Lecture Notes
in Control and Information Sciences 190

Editor: M. Thoma

Raja Chatila and Gerd Hirzinger (Eds.)

Experimental Robotics II

The 2nd International Symposium, Toulouse, France,
June 25-27 1991

Springer-Verlag London Ltd.

Series Advisory Board

A. Bensoussan · M.J. Grimble · P. Kokotovic · H. Kwakernaak
J.L. Massey · Y. Z. Tsypkin

Editors

Raja Chatila
Centre National de la Recherche Scientifique, Laboratoire d'Automatique
d'Analyse des Systemes, 7 Avenue du Colonel-Roche, 31077 Toulouse Cedex,
France

Gerd Hirzinger
DLR, Institute for Robotics and System Dynamics, Oberfaffenhofen,
D-8031 Wessling, Germany

ISBN 978-3-540-19851-2   ISBN 978-3-540-39323-8 (eBook)
DOI 10.1007/978-3-540-39323-8

© Springer-Verlag London 1993
Originally published by Springer-Verlag London Limited in 1993.

The publisher makes no representation, express or implied, with regard to the accuracy of the information contained in this book and cannot accept any legal responsibility or liability for any errors or omissions that may be made.

Typesetting: Camera ready by contributors

69/3830-543210 Printed on acid-free paper

Preface

Experimental Robotics II - The Second International Symposium took place in Toulouse, France in June 1991. It was the second in a series after the first meeting in Montréal in June 1989, organized by Vincent Hayward and Oussama Khatib. The next conference will be held in 1993 in Kyoto, Japan, and chaired by T. Yoshikawa and F. Miyazaki.

Hence the meetings circulate every two years around North America, Europe, and Asia. The objective of this series of symposia is to present and discuss in depth research results and on-going developements in Robotics which have theoretical foundations *and* experimental validations.

Indeed, a robot is a machine in permanent interaction with the environment. The influence of this interaction on its actual behavior is fundamental and is often ill known in advance (e.g., external friction), and cannot be simulated correctly *a priori* (e.g., vision) because of its non deterministic nature or various errors, uncertainties and artefacts.

Robotics is a domain in which the implementation of the theories, methods and algorithms on a physical machine with its sensing, acting and computing capacities - and its real-time and real world constraints - is a fundamental part of the research work. It is not only an important step to sort out what is theoretically possible from what is actually feasible. The concepts themselves must be designed with their implementation in mind. Robotics without robots is maybe Control Theory, Computer Science, Mechanics, Geometric Reasoning, but not Robotics.

The papers presented at ISER and collected in this volume are hence in the very core of robotics research. A total of 38 papers from 13 countries are included in this volume covering the field of Robotics. The book is divided into ten sections corresponding to the ten sessions of the Symposium. The limited number of participants (about eighty) permitted very fruitful and high quality exchanges and discussions.

The meeting was opened by Prof. Alain Costes, Director of LAAS. The keynote lecture was given by Prof. Bernie Roth from Stanford University who got the participants immediately to work on the design of a mechanical system, elegantly showing the important role of experimentation. It was concluded by a closing address by Dr. Georges Giralt from LAAS, who stressed his creed on the importance of experimentation in Robotics.

In addition to the proceedings, a compilation of video segments illustrating the reported research is available.

The international permanent program committee was composed of R. Chatila, *LAAS-CNRS*, France, J. Craig, *Silma*, USA, P. Dario, *Scuola Superiore S. Anna*, Italy, B. Espiau, *ISIA/Ecole des Mines*, France, G. Hirzinger, *DFVLR*, Germany, V. Hayward, *McGill University*, Canada, O. Khatib, *Stanford University*, USA, F. Miyazaki, *Osaka University*, Japan, K. Salisbury, *MIT*, USA, T. Yoshikawa, *Kyoto University*, Japan, and chaired this year by R. Chatila and G. Hirzinger. J-P. Merlet, *INRIA*, France, has since replaced B. Espiau.

The meeting was hosted by LAAS-CNRS and supported by several intitutions and companies: Canadian Space Agency / Agence Spatiale Canadienne (C.S.A. / A.S.C., Ottawa, Ontario), Centre National de la Recherche Scientifique (C.N.R.S., Paris), Commissariat à l'Energie Atomique (CEA, Fontenay-aux-Roses), Conseil Général de la Haute-Garonne (Toulouse), Centre de Robotique Intégrée d'Ile de France (Paris), Laboratoire d'Automatique et d'Analyse des Systémes (L.A.A.S.-C.N.R.S., Toulouse), Matra-Espace (Vélizy), National Science Foundation, Washington D.C., Mairie de Toulouse, Région Midi-Pyrénées (Toulouse).

The meeting could not have taken place without the organizational talent of Mrs Marie-Thérèse Ippolito, and the work and kind availability of Mrs Jackie Som and Nicole Vergriette to whom we are thankful.

Raja Chatila (LAAS-CNRS)
Gerd Hirzinger (DLR)

List of Contributors

M. Accordino
University of Genoa
Dept of Comp. & Systems Science
Via Opera Pia 11A
116145 Genoa
Italy

M. Adams
Katholieke Universiteit Leuven
Depart. Werktuigkunde
Celestijnenlaan 300 B
3001 Heverlee
Belgium

D. S. Ahn
Korea Advanced Inst.
of Science and Technology
P.O.B. 150
Chongryngi, Seoul
Korea

Rachid Alami
LAAS-CNRS
7, Ave. du Colonel Roche
31077 Toulouse Cedex
France
rachid@laas.fr

Benedetto Allotta
Scuola Superiore S. Anna
Via G. Carducci 40
56100 Pisa
Italy

Nils A. Andersen
Inst. of Aut. Control Syst.
Tech. Univers. of Denmark
DK-2800 Lyngby
Denmark

J. L. Arocena
Dpt. of Eng. Science
Univ. of Oxford
Oxford OX1 3PJ
United Kingdom
inaki@robots.oxford.ac.uk

Haruhiko Asada
MIT - Dept. of Mech. Eng.
Room 3-350
Cambridge MA 02139
USA

Christine Bellier
LIFIA IMAG
46, Ave. Félix Viallet
38031 Grenoble Cedex
France
bellier@lifia.imag.fr

Antonio Bicchi
University of Pisa
Centro "E. Piaggio"
56125 Pisa
Italy
lfcmast@icnucevm.bitnet

Philippe Bidaud
Université Paris 6
4 Place Jussieu
75232 PARIS Cedex 05
France

J.E. Bobrow
Dpt. of Mech. & Aerospace Eng.
University of California
Irvine, CA 92717
USA

Yann Bouffard-Vercelli
LAMM
Place Bataillon
34095 Montpellier Cedex
France

Benoît Boulet
McGill Univ. RCIM
3480 University Street
Montreal Quebec
Canada H3A 2A7

David L. Brock
Center for Inf.-Driven
Mech. Systems
M I T
Cambridge MA 02139
USA

Bernhard Brunner
DLR FF-DF/A,
Oberpfafenhofen
Muenchenerstr. 20
D-8031 Wessling
Germany

Nils Bruun
IBM T.J. Watson Res. Center
POB. 704
Yorktown Heights
NY 10598
USA

Giorgio Buttazzo
Scuola Superiore S. Anna
Via G. Carducci 40
56100 Pisa
Italy

Olivier Causse
LIFIA IMAG
46, Ave. Félix Viallet
38031 Grenoble Cedex
France

Raja Chatila
LAAS/CNRS
7, Ave. du Colonel-Roche
31077 Toulouse Cedex
France
raja@laas.fr

Stefano Chiaverini
Dipart. di Inform. e Sistemistica
Universita "Frederico II"
Via Claudio 21
I-80125 Napoli
Italy

Hélène Chochon
Alcatel Alsthom Recherche
Route de Nozay
91460 Marcoussis
France
chochon@aar.alcatel-alsthom.fr

Joël Colly
LAAS/CNRS
7, av. du Colonel-Roche
31077 Toulouse Cedex
France
colly@laas.fr

James L. Crowley
LIFIA IMAG
46, Ave. Félix Viallet
38031 Grenoble Cedex
France
jlc@lifia.imag.fr

Court B. Cutting
Inst. of Reconst. Plastic Surgery
NYU Medical Center
New York 10016
USA
cutting@mcrips.med.nyu.edu

Laeeque Daneshmend
McGill Univ. RCIM
3480 University Street
Montreal Quebec
Canada H3A 2A7
laeeque@larry.mcrcim.mcgill.edu

R.W. Daniel
Dpt. of Eng. Science
Univ. of Oxford
Parks Road
Oxford OX1 3PJ
United Kingdom

Paolo Dario
Scuola Superiore S. Anna
Via G. Carducci 40
56100 Pisa
Italy
dario@ssupl.sssup.it

Pierre Dauchez
LAMM - Univ. Montpellier II
Place Bataillon
34095 Montpellier Cedex
France

Bernard Degallaix
LAAS-CNRS
7, Ave. du Colonel Roche
31077 Toulouse Cedex
France
degal@laas.fr

Eric Degoulange
LAMM - Univ. Montpellier II
Place Bataillon
34095 Montpellier Cedex
France

Xavier Delebarre
Dassault Aviation
78, quai Dassault
92214 Saint-Cloud
France

Michel Devy
LAAS-CNRS
7, av. du Colonel-Roche
31077 TOULOUSE Cedex
France
devy@laas.fr

J. De Schutter
Katholieke Universiteit Leuven
Depart. Werktuigkunde
Celestijnenlaan 300 B
3001 Heverlee Belgium
joris@mech.kuleuven.ac.be

Olav Egeland
Institutt for tek. kybernetikk
Norges tekniske hogskole
O.S.Bragstads plass 8
N-7034 Trondheim
Norway

P. Elosegui
Dpt. of Eng. Science
Univ. of Oxford
Parks Road
Oxford OX1 3PJ
United Kingdom

Daniel Fontaine
Université Paris 6
4 Place Jussieu
75232 Paris Cedex 05
France

Paul Gaborit
LAAS-CNRS
7, Ave. du Colonel Roche
31077 Toulouse Cedex
France
gaborit@laas.fr

D. Grimm
IBM TJ Watson Res. Center
POB. 704
Yorktown Heights
NY 10598
USA

Betsy Haddad
N.Y.U Medical Center
Inst. of Reconst. Plastic Surgery
New York 10016
USA

Gerhard Hirzinger
DLR FF-DF/A
Oberpfafenhofen
Muenchenerstr. 20
D-8031 Wessling
Germany

Shigeki Iida
Inst. of Inform. Science
and Electronics
Univers. of Tsukuba
Tsukuba Ibaraki, 305
Japan

Oussama Khatib
AI Laboratory
Stanford University
Stanford, CA 94305
USA
ok@flamingo.stanford.edu

Janez Funda
Comp. and Inf. Science Dpt.
University of Pennsylvania
Philadelphia, PA 19104
USA
janez@grasp.cis.upenn.edu

Francesca Gandolfo
DIST - University of Genoa
Via Opera Pia 11A
116145 Genoa
Italy
fran@dist.dist.unige.it

G. Grunwald
DLR FF-DF/A
Oberpfafenhofen
Muenchenerstr. 20
D-8031 Wessling
Germany

Tsutomu Hasegawa
ETL
1-1-4 Umezono
Tsukuba - Ibaraki 305
Japan

Koh Hosoda
Div. of Applied Systems Science
Fac. of Eng. Kyoto Univ. Uji
Kyoto 611
Japan

Nobuya Ikeda
Dpt. of Control Eng.
Osaka Univ.
Toyonaka, Osaka 560
Japan

Deljou Khoramabadi
Inst. of Reconst. Plastic Surgery
NYU Medical Center
New York 10016
USA

Jungi Furusho
Dept. of Mech. & Control Eng.
Univ. of Electro-Communications
Chofu Tokyo 182
Japan

Dave Green
Inst. for Inform. Tech.
National Research Council
Ottawa-Ontario K1A 0R6
Canada

Eugenio Guglielmelli
Scuola Superiore S. Anna
Via G. Carducci 40
56100 Pisa
Italy

Vincent Hayward
McGill Univ. RCIM
3480 University Street
Montreal Quebec
Canada H3A 2A7
hayward@larry.mcrcim.mcgill.edu

Koji Ide
Dept. of M. E.
Faculty of Eng.Science
Osaka University
Toyonaka, Osaka 560
Japan

Alan D. Kalvin
IBM T. J. Watson Res. Center
POB. 704
Yorktown Heights
NY 10598
USA

Pradeep Khosla
Elect. and Comp. Eng. Dpt
CMU
Pittsburgh PA 15213-3890
USA

Yong-yil Kim
IBM T J Watson Res. Center
POB. 704
Yorktown Heights
NY 10598
USA

K. Kitagaki
ETL
1-1-4 Umezono
Tsukuba - Ibaraki 305
Japan

Daniel E. Koditschek
Center for Systems Science
Yale Univ. Dpt of Elect. Eng.
New Haven CT 06520
USA
kod@corwin.eng.yale.edu

Eric Krotkov
C M U
Robotics Institute
Pittsburgh PA 15213-3890
USA
epk@ius1.cs.cmu.edu

Osam Kuwaki
Dept. of Control Eng.
Osaka Univ.
Toyonaka, Osaka 560
Japan

David Larose
IBM T. J. Watson Res. Center
PO Box 704
Yorktown Heights NY 10598
USA

Christian Laugier
LIFIA IMAG
46, Ave. Félix Viallet
48031 Grenoble Cedex
France
laugier@lifia.imag.fr

Jadran Lenarčič
The Robotics Lab.
Inst. J. Stefan
Jamova 39
61111 Ljubljana
Slovenia

Thomas Lindsay
Comp. and Inf. Science Dpt.
University of Pennsylvania
Philadelphia, PA 19104
USA
van@grip.cis.upenn.edu

Ramiro Liscano
Inst. for Inform. Tech.
National Research Council
Ottawa-Ontario K1A 0R6
Canada
liscano@iit.nrc.ca

Sheng Liu
Center for Inf. Driven
Mech. Systems
M I T
Cambridge, MA 02139
USA

Allan Manz
Inst. for Inform. Tech.
National Research Council
Ottawa-Ontario K1A 0R6
Canada

Masaaki Maruyama
Dpt. of Control Eng.
Osaka Univ.
Toyonaka, Osaka 560
Japan

Yasuhiro Masutani
Dept. of M. E.
Faculty of Eng. Science
Osaka University
Toyonaka, Osaka 560
Japan

Hirokazu Mayeda
Dept. of Control Eng
Osaka Univ.
Toyonaka, Osaka 560
Japan

Emmanuel Mazer
LIFIA IMAG
46, Ave. Félix Viallet
38031 Grenoble Cedex
France
manu@lifia.imag.fr

J. M. McCarthy
Dpt. of Mech. & Aerospace Eng.
University of California
Irvine, CA 92717
USA

B. W. McDonell
Dpt. of Mech. & Aerospace Eng.
University of California
Irvine, CA 92717
USA

Tad Mc Geer
Aurora Flight Sciences
3107 Colvin Street
Alexandria Virginia 22314
USA
Tad@hiflight.com

Claudio Melchiorri
DEIS - Universy of Bologna
Via Risorgimento 2
40136 Bologna
Italy
cloc1@ingbo1.cineca.it

Jean-Pierre Merlet
Inria Sophia-Antipolis
2004, Route des Lucioles
06561 Valbonne cedex
France
merlet@alcor.inria.fr

Fumio Miyazaki
Dept. of M.E.
Osaka University
Toyonaka, Osaka 560
Japan
miyazaki@crane.mees.osaka-u.ac.jp

Philippe Moutarlier
LAAS-CNRS
7, Ave. du Colonel Roche
31077 Toulouse Cedex
France
philippe@laas.fr

Hiroki Murakami
Ishikawajima-Harima
Heavy Ind. Co., Ltd
1-15, Toyosu 3-chome Kotoku
Tokyo 135
Japan

Chafye Nemri
McGill Univ. RCIM
3480 University Street
Montreal Quebec
Canada H3A 2A7

Marylin Noz
Inst. of Reconst. Plastic Surgery
NYU Medical Center
New York 10016
USA

Tsukasa Ogasawara
ETL
1-1-4 Umezono,
Tsukuba - Ibaraki 305
Japan
ogasawara@etl.go.jp

Robert Olyha
IBM T. J. Watson Res. Center
POB. 704
Yorktown Heights NY 10598
USA

Richard Paul
Comp. and Inf. Science Dept.
University of Pennsylvania
Philadelphia, PA 19104
USA
lou@central.cis.upenn.edu

Victor Perebaskine
LAAS-CNRS
7, Ave. du Colonel Roche
31077 Toulouse Cedex
France
victor@laas.fr

Sean Quinlan
A I Lab.
Stanford University
Stanford, CA 94305
USA
sean@neon.stanford.edu

Ole Ravn
Inst. of Aut. Control Syst.
Tech. Univers. of Denmark
DK-2800 Lyngby
Denmark

Patrick Reignier
LIFIA IMAG
46, Ave. Félix Viallet
38031 Grenoble Cedex
France

Alfred Rizzi
Center for Systems Science
Dept of Elect. Eng.
Yale Univ.
New Haven CT 06520
USA

J. Kenneth Salisbury
M I T
545 Technology Square
Cambridge, MA 02139
USA
jks@ai.mit.edu

Giulio Sandini
University of Genoa
Dpt of Comp. & Systems Science
Via Opera Pia 11A
116145 Genoa
Italy

Akihito Sano
Dept. of Mechanical Eng.
Gifu Univ.
1-1 Yanagido
501-11 Gifu City
Japan

Hans-Kaspar Scherrer
ETH/Institute of Robotics
Leonhardstr.27
CH-8092 Zurich
Switzerland
scherrer@sys.ife.ethz.ch

Gerhard Schrott
Inst. für Informatik
Tech. Un. of Munich
Postfach 20 24 20
D-8000 Munichen 2
Germany

Bruno Siciliano
Dipart.di Inform. e Sistemistica
Universita "Frederico II"
Via Claudio 21
I-80125 Napoli
Italy

Allan Theill Sorensen
Inst. of Aut. Control Syst.
Tech. Univers. of Denmark
DK-2800 Lyngby
Denmark
sl_ats@sl.dth.dk

T. Suehiro
ETL
1-1-4 Umezono
Tsukuba - Ibaraki 305
Japan

Shooji Suzuki
Univ. of Tsukuba
Tsukuba Ibaraki 305
Japan
ssuzuki@roboken.is.tsukuba.ac.jp

Russell H. Taylor
IBM T. J. Watson Res. Center
POB. 704
Yorktown Heights NY 10598
USA
rth@watson.ibm.com

Jocelyne Troccaz
TIMB - Fac. Médecine
Domaine de la Merci
38700 La Tronche
France
troccaz@timb.imag.fr

Gabriele Vassura
DIEM
Via Risorgimento 2
40136 Bologna
Italy

Koji Yoshida
Dept. of Control Eng.
Osaka Univ.
Toyonaka, Osaka 560
Japan

S. Swevers
Katholieke Universiteit Leuven
Depart. Werktuigkunde
Celestijnenlaan 300 B
3001 Heverlee
Belgium

M. Tistarelli
University of Genoa
Dpt of Comp. & Systems Science
Via Opera Pia 11A
116145 Genoa
Italy

Andreja Umek
The Robotics Lab.
Inst. J. Stefan
University of Ljubljana
Ljubljana
Slovenia

Dieter Vischer
ETH/Institute of Robotics
Leonhardstr.27
CH-8092 Zurich
Switzerland
vischer@sys.ife.ethz.ch

Tsuneo Yoshikawa
Dept. of Mechanical Eng.
Kyoto University
Kyoto 606
Japan
ty@image.kuass.kyoto-u.ac.jp

K. Takase
ETL
1-1-4 Umezono
Tsukuba - Ibaraki 305
JAPAN

D. Torfs
Katholieke Universiteit Leuven
Depart. Werktuigkunde
Celestijnenlaan 300 B
3001 Heverlee
Belgium

H. Van Brussel
Katholieke Universiteit Leuven
Depart. Werktuigkunde
Celestijnenlaan 300 B
3001 Heverlee
Belgium

Richard Volpe
JPL, CalTech
4800 Oak Grove Drive
Pasadena CA 91109
USA
volp@telerobotics.jpl.nasa.gov

Shin'ichi Yuta
University of Tsukuba
Tsukuba Ibaraki 305
Japan
yuta@is.tsukuba.ac.jp

Contents

Section 1: Robot Control   1

The Equivalence of Second Order Impedance Control and Proportional Gain Explicit Force Control : Theory and Experiments. R. Volpe, P. Khosla, CMU, USA.   3

Robot Control in Singular Configurations - Analysis and Experimental Results. S. Chiaverini, B. Siciliano, University Napoli, Italy, and O. Egeland, Inst. Tek. Kybernetik, Norway.   25

Implementation of Control Algorithms for Flexible Joint Robots on a KUKA IR 161/60 Industrial Robot: Experimental Results. M. Adams, J. Swevers , D. Torfs, J. De Schutter, H. Van Brussel, Dpt. of Mechanical Engineering, Katholieke Universiteit Leuven, Belgium.   35

Adaptive and Fault Tolerant Tracking Control of a Pneumatic Actuator. B. W. McDonell, J. E. Bobrow, Dpt. of Mechanical and Aerospace Engineering, University of California, Irvine, USA.   49

Section 2: Learning and Skill Acquisition   59

Experimental Verification of Human Skill Transfer to Deburring Robots. H. Asada, S. Liu, MIT, USA.   61

Learning of Robotic Assembly Based on Force Information. F. Miyazaki, K. Ide, Y. Masutani, D. S. Ahn, Osaka University, Japan.   78

Model Based Implementation of a Manipulation System with Artificial Skills. T. Ogasawara, K. Kitagaki, T. Suehiro, T. Hasegawa, K. Takase, ETL, Tsukuba, Japan.   89

A Learning Control System for an Articulated Gripper. P. Bidaud, D. Fontaine, LRP, Université Paris VI, France.   99

Section 3: Grippers and Articulated Hands   113

Implementation of Behavorial Control on a Robot Hand/Arm System. D.L. Brock, J.K. Salisbury, MIT, USA.   115

Mechanical and Control Issues for Integration of an Arm-Hand Robotic System. C. Melchiorri, G. Vassura, DEIS/DEIM, Bologna, Italy.   136

Experimental Evaluation of Friction Characteristics with an Articulated Robotic Hand. A. Bicchi, University of Pisa, Italy, J. K. Salisbury, D.L. Brock, MIT, USA.   153

Intelligent Robot Gripper for General Purposes. H-K. Scherrer, D. Vischer, ETH / Inst. Robotics, Switzerland.   168

Section 4: Robotic Systems and Task-Level Programming   177

A Model-Based Optimal Planning and Execution System with Active Sensing and Passive Manipulation for Augmentation of Human Precision in Computer-Integrated Surgery. Russell H. Taylor (IBM, T. J. Watson Research Center), Court B. Cutting (Institute for Reconstructive Plastic Surgery, NYU Medical Center), Yong-yil Kim, Alan D. Kalvin, David Larose (IBM T. J. Watson Research Center), Betsy Haddad, Deljou Khoramabadi, Marilyn Noz (Institute for Reconstructive Plastic Surgery; NYU Medical Center), Robert Olyha, Nils Bruun, Dieter Grimm (IBM T. J. Watson Research Center), USA.   179

An Experimental Environment for Task-Level Programming of Robots. G. Schrott, Munich University, Germany.   196

An Architecture for Task Interpretation and Execution Control for Intervention Robots: Preliminary Experiments. R. Chatila, R. Alami, B. Degallaix, V. Perebaskine, P. Gaborit, P. Moutarlier. LAAS-CNRS, France.   207

A Sensor-based Telerobotic System for the Space Robot Experiment ROTEX. G. Hirzinger, G. Grunwald, B. Brunner, J. Heindl, DFVLR, Germany.   222

Section 5: Manipulation Planning and Control   239

Towards Real-Time Execution of Motion Tasks. S. Quinlan, O. Khatib, Stanford University, USA.   241

Force Control of a Two-arm Robot Manipulating a Deformable Object. X. Delcbarre, E. Dégoulange, P. Dauchez, Y. Bouffard-Vercelli, LAMM, France.   255

A Practical System for Planning Safe Trajectories for Manipulator Robots. C. Bellier, C. Laugier, E. Mazer, J. Troccaz, LIFIA, France.   270

Preliminary Experiments in Spatial Robot Juggling. A. A. Rizzi, D. E. Koditschek, Yale University, USA.   282

Section 6: Mobile Robots   299

A Comparison of Real-Time Obstacle Avoidance Methods for Mobile Robots. A. Manz, R. Liscano, D. Green, NRC, Canada   301

Object-Oriented Design of Mobile Robot Control Systems. H. Chochon, Alcatel Alsthom Recherche, France.   317

Action Level Behaviours for Locomotion and Perception. J.L. Crowley, P. Reignier, O. Causse, F. Wallner, LIFIA, France.   329

Implementation of a Small Size Experimental Self-Contained Autonomous Robot - Sensors, Vehicle Control, and Description of Sensor Based Behavior. Sin'ichi Yuta, Shooji Suzuki, and Shigeki Iida, Tsukuba University, Japan.   344

Section 7: Sensing and Perception 359

Mapping Rugged Terrain for a Walking Robot. E. Krotkov, CMU, USA. 361

Localization of a Multi-Articulated 3d Object from a Mobile Multisensor System. M. Devy, J. Colly, LAAS-CNRS, France. 367

Object Understanding Through Visuo-Motor Cooperation. M. Accordino, F. Gandolfo, A. Portunato, G. Sandini, M. Tistarelli, Genoa University, Italy. 379

Real-Time Vision Based Control of Servomechanical Systems. N.A. Andersen, O. Ravn, A. T. Sorensen, Technical University of Denmark, Denmark. 388

Section 8: Compliance and Force Control 403

Contact Operations Using an Instrumented Compliant Wrist. T. Lindsay, J. Funda, R. Paul, University of Pennsylvania, USA. 405

Controlling Contact by Integrating Proximity and Force Sensing. B. Allotta, G. Buttazzo, P. Dario, E. Guglielmelli, Scuela Superiore S. Anna, Pisa, Italy. 420

End Point Control of Compliant Robots. J. I. Arocena, R. W. Daniel, P. Elosegui, University of Oxford, UK. 435

Use of C-Surface Based Force-Feedback Algorithm for Complex Assembly Tasks. J-P. Merlet, INRIA, Sophia-Antipolis, France. 450

Section 9: Legged Locomotion 463

Passive Dynamic Biped Catalogue, 1991. T. McGeer, AFS, USA. 465

Realization of Dynamic Quadruped Locomotion in Pace Gait by Controlling Walking Cycle. A. Sano, J. Furusho, Gifu University, Japan. 491

Section 10: Modelling and Control 503

System Identification and Modelling of A High Performance Hydraulic Actuator. B. Boulet, L. Daneshmend, V. Hayward, C. Nemri, McGill University, Canada. 505

Experimental Evaluation of Human Arm Kinematics. J. Lenarčič, A. Umek, Ljubljana University, Slovenia. 521

Modeling and Control of a Three Degree of Freedom Manipulator with Two Flexible Links. T. Yoshikawa, Kyoto University, H. Murakami, Harima Heavy Industry, K. Hosoda, Kyoto University, Japan. 531

Experimental Examination of the Identification Methods for an Industrial Robot Manipulator. H. Mayeda, M. Maruyama, K. Yoshida, N. Ikeda, O. Kuwaki, Osaka University, Japan. 546

Section 1: Robot Control

This first set of papers deals with different aspects of robot control.

The paper of Volpe and Khosla tries to shed light into the confused variety of terminologies used in force control; in particular they discuss equivalence aspects of impedance and proportional gain explicit force control and present experimental results with the CMU direct-drive arm. As in the whole field of robot control these kind of theoretical and practical comparisons are of crucial importance before commonly acknowledged standards will settle down.

The paper of Chiaverini, Siciliano and Egeland is analytically and experimentally concerned with the control of 6 dof-robots near kinematically singular configurations. The approach makes use of the well-known "damped-least-squares" method, but adds a "user-defined accuracy" technique that allows to define directions of high and lower accuracy in the operational space.

Mechanical flexibiliby in industrial robots in general does not appear in the links but in the joints and may show up considerably adverse effects in case of high accelerations. The paper of Adams, Swevers, Torfs, DeSchutter and van Brussel shows how - by using identification and model-based control - the dynamics (overshoot, oscillations, tracking error) of an industrial robot can be significantly improved. Comparisons are especially related to the PID control systems normally delivered to the robot user.

Adaptive and fault tolerant tracking control of a pneumatic actuator is presented in the paper of McDonell, Bobrow and McCarthy. The authors point out that fixed gain control is not adequate for pneumatic actuators. In contrast they propose (and experimentally demonstrate) a controller based on a recursively identified time-varying linear model which uses full-state feedback and tracking control via feedforward terms.

The Equivalence of
Second Order Impedance Control and
Proportional Gain Explicit Force Control:
Theory and Experiments

Richard Volpe* and Pradeep Khosla†

Abstract

This paper reveals the essential equivalence of second order impedance control and proportional gain explicit force control with feedforward. This is first done analytically by reviewing each control method and showing how they mathematically correspond. For stiff environments the correspondence is exact. However, even for softer environments similar response of the system is indicated. Next, the results of an implementation of these control schemes on the CMU DD Arm II are presented, confirming the predictions of the analysis. These results experimentally demonstrate that proportional gain force control and impedance control, with and without dynamics compensation, have equivalent response to commanded force trajectories.

1 Introduction

There is a whole class of tasks that seem to implicitly require controlling the force of interaction between a manipulator and its environment: pushing, scraping, grinding, pounding, polishing, twisting, etc. Thus, force control of the manipulator becomes necessary in at least one of the degrees of freedom of the manipulator; the other degrees of freedom remain position controlled. Mason formalized this idea and called it Hybrid Control [14]. Simply put, the manipulator should be force controlled in directions in which the position is constrained by environmental interaction, and position controlled in all orthogonal directions.

The Hybrid Control formalism does not specify what particular type of position or force control should be used. It only partitions the space spanned by the total degrees of

---

*Currently at The Jet Propulsion Laboratory, California Institute of Technology, Pasadena, California 91109. This work was completed while the author was a member of the Department of Physics, The Robotics Institute, Carnegie Mellon University, Pittsburgh, Pennsylvania, 15213

†Department of Electrical and Computer Engineering, The Robotics Institute, Carnegie Mellon University, Pittsburgh, Pennsylvania, 15213

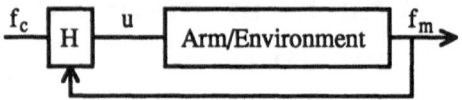

Figure 1: Explicit force control block diagram.

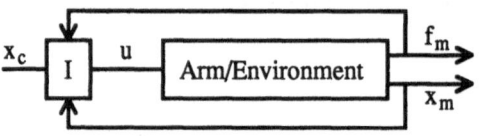

Figure 2: Impedance control block diagram.

freedom into one subspace in which position control is employed, and another in which force control is employed. In the position control subspace simple strategies have proven adequate (eg. PID), while sophisticated enhancements have improved performance (eg. computed torque control, adaptive control). However, in the force control subspace, two main conceptual choices have emerged: *explicit force control* and *impedance control*. Figures 1 and 2 are simple block diagrams of these two types of control schemes. The major difference between these schemes is the commanded value: explicit force control requires commanded force, while impedance control requires commanded position. In order for these to be feedback controllers, explicit force control needs force measurement, while impedance control needs position measurement. In addition, impedance control requires force measurement — without it an impedance controller reduces to a position controller.

Ideally, an explicit force controller attempts to make the manipulator act as pure force source, independent of position. Like position control, the obvious first choice has been some manifestation of PID control (i.e. P, PD, PI, etc.). These have met with varying amounts of success depending on the characteristics of the particular actuators, arm links, sensor, and environment.

Alternatively, impedance control has been presented as a method of stably interacting with the environment. This is achieved by providing a *dynamic* relationship between the robot's position and the force it exerts. A complete introduction to impedance control is beyond the scope of this discussion and the reader is referred to the previous work of other researchers [6, 8]. The basic tenet of impedance control is that the arm should be controlled so that it behaves as a mechanical impedance to positional constraints imposed by the environment. This means that the force commanded to the actuators is dependent on its position: $f = \mathcal{Z}(x)$, where $\mathcal{Z}$ may be a function or an operator. If the impedance is linear, it can be represented in the Laplace domain as $F(s) = Z(s)X(s)$; the order of Z determines the order of the impedance controller. The resultant behavior

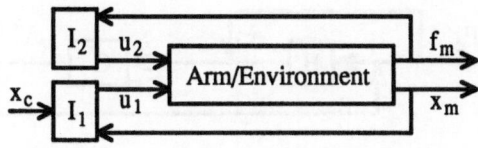

Figure 3: Impedance control block diagram with the controller divided into its position part, I_1, and its force part, I_2.

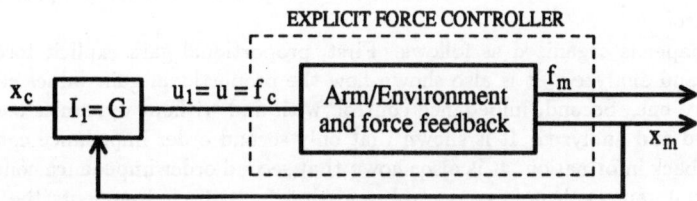

Figure 4: Impedance control block diagram redrawn to show the inner explicit force controller.

of the manipulator is obvious: if it is unconstrained it will accelerate; if it is constrained the forces from the actuators will be transmitted through the arm and exerted on the environment. In this paper we will only consider the second case of physical interaction in which force feedback must be used in the control of the mechanical impedance of the manipulator. For linear impedance relationships, the force feedback loop may be separated as in Figure 3. It can be seen that this figure may be further modified as in Figure 4 to show that the force feedback loop is part of an internal explicit force controller. Thus, an impedance controller that utilizes force feedback contains an explicit force controller. Further, when the arm is against a stiff environment, the feedback term, x_m, is constant and may be ignored. Thus, impedance control reduces to explicit force control.

This paper explores the exact correspondence between explicit force control and impedance control. In particular, it is shown that impedance controllers that utilize force feedback must be second order; lesser order impedance relations are essentially open-loop to force. Analysis of the second order impedance controller reveals that it has an algebraic structure akin to proportional gain explicit force control with feedforward. This correspondence becomes exact when the position feedback is constant. In practice, this criterion is regularly met by stiff environments, or soft environments in equilibrium with the arm.

We have implemented both second order impedance control with and with out dynamics compensation, as well as proportional gain explicit force control with feedforward. These implementations were in six DOF on the CMU DDarm II. The results reveal the

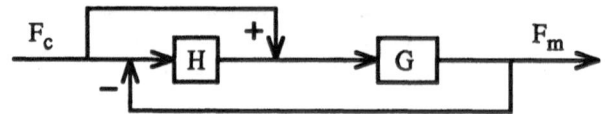

Figure 5: Proportional gain explicit force control block diagram.

equivalent response of the impedance and explicit force control strategies, even for the case of soft environment contact. These results experimentally confirm what is analytically indicated: the equivalence of second order impedance control and proportional gain force control.

This paper is organized as follows. First, proportional gain explicit force control is reviewed and analyzed. It is also shown how the proportional gain values can be as low as negative one. Second, impedance control, with and without dynamics compensation, is reviewed and analyzed. It is shown that only second order impedance control utilizes force feedback information. It is also shown that second order impedance control employs proportional gain explicit force control, and that for stiff environments the two become the same controller. In the last part of this paper, the insights and predictions from this analysis of the controllers is experimentally verified.

2 Proportional Gain Explicit Force Control

The first controller to be discussed is proportional gain explicit force control. The chosen form of this controller is:

$$f = f_c + K_{fp}(f_c - f_m) - K_v \dot{x}_m \qquad (1)$$

where subscripts c and m denote the commanded and measured quantities, respectively. The feedforward term, f_c, is necessary to provide a bias force when the force error is zero — without it the system is guaranteed to have a steady state force error. The velocity gain, K_v, adds damping to the system. A block diagram for this controller is shown in Figure 5, where $H = K_{fp}$ and G is the arm / sensor / environment plant and includes the active damping control loop. It has previously been shown that G can be considered a fourth order transfer function [3], and we have experimentally extracted parameter values for the model [16, 18]. The closed loop transfer function with the feedforward term is:

$$\frac{F_c}{F_m} = \frac{(1 + K_{fp})G}{1 + K_{fp}G}. \qquad (2)$$

This is a Type 0 System and will have a nonzero steady-state error for a step input. The root locus of this system is shown in Figures 6 and 7. The corresponding Bode plots are shown in Figure 8. As can be seen from the root locus, proportional control makes the system more oscillatory and can make it unstable. This instability is contrary to other researchers' predictions which were the result of using a plant model that was not experimentally derived [3, 18]. The Bode plots further illustrate this problem. There is

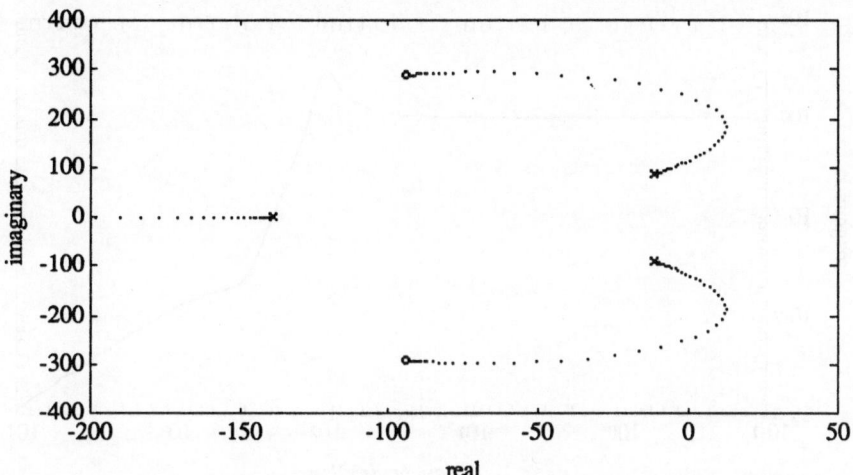

Figure 6: Root locus for the fourth order model under proportional gain explicit force control.

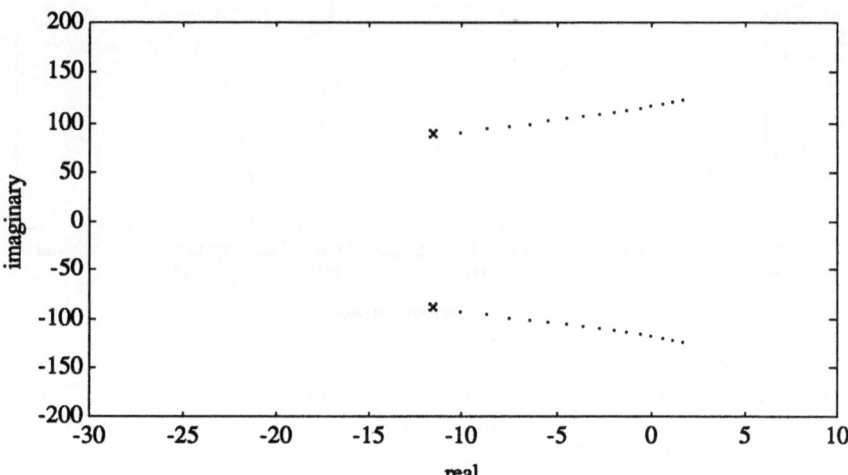

Figure 7: Enlargement of root locus in Figure 6 with K_{fp} values of 0 to 1.5 in steps of 0.1 .

8

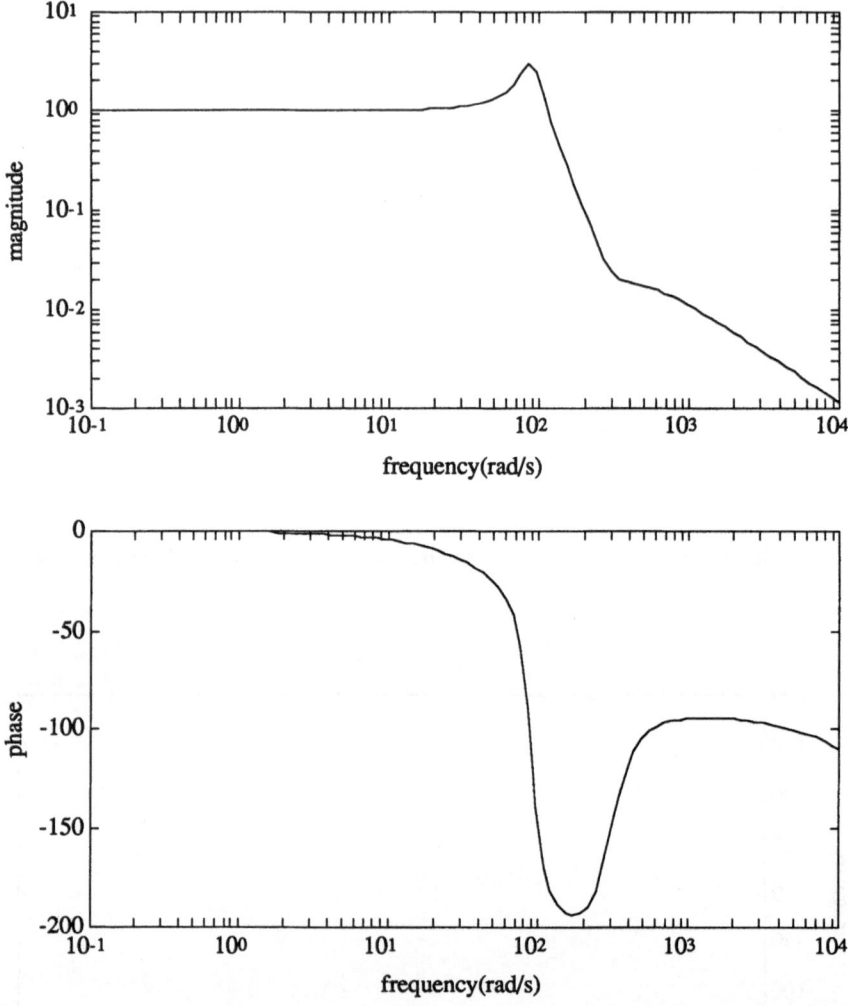

Figure 8: The resonance peak occurs near the natural frequency of the environment. The gain margin is 1.2 at $\omega = 118$rad/s, which corresponds to the root locus crossing to the right half plane in Figure 7.

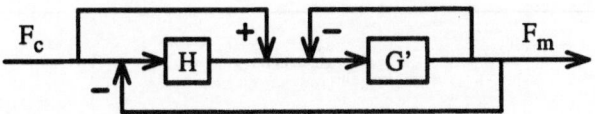

Figure 9: Block diagram a force-based explicit force controller with proportional gain and unity feedforward. The plant G has been replaced by G' and the explicit feedback path of the environmental reaction force.

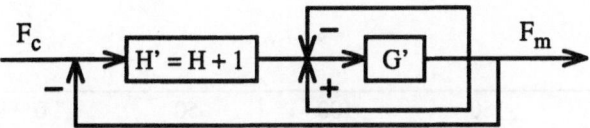

Figure 10: Block diagram a force-based explicit force controller with proportional gain and extra feedback for reaction force compensation. The plant G has been replaced by G' and the explicit feedback path of the environmental reaction force.

a resonance peak from the environment dynamics at approximately 100 rad/s. After this peak there is a 40 dB/decade drop-off which gives a minimum phase margin of $\sim 15°$ at $K_{fp} \approx 1$.

The addition of a lowpass filter in the feedback loop can improve the response by introducing a dominant pole on the real axis [1, 16]. However, this pole placement and the behavior of the system closely match that provided by integral control. A discussion including integral force control is outside the scope of this paper [16]. Therefore, lowpass filtering will not be considered further.

It will prove useful later (in the discussion of impedance control) to now discuss the feedforward term in more detail. It is usually desirable that the feedforward term be unity so that the environmental reaction force will be canceled during steady state. Figure 9 shows the block diagram of the system. The plant G' differs from G in that the reaction force has be extracted and shown explicitly in the block diagram. The transfer function of this system is:

$$\frac{F_m}{F_c} = \frac{(H+1)G'}{1+(H+1)G'} \tag{3}$$

$$= \frac{H'G'}{1+H'G'} \tag{4}$$

where $H' = H + 1$. It is seen directly that an equivalent block diagram of the system may be constructed as in Figure 10. Viewed in this way, the reaction force is negated explicitly, and the proportional gain may have values $H' = K'_{fp} \geq 0$ or $H = K_{fp} \geq -1$. Thus, the proportional gain of the original controller may be as small as negative one.

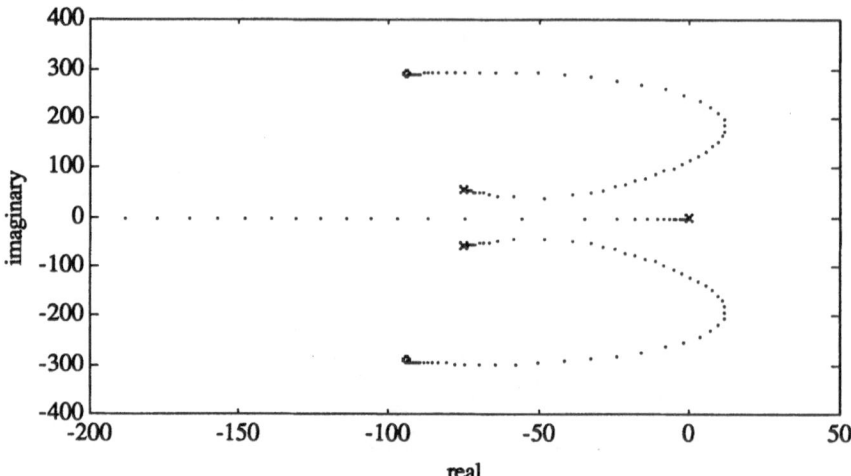

Figure 11: Root locus for the fourth order model for $-1 \leq K_{fp} < \infty$ or $0 \leq K'_{fp} < \infty$.

Figure 11 shows the proportional gain force control root locus for gains as low as negative one. The use of negative gains like this has appeared in the literature previously [5, 7]. However, this result is usually presented within the framework of impedance control. As will be seen in the following sections, the impedance controllers for which this result was obtained actually contain proportional gain explicit force control, which mandates the result.

3 Review of Impedance Control

Impedance control is a strategy that controls the dynamic relation between the manipulator and the environment. The force exerted on the environment by the manipulator is dependent on its position and its impedance. Usually this relation is expressed in Cartesian space as:

$$f = \mathcal{Z}(x). \tag{5}$$

where f, x, and $\mathcal{Z}$, are force, position, and impedance. The impedance consist of two components: that which is physically intrinsic to the manipulator, and that which is given to the manipulator by active control. It is the goal of impedance control to mask the intrinsic properties of the arm and replace them with the target impedance.

The impedance relation can have any functional form. It has been shown that general impedances are useful for obstacle avoidance [6, 10, 17]. However, it will be made clear in this section that sensor based, feedback controlled interaction with the environment requires the impedance to be linear and of second order at most. This is for two reasons. First, the dynamics of a second order system are well understood and familiar. Second, for higher order systems it is difficult to obtain measurements corresponding to the higher order state variables. These measurements are required for closed loop control.

To implement impedance control, model based control can be used. This type of scheme relies on the inverse of the Jacobian. A second type of controller which uses the transpose of the Jacobian is sometimes employed. Both forms of impedance control will be shown to contain proportional gain explicit force control (with feedforward). Also, when in contact with an environment of any appreciable stiffness, the position feedback is essentially constant and impedance control reduces directly to proportional gain force control. The role of proportional gain force control in impedance control, and their equivalence when in contact with stiff environments, has not been recognized or demonstrated previously.

The next sections are organized as follows. First, the order of the desired impedance will be discussed and the implications for implementation will be shown. Second, model based impedance control will be reviewed. The reduced form of impedance control without dynamics compensation will also be presented. Third, it will be shown how each of these schemes contain an internal proportional gain force control loop. This control loop becomes the only active one if the position feedback is constant, as in the case of contact with a stiff environment.

3.1 Zeroth, First, and Second Order Impedance

A linear impedance relation may be represented in the Laplace domain as:

$$F = Z(s)X. \tag{6}$$

The order of the polynomial $Z(s)$ is the order of the impedance.

The simplest form of an impedance controller has a zeroth order impedance. In this case Z is a constant and

$$F = KX. \tag{7}$$

The impedance parameter K is the desired stiffness of the manipulator. When a manipulator has no intrinsic stiffness, K dictates the apparent stiffness of the arm. This is accomplished with active control that uses position feedback. The value of the active stiffness is the position feedback gain.

A more typical form of an impedance controller is a first order impedance. In this case:

$$F = (Cs + K)X. \tag{8}$$

The added parameter C is the desired damping of the manipulator. It is equal to the sum of the active and natural damping. The active damping is accomplished by velocity feedback in a position controlled system. The value of the active damping is the velocity feedback gain. Since the active damping can be modified, C can take on any value which maintains stability. In fact, negative active damping can be used to eliminate the appearance of any damping in the arm. This is rarely desirable, since damping has a stablizing effect.

The last form of impedance control that shall be considered here is a second order type. The second order impedance controller has the form:

$$F = \left(Ms^2 + Cs + K\right)X. \tag{9}$$

The parameter M is the desired inertia of the manipulator. While the arm has an intrinsic inertia due to its mass, this can be modified by active feedback. It follows from the previous two cases that acceleration feedback can be used for this purpose. In this case, the value of active inertia is the acceleration feedback gain. Its value can be used to adjust M. Few researchers have proposed such acceleration feedback schemes for impedance control [4]. This is because an acceleration measurement typically requires a second derivative, which will be extremely noisy. Alternatively, the force may be measured and the acceleration commanded. This is typically the method employed, as will be shown.

3.2 Model Based Control

Model based control involves the use of a dynamic model of the manipulator to determine the actuation torques [2]. Model based impedance control may be summarized by the following equations [6, 16]:

$$\tau_A = Du + h + g + J^T f_m \tag{10}$$

$$u = J^{-1}\left(\ddot{x}_u - \dot{J}\dot{\theta}_m\right) \tag{11}$$

$$\ddot{x}_u = M^{-1}\left[(C\Delta\dot{x} + K\Delta x) - f_m\right] \tag{12}$$

$$\Delta x = x_c - \mathcal{F}(\theta_m) \tag{13}$$

$$\Delta\dot{x} = \dot{x}_c - J\dot{\theta}_m. \tag{14}$$

Equation (10) describes the dynamics of the manipulator with inertia matrix D, Coriolis and centripetal force vector h, and gravitational force vector g. Equation (11) describes the control signal in terms of the desired Cartesian acceleration. Equation (12) specifies the desired second order impedance control relationship. Equations (13) and (14) determine the Cartesian position and velocity errors through the forward kinematics, $\mathcal{F}$, and the manipulator Jacobian, J. The subscripts c and m indicate commanded and measured quantities.

Without force feedback this control scheme is equivalent to position control schemes such as Resolved Acceleration Control [13] and Operational Space Control [9]. These are only first order impedance control schemes since they just modify the stiffness and damping of the arm. Including force feedback information in the controller yields second order impedance control [6]:

$$\tau = DJ^{-1}M^{-1}\left(C\Delta\dot{x} + K\Delta x - f_m\right) - J^{-1}\dot{J}\dot{\theta} + h + g + J^T f_m. \tag{15}$$

or

$$\tau = J^T\Lambda M^{-1}\left(C\Delta\dot{x} + K\Delta x - f_m\right) - J^T\Lambda\dot{J}\dot{\theta} + h + g + J^T f_m. \tag{16}$$

where

$$D(\theta) = J^T\Lambda(x)J \tag{17}$$

and the matrix, Λ, is the Cartesian space representation of the inertia matrix. The first form of the controller is necessary if the inverse dynamics calculations expressed in Equation (10) are used. In this case, the inverse of the Jacobian and the arm inertia must be calculated. The second form is useful as a steady state approximation in which the manipulator inertia does not change or is not known [8]. ($\dot{J}$ as well as h may equal zero also.) In this second case, the inverse of the Jacobian need not be calculated; only its transpose is necessary. The arm inertia need not be calculated either, since only its product with the impedance mass parameter is needed, as will be explained shortly.

Note that the force feedback is used in two places. First, this feedback is used in the physical model of the arm dynamics, Equation (10). This is equivalent to introducing end effector forces into the Newton-Euler dynamics calculations. Second, the feedback is used in the impedance relation, Equation (12). While Equation (10) effectively linearizes the arm, Equation (12) modifies the impedance control signal to compensate for the experienced force.

It can now be seen that it is the force feedback in the control signal which modifies the apparent inertia of the arm [7]. Equation (16) best shows this effect. In this controller, as with the previous ones, the premultiplication of K and C by ΛM^{-1} changes nothing; $\Delta \dot{x}$ and Δx are still multiplied by a gain. However, things are made different by the force feedback signal f_m. It is multiplied by the term ΛM^{-1}, which is a mass ratio that reduces or increases the amount of actuator torque applied. For simplicity sake, it will be assumed that the impedance parameters are diagonal in the Cartesian space defined by the eigenvectors of Λ. In this case, ΛM^{-1} (or its diagonalized counterpart) can be thought of as a matrix of mass ratios along the diagonal. Since Λ is due to the physical inertia of the arm, it is the impedance parameter M which determines each ratio. For $M \to 0$ the ratio becomes very large; for a small measured force, a large accelerating torque is applied to the arm. Thus, the apparent inertia of the arm is reduced. (It is important to remember that the external force does not contribute to the acceleration because it has been effectively negated by the $J^T f_m$ term.) Similarly, for $M \to \infty$ the ratio becomes very small; for a large measured force, a small accelerating torque is applied to the arm. Thus, the apparent inertia of the arm is increased. In this way, second order impedance control not only changes the stiffness and damping properties of the arm, but its inertia as well.

3.3 Explicit Force Control within Impedance Control

The two second order impedance controllers reviewed above can be shown to contain explicit force control. This aspect of impedance control has not previously been recognized. While some correspondence between impedance control and explicit force control has been recognized, the relation was not specifically or clearly stated [7]. A general argument supporting this new interpretation was presented in the introduction. Now, it will be shown explicitly for the impedance controllers described previously in this paper. It has been shown elsewhere how this framework includes both Stiffness Control, and Accommodation Control [16].

Consider the second order impedance controller represented by Equation (16). This can be rewritten in the form:

$$\tau = J^T \left[K'_{fp}(f_c - f_m) + f_m - K'_v \dot{x}_m \right] + g \tag{18}$$

$$f_c = K(x_c - x_m) + C\dot{x}_c \tag{19}$$

$$K'_{fp} = \Lambda M^{-1} \tag{20}$$

$$K'_v = \Lambda M^{-1} C \tag{21}$$

$$\tag{22}$$

This formulation is very similar to the proportional gain explicit force controller discussed previously and corresponds to the block diagram shown in Figure 10 with $K'_{fp} = H'$. Since velocity feedback was also used in the proportional gain explicit force controller, the only major difference here is the use of the feedback position in the calculation of the commanded force. The commanded velocity can be assumed to be zero ($\dot{x}_c = 0$), which is usually the case.

In the above formulation, the impedance parameters K and C determine the commanded force used in an explicit force controller with a proportional gain of $K'_{fp} = \Lambda M^{-1}$. The term $J^T f_m$ can be seen to negate the reaction force that is experienced by the arm. As expressed in the transfer function of Equation (4) the stability of this controller is guaranteed for $K'_{fp} \geq 0$. This is equivalent to the condition:

$$\Lambda M^{-1} \geq 0 \tag{23}$$

(Again, it is assumed that the matrices K'_{fp} and ΛM^{-1} are diagonal. Therefore the inequality is considered to refer to each of the elements individually.) This loosely implies that the open loop pole location of the root locus corresponds to the impedance parameter $M \to \infty$ and the zeros indicate a value of $M \to 0$. Of course, this statement is only true if the principal axes of Λ and M are parallel.

It is also important to note that force control is often used when the manipulator is in contact with a stiff environment; $\dot{x}_m = 0$, and x_m is an arbitrary constant which can be set to zero. Again, it is the usual case that the commanded velocity is zero ($x_c = 0$). Thus, the commanded force reduces to:

$$f_c = K x_c \tag{24}$$

which has no dependence on position feedback. In this case, K acts merely as a scaling factor to the commanded position. This scaled commanded position can be directly replaced by commanded force. With force as the commanded quantity Equation (19) can be eliminated, and the controller becomes simply Equation (18). Since the the gains K'_{fp} and K'_v are independently adjustable, this controller has an identical structure to the explicit force controller in Equation (1). Therefore, is apparent that second order impedance control against a stiff environment is equivalent to explicit force control with proportional gain and feedforward. In the following section it will be seen that experimental results confirm this conclusion.

4   Experimental Results

This section presents the experimental results of implementations of proportional gain force control with feedforward, and second order impedance control, with and without dynamics compensation. It will be seen that in each case, the response and stability of the system are essentially the same.

All experiments were conducted using the CMU DD Arm II and implemented under the Chimera II real time operating system [15]. The experiments presented here were conducted by pushing on a common environment: a cardboard box with an aluminum plate resting on top. The parameters of a second order model of this environmental system were: stiffness $k \approx 10^4$ N/m, damping $c \approx 17$ N·s/m, and mass $m \approx 0.1$ kg. The control rate was 300 Hz, except in the case of dynamics compensation, where it was 250Hz. In all experiments the velocity gain was $K_v = 10$. All graphs of data show the reference values as a dashed line and the measured values as a solid line.

4.1 Proportional Gain with Feedforward Control

The first controller to be discussed is proportional gain force control with the reference force fedforward. Figures 12 (a) through (h) show the response of this controller to the commanded force trajectory. There are several things to note about the response profiles to variations in the proportional gain. First, as predicted by the model, the system exhibits the characteristics of a Type 0 system: finite steady state error for a step input and unbounded error for a ramp input. Second, for an increase in position gain, the steady state error reduces, but at the cost of increasingly larger overshoot. As correctly predicted by the root locus of the system model in Figure 7, this control scheme causes instability at $K_{fp} \approx 1$. Also, the fact that the environmental poles are always off the real axis can be seen in the steady state oscillations that occur at the system's natural frequency (~ 15 Hz), particularly after the step input. Finally, it can be seen that negative proportional gains are increasingly more stable, but the response of the system approaches zero as $K_{fp} \rightarrow -1$.

4.2 Impedance Control

This section presents the results of implementing the second order impedance control schemes presented earlier. The position reference trajectories are chosen such that given the stiffness of the controller, the trajectory should provide the same force profile as commanded for the proportional gain explicit force controller, thus allowing a direct comparison with that controller. As will become apparent, both forms of impedance control (with and without dynamics compensation) respond the same as proportional gain explicit force control with feedforward.

4.2.1 Impedance Control Without Dynamics Compensation

When in contact with a stiff environment, the manipulator will not move very much or very quickly in the direction normal to the environment. As was shown previously, this enables a steady state approximation that eliminates the need to calculate the inverse dynamics and the inverse Jacobian. The control law has the form of Equation (18). For these experiments $K = 150$ and $C = 10$. As discussed in Section 3.3 the mass ratio, ΛM^{-1}, is equivalent to the proportional force gain plus one:

$$\Lambda M^{-1} = K_{fp} + 1. \tag{25}$$

Thus, for this controller the mass ratio is chosen to be diagonal in the desired frame and its components (λ/m) set to correspond to the proportional force gains chosen previously. This allows a direct comparison of impedance and force control schemes.

Figures 13 (a) through (h) show the response of this impedance controller, as well as the commanded position trajectory multiplied by the active stiffness in that direction.

16

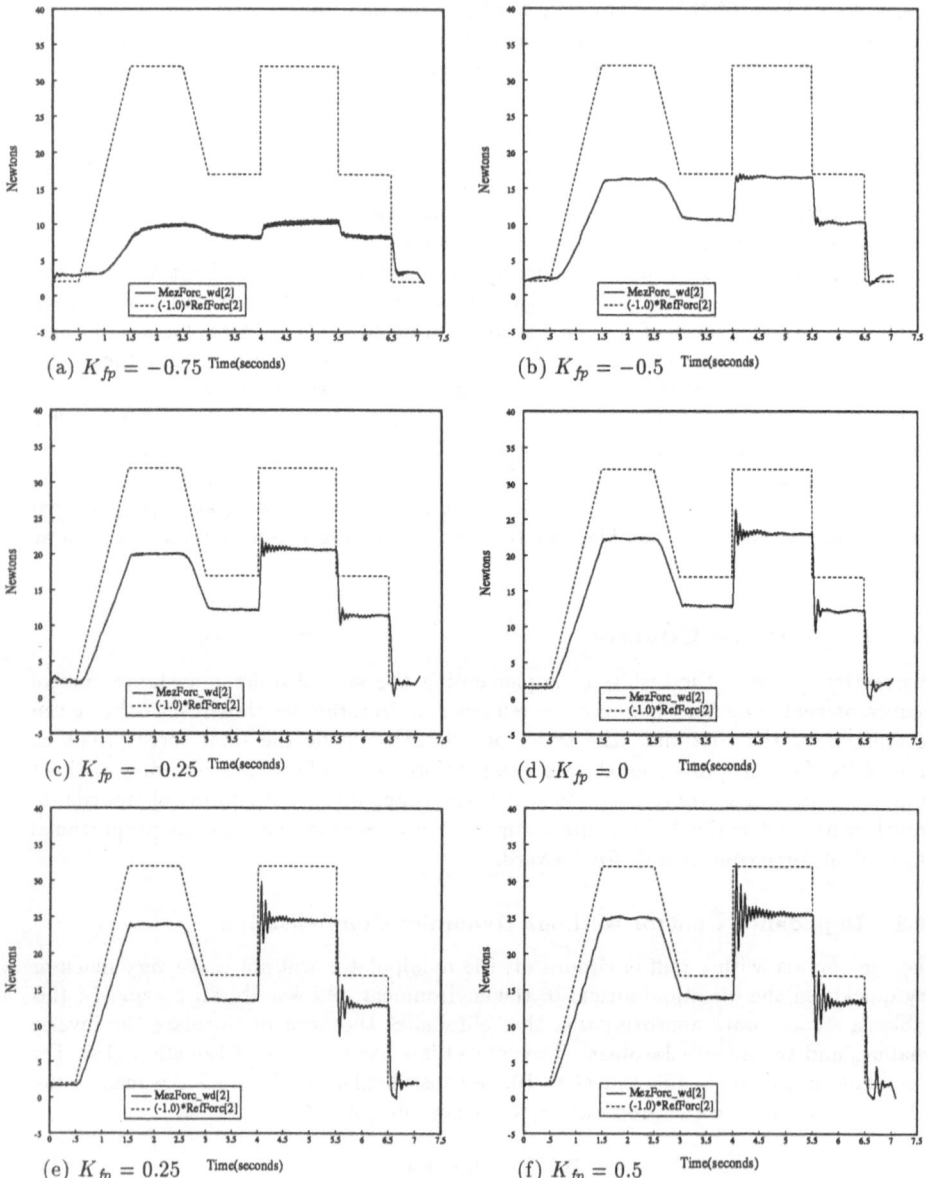

Figure 12: Experimental data of proportional gain explicit force control with feedforward. The proportional gain varies from -0.75 to 1.

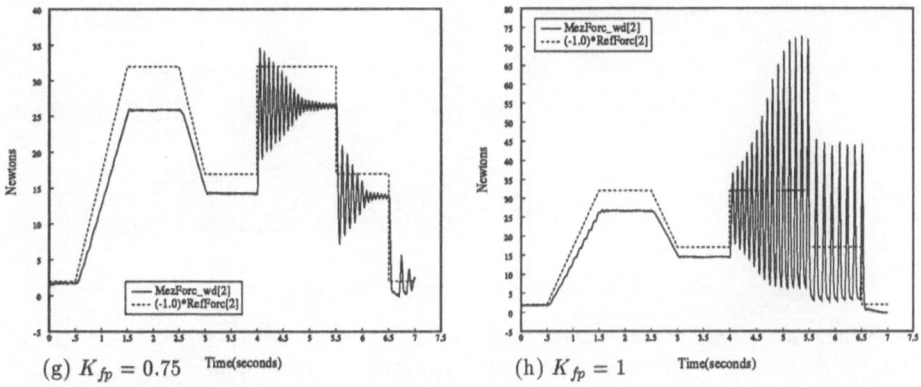

(g) $K_{fp} = 0.75$ Time(seconds)　　　　(h) $K_{fp} = 1$ Time(seconds)

Figure 12: (continued) Experimental data of proportional gain explicit force control with feedforward. The proportional gain varies from -0.75 to 1.

As is readily apparent, the response of this controller is essentially equivalent to that of the proportional gain controller shown in Figures 12 ·(a) through (h). This confirms the previous theoretical assertion that second order impedance control against a stiff environment is equivalent to explicit force control with proportional gain and feedforward compensation.

4.2.2   Impedance Control With Dynamics Compensation

Second order impedance control can also be implemented with dynamics compensation as shown in Equation (15). As can be seen in Equation (16), the mass ratio, ΛM^{-1} can be thought of a proportional force gain. However, it can be seen from Equations (11) and (12) that only M can be specified in this scheme.

Usually M is chosen to be diagonal in the task frame along with K and C. When operating in free space ($f_m = 0$) a diagonal M acts as a simple scaling factor for K and C, thereby preventing coupled motion. If M were nondiagonal, its product with diagonal K and C would be nondiagonal, and coupled motion would result. Further, K and C are usually chosen to be diagonal in some task frame which is aligned with the environment to be contacted. In this way, the manipulator may be made stiff tangential to a surface, but soft normal to it. (The velocity gains are usually chosen for critical damping.)

However, when in contact with the environment ($f_m \neq 0$), the ratio of the inertias, ΛM^{-1}, acts as a proportional force gain which is not diagonal in general, because Λ is not diagonal in general. Therefore, it is necessary to determine the effective value of the mass ratio (force gain). This requires finding the dominant element of Λ for the direction in which the environment is contacted.

Finding the dominant component of the inertia matrix is equivalent to finding the effective mass in the direction of concern. Since it is the force which is being controlled,

18

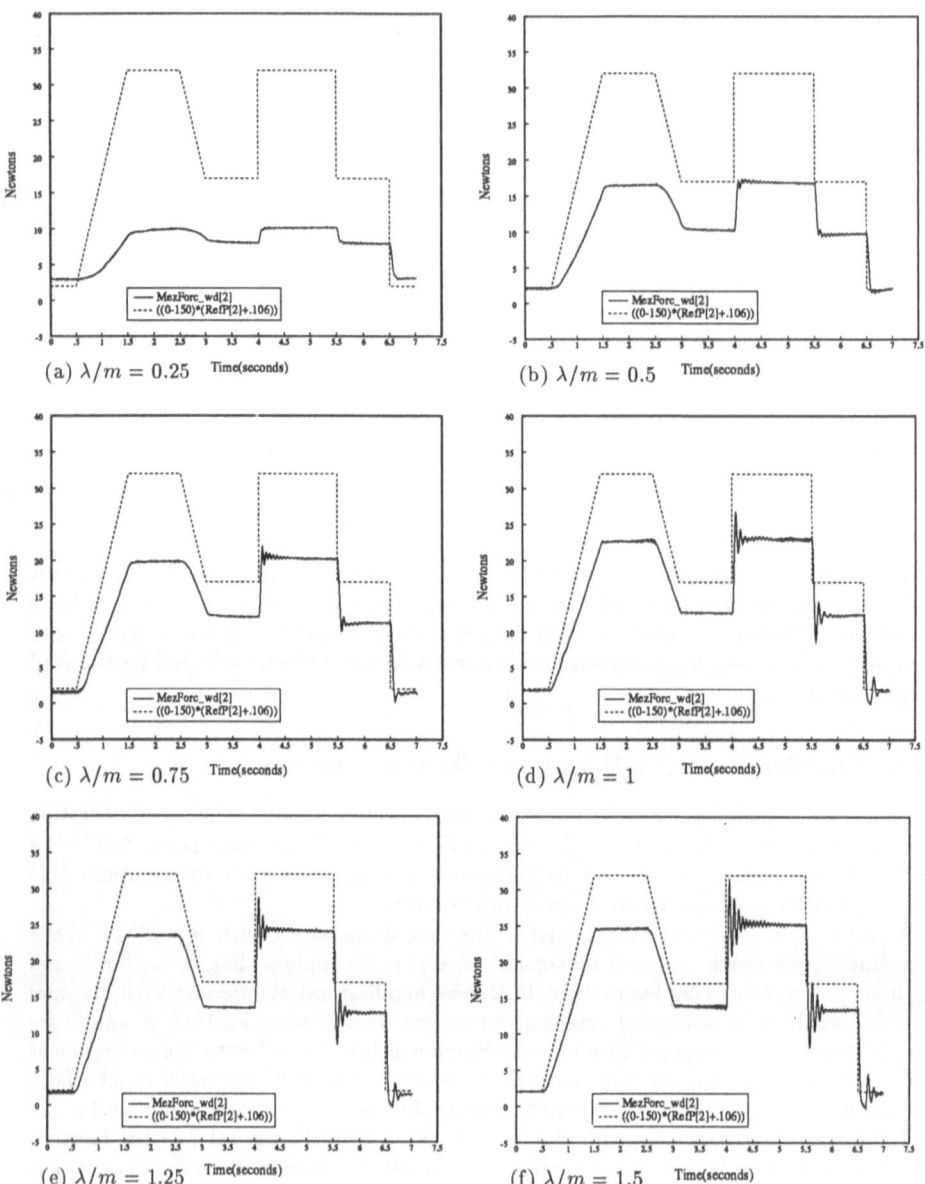

Figure 13: Experimental data of impedance control without dynamics compensation. The mass ratio 'gain' varies from 0.25 to 2.0.

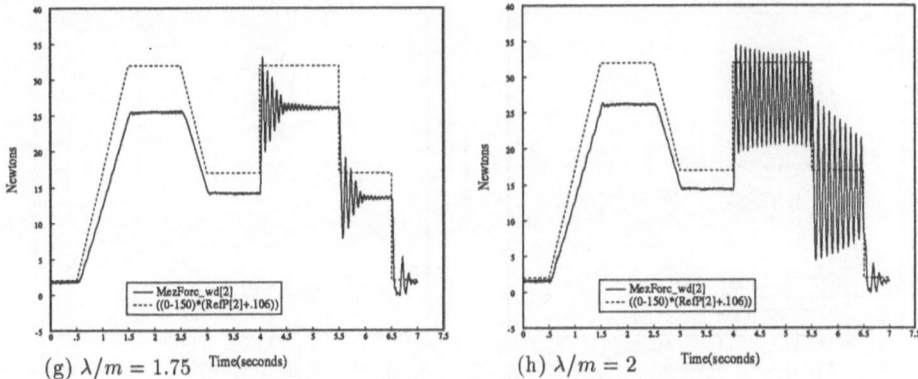

(g) $\lambda/m = 1.75$ (h) $\lambda/m = 2$

Figure 13: (continued) Experimental data of impedance control without dynamics compensation. The mass ratio 'gain' varies from 0.25 to 2.0.

this can only be done by determining the resultant acceleration from an applied force:

$$\ddot{x} = \Lambda^{-1}f \qquad (26)$$

The force may be set to be the unit vector in the direction of the surface. For the experiments performed, the z direction was chosen. The actual values of Equation (26) were:

$$\ddot{x} = \begin{bmatrix} 0.070 & 0 & -0.053 & 0 & 0 & 0 \\ 0 & 1.671 & 0 & -9.723 & -0.049 & 0 \\ -0.053 & 0 & 0.199 & 0 & 0 & 0 \\ 0 & -9.723 & 0 & 59.9 & -1.272 & 0 \\ 0 & -0.049 & 0 & -1.272 & 2.758 & 0 \\ 0 & 0 & 0 & 0 & 0 & 3226 \end{bmatrix} \begin{bmatrix} 0 \\ 0 \\ 1 \\ 0 \\ 0 \\ 0 \end{bmatrix} \qquad (27)$$

It is apparent that for forces applied in the z direction to the arm in this configuration, the dominant acceleration is $\ddot{x}_z \approx 0.2\text{m/s}^2$. Thus, the apparent inverse scalar mass is $\lambda_z^{-1} \approx 0.2\ \text{kg}^{-1}$. This implies that the best scalar approximation of the mass in the z direction is $\lambda_z \approx 5$ kg. This value may then be thought of as a scaling factor applied to the variable gain value M_{33} in Equation (15).

Figures 14 show the response of impedance control with dynamics compensation for $0.1 \leq M_{33}^{-1} \leq 0.45$. The other gains were $K = 150$ and $C = 10$. Using the approximation of $\lambda_z \approx 5$ kg, these gains can be thought of as $0.5 \leq \lambda_z M_{33}^{-1} \leq 2.25$. Thus, a direct comparison can be made between the system response shown in Figures 14 and that shown in Figures 13 and Figures 12. The responses are essentially the same. This comparison supports the analysis above. It also, further supports the conclusion that second order

20

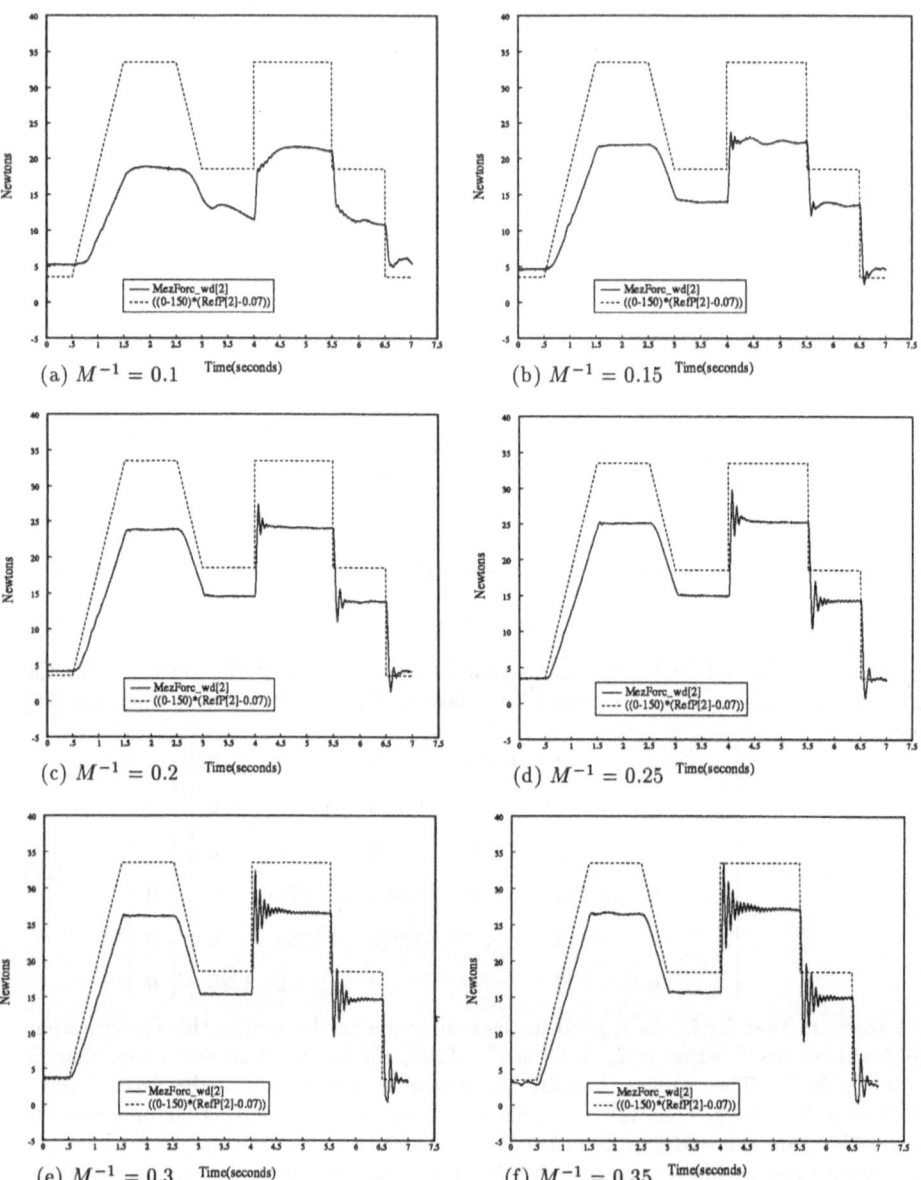

Figure 14: Experimental data of impedance control with dynamics compensation. The commanded inverse mass varies from 0.1 to 0.45. This is approximately the same as having a force gain vary from 0.5 to 2.25.

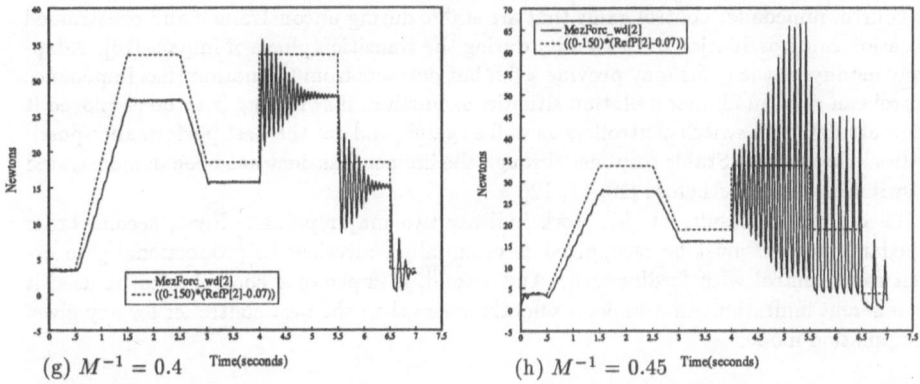

(g) $M^{-1} = 0.4$ _{Time(seconds)} (h) $M^{-1} = 0.45$ _{Time(seconds)}

Figure 14: (continued) Experimental data of impedance control with dynamics compensation. The commanded inverse mass varies from 0.1 to 0.45. This is approximately the same as having a force gain vary from 0.5 to 2.25.

impedance control, with or without dynamics compensation, is essentially equivalent to explicit force control with proportional gain and feedforward compensation.

5   Conclusions

It is apparent from the results presented here that second order impedance control and proportional gain explicit force control with feedforward are essentially the same thing. This leads us to the question the value of impedance control as a unified controller for motion through, and constrained interaction with, the environment. Our conclusion that impedance control is not the best solution for these modes of operation is illustrated by the following discussion.

First, it has been shown the proportional gain force control is not the best force controller; integral gain control provides much better tracking [16]. Therefore, the behavior of the impedance controller while in contact with the environment is not optimal, and not always stable.

Second, impedance control is more cumbersome to use since it requires position reference instead of force reference. Some researchers see this as a strength since there is no need to switch inputs between the modes of free space motion and constrained force application. However, this implies there is knowledge of the position commands necessary for an operation that intrinsically requires force commands.

Third, while not in contact with the environment, impedance control continues to incorporate force feedback information into the control law. Phenomena such as sensor noise or inertial loading can cause nonzero force readings and inhibit the performance of the position control.

Fourth, impedance control gains that are stable during unconstrained and constrained actuation cause oscillation or instability during the transition phase of impact [16]. Adaptively modifying the gains may provide a fix, but detracts from the notion that impedance control can work in all manipulation situations. Further, if switching is to be employed it seems attractive to switch controllers as well as gains, and get the best performance possible from the system. Stable response through the impact transient has been demonstrated by switching control schemes [19, 11, 12].

Therefore, the results of this work indicate two major points. First, second order impedance control must be recognized as essentially equivalent to proportional gain explicit force control with feedforward. And second, if impedance control is to be used it has inherent limitations that make it something less than the best controller for any given manipulation mode.

6 Acknowledgements

This research was performed at Carnegie Mellon University and supported by an Air Force Graduate Laboratory Fellowship (for Richard Volpe), DARPA under contract DAAA-21-89C-0001, the Department of Electrical and Computer Engineering, and The Robotics Institute.

The writing and publication of this paper was supported by the above and the Jet Propulsion Laboratory, California Institute of Technology, under a contract with the National Aeronautics and Space Administration.

The views and conclusion contained in this document are those of the authors and should not be interpreted as representing the official policies, either expressed or implied, of the U.S. Air Force, DARPA, or the U.S. Government. Reference herein to any specific commercial product, process, or service by trade name, trademark, manufacturer, or otherwise, does not constitute or imply its endorsement by the United States Government or the Jet Propulsion Laboratory, California Institute of Technology.

References

[1] C. An and J. Hollerbach. Dynamic stability issues in force control of manipulators. In *Proceedings of the IEEE Conference on Robotics and Automation*, pages 890–896. IEEE, 1987.

[2] A. Bejczy. Robot arm dynamics and control. Technical Memorandum 33-669, Jet Propulsion Laboratory, Pasadena, CA, February 1974.

[3] S. Eppinger and W. Seering. Understanding bandwidth limitations on robot force control. In *Proceedings of the IEEE Conference on Robotics and Automation*, pages 904–909, Raleigh, N.C., 1987. IEEE.

[4] A. Goldenberg. Implementation of force and impedance control in robot manipulators. In *Proceedings of the IEEE Conference on Robotics and Automation*, pages 1626–1632. IEEE, 1988.

[5] W. Hamilton. Globally stable compliant motion control for robotic assembly. In *Proceedings of the IEEE Conference on Robotics and Automation*, pages 1179–1184. IEEE, 1988.

[6] N. Hogan. Impedance control: An approach to manipulation: Parts i, ii, and iii. *Journal of Dynamic Systems, Measurement, and Control*, 107:1–24, March 1985.

[7] N. Hogan. Stable execution of contact tasks using impedance control. In *Proceedings of the IEEE Conference on Robotics and Automation*, pages 1047–1054. IEEE, 1987.

[8] H. Kazerooni, T. Sheridan, and Houpt P. Robust compliant motion for manipulators, parts i and ii. *IEEE Journal of Robotics and Automation*, RA-2(2):83–105, June 1986.

[9] O. Khatib. *Commande Dynamique dans l'Espace Operationnel des Robots Manipulateurs en Presence d'Obstacles*. PhD thesis, Ecole Nationale Superieure de l'Aeronautique et del'Espace (ENSAE), December 1980.

[10] O. Khatib. Real-time obstacle avoidance for manipulators and mobile robots. *The International Journal of Robotics Research*, 5(1), 1986.

[11] O. Khatib and J. Burdick. Motion and force control of robot manipulators. In *Proceedings of the IEEE Conference on Robotics and Automation*, pages 1381–1386. IEEE, 1986.

[12] D. Lokhorst and J. Mills. Implementation of a discontinuous control law on a robot during collision with a stiff environment. In *Proceedings of the IEEE Conference on Robotics and Automation*, pages 56–61. IEEE, 1990.

[13] J. Luh, M. Walker, and R. Paul. Resolved-acceleration control of mechanical manipulators. *IEEE Transactions on Automatic Control*, 25(3):468–474, June 1980.

[14] M. Mason. Compliance and force control for computer controlled manipulators. *IEEE Transactions on Systems, Man, and Cybernetics*, 11(6):418–432, June 1981.

[15] D. Stewart, D. Schmitz, and P. Khosla. Implementing real-time robotic systems using chimera ii. In *Proceedings of the IEEE International Conference on Robotics and Automation*, pages 598–603. IEEE, May 1990.

[16] R. Volpe. *Real and Artificial Forces in the Control of Manipulators: Theory and Experiments*. PhD thesis, Carnegie Mellon University, September 1990.

[17] R. Volpe and P. Khosla. Manipulator control with superquadric artificial potential functions: Theory and experiments. *IEEE Transactions on Systems, Man, and Cybernetics; Special Issue on Unmanned Vehicles and Intelligent Systems*, November/December 1990.

[18] R. Volpe and P. Khosla. Theoretical analysis and experimental verification of a manipulator / sensor / environment model for force control. In *Proceedings of the IEEE International Conference on Systems, Man, and Cybernetics*, Los Angeles, November 1990. IEEE.

[19] R. Volpe and P. Khosla. Experimental verification of a strategy for impact control. In *Proceedings of the IEEE International Conference on Robotics and Automation*, Sacramento, CA, April 1991. IEEE.

Robot Control in Singular Configurations — Analysis and Experimental Results

Stefano Chiaverini   Bruno Siciliano

Dipartimento di Informatica e Sistemistica
Università degli Studi di Napoli Federico II
Via Claudio 21, 80125 Napoli, Italy

Olav Egeland

Institutt for teknisk kybernetikk
Norges tekniske høgskole
O.S. Bragstads plass 8, 7034 Trondheim, Norway

Abstract

This paper describes analytical and experimental work for controlling robotic manipulators in the neighbourhood of kinematically singular configurations. The proposed method is based on a damped least-squares solution with user-defined accuracy. Results are given for a five-joint industrial robot.

Introduction

It has been long demonstrated that singular configurations constitute a serious problem in sensory control of robotic manipulators. Close to kinematic singularities the usual inverse differential kinematics solutions based on Jacobian (pseudo-)inverse become ill-conditioned, and this is experienced in the form of very high joint velocities and large control deviations.

When a preprogrammed reference end-effector trajectory is to be tracked, it is possible to interpolate in joint coordinates close to singular configurations, or to plan motions so that singularities are avoided. Instead, this is not possible in sensory control where the reference trajectory is not known a priori. The same problem is encountered in joy-stick control of a robot if the operator attempts to lead the robot through a singularity using end-effector motion increments.

Several research efforts have been devoted to devise control schemes that allow the robot to handle the occurrence of singularities. Some schemes identify the degenerate directions in the task space associated with a given singularity and eliminate the task velocity components along those directions in a suitable neighbourhood of the singularity [1–4]. Other schemes modify the exact inverse differential kinematic mapping by resorting to approximate mappings that offer robustness to singularities at the expense of reduced tracking accuracy [5–7]. A recent overview of control techniques of robotic systems through singularities can be found in [8].

In this paper we investigate the performance of a scheme for control in singular configurations based on the damped least-squares solution [5,6]. Implementation issues are discussed regarding the tuning of the damping factor for high accuracy with feasible velocities [9,10]. In particular, the use of weighted damped least-squares to achieve user-defined accuracy along given end-effector space directions is presented [11].

The scheme is applied in experiments to the five-joint ABB Trallfa TR400 robot. The results demonstrate the good performance of the system in crossing a singularity, and the influence of the weighting parameters is tested out extensively.

Damped least-squares solution

A robotic manipulator is naturally actuated in the joint space, whereas its end-effector motions are planned in the operational space. Therefore, in order to design effective control schemes in the joint or in the operational space, it is necessary to take into account the differential kinematic mapping between a motion increment $\delta \mathbf{q} \in \mathbf{R}^n$ in the joint space and the corresponding motion increment $\delta \mathbf{x} \in \mathbf{R}^m$ in the operational space, i.e.

$$\delta \mathbf{x} = \mathbf{J}(\mathbf{q})\delta \mathbf{q}, \tag{1}$$

where $\mathbf{J}(\mathbf{q})$ is the $(m \times n)$ Jacobian matrix of the manipulator considered. In practical cases, it is $m \leq n$; when $m < n$ the manipulator is said to be redundant and there exists an $(n - m)$-dimensional subspace of $\mathbf{R}^n$ in which differential joint-space motions give null differential operational space motions.

A configuration $\hat{\mathbf{q}}$ is said to be *singular* if $\mathrm{rank}(\mathbf{J}(\hat{\mathbf{q}})) = r < m$. At a singular configuration the subspace of the differential joint motion space which maps into the null differential motion vector in the operational space increases its dimension, as $\dim(\mathcal{N}(\mathbf{J}(\hat{\mathbf{q}}))) = n - r > n - m$. On the other hand, since $\dim(\mathcal{R}(\mathbf{J}(\hat{\mathbf{q}}))) = r < m$, only an r-dimensional subspace of differential motion in the operational space can be spanned at a singularity; this subspace is the space of feasible motion for the manipulator. In general, at a singular configuration a differential motion vector in the operational space may have both *feasible components*, lying in $\mathcal{R}(\mathbf{J}(\hat{\mathbf{q}}))$, and *degenerate components*, belonging to $\mathcal{R}^\perp(\mathbf{J}(\hat{\mathbf{q}}))$.

The linear mapping (1) can be effectively analyzed in terms of the singular value decomposition of the Jacobian matrix, that is

$$\mathbf{J} = \mathbf{U}\mathbf{\Sigma}\mathbf{V}^{\mathrm{T}} = \sum_{i=1}^{m} \sigma_i \mathbf{u}_i \mathbf{v}_i^{\mathrm{T}} \tag{2}$$

where $\mathbf{U}$ is the $(m \times m)$ matrix of the output singular vectors $\mathbf{u}_i$, $\mathbf{V}$ is the $(n \times n)$ matrix of the input singular vectors $\mathbf{v}_i$, and $\mathbf{\Sigma} = (\mathbf{S} \quad \mathbf{O})$ is the $m \times n$ matrix whose $(m \times m)$

diagonal submatrix $\mathbf{S}$ contains the singular values σ_i of the matrix $\mathbf{J}$. If r denotes the rank of $\mathbf{J}$, the following hold:

a) $\quad \sigma_1 \geq \sigma_2 \geq \ldots \geq \sigma_r > \sigma_{r+1} = \ldots = \sigma_m = 0$

b) $\quad \mathcal{R}(\mathbf{J}) = \text{span}\{\mathbf{u}_1, \ldots, \mathbf{u}_r\}$

c) $\quad \mathcal{N}(\mathbf{J}) = \text{span}\{\mathbf{v}_{r+1}, \ldots, \mathbf{v}_n\}$.

Notice that the $m - r$ output singular vectors associated to the null singular values represent the degenerate directions in the given configuration. The singular value decomposition is continuous and well-behaved not only in singular values but also in the direction of the singular vectors; thus, the $\mathbf{u}_i$ and $\mathbf{v}_i$ vectors will not change appreciably in the neighbourhood of a singularity.

Upon these premises, it can be recognized that the control system of a robotic manipulator should be provided with the capability of handling singularities. Usually, control operates in the joint space generating the driving torques at the joint actuators. The reference trajectory for the joint control servos is to be generated via kinematic inversion of the given end-effector trajectory, and then the occurrence of a singularity may prevent motion along degenerate components. On the other hand, even in the case the robot control system is designed directly in the operational space, still the problem of motion through and close to singular configurations exists since any controller of that kind is based on the use of the manipulator Jacobian.

An effective strategy that permits control of robotic manipulators in the neighbourhood of kinematic singularities is the *damped least-squares* technique originally proposed in [5,6]. The method corresponds to solving the equation

$$\mathbf{J}^{\mathrm{T}}(\mathbf{q})\delta\mathbf{x} = \left(\mathbf{J}^{\mathrm{T}}(\mathbf{q})\mathbf{J}(\mathbf{q}) + \lambda^2\mathbf{I}\right)\delta\mathbf{q} \tag{3}$$

in lieu of eq. (1); in (3) $\lambda \in \mathrm{R}$ is the damping factor and $\mathbf{I}$ is the identity matrix of proper dimension. It can be easily shown that the solution to (3) can be cast in the form

$$\delta\mathbf{q} = \mathbf{J}^{\mathrm{T}}(\mathbf{q})\left(\mathbf{J}(\mathbf{q})\mathbf{J}^{\mathrm{T}}(\mathbf{q}) + \lambda^2\mathbf{I}\right)^{-1}\delta\mathbf{x} \tag{4}$$

which remarkably requires the inverse of an $(m \times m)$ matrix instead of an $(n \times n)$ matrix. Notice that when λ is zero eq. (1) becomes identical to (3) and the damped least-squares solution reduces to the well-known Moore-Penrose pseudo-inverse solution.

It is important to point out that solutions of (3) satisfy the condition

$$\min_{\delta\mathbf{q}} \|\delta\mathbf{x} - \mathbf{J}(\mathbf{q})\delta\mathbf{q}\|^2 + \lambda^2\|\delta\mathbf{q}\|^2 \tag{5}$$

which evidences the possibility of trading off accuracy against feasibility of the joint space motion increment required to match up with the given operational space motion. Therefore, it is essential to select suitable values for the damping factor: Small values of λ give accurate solution but low robustness to the occurrence of singular and near-singular configurations. High values of λ result in low tracking accuracy even where a feasible and accurate solution would be possible.

To gain more insight into the features of solution (4), the above singular value decomposition (2) is helpful as it allows to rewrite the damped least-squares solution (4) as

$$\delta\mathbf{q} = \sum_{i=1}^{m} \frac{\sigma_i}{\sigma_i^2 + \lambda^2} \mathbf{v}_i \mathbf{u}_i^{\mathrm{T}} \, \delta\mathbf{x}. \tag{6}$$

It is clear that the components for which $\sigma_i \gg \lambda$ are little influenced by the damping factor, being

$$\frac{\sigma_i}{\sigma_i^2 + \lambda^2} \approx \frac{1}{\sigma_i}. \tag{7}$$

On the other hand, when a singularity is approached the smallest singular value tends to zero while the associated component of the solution is driven to zero by the factor σ_i/λ^2; this progressively reduces the joint motion required to achieve near-degenerate components of the commanded $\delta\mathbf{x}$.

The value selected for the damping factor represents the index which allows to decide about the closeness of the current configuration to a singularity; moreover, λ determines the degree of approximation introduced with respect to the pure least-squares (pseudo-inverse) solution.

An optimal choice for λ requires consideration of the smallest non-null singular value experienced along the whole trajectory and of the minimum damping needed to ensure feasible joint motion. To achieve good performance in the entire manipulator's workspace the use of a configuration-varying damping factor was proposed in [5]. The natural choice is to adjust λ as a function of some measure of closeness to the singularity at the current configuration of the robot arm; to this purpose manipulability measures [5] or estimates of the smallest singular value [9] can be adopted. A varying damping factor is in general computationally expensive; simplified computation exploiting knowledge of the kinematical structure has been recently proposed in [12] and in [13,10] for the manipulability and for the smallest singular value, respectively.

User-defined accuracy

The above damped least-squares method achieves a compromise between accuracy and robustness of the solution. This is performed without specific regard to the components of the particular task assigned to the manipulator's end effector. Consequently, it would be nice to think of a strategy that allows to discriminate between directions in the operational space where higher accuracy is desired and directions where lower accuracy can be tolerated. This is the case, for instance, of spot welding or spray painting in which the tool angle around the approach direction is not essential to the fulfilment of the task. Below is illustrated a *user-defined accuracy* technique [11] that meets the above requirements.

Let a weighted task increment be defined as

$$\delta\tilde{\mathbf{x}} = \mathbf{W}\delta\mathbf{x}, \tag{8}$$

where $\mathbf{W}$ is the $(m \times m)$ task-dependent weighting matrix that takes into account the anisotropy of the task requirements. The differential mapping relating the motion increment in joint coordinates to the weighted task increment is then given by

$$\delta \tilde{\mathbf{x}} = \tilde{\mathbf{J}}(\mathbf{q})\delta \mathbf{q} \tag{9}$$

where $\tilde{\mathbf{J}} = \mathbf{W}\mathbf{J}$. It is worth noticing that if $\mathbf{W}$ is full-rank, solving (1) is equivalent to solving (9), but with different conditioning in the matrix to be inverted. This suggests to select only the strictly necessary weighting action in order to avoid undesired ill-conditioning of $\tilde{\mathbf{J}}$.

A solution to the inverse kinematics problem for eq. (9) is obtained using the weighted damped least-squares technique [5]. This corresponds to solving the equation

$$\tilde{\mathbf{J}}^{\mathrm{T}}(\mathbf{q})\delta \tilde{\mathbf{x}} = (\tilde{\mathbf{J}}^{\mathrm{T}}(\mathbf{q})\tilde{\mathbf{J}}(\mathbf{q}) + \lambda^2 \mathbf{I})\delta \mathbf{q} \tag{10}$$

which in turn implies the fulfilment of the condition

$$\min_{\delta \mathbf{q}} \|\delta \tilde{\mathbf{x}} - \tilde{\mathbf{J}}(\mathbf{q})\delta \mathbf{q}\|^2 + \lambda^2 \|\delta \mathbf{q}\|^2. \tag{11}$$

Again, the singular value decomposition of the matrix $\tilde{\mathbf{J}}$ comes in support, i.e.

$$\tilde{\mathbf{J}} = \sum_{i=1}^{m} \tilde{\sigma}_i \tilde{\mathbf{u}}_i \tilde{\mathbf{v}}_i^{\mathrm{T}}, \tag{12}$$

and the solution to (10) can be written as

$$\delta \mathbf{q} = \sum_{i=1}^{m} \frac{\tilde{\sigma}_i}{\tilde{\sigma}_i^2 + \lambda^2} \tilde{\mathbf{v}}_i \tilde{\mathbf{u}}_i^{\mathrm{T}} \delta \tilde{\mathbf{x}}. \tag{13}$$

It is clear that the singular values $\tilde{\sigma}_i$ and the singular vectors $\tilde{\mathbf{u}}_i$ and $\tilde{\mathbf{v}}_i$ depend on the choice of the weighting matrix $\mathbf{W}$. While this has no effect on the solution $\delta \mathbf{q}$ as long as $\tilde{\sigma}_m \gg \lambda$, close to singularities where $\tilde{\sigma}_r \ll \lambda$ for some $r < m$ the solution can be shaped by properly selecting the matrix $\mathbf{W}$. To this purpose, two major issues must be considered:

a) The matrix $\mathbf{W}$ modifies the singular values σ_i into $\tilde{\sigma}_i$, and then the relative weight of each singular value is changed with respect to the selected λ; this differently modulates the damping of the near-degenerate components in comparison to the basic solution.

b) The matrix $\mathbf{W}$ modifies the singular vectors; in particular, this can be exploited to obtain that the output singular vectors $\tilde{\mathbf{u}}_i$ associated to degenerate directions be aligned to the task-space components which can tolerate loss of tracking accuracy.

Experimental results

The weighted damped least-squares method was applied in experiments to an ABB Trallfa TR400 industrial robot with hydraulic actuators. The computer system had a VME interface to the robot which was developed in cooperation with ABB Trallfa. The interface runs at 100 Hz. The inverse kinematic solutions were executed on a Motorola 68020/68881 VME board. On the same processor, independent joint control servos were implemented using proportional feedback which is the conventional type of controller for hydraulic drives; a bandwidth of 18 rad/sec was assigned.

As for the kinematic description, the manipulator has five joints with a vertical first axis, horizontal and parallel second, third and fourth axes, and a fifth axis which is orthogonal to the fourth axis; Figure 1 shows the manipulator with its Denavit-Hartenberg parameters.

The workspace of this manipulator is limited by the linear actuators of the rotary joints 2 and 3, so that the elbow and shoulder singularities are outside the workspace. Then, the only internal singularity of the manipulator is associated with the wrist.

The actual configuration of the wrist singularity will depend on the definition of the end-effector coordinates. It is clear that at most five end-effector coordinates can be specified. The orientation around the tool axis was left undefined; thus, the end-effector motion is described by the velocity vector

$$\dot{\mathbf{x}} = (\dot{x}_0 \quad \dot{y}_0 \quad \dot{z}_0 \quad \omega_y \quad \omega_z)^{\mathrm{T}}$$

where $\dot{x}_0$, $\dot{y}_0$, $\dot{z}_0$ are the end-effector velocity components in the base frame 0, and ω_y, ω_z are the angular velocity components along the tool axes y and z. The end-effector motion increment to be used in connection with (1) is $\delta\mathbf{x} = \dot{\mathbf{x}}\delta t$, where δt is a time increment.

With this end-effector coordinate vector, the singularity occurs for $\cos q_5 = 0$. In the neighbourhood of the singularity, it can be shown that the output singular vector corresponding to the smallest singular value is [13]

$$\mathbf{u}_5 = (0 \quad \alpha \quad 0 \quad 1 - \alpha^2 \quad 0)^{\mathrm{T}}, \tag{14}$$

where $\alpha \ll 1$ depends both on the weighting of the matrix $\mathbf{W}$ in (8) and on the actual configuration of the arm. From the above expression, it follows that the effect of damped least-squares is to eliminate a linear combination of the commanded motion in end-effector coordinates 2 and 4, which correspond to a translation in the y_0 direction in the base frame and a rotation around the same axis in the tool frame.

As emphasized in the previous section, the adoption of the weighting matrix $\mathbf{W}$ modifies the singular values and the singular vectors of $\mathbf{J}$ into the ones characterizing $\tilde{\mathbf{J}}$. According to the expression in (14), different weights were assigned to the second and fourth components of the task space end-effector vector. The choice

$$\mathbf{W}_1 = \mathrm{diag}(1, w, 1, 1, 1) \tag{15}$$

provides low task space sensitivity for a translation along y_0 in the base frame, while the choice

$$\mathbf{W}_2 = \mathrm{diag}(1, 1, 1, w, 1) \tag{16}$$

provides low task space sensitivity for a rotation along y in the tool frame; obviously, a scalar $0 < w < 1$ is to be chosen.

The other crucial point for a damped least-squares solution is the selection of the damping factor. To this purpose, a singular region can be defined on the basis of the estimate of the smallest singular value of $\mathbf{J}$; outside the region the exact solution is used, while inside the region configuration-varying damping and weighting factors are introduced to obtain the desired approximate solution. Both factors must be chosen so that continuity of the joint space motion increment $\delta\mathbf{q}$ is ensured in the transition between the solutions. Following the guidelines of [11], the damping and weighting factors take on the following expressions:

$$\lambda^2 = \begin{cases} 0 & \text{when } \hat{\sigma}_5 \geq \epsilon \\ \left(1 - \left(\frac{\hat{\sigma}_5}{\epsilon}\right)^2\right)\lambda_{max}^2 & \text{otherwise,} \end{cases}$$

$$(w-1)^2 = \begin{cases} 0 & \text{when } \hat{\sigma}_5 \geq \epsilon \\ \left(1 - \left(\frac{\hat{\sigma}_5}{\epsilon}\right)^2\right)(w_{min}-1)^2 & \text{otherwise,} \end{cases}$$

where $\hat{\sigma}_5$ is the estimate of the smallest singular value, which can be computed as

$$\hat{\sigma}_5 = \left| |q_5| - \frac{\pi}{2} \right|,$$

and ϵ defines the size of the singular region. The numerical values used in the experiments were $\epsilon = 0.2$, corresponding to a singular region of about 20 degrees wide in q_5, and $\lambda_{max} = 0.15$.

The initial configuration of the arm was

$$\mathbf{q}(0) = (\,0 \quad -0.1 \quad 0.1 \quad -0.1 \quad \pi/2 - 0.4\,)^{\mathrm{T}},$$

and the commanded step velocity was

$$\dot{\mathbf{x}} = (\,0 \quad 0 \quad 0 \quad 0.07 \quad 0.17\,)^{\mathrm{T}}.$$

It can be easily recognized that the resulting motion trajectory passes through the singularity $q_5 = \pi/2$ with significant degenerate components.

Experimental results were obtained adopting four different solutions; namely,

1. pure resolved-rate inverse Jacobian,
2. usual damped least-squares without weighting ($\mathbf{W} = \mathbf{I}$),
3. weighted damped least-squares with $\mathbf{W}_1$ as in (15) and $w_{min} = 0.1$,
4. weighted damped least-squares with $\mathbf{W}_2$ as in (16) and $w_{min} = 0.01$.

This was intended to analyze the performance of the proposed method in comparison to the standard methods.

With the first solution, the motion of the arm in the neighbourhood of the singularity becomes unacceptable; large joint velocities are generated which reach the mechanical limits and the arm breaks down.

With the second solution, a significant deviation in y_0 was observed due to the damping of the motion in the $\mathbf{u}_5$ task space direction. The nonzero $\alpha = 0.064$ in (14) gives a kinematic coupling between the second and the fourth component of the task velocity.

With the third solution, a very large deviation from the given trajectory along y_0 was generated (Fig. 2) due to the fact that the weighting is applied along that same direction (resulting in $\alpha = 0.54$); a maximum error of 0.54 m occurred after 5 sec of motion.

With the fourth solution, the performance along y_0 was much improved at the expense of that along ω_y ($\alpha = 0.0006$). The arm tracked satisfactorily the translational part of the trajectory with a maximum error of 0.03 m (Fig. 2).

Notice that both the resulting joint trajectories in the last two cases (Fig. 3) pass through the singular configuration without any discontinuity. Obviously, the joint motion is different in the two cases.

The motion of the manipulator in the experiments is illustrated in a section of the video proceedings of this conference.

Conclusions

Analytical and experimental results have been presented for a damped least-squares solution with user-defined accuracy. Satisfactory performance was achieved for controlling the motion of a five joint industrial robot manipulator through the wrist singularity. It was shown that suitable weighting of task space directions can be exploited to achieve accurate tracking along specified axes of motion. Smoothness and continuity of the solution was ensured by defining a proper singular region in the neighbourhood of the singularity.

Future developments of this joint research project will be devoted to applying the proposed method to the case of the six joint ABB IRB 2000 robot with both shoulder and wrist singularities.

Acknowledgements

The research work described in this paper was supported by *Ministero dell'Università e della Ricerca Scientifica e Tecnologica* under the 40% funding project, by *Consiglio Nazionale delle Ricerche* under contract 90.00355.PF67, and by *ABB Robotics*.

References

[1] D.E. Whitney, "The mathematics of coordinated control of prosthetic arms and manipulators," *ASME J. of Dynamic Systems, Measurement, and Control*, vol. 94, pp. 303–309, 1972.

[2] E.W. Aboaf and R.P. Paul, "Living with the singularity of robot wrists," *Proc. 1987 IEEE Int. Conf. on Robotics and Automation*, Raleigh, NC, pp. 1713–1717, Mar.–Apr. 1987.

[3] O. Khatib, "A unified approach for motion and force control of robot manipulators: The operational space formulation," *IEEE J. of Robotics and Automation*, vol. 1, pp. 43–53, 1988.

[4] S. Chiaverini and O. Egeland, "A solution to the singularity problem for six-joint manipulators," *Proc. 1990 IEEE Int. Conf. on Robotics and Automation*, Cincinnati, OH, pp. 644–649, May 1990.

[5] Y. Nakamura and H. Hanafusa, "Inverse kinematic solution with singularity robustness for robot manipulator control," *ASME J. of Dynamic Systems, Measurements, and Control*, vol. 108, pp. 163–171, 1986.

[6] C.W. Wampler, "Manipulator inverse kinematic solutions based on vector formulations and damped least-squares methods," *IEEE Trans. on Systems, Man, and Cybernetics*, vol. 16, pp. 93–101, 1986.

[7] P. Chiacchio and B. Siciliano, "Achieving singularity robustness: An inverse kinematic solution algorithm for robot control," *IEE Int. Work. on Robot Control: Theory and Application*, Oxford, GB, pp. 149–156, Apr. 1988.

[8] S. Chiaverini, L. Sciavicco and B. Siciliano, "Control of robotic systems through singularities," *Advanced Robot Control — Proc. Int. Workshop on Nonlinear and Adaptive Control: Issues in Robotics*, C. Canudas de Wit (Ed.), Lecture Notes in Control and Information Sciences, vol. 162, pp. 285–295, Springer-Verlag, Berlin, 1991.

[9] A.A. Maciejewski and C.A. Klein, "Numerical filtering for the operation of robotic manipulators through kinematically singular configurations," *J. of Robotic Systems*, vol. 5, pp. 527–552, 1988.

[10] O. Egeland, M. Ebdrup and S. Chiaverini, "Sensory control in singular configurations — Application to visual servoing," *Proc. IEEE Int. Work. on Intelligent Motion Control*, Istanbul, TR, pp. 401–405, Aug. 1990.

[11] S. Chiaverini, O. Egeland and R.K. Kanestrøm, "Achieving user-defined accuracy with damped least-squares inverse kinematics," *Proc. 5th Int. Conf. on Advanced Robotics*, Pisa, I, pp. 672–677, June 1991.

[12] T. Yoshikawa, "Translational and rotational manipulability of robotic manipulators," *Proc. 1990 American Control Conference*, San Diego, CA, pp. 228–233, May 1990.

[13] A.A. Maciejewski and C.A. Klein, "The singular value decomposition: Computation and application to robotics," *Int. J. of Robotics Research*, vol. 8, n. 6, pp. 63–79, 1989.

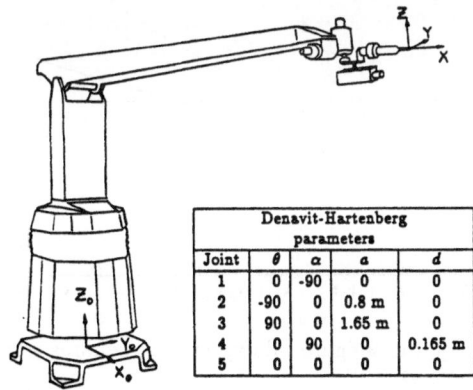

Denavit-Hartenberg parameters				
Joint	θ	α	a	d
1	0	-90	0	0
2	-90	0	0.8 m	0
3	90	0	1.65 m	0
4	0	90	0	0.165 m
5	0	0	0	0

Fig. 1. The ABB Trallfa TR 400 industrial robot manipulator.

34

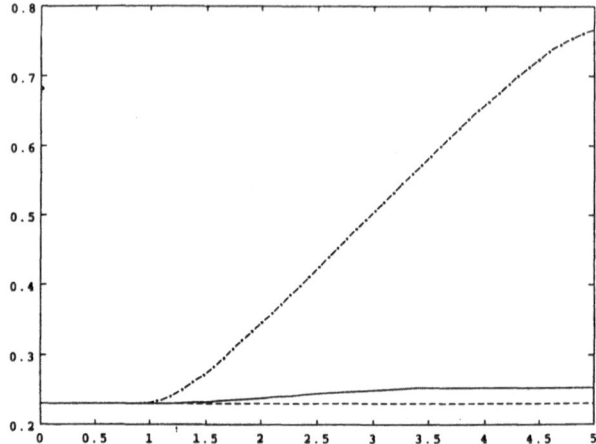

Fig. 2. Motion in y_0 due to commanded motion in the ω_y for the weighted damped least-squares solution. The y_0 reference is constant (dashed curve). The motion in y_0 and ω_y becomes linearly dependent in the singularity, and an error in y_0 is observed. The error in y_0 depends strongly on the weighting. With a weighting of 0.01 on ω_y the maximum error is 0.03 m (solid curve), while with a weighting of 0.1 on y_0 the maximum error is 0.54 m (curve with dots and dashes).

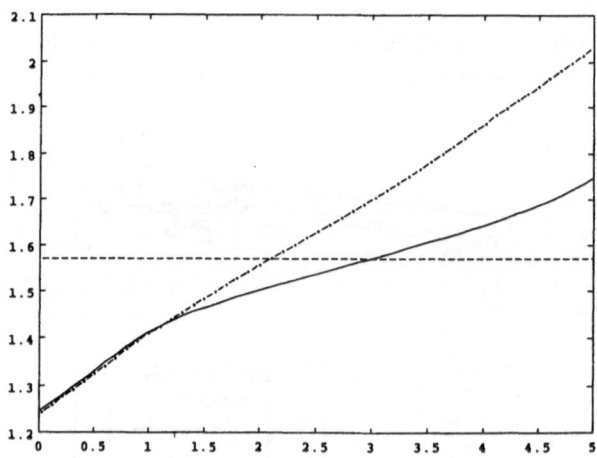

Fig. 3. Motion in q_5 for the weighted damped least-squares solution. The singularity occurs for $q_5 = \pi/2$.

Implementation of Control Algorithms for Flexible Joint Robots on a KUKA IR 161/60 Industrial Robot: experimental results.

M. Adams, J. Swevers , D. Torfs, J. De Schutter, H. Van Brussel

Katholieke Universiteit Leuven,

Division of Production Engineering, Machine Design and Automation (PMA)

Celestijnenlaan 300 B, B-3001 Leuven (Heverlee) Belgium

Abstract

This paper reports on experiments with identification and control of flexible joint robots. It presents two simple modifications of a standard industrial robot controller for a KUKA IR 161/60 robot: improved trajectory generation and flexible control of the first three joints.

The flexible controller consists of a linear position dependent state feedback controller for the first axis and a nonlinear decoupling controller for the second and third axis.

This paper gives a description of modeling, identification and control design of both the linear position dependent state feedback controller and the nonlinear decoupling controller.

Tests show that the smooth trajectory gives the largest contribution to the improved dynamic performance (overshoot, oscillations tracking error), but that there is also a significant improvement of a flexible controller over a classical PID controller in terms of tracking accuracy. In addition the static error due to gravity is compensated.

1   Introduction

Flexibility of joints or links limits the performance of robots. Traditional control designed for rigid robots must be supplemented. First, to prevent or damp out undesired oscillations resulting from excitation of structural dynamics and secondly, to improve the tracking of a reference trajectory designed for a rigid robot.

During the last two decades, a lot of complex control algorithms for multi degree of freedom flexible robots have been developed. Most algorithms are tested on idealised and specially developed test setups, because they require an accurate model of the robot. Industrial robots are avoided because they are too far from ideal and therefore difficult to model. In addition flexibility is avoided in industrial robots. This questions the applicability of these algorithms to industrial robots.

This paper presents experimental results of flexible control algorithms implemented on a KUKA IR 161/60 industrial robot. The flexibility of the KUKA robot is limited.

The main flexibilities are located in the first three joints. The dynamics of second and third axes are decoupled from the dynamics of the first axis. Therefore, the first axis is controlled with a linear position dependent state feedback controller and the second and third axis are controlled with a nonlinear decoupling controller. The nonlinear control method is generally formulated in [1]. It consists of:

- a nonlinear feedforward to cancel out nonlinear terms in the rigid body dynamics.

- linearization of the remaining nonlinearities, and linear feedback of all state variables (rigid and flexible coordinates). This ensures proper stability of the system about the desired state.

- Compensation of static and dynamic deflection improves the tracking of a reference trajectory. This compensation is achieved by calculation of the reference values of all state variables based on the model of the flexible robot.

The last three joints of the robot are rigid. They are controlled with a PID controller in velocity command.

The experimental results of the flexible controller are compared to the performance of the standard industrial robot controller which is normally delivered with the robot. Besides taking into account the flexibility in the control design a second improvement of the standard controller is the use of a smoother trajectory. Tests show that the smooth trajectory gives the largest contribution to the improved dynamic performance (overshoot, oscillations, tracking error), but that there is a significant improvement of a flexible controller over a classical PID controller in terms of tracking accuracy. In addition the static error due to gravity is compensated. At very high velocities and accelerations, there is also a significant improvement of a flexible controller over a classical PID controller in terms of overshoot and oscillations. Therefore is concluded that even for an industrial robot with limited flexibility, some improvements can be obtained.

Section 2 describes the KUKA robot, the sensor equipment, the control hardware and the standard industrial robot controller. Section 3 describes modeling, identification, control design and the test results of the linear position dependent state feedback controller of the first axis. Section 4 describes modeling, identification and control design of the nonlinear decoupling controller of the second and third axis. It also reports and interprets the test results of this controller.

2 Description of the test setup

Fig. 1 shows the setup for one of the tests. The robot is a KUKA IR 161/60 industrial robot (with an arm extension AV400): it has 6 axes, it has a working envelope with radius 3150 mm, and it is equiped with a spot welding tool of 45 kg. Three extra encoders are attached to the first three links of the robot. They measure the link position in addition to the motor position, which is measured with an encoder mounted on the motor shaft. The difference between the motor encoder signal and the corresponding extra encoder signal is a measure for the deformation of the joint. An experimental modal analysis of this robot has shown that the main flexibility in this robot is caused by the first three joints [12], such that these extra encoders suffice to measure the flexibility. A digitizing

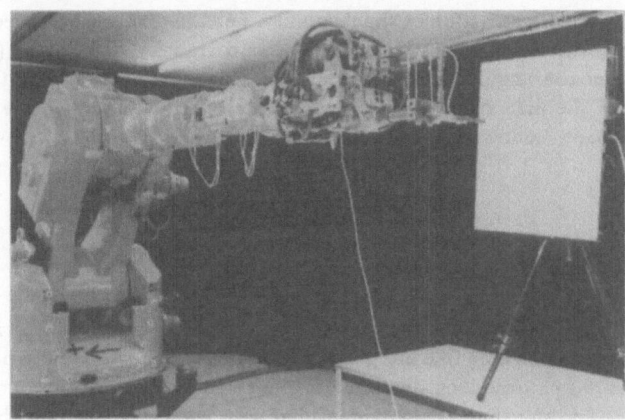

Figure 1: KUKA IR 161/60 industrial robot equiped with spot welding tool

pen and tablet together with the software package RODYM [11] measure the end effector movements and positioning. The digitizing pen is attached to the tool of the robot.

A standard robot controller is delivered with the robot. It controls all axes in velocity command with an independent PID controller. Velocity feedforward is added, and a trajectory based on a trapezoidal velocity (constant acceleration) profile is used.

The improved trajectory generation and the control algorithms are programmed on a VME–system. This system can read encoder signals and send out analog voltages in the range of $\pm 10\,volt$. These voltages are linearly converted by the power supply of the motors into a motor current, when the power supply is working in torque command, or a motor velocity, when the power supply is working in velocity command. The command mode can be altered for each axis separately.

3 Linear position dependent state feedback control of the first axis

3.1 Modelling and identification

The controller for the first axis is based on 4 fourth order discrete time linear models of the robot. Each of these models has been identified experimentally [10] for a certain configuration of axes 2 and 3. These configurations lie between the fully extended configuration and fully contracted configuration of the robot. The models take into account the flexibility of the first joint with one resonance and one anti–resonance frequency.

The identification of each fourth order state space models is based on the identification of two discrete time transfer functions. The division of the total model into two submodels, and the separate identification of these two submodels based on two frequency response functions, results in a more accurate system model than the direct identification of the total model based on one frequency response function [9]. The first transfer function, $H_1(z)$, relates the input command for the motor of the first axis (proportional to the motor torque), to the deformation of the flexibility, i.e. the difference between the angular link and motor position of the first axis. The second transfer function $H_2(z)$ relates the difference between the angular link and motor position of the first axis to the angular motor position of the first axis. These discrete time transfer functions are used to build the total state–space models. They are given by:

$$H_1(z) = \frac{b_{11}z}{z^2 + a_{11}z + a_{12}}, \tag{1}$$

$$H_2(z) = \frac{b_{20}z^2 + b_{21}z + b_{22}}{z^2 - 2z + 1}.$$

$H_1(z)$ contains a pair of complex conjugated poles which describes the resonance frequency. $H_2(z^{-1})$ contains a double integration (two poles at $z = 1$), and a pair of complex conjugated zeros which describes the anti–resonance frequency. The parameters of these two discrete time transfer functions are identified with the frequency domain method presented in [10]. The method calculates the parameters of the models using stepped–sine measured frequency response functions of the system. Figure 2 shows the amplitude and the phase of identified models $H_1(z)$ and $H_2(z)$ and of the measured frequency response functions for the robot configuration which is closest to the fully extended configuration.

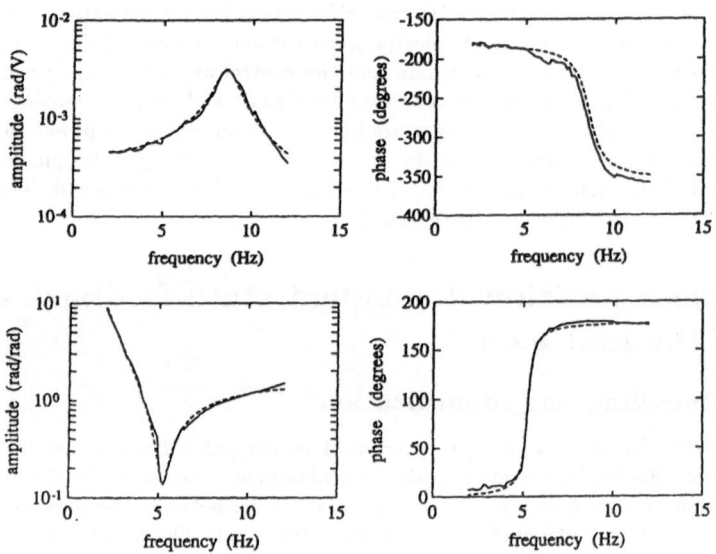

Figure 2: Comparison between the measured frequency response functions (solid line) and the identified transfer functions $H_1(z)$ (upper figures) and $H_2(z)$ (bottom figures) (dashed line), for the robot configuration which is closest to the fully extended configuration.

3.2 Control design

The control signal $u[k]$ for the first axis is a weighed sum of four control signals $u_j[k]$.

$$u[k] = \sum_{j=1}^{4} w_j[k] u_j[k]. \tag{2}$$

$w_j[k]$ is the weight for $u_j[k]$. The control signals $u_j[k]$ consist of feedforward, and feedback of the difference between the desired and the measured states:

$$u_j[k] = u_{ff\,j}[k] - \mathbf{K}_{fb\,j}(\mathbf{x}[k] - \mathbf{x}_{d\,j}[k]). \tag{3}$$

The design of the state feedback controllers and the calculation of the feedforward signals is based on discrete time state space models. Each state space model is based on two frequency domain identified discrete time transfer functions $H_1(z)$ and $H_2(z)$, and the transfer function of a second order digital low pass Butterworth filter with a cut off frequency of $10Hz$. This filter is added to the models because both encoder signals are filtered in order to reduce the spillover problem. The combination of the three submodels into the total state space models is such that all state variables are measurable or exactly known. A closed–loop state estimator in unnecessary, which simplifies the implementation of the controller considerably. The calculation of the feedforward signal $u_{ff\,j}[k]$ is a simulation of the inverse state space model corresponding to the j-th robot configuration, with the desired trajectory. The inverse simulations are stable, because all state space models are of minimal phase. They require a smooth trajectory, based on a ninth order polynomial, to prevent peaks in the feedforward signals. The desired state trajectories $\mathbf{x}_{d\,j}[k]$ result from a simulation with the j-th state space model using $u_{ff\,j}[k]$. The state feedback gains are calculated using pole placement.

Each of these state feedback controllers is stable in the whole working space of the robot, and gives a good dynamic behaviour in an area around the robot configuration for which they are designed. The weights $w_j[k]$ are function of the distance of the end effector from the first axis of rotation. Figure 3 shows the four weights $w_j[k]$ as a function of $d[k]$. In this figure, d_1, d_2, d_3, and d_4 denote the distances of the end effector from the first axis of rotation, corresponding to the four robot configurations in which the robot is identified.

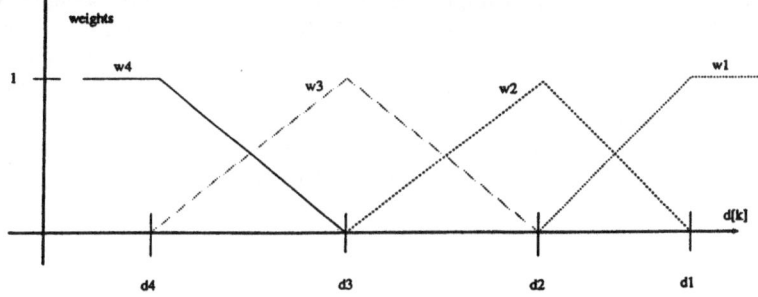

Figure 3: The four weights $w_j[k]$ as a function of $d[k]$

3.3 Experimental Results

Three controllers are implemented:

1. The standard robot controller: PID rigid controller in velocity command with appropriate feedforward. A trajectory based on a trapezoidal velocity profile is used.

2. A PID rigid controller in velocity command with appropriate feedforward. A smooth trajectory is used.

3. The flexible controller in torque command. A smooth trajectory is used.

All controllers are optimized (appropriate choice of feedback gains). Integration is added. A sample period of $6ms$ is used.

The experiments leads to the conclusion that a smooth trajectory gives the most important contribution to an improved dynamic behaviour (comparison of controller 1 and 2). Tests performed with controller 2 and 3 show no vibrations in both cases. Only at extreme high velocities, there is a significant difference between the flexible controller and the PID controller. In that case there is still an excitation of the structural dynamics if the controller does not take the flexibility into account.

Figure 4 evaluates the dynamic performance of the three controllers for a test motion at extreme high velocity. Only the first axis is controlled here, because the first joint is the most flexible part of the robot. The robot is put in its fully extended configuration. This results in a minimal eigenfrequency and a maximum end point oscillation. In this test, the first axis moves at maximum achievable velocity over 90 degrees while all other axes have their brakes on. The end effector positioning is measured with the RODYM measurement system [11]. For controller 1, there is an overshoot of 0.65 mm and the settling time is 1.60 seconds. For controller 2, there is an overshoot of 0.16 mm and the settling time is 0.56 seconds. For controller 3, there is no overshoot and the end-effector does not leave the repeatability zone. This corresponds to a settling time of 0.0 sec. The settling time is defined as the period of time which elapses between the instant at which the end effector enters a specified limit band around the desired position for the first time and the time instant of staying inside this band.

Figure 5 show the tracking test results of the first three controllers for a test contour which is a circle with radius 80 mm, lying in a horizontal plane. The test contour is executed at low and at high velocity, i.e. at 10% and at 100% of the maximum velocity allowed by the standard controller. The end effector path measured at low velocity is considered as the reference path, with which the maximum velocity path is compared. The solid line is the result obtained at low velocity, and the dashed line is the result obtained at high velocity. The bottom figure is a detail of the top figure, taken at the point where the velocity reaches its maximum. The maximum deviation for controller 1 is 1.5 mm, for controller 2 0.7 mm, and for controller 3 0.5 mm.

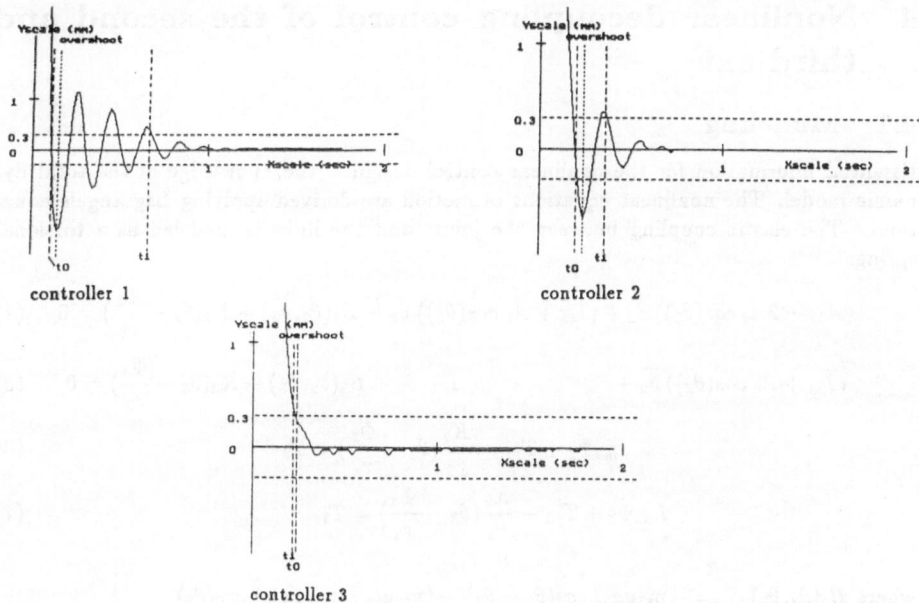

Figure 4: End effector response measured for a test motion at extreme high velocity with controller 1, 2 and 3.

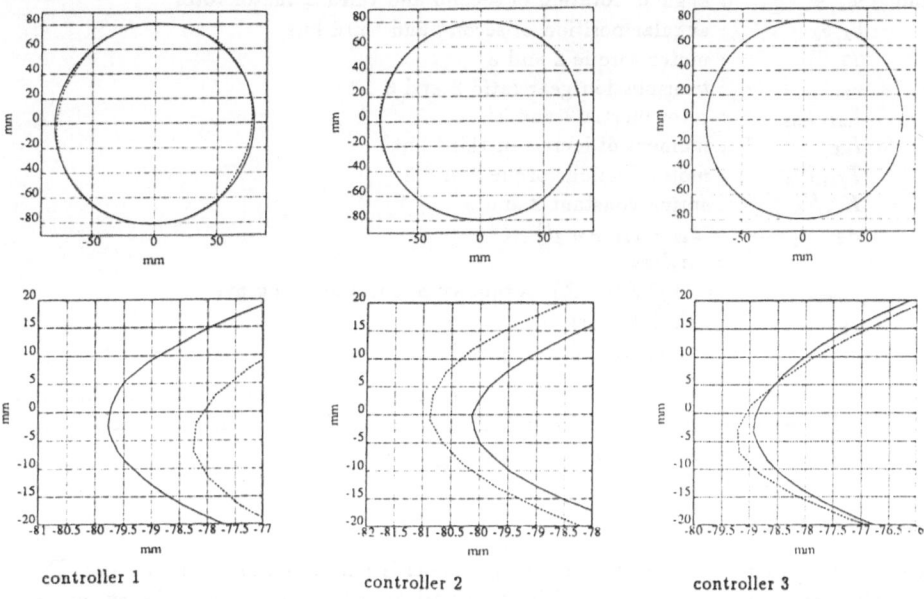

Figure 5: End effector response measured during circle tracking test with controller 1, 2 and 3; low velocity: solid line, high velocity: dashed line.

4 Nonlinear decoupling control of the second and third axis

4.1 Modeling

Essential information for the nonlinear control design is the knowledge of the total dynamic model. The nonlinear equations of motion are derived applying Lagrange's equations. The elastic coupling between the joints and the links is modeled as a torsional spring.

$$(A_{22} + 2A_c \cos(\theta_3)) \ddot{\theta}_2 + (I_{z3} + A_c \cos(\theta_3)) \ddot{\theta}_3 + B_2(\theta_2, \theta_3) + K_2(\theta_2 - \frac{\Phi_2}{r_2}) = 0 \quad (4)$$

$$(I_{z3} + A_c \cos(\theta_3)) \ddot{\theta}_2 + I_{z3} \ddot{\theta}_3 + B_3(\theta_2, \theta_3) + K_3(\theta_3 - \frac{\Phi_3}{r_3}) = 0 \quad (5)$$

$$I_{m2}\ddot{\Phi}_2 + T_{f2} - \frac{K_2}{r_2}(\theta_2 - \frac{\Phi_2}{r_2}) = T_2 \quad (6)$$

$$I_{m3}\ddot{\Phi}_3 + T_{f3} - \frac{K_3}{r_3}(\theta_3 - \frac{\Phi_3}{r_3}) = T_3 \quad (7)$$

where $B_2(\theta_2, \theta_3)$ $= -m_3 g x_{3c} \cos(\theta_2 + \theta_3) - (m_3 g l_2 + m_2 g x_{2c}) \cos(\theta_2)$
$\qquad\qquad\quad -A_c \dot{\theta}_3(2\dot{\theta}_2 + \dot{\theta}_3)\sin(\theta_3) - T_{comp}$

$\qquad B_3(\theta_2, \theta_3)$ $= -m_3 g x_{3c} \cos(\theta_2 + \theta_3) + A_c \dot{\theta}_2^2 \sin(\theta_3)$

and $\quad \Phi_2, \Phi_3$: angle of rotation of second and third actuator rotor

$\qquad \theta_2, \theta_3$: angular position of second and third link

$\qquad T_2, T_3$: motor torque 2 and 3

$\qquad r_2, r_3$: transmission gear ratio 2 and 3

$\qquad I_{m2}, I_{m3}$: rotor inertia 2 and 3

$\qquad I_{z3}$: moment of inertia of third link

$\qquad T_{f2}, T_{f3}$: motor friction torque 2 and 3

$\qquad K_2, K_3$: spring constant 2 and 3

$\qquad A_{22}$ $= I_{z2} + I_{z3} + m_3 l_2^2$

$\qquad A_c$ $= m_3 l_2 x_{3c}$

$\qquad T_{comp}$ $= -\frac{3690}{r_2}(\theta_2 + \frac{\pi}{2})$: torque compensating the gravity

$\qquad m_i$: i th link mass

$\qquad x_i$: i th center of mass position

The second link position is measured with respect to the horizontal axis. The third link position is measured with respect to link 2. The motor friction is modeled as a sum of linear viscous friction and Coulomb friction.

4.2 Identification

The identification of the parameters in this model is performed in different steps: First, the parameters related to the rigid body model are estimated. Then the spring constants are estimated by using static, quasi-static, noise measurements and stepped sine measurements.

In the different steps the following identification procedure is applied: a known input torque is applied to the system. The corresponding angle of rotation of the actuator rotors and corresponding angular position of the links are measured. Introducing the measured values and their first and second derivatives in the non linear equations gives rise to an overdetermined set of linear equations. Solving this set of equations in a linear least squares sense gives an estimate of the unknown parameters.

4.2.1 Identification of inertial and gravitational coefficients.

The identification method requires motion of a single axis, while the positions of the others are fixed. A test motion with high accelerations but small changes in position is chosen. In this way there are no coupling torques acting on the moving axis, and the gravitational forces can be regarded as constant. The main component of the motor torque can thus be related to the effective inertia of the moving axis and to the constant gravitational forces. The coupling inertias are determined from the same measurements by considering the control torques needed to maintain the fixed position at the other axes. By repeating the procedure in different positions, sets of unknown inertial and gravitational coefficients are obtained as a function of the joint position. From this set of coefficients the parameters A_{22}, A_c and I_{z3} are solved.

4.2.2 Identification of the spring constants.

- Static measurements: In four different positions , the endeffector (without spot welding tool) is loaded with a mass $m_l = 96.7\ kg$. The deformation before and after loading is measured. From the difference between these two deformations Δq_2 and Δq_3, the spring constants are calculated:

$$K2 = \frac{l_3 m_l g\ cos(\theta_2 + \theta_3) + g m_l l_2\ cos(\theta_2)}{\Delta q_2}$$

$$K3 = \frac{l_3 m_l g\ cos(\theta_2 + \theta_3)}{\Delta q_3}$$

- Quasi-static measurements: The same measurement data as in 4.2.1 are used. The deflection $q_2 = \theta_2 - \frac{\Phi_2}{r_2}$ and $q_3 = \theta_3 - \frac{\Phi_3}{r_3}$ is considered. K_2 and K_3 are calculated from the relationship

$$\Delta q_2 = -\frac{r_2}{K_2}(\Delta T_2 - I_{m2}\ddot{\Phi}_2)$$

$$\Delta q_3 = -\frac{r_3}{K_3}(\Delta T_3 - I_{m3}\ddot{\Phi}_3)$$

In order to eliminate the influence of the extra encoder offsets, only variations with respect to the initial position are considered. Figure 6 gives the measured and identified deflection at each joint for a position step of $5°$ for link 2, resp. link 3.

- Bandlimited noise measurements: The spring constants are determined from measured data obtained by applying bandlimited noise to the system.

- Stepped sine measurements: Transfer functions are measured using stepped sine measurements. The excitation is generated by the second and third motor separately while the not excited motors are clamped by brakes. The spring constants are determined by comparing the analytical expression for the resonance and anti-resonance frequencies with the measured resonance and anti-resonance frequencies.

In table 1 , the identified parameters are compared with the analytical parameters. For control design, the quasi-static values are used. Using these values, the smallest differences between feedforward and actual torques and between measured and calculated state variables are obtained.

		analytical	quasi-static	noise	static	stepped sine
A_{22}	(kgm^2)	597	576	–	–	–
A_c	(kgm^2)	91.3	115	–	–	–
I_{z3}	(kgm^2)	214	199	217	–	–
I_{m2}	(kgm^2)	0.013	0.0158	0.0126	–	–
I_{m3}	(kgm^2)	0.013	0.0158	0.00899	–	–
$m_3 g x_{3c}$	(Nm)	835	782	779	–	–
K_2	$(10^6 \, Nm/rad)$	–	1.35	1.328	3.183	1.722
K_3	(Nm/rad)	–	0.612	0.582	1.651	0.626

Table 1: Comparison between the identified and analytical parameters of the model.

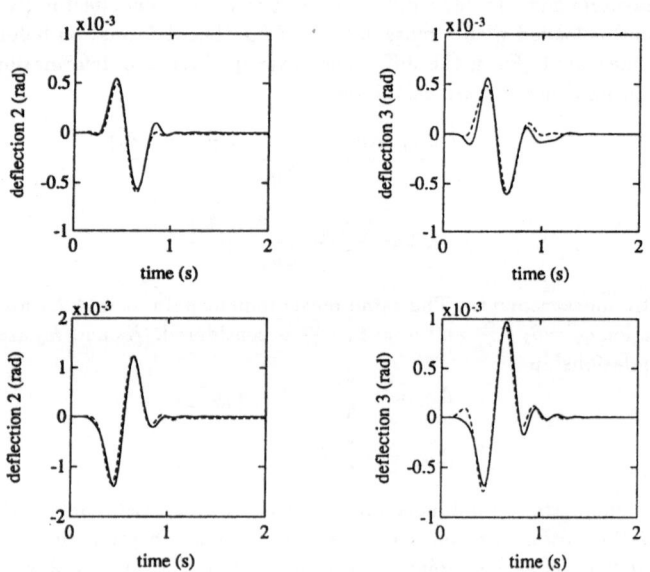

Figure 6: Measured (solid line) and identified (dashed line) deflection at joint 2 and 3 for a position step of 5° for link 2 (upper figures), resp. link 3 (bottom figures).

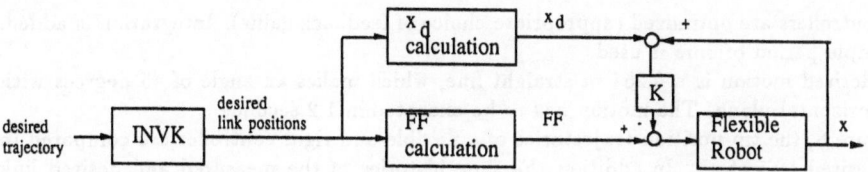

(INVK: Inverse Kinematics , FF: feedforward torque, x_d : desired state (rigid + flexible coordinates),   K: feedback gain)

Figure 7: Block diagram of the nonlinear controller.

4.3   Control System Design.

The second and third axis are controlled with the nonlinear decoupling method for flexible robot structures presented in [1]. The control law is split up into two stages. The first stage is a nonlinear feedback to cancel out nonlinear terms in the rigid body dynamics. In this paper the nonlinear feedback is replaced by a nonlinear feedforward. This means that the nonlinear decoupling terms are calculated based on the desired state variables instead of the actual values. The advantage is that the nonlinear decoupling torque can be calculated off-line, and that using the noise free desired values results in more stable behaviour of the controller.

The second stage is a linear feedback of all state variables (motor position 2 and 3, deflection at joint 2 and 3 and their derivatives). In order to use linear techniques (pole placement, optimal control) for feedback gains computation, the model is linearized to the first order about the desired trajectory. The feedback gains are calculated in discrete time. To limit the computation time, the feedback gains are chosen constant: they are computed with respect to the final desired position.

In the second stage the tracking error between measured and desired state is fed back. Therefore the reference values of all state variables are calculated, based on the model of the flexible robot. This results in a compensation of static and dynamic deflection. The trajectory is based on a 9 th order polynomial. This smooth trajectory is required to prevent peaks in the feedforward signals.

Instead of using a nonlinear state estimator [1], the velocities are calculated by differentiation. To overcome the problem of sensitivity to high frequency noise of the differentiator, the measured signals are filtered with digital low pass filters. In addition the filters reduce the spillover problem. The filters are added in the model. This extends the model order.

Figure 7 gives the block-diagram of the controller.

4.4   Experimental Results

The experimental results of the controller presented above, are compared to the performance of a PID rigid controller in velocity command with appropriate feedforward. For both controllers the same smooth trajectory is used.

All controllers are optimized (appropriate choice of feedback gains). Integration is added. A sample period of $6ms$ is used.

The desired motion is a 0.254 m straight line, which makes an angle of 45 degrees with the horizontal plane. The motion has to be executed in 1.2 seconds.

In figure 8, the tip position trajectories of a flexible and rigid controller are compared to the desired trajectory. In addition the time histories of the measured and desired link positions are plotted. Figure 9 compares the dynamic tip position error normal and tangential to the desired trajectory of the rigid controller and the flexible controller. In both cases there are no vibrations. The reason is the use of smooth trajectories.

The following improvements are obtained in comparison with rigid control:

- The maximum dynamic tracking error normal to the desired trajectory is reduced from 1.6 mm to 0.8 mm.

- The maximum dynamic tracking error tangential to the desired trajectory is reduced from 2.45 mm to 2.0 mm.

- The normal and tangential static errors due to gravitation are compensated.

5 Conclusion.

The tests have shown that the performance of industrial robots, even with limited flexiblity, can be improved by simple modifications of the industrial robot controller. The change from a trajectory based on a trapezoidal velocity profile to a smooth trajectory gives the largest contribution to the improved dynamic performance (overshoot, oscillations, tracking error), but there is a significant improvement of a flexible controller over a classical PID controller in terms of tracking accuracy. In addition the static error due to gravity is compensated. At very high velocities and accelerations, there is also a significant improvement of a flexible controller over a classical PID controller in terms of overshoot and oscillations.

6 Acknowledgement

This paper presents work carried out in the framework of ESPRIT Project 1561 ("SACODY"), and the Interuniversity Attraction Pole No 13, initiated by the Belgian State, Prime Minister's Office, Science Policy Programme.

References

[1] M. Adams , J. De Schutter, *Experiments on nonlinear control of flexible robots.* Proc. 15 th Int. seminar on modal analysis, 1990.

[2] S. A. Bortoff and M. W. Spong, *Feedback linearization of flexible joint robot manipulators.* Proc. 26th Conf. on Desicion and Control, Los Angeles, 1357-1362, 1987.

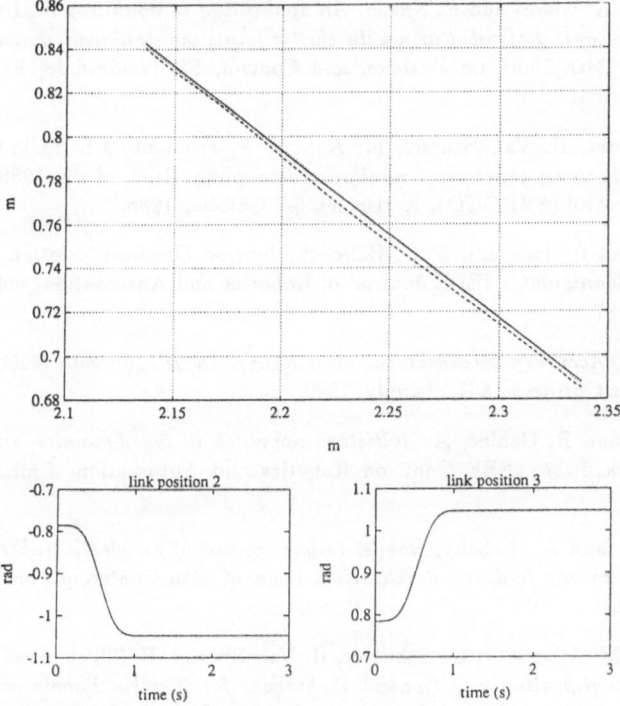

Figure 8: The tip position trajectory and link position trajectories of a generalized non-linear decoupling controller (dotted line) and a rigid controller (dashed line) compared with the desired trajectories (solid line).

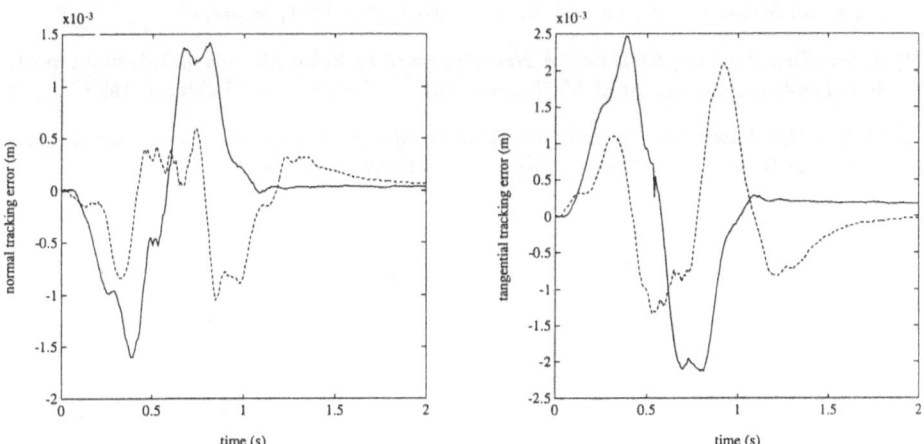

Figure 9: Dynamic tracking error normal and tangential to the desired trajectory in function of time for a rigid controller (solid line) and a flexible controller (dashed line).

[3] A. De Luca, A. Isidori and F. Nicolo, *An application of nonlinear model matching to the dynamic control of robot arm with elastic joints via nonlinear dynamic feedback.* Proc. IEEE 24th Conf. on Desicion and Control, Ft. Lauderdale, FL, 1671-1679, 1985.

[4] J. De Schutter, H. Van Brussel, M. Adams, A. Froment, J.L. Faillot, *Control of flexible robots using generalized nonlinear decoupling.* Proc. of the 1988 Symposium on Robot control (SYROCO), Karlsruhe, 5-7 October 1988.

[5] M. G. Forrest-Barlach and S. C. Babcock, *Inverse Dynamic position Control of a Compliant Manipulator* IEEE Journal of Robotics and Automation, vol. RA-3, NO. 1, 75-83 , 1987.

[6] M. Gerung , *Accuracy Measurement on a KUKA IR 161/60 with RCM3.* Technical Report, Esprit project 1561: Sacody, 1990.

[7] F. Pfeiffer, and B. Gebler, *A Multistage approach to the dynamics and Control of elastic robots.* Proc. IEEE Conf. on Robotics and Automation, Philadelphia, 2-8, 1988.

[8] S. N. Singh, and A. A. Schy, *Robust torque control of an elastic robotic arm based on invertibility and feedback stabilization.* Proc. of 24th Conference on Decision and Control, 1985.

[9] J. Swevers, M. Adams, J. De Schutter, H. Van Brussel, H. Thielemans, *Limitations of Linear Identification and Control Techniques for Flexible Robots with Nonlinear Joint Friction.*, Experimental Robotics 1, Lecture Notes in Control and Information Science, Springer–Verlag, 1990.

[10] J. Swevers, B. De Moor, and H. Van Brussel, *Stepped sine system identification, errors–in–variables, and the quotient singular value decomposition.* Mechanical Systems and Signal Processing, vol. 6, no. 5, November 1991, in press.

[11] J. Van Den Bossche, *RODYM: A New Approach to Robot Metrology.* Internal report, K.U.Leuven, Department of Mechanical Engineering, Leuven, Belgium, 1990.

[12] H. Van Der Auweraer, *Off–line identification of a KUKA IR 161/60 industrial robot.* Technical Report, task 4.8.5.1., Esprit project 1561: Sacody, 1990.

Adaptive and Fault Tolerant Tracking Control of a Pneumatic Actuator *

B.W. McDonell and J.E. Bobrow
Department of Mechanical and Aerospace Engineering
University of California, Irvine
Irvine, California 92717

Abstract

Pneumatic actuators are of interest for robotic applications because of their large force output per unit weight and their low cost. A pneumatic actuator with high bandwidth is difficult to obtain because of the compressibility of air and the nonlinear characteristics of air flowing through a variable area orifice. In this paper a controller for a one degree of freedom pneumatic system is presented which uses full-state feedback for simultaneous parameter identification and tracking control. Experimental results indicate that trajectory tracking comparable to electric motor driven robot systems can be achieved.

1 Introduction

Air power has been used for many years in factory environments for open loop actuation of motion between two fixed positions. Pneumatic actuators have the advantages of producing large output forces per unit weight, of being clean and easy to work with, and of being relatively inexpensive. In addition, compressed air is readily available at nearly every shop. Unfortunately, position and force stabilization of a pneumatic actuator is difficult if a high bandwidth closed-loop system is desired. In this paper a controller is described which has performed well in laboratory experiments. The controller is based on a recursively identified time-varying linear model for the system and uses full-state feedback plus a feedforward term to provide trajectory tracking.

Several controllers have been used in the past for pneumatic systems. Most have been fixed gain linear controllers based on a nominal model usually obtained by linearizing the air flow dynamics about the cylinder midstroke position (Shearer[6], Burrows and Webb[1], Vaughan[7], Liu and Bobrow[4]). Fixed gain controllers are inadequate for robot applications because the rigid body dynamics of the robot change when the payload varies

---

*This research was supported by the Rockwell International Graduate Fellowship Program and by Parker-Hannifin, Parker-Bertea Aerospace Division.

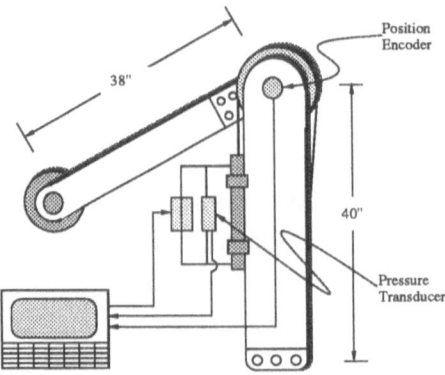

Figure 1: A cable driven pneumatic arm with a double acting cylinder.

and the air flow dynamics change drastically near the cylinder end positions. In addition, complications in modeling arise due to the dynamic characteristics of the servovalve which meters the air flow and from the assumption that the air pressure increases uniformly in the chambers. The actual air pressure response occurs during the time it takes an acoustic wave to travel the length of the actuator (Mannetje[5]), so unmodeled dynamics are present in the system.

2 System Dynamics

The experimental test bed uses a double acting pneumatic cylinder to drive a rotary joint via cables as shown in Figure 1. Since the system operates in the vertical plane gravity must be compensated for by the controller. A high resolution position encoder (432,000 counts/revolution) measures the joint angle, and a differential pressure transducer with a 16 bit A/D converter measures the pressure difference between the two sides of the cylinder. A four way servovalve operated by a 16 bit D/A converter controls airflow to both sides of the cylinder.

The equation of motion for the arm is

$$J\ddot{\theta} + mgl\cos\theta = Fr \tag{1}$$

where J is the moment of inertia of the arm about the pivot, $mgl\cos\theta$ is the torque due to gravity, F is the force in the cable, and r is the moment arm.

A schematic representation of the pneumatic actuator and servovalve is shown in Figure 2. Using conservation of energy, the relationship between mass flow rate of air and the change of pressure in chamber A (or chamber B) (Shearer[6], Liu and Bobrow[4]) is

$$\dot{m}_a T_s - \frac{p_a \dot{v}_a}{c_p} + \frac{\dot{Q}}{c_p} = \frac{c_v(p_a v_a)\dot{}}{c_p R}, \tag{2}$$

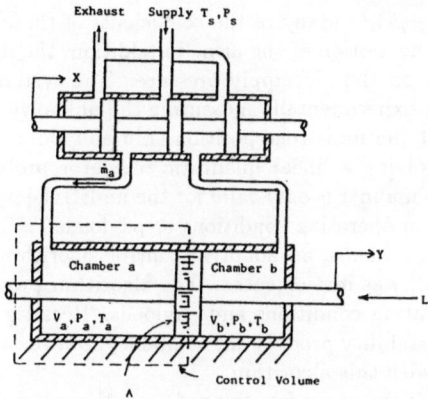

Figure 2: A four-way servovalve controls air flow to cylinder.

where $\dot{Q}$ is the rate of heat transfer to the cylinder, $\dot{m}_a$ is the mass flow rate of air, c_v and c_p are the constant volume and constant pressure specific heats of air, R is the universal gas constant, v_a is the volume of chamber A, T_s is the air supply temperature, and p_a is the cylinder pressure. The mass flow rate $\dot{m}_a$ is a nonlinear function of pressure and the spool valve position, which is assumed to be directly proportional to the valve input current. Hence, the flow equations have the form

$$\dot{m}_a = f_a(u, p_a), \tag{3}$$

$$\dot{m}_b = -f_b(u, p_b), \tag{4}$$

where u is the valve input current and f_a and f_b have the property that $f(0, p) = 0$. The cylinder volumes are related to the arm position by

$$v_a = A(l/2 + r\theta) \tag{5}$$

$$v_b = A(l/2 - r\theta), \tag{6}$$

where A is the cylinder cross sectional area, l is the stroke, and r is the radius of the pulley that rotates the arm. Substituting (3) and (5) into (2) and rearranging gives

$$\dot{p}_a = \frac{f_a(u, p_a)T_s R c_p}{A(l/2 + r\theta)c_v} - \frac{p_a \dot{\theta} c_p}{c_v(l/2 + r\theta)}, \tag{7}$$

where the relation $R + c_v = c_p$ has been used and $\dot{Q}$ is assumed to be zero. Similarly for side b,

$$\dot{p}_b = -\frac{f_b(u, p_b)T_s R c_p}{A(l/2 - r\theta)c_v} + \frac{p_b r \dot{\theta} c_p}{(l/2 - r\theta)c_v}. \tag{8}$$

3   The Control Algorithm

The above equations (7), (8), and (1) can be cast in the form

$$\dot{x} = \bar{A}(x)x + \bar{b}(u) + \bar{d}(x), \tag{9}$$

where $x^T = [\theta \quad \dot{\theta} \quad (p_a - p_b)]^T$. Many of the coefficients of these equations are either not known or vary during the motion of the arm. In addition, the flow equations (3) and (4) are strongly dependent on the air supply pressure. The system can be controlled with good performance after experimentally measuring the unknown coefficients to obtain the linearized system about the midstroke position (Liu and Bobrow[4]). The linear control law was obtained by solving a linear quadratic regulator problem. Unfortunately, the control obtained in this manner is only valid for the midstroke position and is not capable of adapting to changes in operating conditions or payloads.

To overcome these problems, an adaptive control algorithm similar to that used by Kim and Gibson 1991[3] was implemented. The algorithm provided accurate trajectory tracking for many operating conditions and payloads. Because of the complexity of the equations of motion, a stability proof is not available, however an experimental system is operating successfully with this algorithm.

Because the nonlinear terms are functions of time, the system (9) can be interpreted as linear with time-varying coefficients. The benefit of this interpretation is that the model becomes

$$\dot{x} = \tilde{A}(t)x + \tilde{b}(t)u + \tilde{d}(t), \tag{10}$$

with the corresponding discrete-time model

$$x_{k+1} = A_k x_k + b_k u_k + d_k, \tag{11}$$

where the index k refers to the sampling instant. For the following controller to operate successfully, the coefficients in (11) must change slowly relative to the dynamics of the closed loop system.

Since full state feedback is available, x_k is known at each instant. Hence, all coefficients of A_k, b_k, and d_k can be identified as coefficients of known parameters using a recursive least squares algorithm (Goodwin and Sin[2]) with a forgetting factor to allow for the time-variation of these parameters. Three independent scalar least squares regressions, one for each component of the state, are updated after each sample.

In order to achieve trajectory tracking, define the error

$$z_k = x_k - x_k^d \tag{12}$$

where x_k^d is a desired trajectory to be followed. Next, break the control into two parts

$$u_k = u_k^o + u_k^f, \tag{13}$$

where u_k^o is the nominal feedback control and u_k^f is a feedforward term. Using (12) and (13) with (11) the state equations become

$$z_{k+1} = A_k z_k + b_k u_k^o + \epsilon_k, \tag{14}$$

where ϵ_k satisfies

$$\epsilon_k = A_k x_k^d - x_{k+1}^d + d_k + b_k u_k^f. \tag{15}$$

The performance index used is

$$J_k = \sum_{j=k}^{k+N} (z_j^T Q z_j + u_j^T R u_j), \tag{16}$$

where Q is a positive semidefinite matrix and R is a positive definite matrix. The control law is obtained by minimizing (16) subject to (14). If the system parameters were known, and if $\epsilon_k = 0$, then the unique minimum u_k is known by solving a regulator problem with time-varying coefficients. For real time control, it is not feasible to solve this problem for each trajectory and plant. The gains obtained from the Riccati update equation were found to provide a stable control. At each instant, the identified plant parameter estimates are assumed to be the true parameters and the equations

$$R_k = R + b_k^T P_k b_k \tag{17}$$

$$F_k = R_k^{-1} b_k^T P_k A_k \tag{18}$$

$$u_k = -F_k z_k + u_k^f \tag{19}$$

$$P_{k+1} = A_k^T [P_k - P_k b_k R_k^{-1} b_k^T P_k] A_k + Q, \tag{20}$$

are solved where P_0 is initialized with the value corresponding to the solution of the algebraic Riccati equation for a system with A and b fixed at some nominal values.

To apply the control in (19) the feedforward signal u_k^f is needed. In addition, the third component of the state vector $x_{3_k}^d = p_{a_k}^d - p_{b_k}^d$ must be also be defined. Note that the first two components of the desired state x_k^d are known because the reference trajectory $\theta(t)$ and $\dot{\theta}(t)$ is known. Because the solution to the above regulator problem assumes $\epsilon_k = 0$, u_k^f and $x_{3_k}^d$ are chosen to minimize ϵ_k in (15). Several methods of choosing the desired pressure and feedforward control were tried with the following approach yielding the best performance.

The first two components of b_k are forced to be zero in the least squares model. That is, $b_k^T = (0 \ 0 \ b_{3_k})^T$, so the linear regression model (14) does not include second order effects of the control on the state ($\tilde{b}(t)$ in (10) has the form $\tilde{b}(t)^T = (0 \ 0 \ b_3(t))^T$). Next, the second state equation in (15) and the current parameter estimates are used to compute the desired pressure from the desired position and velocity. The second state equation has the form

$$x_{2_{k+1}}^d = a_{21} x_{1_k}^d + a_{22} x_{2_k}^d + a_{23} x_{3_k}^d + d_{2_k}, \tag{21}$$

where all terms except for $x_{3_k}^d$, the desired pressure, have either been estimated or are specified from the desired trajectory. Hence, at instant k, (21) is used to solve for $x_{3_k}^d$, and the same equation is used to solve for $x_{3_{k+1}}^d$ by shifting the index k forward by one. Finally, u_k^f is computed by solving the third state equation for the control given the desired position, velocity, and pressure

$$u_k^f = (x_{3_k+1}^d - (a_{31} x_{1_k}^d + a_{32} x_{2_k}^d + a_{33} x_{3_k}^d + d_{3_k}) / b_{3_k}. \tag{22}$$

4   Implementation and Experiments

The adaptive controller is implemented on a 20-MHz 80386 PC equipped with a 33-MHz 68030 parallel processor board. All control calculations are performed on the 68030 board with the PC performing only I/O and user interface functions.

The adaptive control algorithm is loaded onto the 68030 board and started, after which control is returned to the PC. A resident interrupt server is then loaded on the PC to provide I/O and an interface to the 68030 board. Once control is started, a clocked hardware interrupt is generated on the 68030 board to initiate the control routines. At each interrupt the following sequence of operations are performed:

- Read the arm position and pressure.

- Compute and output the control from (19) with previously calculated u_k^f.

- Update the recursive least squares estimates of the system parameters in (11).

- Update discrete LQR gains (17), (18), and (20).

- Compute next reference state (21) and feed forward control (22).

Numerous test cases were examined in order to test the control algorithm. The results of three tests, revealing the tracking as well as adapting ability of the controller, are shown below. To demonstrate the tracking of the system, a polynomial curve with continuous first, second, and third derivatives was used to blend motion between two fixed positions in a specified time.

Figures 3 and 4, respectively, show the response and the error of the arm while tracking four pairs of 60 degree motions with rise times of 0.8, 0.7, 0.6, and 0.5 seconds. The tracking is observed to be deteriorating for the 0.6 and 0.5 second motions. As is shown in Figure 5, this is due to reaching the limit of the valve operation.

The adapting ability of the controller is demonstrated by executing the previously mentioned polynomial curve, then adding an additional 25 pounds to the initial 10 pound load at the tip of the four foot arm. The arm is then allowed to shake six times to reidentify the system, after which another polynomial curve is executed. Figures 6 and 7 show the response and the error, respectively. Figure 8 shows the (2,3) element of the A matrix as it adapts to the added weight. This element should be nearly proportional to the reciprocal of the moment of inertia of the arm. The identification is seen to be essentially complete after just one cycle.

The fault tolerance of the controller is demonstrated by completely severing the 1/4 inch air hose (the supply line is only 5/16 inch) between the cylinder and one side of the pressure transducer resulting not only in a large leak but also in the loss of one side of the delta pressure input. This was performed while the system was attempting to track a 30 degree, 1 Hz sine wave. The results and error are shown in figures 9 and 10.

5 Conclusion

A pneumatic system has been controlled using a time-varying third order model. An adaptive controller is presented which estimates all of the system parameters using full state feedback and recursive least squares parameter estimation. The feedback control gains are obtained by updating a discrete-time Riccati equation, and a feedforward control is computed that provides trajectory tracking. Experimental results demonstrate the ability of the controller to follow a trajectory and to adapt to large changes in the system.

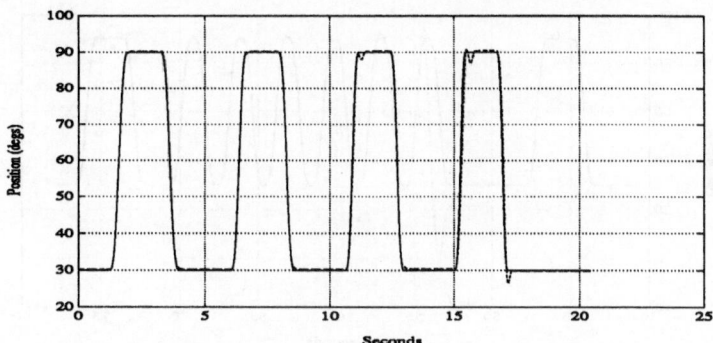

Figure 3: Tracking of 0.8, 0.7, 0.6, 0.5 second step approximations.

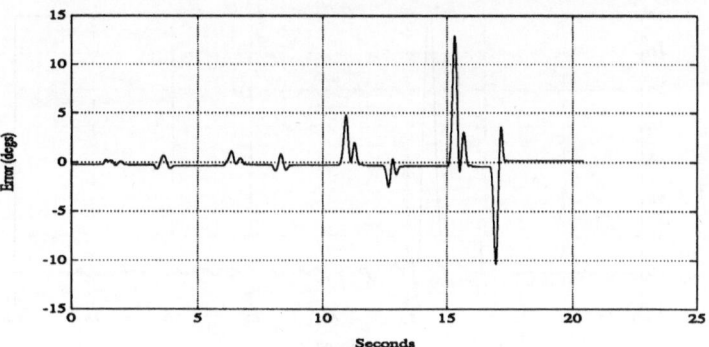

Figure 4: Error of Step Tracking

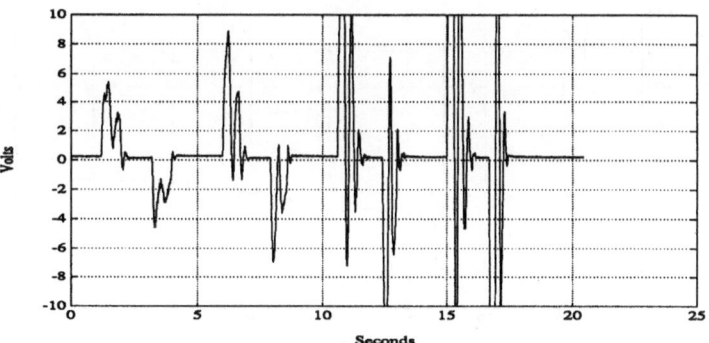

Figure 5: Control Signal for Step Tracking

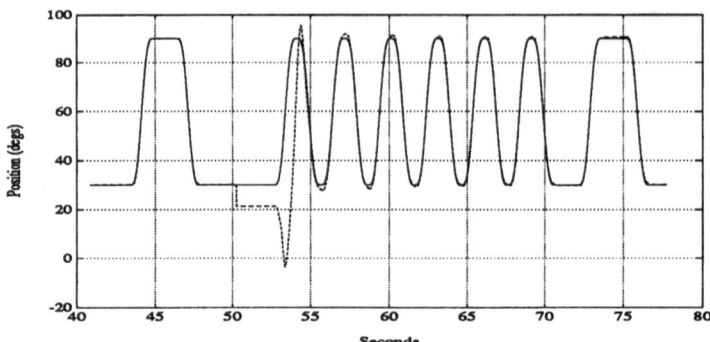

Figure 6: Tracking of Steps with addition of 25 pounds.

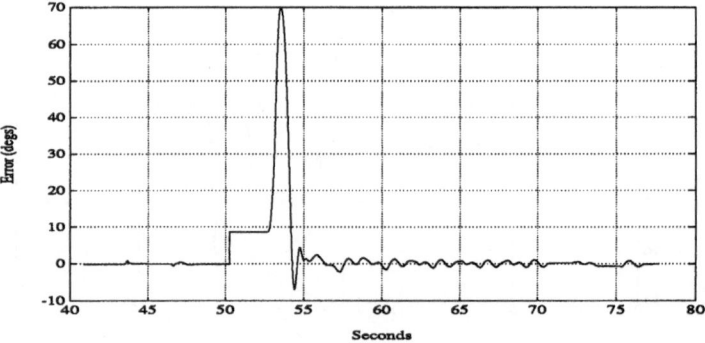

Figure 7: Error of Step Tracking

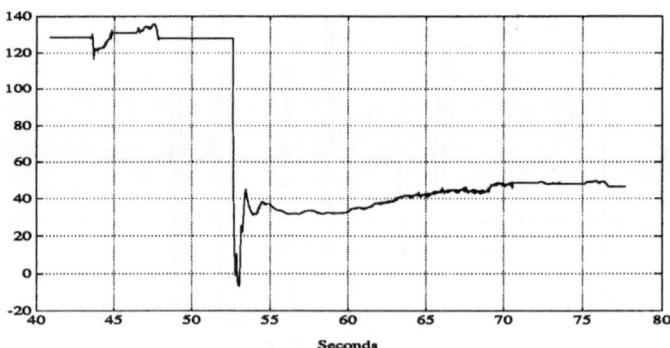

Figure 8: Element a_{23}

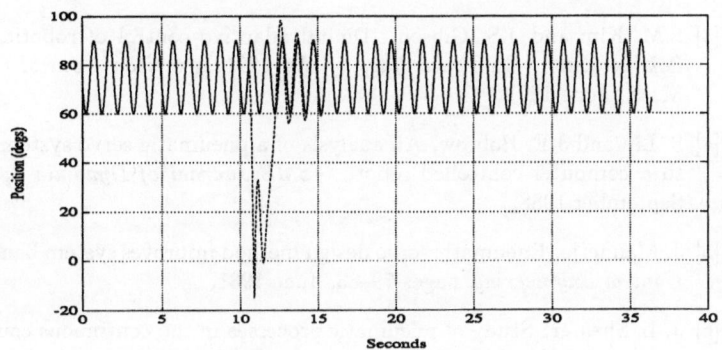

Figure 9: Tracking of Sine wave with cut hose.

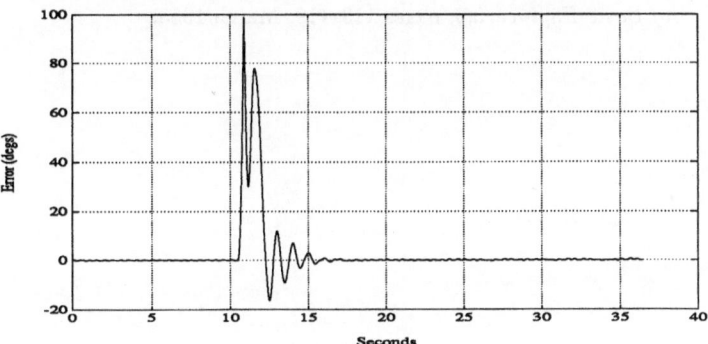

Figure 10: Error of Sine Tracking

References

[1] C.R. Burrows and C.R. Webb. Use of root loci in design of pneumatic servo motors. *Control*, pages 423–427, August 1966.

[2] G.C. Goodwin and K.S. Sin. *Adaptive Filtering Prediction and Control*. Prentice-Hall, 1984.

[3] S.M. Kim and J.S. Gibson. Digital adaptive control of robotic manipulators with flexible joints. In *Proceedings of the 1991 American Control Conference*, Boston, Mass., June 1991.

[4] S. Liu and J.E. Bobrow. An analysis of a pneumatic servo system and its application to a computer-controlled robot. *ASME Journal of Dynamic System Measurement*, September 1988.

[5] J. Mannetje. Pneumatic servo design method improves system bandwidth twenty-fold. *Control Engineering*, pages 79–83, June 1981.

[6] J. L. Shearer. Study of pneumatic processes in the continuous control of motion with compressed air - I,II. *Transactions of ASME*, pages 233–242, February 1956.

[7] D.R. Vaughan. Hot-gas actuators: Some limits on the response speed. *ASME Journal of Basic Engineering*, pages 113–119, March 1955.

Section 2: Learning and Skill Acquisition

Robot learning is an area which is steadily gaining more interest, partly due to the fact that modelling all details in and around a robot is just impossible, partly due to the fact that robots improving their performance like a learning child corresponds to an old dream that gains new hopes e.g. from some progress in the area of neural nets.

Indeed the first paper of Asada and Liu addresses the problem of skill transfer from human operator to robot using neural nets. The application given, namely deburring, is probably one of the most typical robot applications where it is completely unclear how e.g. a conventional programming language might be used to instruct the robot - in a task where usually even the human worker is not able to express verbally why he/she reacts in such or such way to his tactile and visual sensors.

The paper of Miyazaki, Ide, Masutani and Ahn addresses the application-oriented problem of part-mating in assembly. The authors propose an unsupervised learning scheme based on a self-organizing, binary tree type database for representing and generating task strategies. Their approach is less continuous-feedback control-oriented, instead it has much similarity with search strategies implying successful and non-successful nodes and evaluation criteria.

Showing a robot in simulated environment how to perform a task so that in a later phase it is capable of executing it autonomously in the real world corresponds to a philosophy that gains more and more interest in telerobotics, i.e. remotely controlled systems. One of the main problems hereby is to set up a library of elementary skills involving sensory feedback and use them as ingredients for higher level task commands. Such an approach is the basis of a paper by Ogasawara, Kitagaki, Suehiro, Hasegawa and Takase ; their testbed is a model-based telerobot system.

The paper of Bidaud and Fontaine introduces the concept of "intelligent effectors", into which robot systems break down. Their work basically deals with the use of knowledge based techniques to construct an assembly plan, subdividing the task into elementary actions, and for these actions to run successively appropriate rules for triggering them or deciding whether the determination predicates are satisfied or whether exceptions occur.

Experimental Verification of Human Skill Transfer to Deburring Robots

Haruhiko Asada and Sheng Liu
Center for Information-Driven Mechanical Systems
Massachusetts Institute of Technology
Cambridge, MA. 02139, U.S.A.

Abstract

A new adaptive controller for deburring robots is presented in this paper. The design of the control system is entirely based on a human skill model. With the use of the human skill model, the performance skills of a human expert can be acquired and transferred to robot control systems that allow robots to mimic human skills in performing deburring tasks. The human skill model comprises a tool manipulation strategy, a process perception model and associative memories. The tool manipulation strategy describes the dynamic interaction between the tool, held by the human expert, and the environment. Teaching data taken from human demonstration motions shows that the human expert changes his tool manipulation strategy with respect to the varying process condition. The process model, which characterizes the process conditions, provides a way to detect variations in process conditions. The associative memories, represented by mapping functions, reflect how a human expert modifies his tool control strategy with respect to changing process conditions. These mapping functions can be identified by using human teaching data. The consistency of the mapping and the transferrability of human skills are analyzed by using Lipschitz's condition. A robot controller is constructed based on the human skill model that involves associative memories derived from human teaching data. The control system is implemented on a direct-drive robot. The experimental results show that the robot can adapt itself to the deburring process in a manner very similar to a human expert.

1 Introduction

Today's robots require task-specific, detailed instructions substantially different from what is usually designated as skill and dexterity. Skill is the ability to use knowledge in performing and executing learned physical tasks. Rather than giving the robot instructions for performing a specific task, we should teach the robot skills which will enable it to exhibit dexterity and intelligence.

In this paper a new method based on task process models for acquiring manipulative skills from human experts is presented. In performing manipulative tasks such as deburring, a human expert moves a tool at an optimal feedrate and cutting force as well as with an appropriate compliance for holding the tool. An experienced worker can select the correct strategy for performing a task and change it dynamically in accordance with the task process state. In this paper, the human expertise for selecting a task strategy that accords with the process characteristics is modeled as an associative mapping, and represented and generated by using a neural network.

First, the control strategy for manipulating a tool is described in terms of feedforward inputs and tool holding dynamics. The parameters and variables representing the control strategy are then identified by using teaching data taken from demonstrations by an expert. The task process is also modeled and characterized by a set of parameters, which are identified by using this same teaching data. Combining the two sets of identified parameters, we can derive an associative mapping from the task process characteristics to the task strategy parameters. The consistency of the mapping and the transferrability of human skills are analyzed by using Lipschitz's condition. The method is applied to deburring, and implemented on a direct-drive robot. It is shown that the robot is able to associate a correct control strategy with process characteristics in a manner similar to that of the human expert.

2 Manipulative Skills

Figure 1 shows a schematic model of the manipulative skills discussed in this paper. Given a task goal and *a priori* information about the task, a human expert selects an appropriate strategy for performing the task. He also perceives the state of the task process, using acute senses such as visual, acoustic, and tactile. Depending on the actual process state, the human modifies the task performance strategy. The skill addressed in this paper is defined as the ability to associate an appropriate strategy with the process state and a priori information about the task. A particular motion or a feedback law is not a skill if it is valid only for a specific task to be performed under particular conditions. Skill, as used in this paper, refers to the generic ability to perform a class of tasks; this includes association and generation of strategies based on the perception of the task process state and a priori task information. An experienced worker has already acquired techniques and built up a certain scheme for generating an optimal strategy based on his perception of the process state and knowledge about the task. The major objective of this paper is to model such learned skills for generating control strategies and to transfer these skills to a robot. As shown in Figure 1, we represent skills as an associative mapping from the process characteristics and task information to the task strategy, which is described by a set of variables.

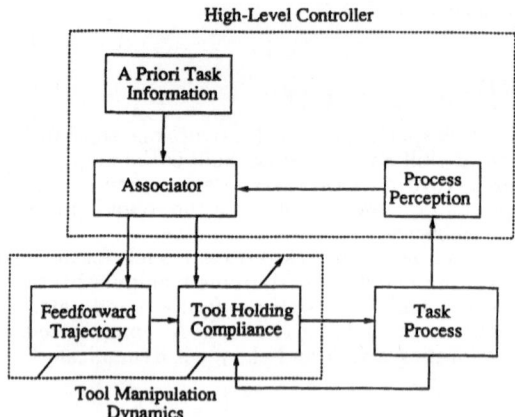

Figure 1: Schematic model of human skills

The tasks considered in this paper are generally performed by manipulating tools, e.g. deburring and grinding. In these tasks, the motion of a human expert and his strategy for performing a task are all reflected to the motion of the tool and its dynamics characteristics. In consequence, we describe an expert's task strategy in terms of variables describing the tool motion as well as parameters characterizing the tool dynamics. For example, the tool dynamics are represented by the compliance or the impedance with which the human holds the tool. The tool motion variables, in the case of deburring and grinding, are the feedrate and cutting force that the human intends to achieve. The former is modeled as feedback gains, or more generally, as feedback laws, while the latter is provided by feedforward inputs, as shown in Figure 1.

The model of tool manipulation skills is thus described as a process of associating process characteristics and task information with a set of variables and parameters for controlling the tool motion. With this framework of skill representation, we will develop a method for transferring expert skills to a robot. Namely, the high-level controller that modifies the tool holding compliance and the feedforward trajectories, as shown in Figure 1, is generated from teaching data.

One of the difficulties in acquiring human skills is that humans often lack awareness concerning their skills. As they train over a long period of time, they build up some memory or association mechanisms, which are often on a subconscious level. In particular, the manipulative skills that we are discussing in this paper are difficult to explain explicitly; a human expert can not quantitatively describe the compliance with which he holds the tool, how much force he applies to the tool, and how he modifies the compliance and force in accordance with the process state. These actions are performed mostly at a subconscious level. As a result, a verbal picture of the process is not sufficient for identifying manipulative skills. For this reason we take quantitative data from demonstrations by human operators, and then process the data in order to identify the associator that elucidates the manipulative skills.

3 Acquisition of Deburring Skills

(1) Dynamic Models of Tool Manipulation

In this section, let us consider the transfer of deburring skills as a metaphor of the skill transfer technique described in the previous section. First we need to identify how a human expert manipulates the deburring tool against the workpiece while accommodating the interactive force. In this paper, we model the human behavior of manipulating a tool in terms of equivalent impedance and feedforward inputs. Namely, it comprises mass, damper and spring terms given by

$$M\ddot{X} + B\dot{X} + KX = F + U \tag{1}$$

where X is the position of the tool, F is the force acting on the tool, as defined previously, and M, B and K are coefficient matrices having a dimension consistent with the position and force vectors. The last term U represents a feedforward term generated by the human. In this paper, we model the feedforward term as reference feedrate and cutting force conceived by the human. The impedance model parameterizes the human motion by the set of parameters $a = (M, B, K, U)$. If the tool motion is within a two-dimensional plane, as shown in Figure 2, and the off-diagonal terms of the coefficient matrices are negligible, then the above model is simplified as

$$m_x\ddot{x} + b_x(\dot{x} - V_o) = -F_t \tag{2}$$
$$m_y\ddot{y} + b_y\dot{y} + k_yy = F_n - F_o \tag{3}$$

where F_t and F_n are the forces acting on the tool in the tangential and normal directions, respectively (Figure 2). Velocity V_o is the reference feedrate in the tangential direction, while F_o is the feedforward reference force in the normal direction. Note that a downward reference force F_o is defined to be negative. The tool control model may be further simplified by matching actual data of human motions.

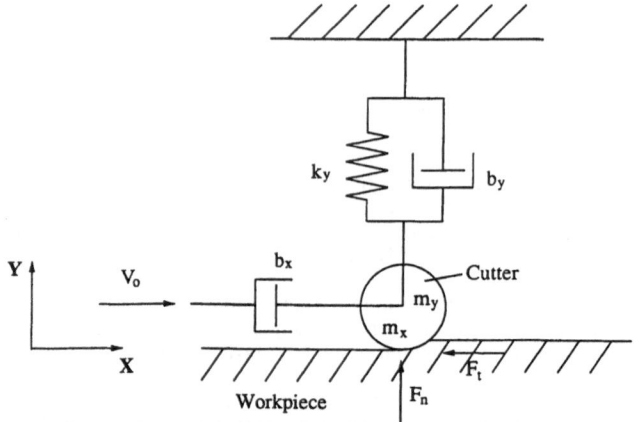

Figure 2: Tool manipulation dynamics

Some preliminary observations of human teaching motions reveal that there is no clear correlation between tangential cutting force and tool acceleration as well as its displacement in the tangential direction. However, the tool velocity, $\dot{x}$, is in direct proportion to the tangential cutting force, F_t. The relationship between the tool velocity and the cutting force in the tangential direction is governed by a metal cutting mechanism, which will be discussed shortly. Since this relationship is not a result of an active motion of the human arm, it can not be considered as a tool control strategy. Therefore, based on human teaching motions, the tool control strategy of a human expert in the tangential direction can be simply modeled as a velocity sorce that can specify any feedrate for the deburring tool. Equation (2) then is reduced to

$$\dot{x} = V_o \qquad (4)$$

Based on observation of human motions, it was found that the inertia term and the damping term in equation (3) are also negligible. The simplified tool manipulation model in the normal direction can be shown as

$$k_y y + F_o = F_n \qquad (5)$$

Namely, in the normal direction, the human worker manipulates the tool simply like a spring, with a varying spring constant and reference applied force.

According to the skill model addressed in the previous section, a human expert changes the tool manipulation parameters in accordance with changes in the task process. Therefore, the parameters in equations (4) and (5) are constant only for a certain interval where the task conditions do not change significantly. In the following discussion, we assume that the tool manipulation parameters are constant for an interval in the task process where the task conditions remain the same. Namely, the expert uses constant feedrate and normal force as feedforward inputs, V_o and F_o, and holds the tool with a constant compliance under the same process conditions. These parameters are to be identified by using teaching data.

While a human expert demonstrates an operation, displacement, velocity and the interacting force are measured. The parameters involved in equations (4) and (5) are then identified by using a standard curve fitting technique [Asada and Asari, 1988]. Namely, these data are substituted into the model given by equations (4) and (5), and the set of parameters that minimize the error are determined. For a certain number of data points, where the task conditions are the same, the parameter values that minimize the mean squared error between the calculated and measured values are considered as the optimal values for that interval of the grinding process.

(2) Modeling of Deburring Processes

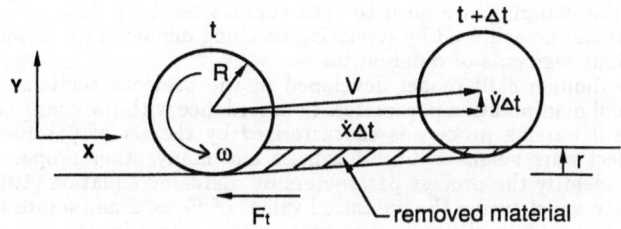

Figure 3: Surface metal cutting schematics

To characterize the task process condition, we use a standard process model of metal cutting. Generally, the constitution of the process model is based on the conservation laws of energy and mass. Figure 3 shows a surface metal cutting schematic to illustrate the process model. When the tool is in contact with the workpiece for removing the burr, the material removal rate (MMR) is in proportion to the power consumption rate at the point of contact [King and Hahn, 1986]. Namely,

$$(MMR) \propto (P - P_{th}) \tag{6}$$

where P_{th} is a threshold power. At the point of contact, the power consumption rate is calculated by

$$Power = F_t \times \omega \times R \tag{7}$$

where F_t is the interacting force in the tangential direction, ω is the spindle speed of the grinding tool, and R is the radius of the grinding wheel. The following equation can be derived from geometrical analysis.

$$(MMR) \propto g(r)\dot{x} - h(r)\dot{y} \tag{8}$$

where $g(r)$ and $h(r)$ are functions of burr height r. Combining equations (6), (7) and (8), we have the following equation

$$F_t \omega R - P_{th} = \lambda(g(r)\dot{x} - h(r)\dot{y}) \tag{9}$$

where λ is a constant of proportionality.

For the deburring process over a short section of the workpiece, the burr can be considred as of constant height. Thereby, $g(r)$ and $h(r)$ can be replaced by constant values. Assuming that ω and R are kept constant, then equation (9) reduces to a linear form

$$F_t \omega = g\dot{x} - h\dot{y} + P' \tag{10}$$

where $P' = P_{th}/R$.

For a short interval in the deburring process, we use the time-invariant function given by equation (10) in order to model the process dynamics that constrains the tool motion. Namely, we parameterize the process characteristics by the set of parameters g, h and P' involved in this process model. These parameters are identified by using the same set of measured data as the ones for identifying the tool manipulation parameters. The least squared error approach is used again to obtain optimal values.

(3) Data Processing and Generation of Associative Mappings

Now that both the dynamic model of tool manipulation and the deburring process model have been derived, let us determine the parameters involved in both models using teaching data and then generate an associative mapping between the two sets of parameters. To generate the mapping, we need to take copious teaching data under various task conditions. This can be achieved by repeating teaching demonstrations for different workpieces with various segments of different burrs.

According to the human skill model developed in the previous sections, a human expert changes the tool manipulation parameters in accordance with the change of process characteristics. The deburring process is characterized by the set of parameters, $b = (g, h, P')$, which reflect burr height r, burr hardness and many other properties of the process. Let us first identify the process parameters by matching equation (10) with the teaching data. Namely, substituting the measured values of F_t, ω, $\dot{x}$ and $\dot{y}$ into both sides of equation (10), we can obtain the three parameters that minimize the squared error. The process parameters are determined for each segment of the teaching data. Note that the error increases as the length of the segment increases. Particularly, when the task process varies significantly, the constant parameters do not fit the data, resulting in a large error. We partition the data into several segments so that the error may be kept lower than an appropriate threshold level. In this way, we perform the segmentation of the teaching data and characterize the individual segment of each task process with the three parameters.

The manipulation parameters involved in equation (1), or equations (2) and (3), are then determined for each segment of the teaching data. Namely, a pair of parameter vectors, a v.s. b, are obtained for each segment, describing how the human selected the manipulation parameters in accordance with the process characteristics. Let us assume that the teaching data are divided into N segments, which provide N pairs of corresponding parameter vectors, $a^i = \{V_o^i, k_y^i, F_o^i\}$, and $b^i = \{g^i, h^i, P'^i\}$. Our problem is to derive an associative mapping from the N pairs of samples, $\{(a^i, b^i), i = 1, ..., N\}$.

The mapping can be generated in a straightforward way, if the relationship is linear. Namely, each manipulative parameter, V_o, k_y, or F_o is treated as a linear function of variables g, h and P', determined by using a standard least squared error method. However, human behavior that associates the process characteristics with the tool manipulation strategy is presumably nonlinear. We need to use a nonlinear mapping technique, if the linear mapping does not elucidate the human skills, having a significant mismatch between the model and the data. In this paper, we use a multi-layer neural network to represent and generate the nonlinear mapping. If an appropriate structure is provided, the neural network may produce a correct correlation between the inputs b and the outputs a. Given the N pairs of parameter vectors (a^i, b^i), we can train the network to generate a correct mapping. In the following section, the results of using both linear and nonlinear mapping functions will be compared.

4 Validity of Acquired Skill Models

In this section, let us consider the validity of the skill models obtained through the data acquisition described in the previous section. To transfer human skills to a robot, we need to measure the human motion and monitor the task process using available sensors. A question arises from this measurement since the available sensors for acquiring teaching data are inevitably different from actual human senses. A human can take a control action based on the perception of the task process using acute senses such as the visual and acoustic. The physical sensors we use for the measurements are not necessarily able to detect all the process phenomena that a human can detect. What the available sensors can detect are similar to those of the human senses but are different in many aspects. This is an unavoidable problem in transferring human skills to robots. Human perception is so complicated and intricate that we can hardly identify how a human recognizes the task process characteristics. It is not our goal in this paper to exactly replicate human perception but to acquire useful knowledge and techniques that can be utilized for performing tasks by robot manipulators.

In the previous section we employed a process dynamics model to characterize the task process. Namely, the parameters involved in a process dynamics equation are used to characterize the task process. These parameters, denoted b, provide a compact representation of the process characteristics that is directly identifiable from data. One should note, however, that this set of parameters reflects only characteristics such as material properties and workpiece geometry, which are involved in the process dynamics equation. There might be some features about the task process that are essential to human perception but are not involved in the process dynamics model. When essential task features are missing from the process parameters b, the human skills can not be transferred correctly to robot manipulators.

The question is whether critical information on the task process that has a significant influence upon human decision making is missing from the process parameters b. The associative mapping from the process parameter space to the manipulation parameter space can not represent a correct relationship, if the process parameters do not contain sufficient information about the task process. Namely, manipulation parameters a might be determined depending on some parameters other than the process parameters b. As a result, the mapping from b to a may produce inconsistent results.

Figure 4-(a) illustrates an inconsistent mapping from the space of process parameters, B, to the space of manipulation parameters, A. The same point in the process parameter space corresponds to multiple points in the manipulation parameter space. The human selects a^i at one time and a^j at another time, while the process parameters b are the same. The human behavior looks erratic since two different motions are observed for the same process parameters. Note that the lack of critical information always incurs this inconsistent mapping as long as the missing information is significant. If the mapping is consistent for all the human motions, the missing information does not influence the human decision, hence it is negligible. Therefore, the inconsistent mapping occurs whenever significant information is missing in the process parameter vector, b. By contraposition, we can conclude that no significant information is missing if the inconsistent mapping does not occur. We can validate an acquired skill model by examining the inconsistency of the mapping from space B to space A.

A difficulty in this validation method is that it is necessary to examine all possible cases. Strictly speaking, an infinite number of teaching data must be examined in order to prove that inconsistency does not occur, and this is practically impossible to do. We need to deal with a finite number of data that can be acquired from a human without causing fatigue.

Let B^* be a subset of B, representing a set of sample data. By covering the majority of common cases, we may assume that the samples are large and copious enough to represent a human skill. Our goal is to examine the consistency of the finite number of data; whether they are erratic or meaningful. Since we are dealing with a finite number of data, the same value of variables b is seldom obtained twice, as shown in Figure 4-(a). However, the situation shown in Figure 4-(b) should occur, as we examine a sufficient number of data. Namely, if significant information is missing in b, totally different actions are observed for almost the same two data points.

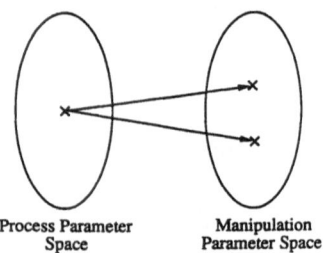

Process Parameter
Space

Manipulation
Parameter Space

Figure 4-(a): Inconsistent mapping from process parameter space B to manipulation parameter space A

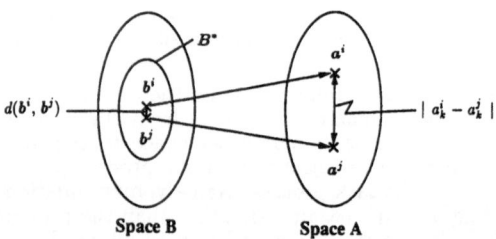

Space B

Space A

Figure 4-(b): Condition where the quotient $\mid a_k^i - a_k^j \mid / d(b^i, b^j)$ exceeds the given maximum value L

To formulate this, we need to use a norm to evaluate the difference between two data points. Let $d(b^i, b^j)$ be the distance between points b^i and b^j in space B^*, and $d(a^i, a^j)$ the distance between points a^i and a^j that correspond to b^i and b^j respectively. To be consistent, the quotient of the two distances must be bounded:

$$q_{ij} \equiv \frac{d(a^i, a^j)}{d(b^i, b^j)} \leq L, \quad \forall b^i, \forall b^j \in B^* \tag{11}$$

where L is a bounded value. Rewriting the above equation, we obtain

$$d(a^i, a^j) \leq L \times d(b^i, b^j), \quad \forall b^i, \forall b^j \in B^* \tag{12}$$

which is equivalent to Lipschitz's condition for continuity of a function. In this paper, we use Lipschitz's condition for examining the consistency of the teaching data.

Let us assume that a human expert repeats teaching operations many times so that similar process parameters may be observed. If some unmodeled process characteristics have a significant influence, then manipulation parameters a must vary as the unmodeled characteristics vary, resulting in the erratic behavior as shown in Figure 4-(b). Thus, along with sufficient teaching data, the deficit of significant process characteristics can be examined by using Lipschitz's condition.

5   The Control System

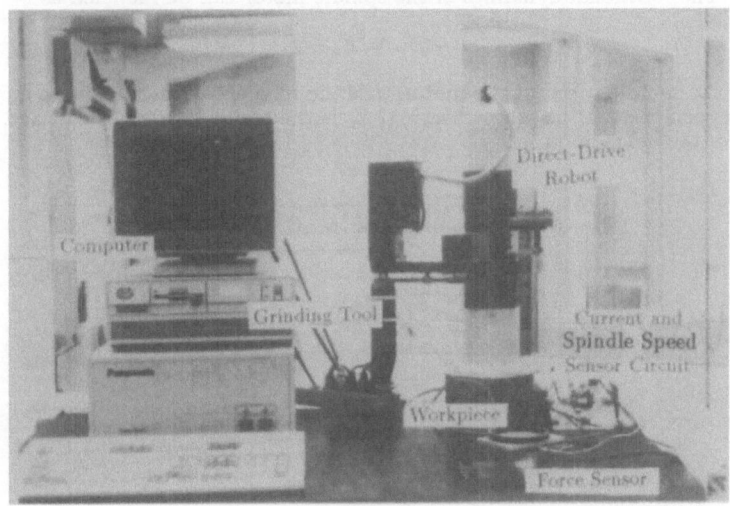

Figure 5: The two degree-of-freedom, deburring robot

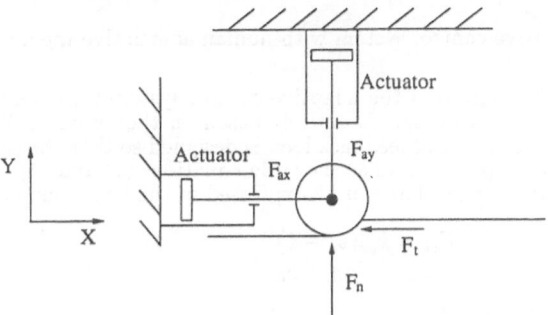

Figure 6: Schematic of the equivalent two degree-of-freedom, deburring robot

In this section, the human skill model will be implemented on a two degree-of-freedom, direct-drive robot, as shown in Figure 5. To simplify the analysis, the system dynamics can be studied in a Cartesian coordinate. For a direct-drive robot with a decoupled parallelogram design, the translation from joint torques to end point actuating forces is easy to achieve. Therefore, as shown in Figure 6, it can be considered that the deburring tool is driven by two linear actuators that output actuating forces in x and y directions respectively. The actuators exert forces on the deburring tool at its center of mass, while the deburring wheel is driven by a spindle motor. The dynamics of this system can be simply formulated by the following equations in the tangential and normal directions respectively.

$$M\ddot{x} = F_{ax} - F_t \qquad (13)$$
$$M\ddot{y} = -F_{ay} + F_n \qquad (14)$$

where M is the inertia of the tool, F_{ax} and F_{ay} are actuator forces in the tangential and normal directions, and F_t as well as F_n are interacting forces between the tool and the workpiece. The rotational dynamics of the spindle motor can be modeled as

$$J\dot{\omega} + B\omega = E_a - F_t r \tag{15}$$

where ω is the spindle speed of the motor, J is the moment of inertia of the motor, B is the friction coefficient in the motor bearing, E_a is the applied armature voltage, and r is the radius of the deburring tool.

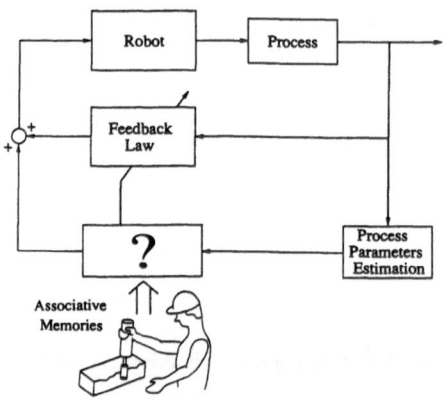

Figure 7: Adaptive control system with human associative memories

Figure 7 shows a block diagram of the adaptive control system for deburring robots. The design of the controller architecture is entirely based on the human skill model depicted in Figure 1. In Figure 7, a local feedback loop is designed so that the robot holding the tool will display closed-loop dynamics as shown in equations (4) and (5). For the two degree-of-freedom robot, the control law can be expressed in the Cartesian coordinate as

$$F_{ax} = k_{vx}(v_o - \dot{x}) \tag{16}$$
$$F_{ay} = k_y y + F_o \tag{17}$$

where k_{vx} is a velocity feedback gain that we specify.

The control system uses the same set of available sensors as the one used for the acquisition of human skills to monitor process states. Using sensor data, process characteristics can be represented by parameters in the process model (equation (10)). To execute deburring tasks based on human skill model, the robot must be able to estimate the process condition in real time and then select an appropriate control action by referring to the associative memories. For real time process identification, parameters in equation (10) need to be estimated on-line during execution. A recursive least squares estimation technique can be applied for on-line parameter estimation. For a deburring operation, the process characteristics vary from time to time. That is, the process parameters are time-varying. To deal with systems whose parameters are time-varying, we can introduce an exponential forgetting factor to the least squares algorithm. The exponential forgetting factor makes the algorithm assign high weighting to the present measurement and exponentially decaying weighting to the past measurements. The complete recursive least squares algorithm with exponential forgetting can be found in [Goodwin and Sin, 1984] and [Åströn and Wittenmark, 1989].

As discussed earlier, one unique feature of human skills is the ability to adapt the control strategy to the current process condition. The adaptation scheme in the human skill model is described by the associative memories, which can be identified entirely based on human teaching data. By applying the acquired associative memories in its adaptation loop, the control system enables the robot to perform deburring tasks in an adaptive manner similar to a human expert. Namely, based on the associative memories, reference inputs and control gains in the feedback loop will be modified with respect to the changing process characteristics. By the 'certainty equivalence principle', the estimated process parameters are used as if they were actual system parameters for computing control parameters. Systems with this architecture are recognized as self-tuning, adaptive control systems. However, there is a major difference between a common self-tuning controller and this new adaptive controller. For commonly used self-tuning controllers, the calculation of control parameters is based on a specific controller design technique, e.g. pole-placement, linear quadratic regulator or minimum variance control. In the new control system, all control parameters are determined by the associative memories, which can be trained by using human teaching data. Therefore, the system can learn how to perform tasks from human experts. Namely, we can teach human skills to robots. Actually, this method can be applied to transfer human skills not only in deburring or grinding, but also in other general tasks. This teachable adaptive controller is particularly useful when task specifications and control objectives are not exactly known or are difficult to formulate in an explicit mathematical form. Instead of writing a program or designating a fixed algorithm to construct a completely autonomous adaptive control system, the approach we are proposing will allow us to construct an adaptive control system by simply acquiring demonstration data from a human expert.

6 Implementation and Experimental Results

6.1 Identification of Human Deburring Skills

As shown in Figure 5, an experimental set up was developed in order to measure deburring operations of a human expert. A deburring tool instrumented with a spindle speed sensor is attached to the end of a three-axis, direct-drive robot. Since there is very little friction in any joint, the robot is highly back-drivable. In our experiments , the robot is used as a positioning device that measures the displacement of the tool. The human expert grasps the deburring tool and moves it against a workpiece to demonstrate the deburring operations. The workpiece is secured to a fixture under which a six-axis force sensor is placed to measure the force acting on the workpiece. Since we used workpieces with relatively low inertia, the measured force is almost the same as the cutting force. The workpieces are rectangular plates that have burrs of varying height and thickness. We used both stainless steel and aluminum plates for taking data. The deburring tool has a cutter made of high speed steel.

A task coordinate frame is placed at the initial point of the deburring tool so that the x-axis is aligned with the workpiece surface as shown in Figure 3. During the deburring demonstration, both displacement and force signals are recorded at a sampling rate of 6 miliseconds. Velocity of the tool was derived by numerical differentiation.

Based on the acquired data, we identified the tool manipulation parameters, V_o, k_y, F_o and process parameters g, h, P'. In both the tool manipulation model and the process model, we have assumed that the parameters involved are all constants for a limited segment in deburring processes. To process the teaching data, it is necessary to divide it into several segments so that the time-invariant models may be valid for the individual intervals of the data. To this end we used the following segmentation algorithm. First, we divided the data into many short intervals, and then combined adjacent intervals by examining whether the error of the combined segments is less than a certain tolerance. Namely, we began with very small segments for calculating the manipulation and process parameters. If the mean squared error is smaller than a specified value, then we increase the size of that data segment until the mean squared error reaches the allowable limit.

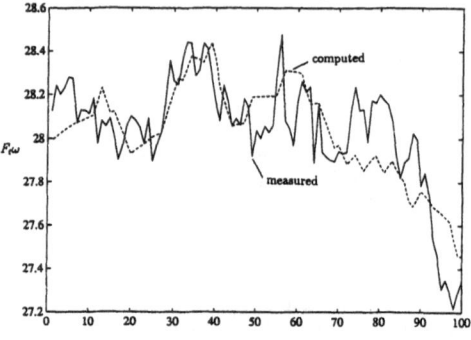

Time(6 ms/div.)

Figure 8: Measured $F_t\omega$ values vs. computed values based on the process model (equation (10))

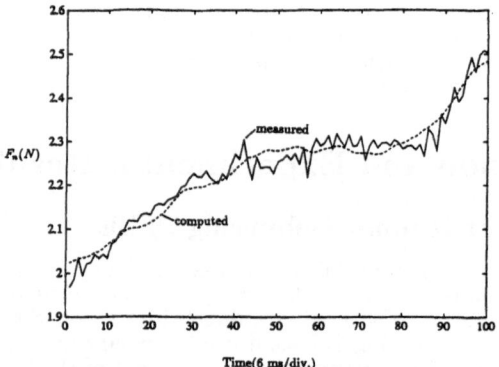

Time(6 ms/div.)

Figure 9: Measured F_n values vs. computed values based on the tool manipulation dynamics (equation (5))

Figure 8 shows one of the results of the process identification using the above segmentation technique. The product of the measured tangential force F_t and the spindle speed ω, that is, the left hand side of equation (10), is compared with the predicted value using the identified model, that is, the right hand side of equation (10). The mean squared error is less than 5 percent of the average value. Figure 9 shows a comparison between the normal force F_n and the predicted values based on the manipulation model in the normal direction (equation (5)). The mean squared error is also less than 1 percent of the average value of the normal force.

We repeated the data acquisition and processing for several workpieces using different materials and various burr sizes and shapes. From these measurements, we obtained 40 pairs of manipulation parameters and process parameters. Table 1 shows 10 of the 40 pairs of manipulation and process parameters. All these parameters have been normalized by the variance of each individual parameter. Before using these parameter data to obtain a smooth and continuous mapping function, we need to examine their consistency. As pointed out in the previous sections, if the process parameters vector b^i does not cover some critical information about the process characteristics, the mapping from b^i to a^i will be inconsistent. This inconsistent mapping can be examined by evaluating Lipschitz's condition (equation (11)) for all the parameter pairs.

Table 1: Teaching data acquired from human deburring operations

Sample	Process	Parameters	(Inputs)	Manipulation	Parameters	(Outputs)
	g	h	P'	V_o	k_y	F_o
1	-1.73	-1.09	3.75	2.38	1.78	-2.04
2	0.30	1.09	2.68	2.82	2.21	-2.99
3	-0.05	-1.29	2.96	2.03	2.32	-3.01
4	2.08	0.70	2.02	1.72	2.21	-2.91
5	1.01	0.26	2.75	2.28	1.68	-2.54
6	1.11	0.22	2.99	2.55	1.17	-1.78
7	1.65	0.70	2.71	2.95	1.31	-2.32
8	1.99	1.17	4.62	4.21	-1.26	-1.17
9	-0.10	0.22	5.57	3.60	1.02	-1.24
10	2.05	1.82	4.46	4.29	2.72	-2.81

The quotient q_{ij} of every data point was computed (equation (11)). For each data point, we take the average of the sum of the quotients, using the following equation.

$$\bar{q}_i = \sum_j \frac{d(a^i, a^j)}{d(b^i, b^j)} / n \tag{18}$$

where n is the total number of data points.

Table 2 shows the computed quotients for the first 10 data points. The pair of data (a^5, b^5), and (a^6, b^6), for instance, have a large quotient. To obtain a consistent mapping, we eliminated some data which has caused the mapping to be inconsistent and erratic. The average quotients q_i of several (a, b) pairs are listed in the last column of Table 3. Data points 4 and 9 have significantly higher average quotients than the others. Setting up a value of 1.6 as the upper limit of average quotients, we eliminated these data points from the list. Out of 40 parameter pairs, 10 pairs were found to have average quotients above the 1.6 limit value, and thus were deleted from the training set. The other 30 parameter pairs were considered adequate to define a consistent mapping. From the 30 sets of consistent data, we chose 27 parameter pairs as the training samples to identify the mapping function. The remaining 3 pairs of data were retained to examine the effectiveness of the mapping function. It is expected that the mapping between process parameters and normal manipulation parameters is nonlinear. We used a neural network to generate a nonlinear mapping that fits the selected 27 pairs of training data. After trying several structures, we found that a network with six hidden units along with three input units and three output units was satisfactory. Table 3 shows 3 pairs of parameters (a, b) that were not used for training the network. The table also shows the estimated values of V_o, k_y and F_o using the neural network. The nonlinear mapping generated by the neural network yields an average error of about 10 percent.

Table 2: Sample quotient values for 10 sets of training data

	2	3	4	5	6	7	8	9	10	$\bar{q}$
1	0.30	0.52	1.00	0.11	0.40	0.06	1.05	2.85	0.36	0.72
2		0.06	2.11	0.96	1.46	0.82	0.87	0.96	0.99	0.72
3			1.52	0.54	1.01	0.45	2.02	2.14	0.94	0.84
4				1.97	1.96	2.14	0.97	1.74	1.31	2.09
5					5.98	0.67	0.56	1.42	0.37	0.86
6						1.54	0.58	1.63	0.70	1.04
7							0.32	1.20	0.23	0.64
8								4.80	0.56	1.54
9									2.29	2.53
10										1.02

Table 3: Comparison between identified and computed values of parameters (a, b)

Sample No.	Inputs			Outputs			Neural Nets		
	g	h	P'	V_o	k_y	F_o	V_o	k_y	F_o
1	1.02	2.80	2.75	2.21	1.97	-2.56	2.07	1.67	-2.23
2	0.41	-6.75	5.72	4.43	0.85	-1.59	4.93	1.12	-1.67
3	0.15	-0.22	5.67	5.13	3.26	-1.17	4.58	3.57	-1.00

6.2 Application of the Acquired Associative Mappings

In this section, we will examine the closed-loop response of the underlying control system by simulations and simple experiments. For simplicity, we study the system responses only in the tangential direction. In the normal direction, we simply regulate the tool at a fixed displacement along the y axis.

6.2.1 Evaluating the Recursive Least Squares Algorithm

Before we simulate the tool motion, we need to study how well the recursive least squares algorithm can estimate the process parameters. Since the tool motion in the y axis direction is constrained, the $h\dot{y}$ term in equation (10) can be neglected, and equation (10) can be reduced to

$$F_t \omega = g\dot{x} + P' \tag{19}$$

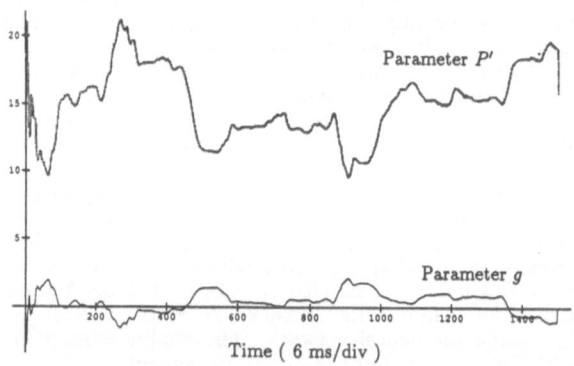

Figure 10: Specified values of process parameters

With this simplified process model, the process parameters that need to be estimated in a deburring process are g and P. The performance of the recursive algorithm can be examined in the following way: As shown in Figure 10, a time series of process parameters, $g(t)$ and $P'(t)$, were specified first. This set of process parameters was used for the mapping functions to compute the manipulation parameters. Given the time history of process parameters and manipulation parameters, the output of the system can be calculated, based on equations (16) and (19). The data generated in this manner simulate the system responses under the process conditions specified by the time series of process parameters shown in Figure 10. These force and motion data were used as if there were real data measured by the force and motion sensors in an actual demonstration. Based on these data, the process parameters can be estimated by applying the process dynamics model and the recursive least squares algorithm. Figure 11 shows the comparison between the estimated values and the specified values of the parameter P'. The two values match very well except at some points where the specified values change abruptly. The reason for the discrepancy is that as the predefined parameter values change very quickly, the recursive estimation algorithm may not converge fast enough to catch up with the variation algorithm may not converge fast enough to catch up with the variation of specified values. And this can cause the estimated values deviate from the correct values. The comparison shown in Figure 11 indicates that the recursive least squares algorithm can provide good estimates of process parameters except at some instants when process condition changes abruptly. As abrupt changes of process characteristics take place, all process parameters are reset to new values. Since the least squares algorithm converges so fast, it only takes a short while for the estimation to follow that step change and converge to the new set of process parameters. Therefore, from this simulation, it is verified that the least squares algorithm is effective for estimating process parameters in a deburring task.

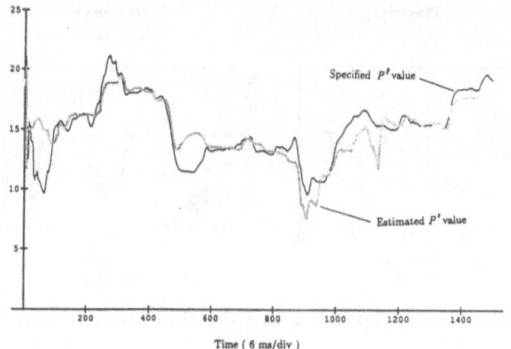

Figure 11: Specified values of P' vs. estimated values

6.2.2 Experimental Results

The main purpose of this experiment is to show the closed-loop behavior of the adaptive control system with the mapping function involved in the adaptation loop. The velocity feedback gain, k_{vx}, was set at a high value so that the actual tool velocity can closely follow the reference feedrate trajectory generated by the mapping function $v_o(\hat{g}, \hat{P})$. The experiment set-up is depicted in Figure 5. The set-up is exactly the same as the one used in [Liu and Asada, 1991] for acquiring teaching data from a human expert, except that the direct-drive robot now is servoed to manipulate the deburring tool. The deburring tool we used is a rotary file made of high speed steel. The workpiece material is 2014-T6 aluminum alloy. The average thickness of the burr is about 0.5 mm, and the height of the burr ranges from 0.5mm to 4mm. Figure 12 shows the system response with adaptive feedrate. Figure 12-c is the plot of the reference feedrate generated by the mapping function, $v_o(\hat{g}, \hat{P})$. Figure 12-d is the plot showing the reaction force in the tangential direction. When the tool is cutting a small burr, the feedrate is high and the cutting force is low. While the tool encounters a big burr, the reaction force increases and the feedrate decreases. This result reveals that the tool responds to process conditions in a way similar to a human expert. That is, when cutting a small burr, the feedrate is high; whereas for a big burr, the feedrate is low. It should be noted that in programming the control system, no explicit command is given to adjust the tool feedrate with respect to the reaction force. The computation of the feedrate is entirely based on the mapping function, $v_o(\hat{g}, \hat{P})$, which is obtained from human teaching data. Figure 13 shows the result

76

when the traditional controller is used with a fixed feedrate. Figure 13-a shows that the tool was moving with almost a constant feedrate regardless the size of the burr. Figure 13-d shows that when the tool hit a big burr, the reaction force increased substantially so that it caused the tool to stall (Figure 13-b).

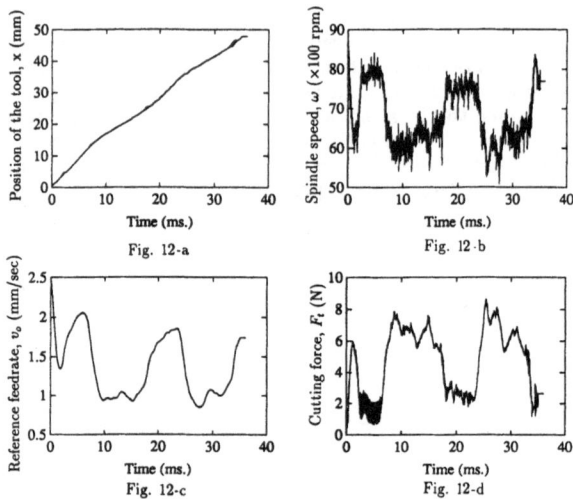

Figure 12: System response using the skill-based adaptive controller

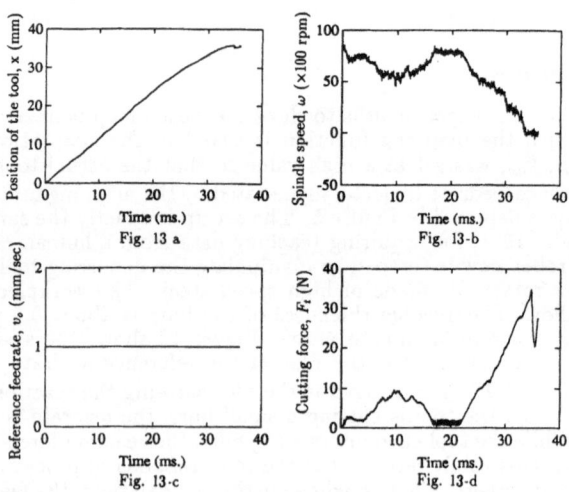

Figure 13: System response using the traditional feedrate controller

7 Conclusion

A human skill model that describes the performance skills of a human expert has been presented. In the human skill model, a metal cutting model is used to characterize the process condition in a deburring task. It is found that a human expert modifies his tool control strategy with respect to the changing process condition. Using teaching data acquired from human demonstrations, the associative mappings between the expert's tool control strategy and the process characteristics can be identified. A neural network is used to learn and represent the associative mappings. Test results show that the trained neural network provides an effective way to estimate control gains of the human teacher based on task feature information.

A new controller for robotic deburring was developed based on the human skill model. The architecture of this controller is a self-tunning adaptive control system. The associative mappings acquired from human teaching motions is used for the gain tunning mechanism. The experimental results of the system response show that the adaptive controller can adjust the control parameter in accordance with the process conditions in the manner similar to a human expert.

Acknowledgements

The authors wish to acknowledge the Leader for Manufacturing Program for its partial support of this work.

References

[Asada and Asari, 1988] H. Asada and Y. Asari, *The Direct Teaching of Tool Manipulation Skills Via the Impedance Identification of Human Motions*, Proc. 1988 IEEE Int. Conf. on Robotics and Automation, pp. 1269 - 1274, 1988.

[Åströn and Wittenmark, 1989] K. Åströn and B. Wittenmark, *Adaptive Control*, pp. 68, 1989

[Goodwin and Sin] G. C. Goodwin and K. S. Sin, *Adaptive Filtering, Prediction and Control*, 1984.

[Liu and Asada, 1991] S. Liu and H. Asada, *Transferring Manipulative Skills to Robots: Representation and Acquisition of Tool Manipulative Skills Using a Process Dynamics Model*, accepted for publication in the ASME Journal of Dynamic Systems, Measurement and Control, 1991.

[King and Hahn, 1986] R. King and R Hahn, *Handbook of MODERN GRINDING TECHNOLOGY*, pp. 34 - 38, 1986

Learning of Robotic Assembly
Based on Force Information

*Fumio MIYAZAKI, *Koji IDE, *Yasuhiro MASUTANI, and **D. S. AHN

* Department of Mechanical Engineering
Faculty of Engineering Science, Osaka University
Toyonaka, Osaka, 560 JAPAN

** Korea Advanced Institute of Sience & Technology
P.O.Box 150, Chongryngri, Seoul, KOREA

Abstract

This paper treats a practical method to generate assembly strategies applicable to part-mating tasks that are of particular interest. The difficulties in devising reliable assembly strategies results from various forms of uncertainty such as an imperfect knowledge of the parts being assembled and limitations of the devices performing the assembly. Our approach to cope with this problem is to have the robot learn the appropriate control response to measured force vectors, that is, the mapping between sensing data and corrective motion of robot, during task execution. In this paper, the mapping is acquired by using a learning algorithm and represented with a binary tree type database. Remarkable features of the proposed method are the use of a priori knowledge and accomplishment of the task with little human trouble. Experiments are carried out by taking account of practical production facilities. It is shown by experimental results that an ideal mapping is acquired effectively by using the proposed method and the assembly task is carried out smoothly.

1 Introduction

A number of work have been done on the mating of tightly-fitting parts. Especially, practical methods of utilizing passive compliance such as RCC(Remote Compliance Center) have been devised to solve the chamfered peg-in-hole problem [1] [2] [3]. From the system theoretic point of view, we can regard the force/moment exerted from a hole into a peg as the input and the motion of the peg caused by the compliant device as the output. That is, the passive compliant device is a hardware to implement a given input-output relation, "mapping". Of course, this mapping should be designed considering a specified pair of peg and hole.

In the unchamfered peg-in-hole problem, however, a multiple mapping relation is necessary to carry out the insertion task as a whole because the mapping varies in accordance with the contact configuration between the peg and hole [9]. A selection matrix in the hybrid control, for example, must be switched corresponding to the change of contact. Work on high-precision parts mating done so far by taking multiple mapping into consideration is classified into the following two approaches:

1. Analytic Approach
   Mappings necessary to perform the insertion task are derived by analyzing the model that expresses a variety of contact configurations [5] [10].

2. Experimental Approach Based on Learning
   Mappings are acquired through practice in a real task environment [7].

In the first approach, it is most important to come up with an appropriate model that represents the real environment and derive strategies robust to the uncertainties involved in the model. The second approach lessens this kind of burden because the mappings that overcome the uncertainties are automatically acquired owing to the learning ability, provided that there exists at least a correct mapping between the input and the output. From a practical point of view, the efficiency of learning, that is, the number of trials in the learning process should be taken into consideration.

The key issue addressed in this paper is to associate sensor signals with appropriate actions of a robot and generate task strategies via iterative learning in an actual physical environment. Task strategies are represented as the mapping between the perception space and the action space. Skills are denoted in terms of IF(the state of sensor signals) and THEN(the corresponding desired action). Vaaler and Seering proposed assembly algorithms that use self-generated corrective response based on sensor measurement. Simons et.al. proposed the stochastic automata theory for high precision assembly operations performed by force sensing robot [7]. Both methods are based on transition tables determined through a lot of experiments. The dimension of transition tables largely depends on the quantization level of the signal. Hence these methods require a large amount of memory size that cannot be physically realized for most real problems.

In this paper, we propose an unsupervised learning scheme based on the self-organizing database for representing and generating task strategies. Task strategies, in other words, skills are acquired via iterative learning and expressed in terms of a binary tree type database [11] [8]. Unsupervised means here that whether the database is updated or not is decided automatically depending on the evaluation of criteria given beforehand. A priori knowledge such as a rough estimate of the input/output relation is also available to improve the efficiency of learning.

In the following sections, we explain the details of the learning algorithm and how to implement it to the whole peg-in-hole assembly task. It is also shown that our proposed scheme are able to successfully insert pegs into holes at a clearance level of only 10 μm under a task environment given by taking into consideration of the uncertainties inevitable in the actual assembly process.

2 Learning Algorithm for The Peg-in-Hole Task

In general the whole peg-in-hole task is divided into two subtasks, searching task and inserting task. At the searching stage, the peg is so moved that the bottom surface of the peg is completely in the upper end of the hole. At the inserting stage, the peg enters the hole and reaches its destination, a certain depth. Strategies to perform the searching task is quite different from those for the inserting task due to the different contact configurations. However, it is desirable to learn them by using the same frame of algorithm.

2.1 How to Acquire The Mappings

Fig.1 shows the basic algorithm to learn the mappings at both searching and inserting stages. The input and the output to the learning system are the force information picked up with a 6-axes force sensor and the corrective motion of a robot at each stage, respectively. Let the perception space S and the action space A be $n-$ and $m-$ dimensional vector spaces, respectively. Any $s \in S$ represents a situation in the perception space and is also an input to the learning system. Any $a \in A$ represents an action in the action space and is also an output of the learning system corresponding to a s. When a corrective motion is needed to perform the peg-in-hole task, sensing data detected then are input to the learning system as a situation s_{now}. s_{now}'s nearest neighbor s_{base} is retrieved in the database. Since a corrective motion a_{base} corresponding to s_{base} is memorized as a pair in the database, we can directly get a_{base} as an output of the learning system. As a result of the actual corrective motion, the action performed is evaluated based on a performance criterion. If it is successful, the system goes on to the next step. Otherwise, the system randomly select another action a_{random} and try to use it after returning the contact configuration to its original situation and getting force information s'_{now} again. This process is carried out until a successful action can be chosen. Then the database is updated by adding a pair of s'_{now} and a_{random} to it.

2.2   How to Represent The Mappings

Experience, as a collection of situation-action pairs (s_i, a_i), can be organized in a tree type database. This learning system analyzes the samples of $(s, a) \in S \times A$ with a descrimination function d used to quantize the distance between two situations and memorizes its results in a self-organizing way. The followings are the algorithm of unsupervised learning database.

[Step 1](Initialization)
Set $i = 1$. Let $(s[root], a[root]) = (s, a)$. Here $s[\ ]$ and $a[\ ]$ are the memory variables assigned for respective nodes to memorized samples.

[Step 2](Finding terminal node)
Increase i by 1, and put s_i in. After resetting a pointer n to the root node, repeat the following until the pointer arrives at some terminal node. If $d(s_i, s[n_l]) < d(s_i, s[n_r]), n = n_l$. Otherwise, $n = n_r$. Here n_l and n_r mean the successor nodes of n.

[Step 3](Evaluation)
Move a robot as the action memorized in $a[terminal\ node]$ and evaluate the action to judge a success or a failure with an evaluation function. If evaluation results in a success, back to [Step 2]. Otherwise, go to [Step 4].

[Step 4](Selecting better action)
Let $a^i = a[terminal\ node]$. Choose $a \neq a^i$ randomly, and execute the action a and evaluate. If successful, go to [Step 5]. Otherwise, return to the original state and repeat [Step 4] until success.

[Step 5](Expanding the database)
Regard and establish new successor nodes as follows:

$$(s[n_l], a[n_l]) = (s[n], a[n]), (s[n_r], a[n_r]) = (s^i, a)$$

Finally back to [Step 2].

Because of backlash or stick motion of mechanical systems, it is difficult to expect that the returned situation will remain in the same action region during [Step 4]. In fact, after returning to the original position during [Step 4], the returned position may be in the same action region or not. In case of returning to any other action region, we can heuristically assume that the correct relation pair (s, a) will be expanded in $s \times A$ space incrementally.

3   Action Space and Evaluation Function

Whether or not correct mappings for the peg-in-hole task can be acquired through learning operation largely depends on the action space. If there is no appropriate action corresponding to a percepted signal in the action space, the robot is unable to learn the skill. Besides, learning becomes meaningless if correct mappings vary due to the change of circumstances. Evaluation Function plays another important role to carry out effective learning. In the followings we explain what actions and Evaluation Functions are introduced in the searching and inserting stages respectively.

3.1   Action and Evaluation Function in Searching Stage

When the peg contacts with the surrounding of the hole, contact configuration can be classified into two groups, 1-point contact and 2-point contact, depending on positional and angular errors. Fig.2 shows the force and moment applied to the peg and the corresponding reaction forces at the contact points. In our models, we assume the parts to be infinitely rigid and massless, and we use the dry Coulomb model to represent friction. We represent the normal component of the reaction force at a contact point as f and the static coefficient of friction as μ. We represent the angle of the peg's axis with respect to the hole's axis as θ. Assuming that there is a certain error only in x-direction and 1-point contact, the resulting equilibrium relations are expressed in the coordinates of the peg frame as follows.

$$F_z + f\cos\theta + \mu f \sin\theta = 0 \tag{1}$$

$$F_x - f\sin\theta + \mu f \cos\theta = 0 \tag{2}$$

$$M_y - fr\cos\theta - \mu fr \sin\theta = 0 \tag{3}$$

Here F_i and M_i (i=x,y,z) represent components of measured force and moment transformed by the peg coordinate frame respectively, and r is the radius of the peg. The similar analysis can be applied to other directions by the transformation of coordinate frame. The direction of the peg's center, which is calculated by this analysis, coincides with that of the hole center provided that there is no positional error in y-direction. However, such an ideal situation rarely happens because there is generally a certain positional error in y-direction as well as x-direction. Taking into consideration of the fact that the range of the initial positional error is small, this analysis can be employed to roughly estimate the hole center. Table 1 shows the final result usable to estimate the relations between the direction of positional error and the force sensor information. For the given direction of positional error, the corresponding corrective action can be decided by using Table 1, as a priori knowledge. It is clear that translational motions as a set of corrective action are insufficient for successful searching due to the uncertainty of Table 1. This is the reason we introduce a rotational motion as one of the corrective actions. To explain the effectiveness of a rotational motion, first we define two different areas in the configuration space.

1. Goal Area
   Goal Area is the region in the configuration space where the searching task is completed successfully.

2. Sub-Goal Area
   Sub-Goal Area is the region in the configuration space where the peg can easily move into Goal Area with only the rotational action.

Assuming that the positional error is extremely small, in case of 2-point contact, if we rotate the peg clockwise under the situation depicted in Fig.2, either of the contact points remains fixed on the edge of the hole. If the point B_1 is fixed, the peg can move toward the Goal Area. On the other hand, if the point B_2 is fixed, the peg goes away from the Goal Area. Either way we can make sure whether the result is successful or not by executing small rotation of the peg. Hence in unsuccessful case, trying to rotate in reverse leads to success. This strategy takes effect in the Sub-Goal Area. Of course, if the peg are connected with the robot in one rigid body, the movement as shown in Fig.2 will be impossible because the motion between the peg and the edge of the hole will become sliding motion. With the aid of compliant structure installed between the end effector of the robot and the peg, the motion between the peg and the edge of the hole becomes rotation about a contact point.

To acquire task strategies automatically, the learning system requires some criteria for evaluating whether the executed action is successful or not. We made simulated experiments to obtain them. The following is the criterion to discriminate whether the peg is in the Sub-Goal Area or not.

$$\frac{|F_{after}| - |F_{before}|}{|F_{before}|} < -0.20 \tag{4}$$

or

$$|F_{after}| < 1500(gf) \tag{5}$$

where F_{before} and F_{after} represent, respectively, F_z signals before and after the executed action. Next in order to discriminate whether the searching task is completed or not, we employ the following heuristic function based on the preliminary experiments.

$$\frac{|F_{after}| - |F_{before}|}{|F_{before}|} < -0.95 \tag{6}$$

or

$$|F_{after}| < 50(gf) \tag{7}$$

3.2 Action and Evaluation Function in Inserting Stage

The action space at the inserting stage is defined by any possible combination of motion of the peg. Considering the positioning accuracy of robot (repeatability $\pm 0.05mm$), we limit actions with $\{\pm x, \pm y, \pm \theta_z\}$ where x, y, θ_z are the linear movement in x and y directions, and the rotation about z axis, respectively. Due to backlash or stick motion, it is difficult to make small movement of the peg. To assure an action is performed accurately in a selected direction, the action $\{+x\}$, for example, is defined as the motion in $+x$ direction until the force F_x changes to some level of magnitude of force. The other actions are defined in a similar manner. The evaluation criterion in the inserting stage is heuristically defined as "The force level is in the possible insertion range?"

4 Experimental Results

Fig.3 Shows the configuration of the robotic assembly system. It consists of a SCARA type robot, 6-axis force sensor, a compliant mechanical device(RCC), and a personal computer. The computer communicates with the force sensor by GP-IB and the robot by RS-232-C. The learning algorithm and control schemes of circumferential equipments are implemented using C language. The diameter of the peg is $20mm$, clearance is $10\mu m$, and both the peg and hole are unchanfered. The swivel table is used to adjust the angular error.

Fig.4 shows the transition of the success rate and the number of nodes during the learning process with/without a priori knowledge. The initial position of the peg relative to the hole is randomly given every trial in the learning process. The success rate is computed as the rate of successful times in 30 trials right before the action. The success rate comes to nearly 100% after about 100 trials. Though the learning system with a priori knowledge is more effective than that without a priori knowledge, the difference is small. The node number in Fig.4 represents the amount of generated nodes in database.

About 30 byte memory size is enough for each node to contain 6-component of percepted signal data, action data, and data for the connection of binary tree type database. Therefore the memory quantity for the database and the time to handle the database is negligible.

In actual assembly tasks, there must be a variation in the part size or the angular error between the peg and the hole. To investigate the robustness of the learning system to this variation, we made several experiments. In batch-type manufacturing environment, the size of parts varies within the design tolerance. First, to investigate the robustness to the variation of the part clearance, we used pegs with different tolerances. The clearnace was changed from $10\mu m$ to $50\mu m$ at 100th trial. Fig.5 shows that the learning system is robust to the variation of the part size. Fig.6 shows success rate vs trial number while x component of the angular error is varying from 0 to 0.1 through 0.2 degree. In summary the proposed learning system is effective for the acquisition of the strategy in a given actual assembly task and robust to the variation of angular error or part size.

5 Conclusion

We have discussed the assembly system which can acquire the mapping between the force information and the corrective action of the robot through the iterative learning in actual task environments. Though the acquisition of the mapping can be done without the assistance of the experts, if possible, kinematical analysis is usable as the form of a priori knowledge to accelerate the learning. It is important to select inputs and outputs of the learning system based on the effectiveness in given tasks. We have introduced the rotational corrective motion at the searching stage, in addition to the conventional translational corrective motion, and have shown the effectiveness of it. Our proposed learning system has been applied to all stages of assembly task in actual environments to acquire the task strategy automatically. Also the learning system has been experimented and confirmed to be robust to the varying part size or angular error.

References

[1] James L. Nevins and Daniel E.Whitney *"Computer-controlled Assembly"*, Scientific Am. , Vol.238, No.2, 62/74, (1978)

[2] J. L. Nevins and D. E. Whitney *"Assembly Research"*, Automatica, Vol.16, 595/613, (1980)

[3] D. E. Whitney *"Quasi-Static Assembly of Compliantly Supported Rigid Parts"*, Robot Motion, MIT Press, 439/471, (1981)

[4] K.W. Jeong and H.S. Cho *"Development of a Pneumatic Vibratory Wrist for Robotic Assembly Robotics, Vol.7, 9/16, (1989)

[5] T. Goto, K. Takeyasu, T. Inoyama *"Control Algorithm for Precision Insert Operation Robots"*, IEEE Trans. on System, Man, and Cybernetics, Vol. SMC-10, No. 1, January, 19/25, (1980)

[6] Yangsheng Xu, Richard P. Paul *"A Robot Compliant Wrist System for Automated Assembly"*, Proc. IEEE Int. Conf. Robottics and Automation, 1750/1755, (1990)

[7] J. Simons, H. Van Brussel, I. De Schutter, J. Verhaert *"A Self-Learning Automation with Variable Resolution for High Precision Assembly by Industrial Robots"*, IEEE Trans. on Automatic Control, Vol. AC-27, No. 5, Octber, 1109/1113, (1982)

[8] D. S. Ahn, H. S. Cho, F. Miyazaki and S. Arimoto *"Automated Robotic Assembly with Unsupervised Learning Machine"*, Proc. 28th SICE, 1385/1388, (1989)

[9] Haruhiko Asada *"Teaching and Learning of Compliance Using Neural Nets: Representation and Generation of Nonlinear Compliance"*, Proc. IEEE Int. Conf. Robotics and Automation, 1237/1244, (1990)

[10] S.N. Gottachlich and A.C. Kak *"A Dynamic Approach to High-precision Parts Mating"*, Proc. IEEE Int. Conf. Robotics and Automation, 1246/1253, (1988)

[11] H. Suzuki *"Study of Universal Learning Machine Based on a Self-organization Algorithm"*, Ph.D Thesis in Osaka University, (1988)

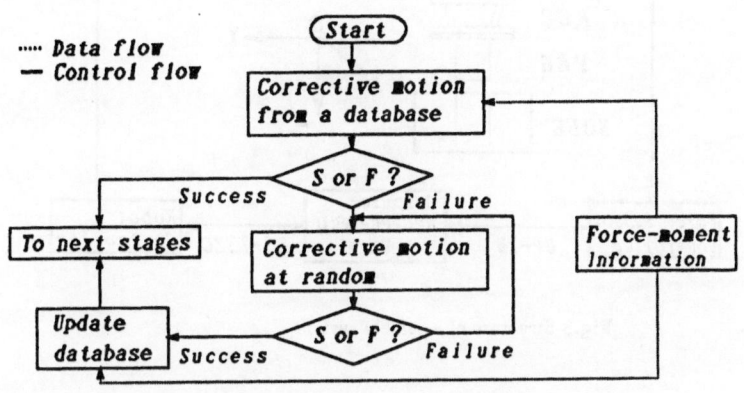

Fig.1 Algorithm for mapping acquisition

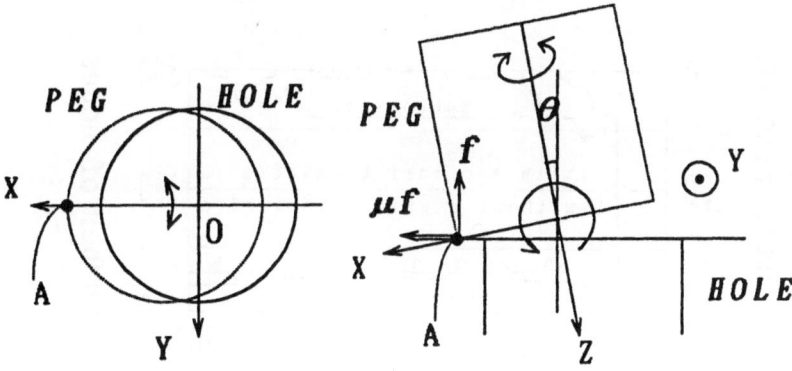

Fig.2 Configuration of one-point contact

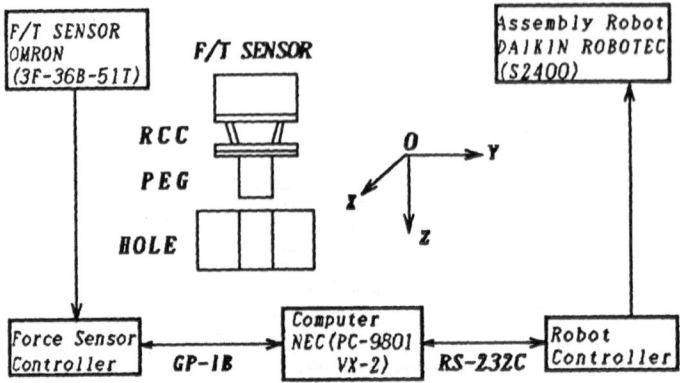

Fig.3 Structure of assembly system

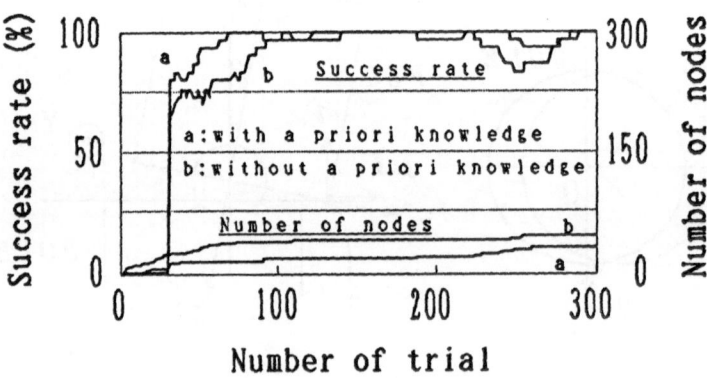

Fig.4 Comparison of the success rate and the number of
nodes with/without a priori knowledge for searching
stages

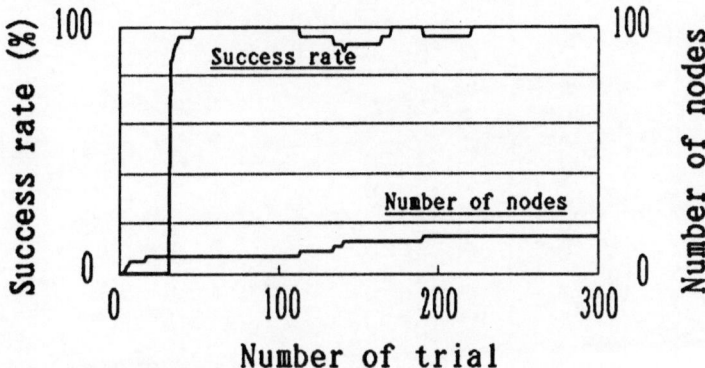

Fig. 5 Transition of the success rate and the number of
nodes in the case that the initial angular error(θ_x)
is changed as follows: $0.0 \rightarrow 0.2 \rightarrow 0.1$degree

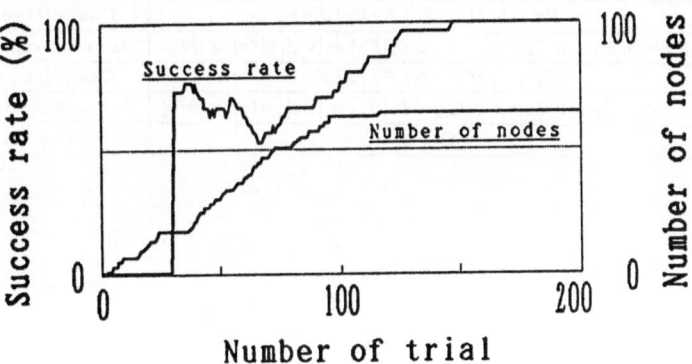

Fig. 6 Transition of the success rate and the number of
nodes in the case that the clearance is changed from
10μm to 50μm

Table 1 A priori knowledge for mapping between sensor information and direction of lateral error

Sensor Information	Contact configuration Direction of lateral error
$Fx < 0, My < 0, \|Fx\| \ggg \|Fy\|$	One-point contact, $+X$
$Fx < 0, Fy < 0, Mx > 0, My < 0, \|Fx\| \approx \|Fy\|, \|Mx\| \approx \|My\|$	One-point contact, $+X+Y$
$Fy < 0, Mx > 0, \| Fy \|\ggg\| Fx \|$	One-point contact, $+Y$
$Fx > 0, Fy < 0, Mx > 0, My > 0, \|Fx\| \approx \|Fy\|, \|Mx\| \approx \|My\|$	One-point contact, $-X+Y$
$Fx > 0, My > 0, \| Fx \|\ggg\| Fy \|$	One-point contact, $-X$
$Fx > 0, Fy > 0, Mx < 0, My > 0, \|Fx\| \approx \|Fy\|, \|Mx\| \approx \|My\|$	One-point contact, $-X-Y$
$Fy > 0, Mx < 0, \| Fy \|\ggg\| Fx \|$	One-point contact, $-Y$
$Fx < 0, Fy > 0, Mx < 0, My < 0, \|Fx\| \approx \|Fy\|, \|Mx\| \approx \|My\|$	One-point contact, $+X-Y$
Others	Two-point contact

Model Based Implementation of a Manipulation System with Artificial Skills

T.Ogasawara, K.Kitagaki, T.Suehiro, T.Hasegawa, and K.Takase
Electrotechnical Laboratory
1-1-4 Umezono, Tsukuba, Ibaraki 305, Japan

Abstract

This paper describes a manipulation system integrating a geometric model and manipulation skills. The model provides the geometric structure and physical properties of the objects in the environment. The manipulation skills enable reliable task execution in the presence of unavoidable errors and uncertainties. The geometric modeler enables high level programming, using skill based task commands, for assembly tasks. Based on the model and skills, the system autonomously executes specified tasks. As a result of these features, robustness and reliability in remote task execution has been achieved.

1 Introduction

The conventional industrial robots are controlled in the framework of direct teaching of motion and subsequent replay. This framework is valid for repetitive tasks in a structured environment, such as a manufacturing plant, where errors and uncertainties are reduced to effectively zero by carefully adjusting the setups and refining manipulator motion. In assembly tasks, however, the clearance is generally smaller than the residual position control errors. Therefore, error and uncertainties are not negligible. To overcome this, dedicated mechanical compliant wrists have been userd for insertion of particular parts. Since this approach to insertion requires a different compliant wrist for each part, it cannot be applied to general assembly tasks. A new framework is necessary to generally enable reliabe task execution in the presence of errors and uncertainties. We have proposed a model-based telerobot system [1] as a new framework for robots in unstructured environments. Remote execution of assembly and disassembly of mechanical components has been successfully achieved whith this system.

This paper describes a manipulation system integrating a geometric model and manipulation skills which is the nucleus of the model-based telerobot system. The model provides the geometric structure and physical properties of the objects in the environment. The manipulation skills enable reliable task execution in the presence of unavoidable errors and uncertainties. The geometric modeler enables high level programming, using skill based task commands, for assembly tasks. Based on the model and skills, the system autonomously executes specified tasks. As a result of these features, robustness and reliability in remote task execution has been achieved.

2 Model based approach to manipulation skills

The conventional manipulation system is composed of two layers; task level layer and servo level layer. Manipulation tasks are described using task level programming language, and task planning system is implemented in the task level layer. On the other hand, manipulator control system is implemented in the servo level layer. It is difficult to combine these two layers directly. We have introduced a new layer which is called *skill level* between two levels to make it easy to integrate task level system and servo level system (Figure 1). The *skill level* is based on the concept of manipulation skills.

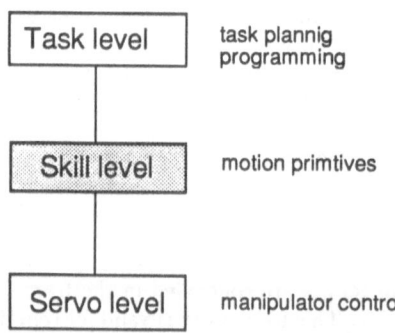

Figure 1: Hierarchy of skill based manipulation system

The manipulation skills are motion primitives. Based on the analysis of assembly sks, a sequence of motions and sensor actions are grouped into a skill for assembly. ich skill achieves a transition of one state to another state in terms of an assembly [2]. Typical examples of the state of a relation between two objects in assembly are vertex-to-face contact, edge-to-face contact, and face-to-face contact. Motion of the object to be manipulated is specified in the Cartesian task coordinate system.

To enable reliable task execution in the presence of unavoidable errors and uncertainties, skills are implemented within the hybrid (position/velocity/torque) control scheme and are incorporated with sensing procedures to detect the state to be achieved.

Three primitive skills; move-to-touch skill, rotate-to-level skill, and rotate-to-insert skill; are implemented using internal position/velocity sensor and hybrid position/force control strategy. These primitive skills are shown in Figure 2.

The strategy of skill depends on the target object, task situation, manipulators, and sensors. A variety of skills are needed to implement a general purpose manipulation system. Sensory information from different sensors must be integrated and fused to detect the state transition of the target object. Internal position sensors and a force/torque sensor are used to realize sensing functions. The complex skills are a composition of primitive skills and active sensing functions.

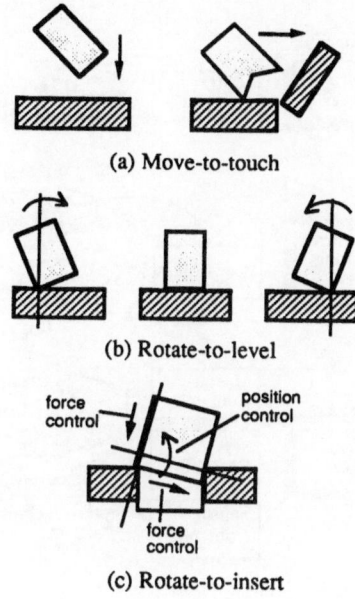

(a) Move-to-touch

(b) Rotate-to-level

(c) Rotate-to-insert

Figure 2: Primitive skills for assembly

3   Hardware configuration of the task execution system

The hardware configuration of the manipulation system is shown in Figure 3. The system is composed of three parts; an interactive teaching system, an off-line programming system, and a task execution system.

The task execution system consists of the ETA-3 direct-drive (DD) manipulator [3], a task-coordinate software servo system, a skill control system, and a task program control system.

The ETA-3 DD manipulator is being developed at ETL, and has favorable characteristics for a skill based manipulation. A force-torque sensor is attached to the wrist of the manipulator to detect force precisely. The task-coordinate servo system enables any coordinate frame to serve as the control frame for the hand position and orientation. This servo system is implemented on Plessey's Miproc16 microprocessor. The skill control system is implemented on a MicroVAX II. The MicroVAX is connected to the Miproc via a DMA line.

The task execution system is the main part of the runtime monitoring system [4] of an integrated telerobotic system for remote execution of assembly and disassembly of mechanical components.

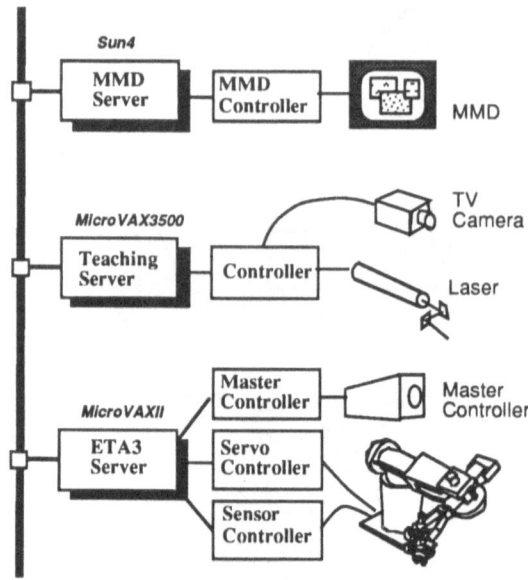

Figure 3: Hardwareconfiguration of the system

4   Implementation of skills on task execution system

The hierarchical structure of the manipulation system [5] is shown in Figure 4, and software details are shown in Figure 5. Task level programming and planning are performed on the top level system to execute complex tasks. Tasks are executed by integration of skills without knowledge of their underlying complex control strategy.

The control scheme of skills is accomplished using the geometric model. Skills are classified into three motions; move-to-touch, rotate-to-level, and rotate-to-insert. The skills are provided as Lisp functions. Typical skill level Lisp functions are shown in Figure 6. Primitive functions are provided to implement the skills. These primitivs are a set of control functions for the DD Manipulator; such as setting parameters of hybrid control and task-coordinate servo, and changing control mode. Several primitive funtions are shown in Figure 7.

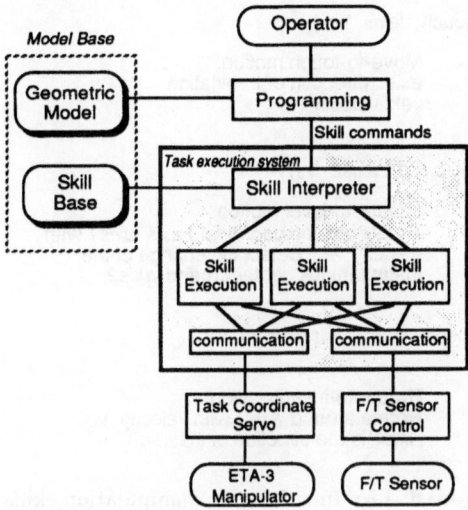

Figure 4: Skill based task execution system

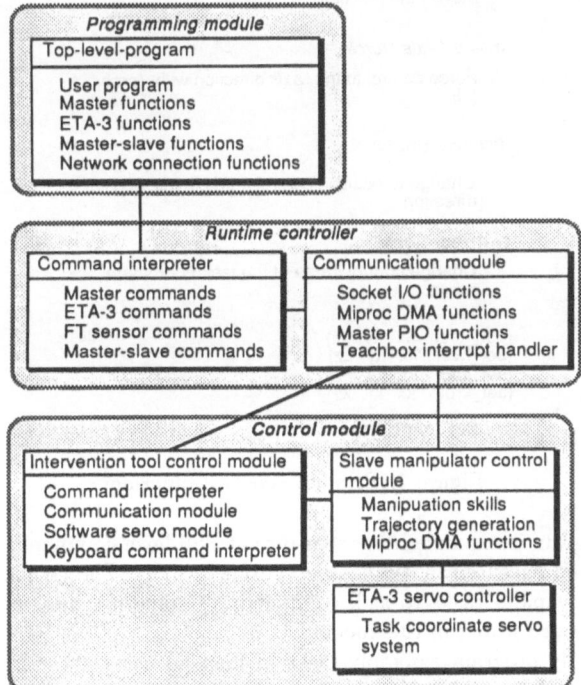

Figure 5: Software constructs of teleoperation system

(touch *'axis 'vel*)

> Move-to-touch motion.
> axis : direction of translation
> vel : velocity

(fit *'axis1 'axis2 'vel*)

> Rotate-to-level motion.
> Rotate around coordinate axis *axis1* with
> velocity *vel*. Detect the change of the
> instantanious center around *axis2*.

(insert *'axis1 'rot 'axis2*)

> Rotate-to-insert motion.
> Rotate around *axis1* with velocity *vel*.
> *Axis2* is the direction of insertion.

Figure 6: Lisp functions for manipulation skills

(move *'pos*)

> Move t the position.

(press *'axis 'force*)

> Force control to the *axis* direction with *force*
> (Nm)

(fix-pos *'axis 'k*)

> Change to position control mode in the *axis*
> direction.

(set_tool *'tool*)

> Change control frame of the task-coordinate
> servo.

(set_kpos *'kx 'ky 'kz*)
(set_kvel *'kx 'ky 'kz*)
(set_kfig *'kx 'ky 'kz*)
(set_krot *'kx 'ky 'kz*)

> Set control gains.

Figure 7: Primitive control functions

The task program is expanded to a sequence of skill level commands. The commands are interpreted and executed by the task execution system. The geometric environment model is used to expand tasks to skill commands. Geometric data, position data, and coordinate frames are stored in this model.

Users can easily program assembly tasks in Lisp [6] using these functions. Typical example motions, **pick** and **place**, are shown in Figure 8. Several skill functions, such as **touch**, and primitive functions, such as **move** and **press**, are used to describe motions. Furthermore, functions to access environmet model are necessary to program tasks. The attributes of objects are stored in the environment model. The `mass-of` function obtains the mass of objects, the `center-of-grabity-of` funtion obtains the center of the grabity of objects, and the `approach-position-of` function acquires the approach point of objects.

The task execution system of the skill based manipulation is connected to other modules via local area network. The execution system performs skill functions communicating with the programming system by Unix's *socket* mechanism. This distributed architecture increases modularity and reliability of the manipulation system. The skill interpreter refers the model base and selects the skill according to the current task.

```
(defun pick (object)
    (reset_object)
    (move (approach-position-of object))
    (f-width (grasping-width-of object))
    (move (grasping-position-of object))
    (grasp 4)
    (grasp 10)
    (set_object  (mass-of object)
                 (center-of-gravity-of object))
    (set_tool (control-frame-of object))
)
```

(a) Pick motion

```
(defun place (object place-position)
    (move (approach-position-of place-position))
    (compensate-position)
    (touch 'z -15.0 0 0.7 0.1)
    (rx-free) (ry-free)
    (press 'z -3) (press 'z -3) (press 'z -3)
    (move (read_frm))
    (tr-fix)
    (assign-current-position-to object)
    (f-width (grasping-width-of object))
    (reset_object)
    (move (approach-position-of object))
)
```

(b) Place motion

Figure 8: Example program

5 Experiment

A diaphragm exchange task has been completed using this system. The experimental task is expanded into the following subtasks.

(1) Unfasten bolt using wrench
(2) Remove bolt by gripper
(3) Remove old diaphragm
(4) Hook new diaphragm and insert into valve

Some scenes from the task are shown in Figure 9.

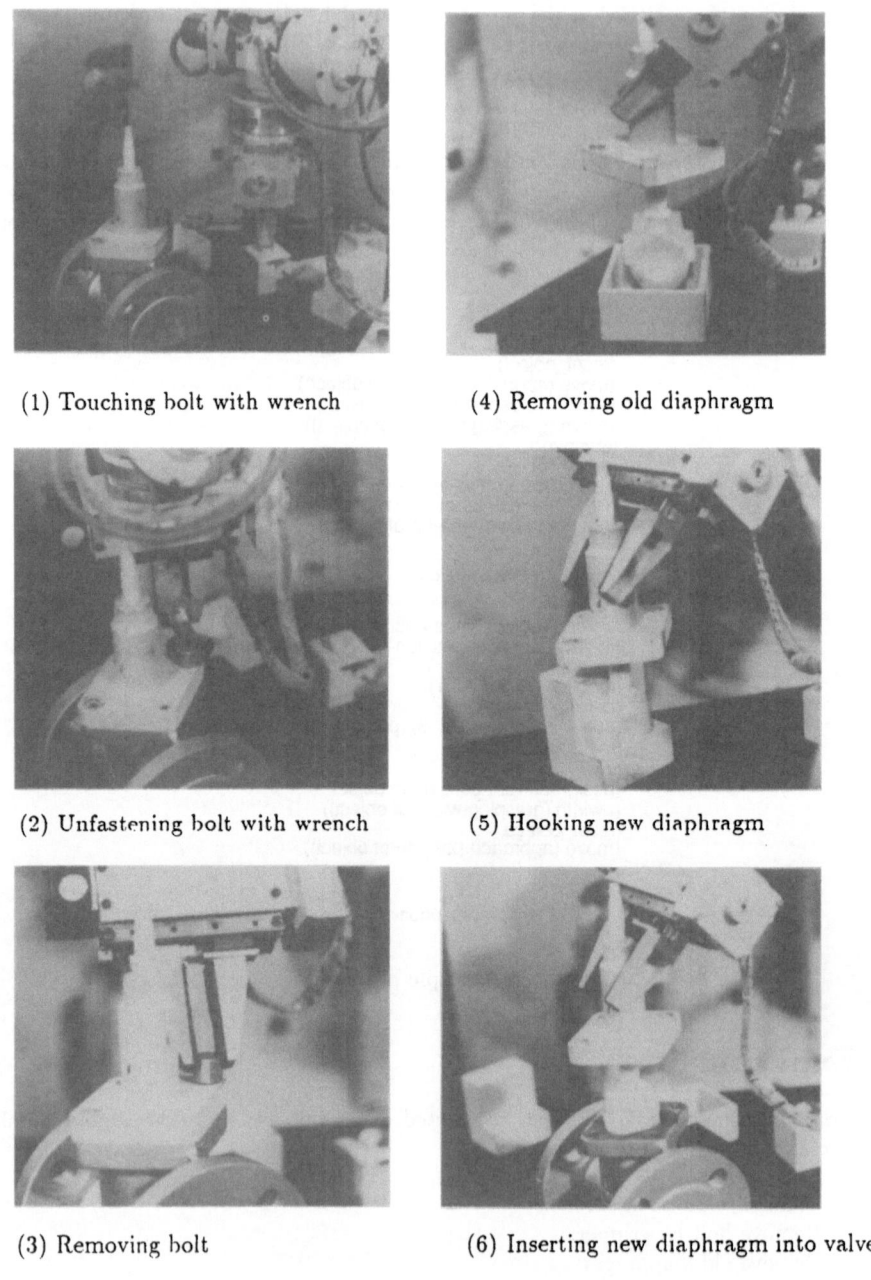

(1) Touching bolt with wrench

(2) Unfastening bolt with wrench

(3) Removing bolt

(4) Removing old diaphragm

(5) Hooking new diaphragm

(6) Inserting new diaphragm into valve

Figure 9: Diaphragm exchange task

Skill based motions are necessary to complete this task. Edge mating motion is implemented by detecting a change of situation of a target object. The task which fits a wrench to a bolt is accomplished using move-to-touch and rotate-to-insert skills. The move-to-touch skill is frequently used to place an object, such as a valve stand, bolt, and diaphragm.

In the event of unexpected trouble, robot systems sometimes fail to complete the specified task. An intervention tool enables recovery from such failures. When an error occurs, the operator issues a command to switch manipulator control from autonomous mode to master-slave mode. In master-slave mode, rotation of the intervention tool is converted to 1-DOF motion in the control frame. Assignment of this dial rotation to any of the 6 degrees of freedom in task coordinates or world coordinates can be achieved by typing one of a number of specific keys. After manual recovery, the control mode of the manipulator can be switched back to resume autonomous execution of the task. The error recovery process is shown in Figure 10. This mechanism is required to hook the new diaphragm and to insert the diaphragm into valve body. The experimental task has been successfully completed using this feature.

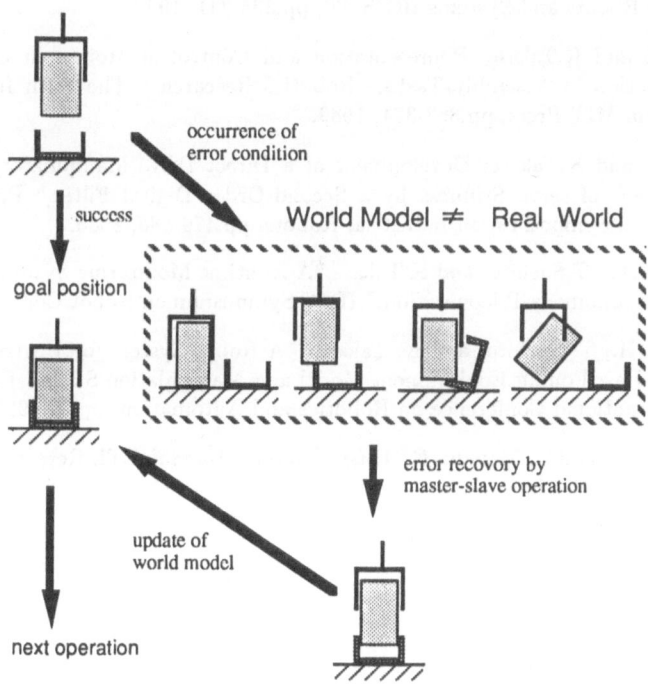

Figure 10: Error recovery process by operator's assistance

6   Conclusion

The concept of manipulation skills has been implemented using a direct-drive manipulator. The geometric environmental model and task execution system based on manipulation skills have been integrated as a model-based telerobot system. Implementation of manipulation skills and task execution system is described.

The remote execution of assembly and disassembly of mechanical components has been successfully achieved with this system.

ACKNOWLEDGMENTS

The authors express their gratitude to Dr. T.Yuba to his encouragement and support of this works, and to the members of the Robotics Group of the Elecetrotechnical Laboratory for their valuable comments.

References

[1] Hasegawa,T., et al. :"An Integrated Tele-Robotics System with a Geometric Environment Model and Manipulation Skills," Proc. of IEEE International Workshop on Intelligent Robots and Systems IROS '90, pp.335-341, 1990.

[2] Suehiro,T. and K.Takase:"Representation and Control of Motion in Contact and Its Application to Assembly Tasks," Robotics Research : The Fifth International Symposium, MIT Press, pp.367-374, 1989.

[3] Suehiro,T. and K.Takase:"Development of a Direct Drive Manipulator ETA3 and Enhancement of servo Stiffness by a Second-Order Digital Filter," Proc. of 15th International Symposium on Industrial Robots, pp.479-486, 1985.

[4] Ogasawara,T. , T.Suehiro, and K.Takase:"A Runtime Monitoring System for Hybrid Manual/Autonomous Teleoperation," IFAC Symposium on Robot Control '88, 1988.

[5] Hasegawa, H., T. Suehiro, and K. Takase: "A Robot System for Unstructured Environmets Based on an Environment Model and Manipulation Skills," Proc. of 1991 IEEE International Conference on Robotics and Automation, pp.916-923, 1991.

[6] Ogasawara, T. and T. Matsui: *ETALisp ReferenceManual*, ETL Research Memorandum, RM-87-08E, 1986.

A LEARNING CONTROL SYSTEM FOR AN ARTICULATED GRIPPER

P. Bidaud   D. Fontaine
Laboratoire de Robotique de Paris
Université de Paris 6 - CNRS - ENSAM
Tour 66 - 4 Place Jussieu
75252 PARIS Cedex 05

Abstract - This paper presents an advanced control system we developed for an articulated gripper. This articulated gripper was previously designed to achieve stable grasp of objects with various shapes and to impart compliant fine motions to the grasped object. In the control system of this device we introduced autonomous reasoning capabilities. Fine motion strategies, needed for mating or grasping, use inductive learning from experiments to achieve uncertainty and error recovery (on sensing, control and model). An overview of the articulated gripper's capabilities is provided for a better understanding of the programming environment we propose. For solving the problem of synthesis programs for fine motion planning we introduce declarative programming facilities in the controller through a time-sensitive expert system. The paper gives some details on the implementation of this expert system. Then we develop an heuristic procedure to obtain an implicit local model of contacts in complex assembly tasks. Finally, a specific example of this approach -- a peg-in-hole operation -- is outlined.

I. INTRODUCTION

The initial need for robot grippers was simply to secure objects. Mechanical devices and contact surfaces of grippers are often designed for specific applications, so that they can fit the object's shape to maximize the contact area and friction grasping forces. Limitations inherent in this simple form account for the growing interest in multi-fingered grippers. The added functionalities of general purpose dexterous multi-fingered grippers, which can grasp and manipulate objects with different shapes and sizes as well and to sense informations about the local environment seem useful for performing complex tasks. Several mechanisms illustrating these capabilities have already been built.

These grippers are constituted of several individually controlled articulated fingers, equipped with tactile sensors which are used for guiding grasp and fine manipulation operations. A great deal of work has been done on analyzing robust tip prehension grasp and kinestatics of finger mechanisms for mechanical design or grasp configuration pre-planning. With respect to this, we consider that finding optimal grasp configuration and grasping forces for a given set of external forces and desired displacement of the object is a key problem. Screw theory is well adapted to formalize this problem. Kerr [1], Ji and Roth [2], Romdhane and Duffy [3], Podhorodesky [4], Li and Canny [5] used it to find some theoretical solutions to the problems of predicting the stability of grasps and optimizing internal forces for a general multi-point friction grasp.

In these studies, assumptions are made about the geometry of contact surfaces, material properties, friction and other physical phenomena to simplify the analytical solution. Besides, real objects and mechanisms introduce errors and uncertainties. Therefore, the resulting method for individual and coordinated joint fine motion planning can't be used practically. However, it may used in conjunction with force sensor informations, to take into account interaction with the real world.

The use of such a complex system requires a task-level programming environment to achieve a level of abstraction and autonomy. Different approaches have been proposed to synthesize fine motion programs in order to achieve part mating with errors and uncertainties recovery. One of these methods uses backprojections [6] [7]. It involves the providing, through a purely geometric reasoning, of a starting configuration space (strong pre-image concept) from which any basic action will be successful, regardless of the presence of uncertainties.

Another method, developed mainly by C. Laugier [8] uses explicit processing of uncertainties. It involves the developing of a model for the propagation of uncertainties during the execution of a task. This model can be used in two ways : The first way uses the set theory. It is based on work whose purpose is to evaluate the uncertainties in the relative position of the parts at various steps of the assembly process [9]. The major drawback of this representation lies in the fact that it leads to a reasoning based on maximum errors and doesn't take into account error compensation phenomena. The second way is based on a probability method. A probability law is associated with each value subject to uncertainty. However, this becomes difficult to use when contact pairs are complex and when possible initial relative configurations are various.

In this paper, we propose a way to produce a relatively simple and fast planner developed using a knowledge-bases approach. This approach, inspired largely by Dufay and Latombe's work [10] and more recently, by Xiao and Voltz work [11], allows to develop fine motion strategies with inductive learning from experiments.

The method use the execution traces of many attempts to generate a modified plan able to deal with uncertainties.

In section II, we illustrate the concept of an " intelligent effector", we use for automatic error detection and recovery. Section III introduces a programming methodology for fine motion synthesis, and section IV presents the implementation and experiments of force/motion strategies, using LRP's articulated gripper.

II. CONSIDERATIONS ON PLANNING MOTIONS WITH UNCERTAINTIES

The execution of a task by a robot operator requires a complete, exact and coherent model of the environment in which the task takes place. However, it is difficult to define a parametric model for a physical environment subject to variations (such as variations in the dimensions, geometry, and physical properties of the parts to be manipulated) and in which effectors and observable objects are imperfect (physical measurement errors, non-linearities in the command,.....). Therefore, effectors need to be adaptable, which means that they must have decision-making capabilities.

The usual organization of hardware and software for the control of robot systems breaks down into four hierarchical levels : the actuator level, the effector level, the object level, and the objective level. However, this analytical breakdown has some drawbacks, mainly with respect to inter-module communications. Indeed, the planning of actions for a particular effector is done at a higher level. Consequently, when changes are made to an effector, software and sometimes hardware modifications need also made at all other hierarchical levels. Therefore, it seems preferable to use decentralized decision-making, so that each effector can control its own tasks. In this case, a robot system will break down into several "intelligent effectors".

II.1. Definition of an "intelligent effector".

By "intelligent effector", we mean any physical system capable of :
• producing, or executing a task using its own physical resources,
• maintaining the physical integrity of own resources and any resource attached to it (manipulated object, tools, ...),
• exchanging information with other effectors.

Thus, an "intelligent effector" is a complex system, which is a self learning system, whose representation by a graph under the form of a finite states robot would generate a complex structure. Therefore, we choose to represent it as follows :

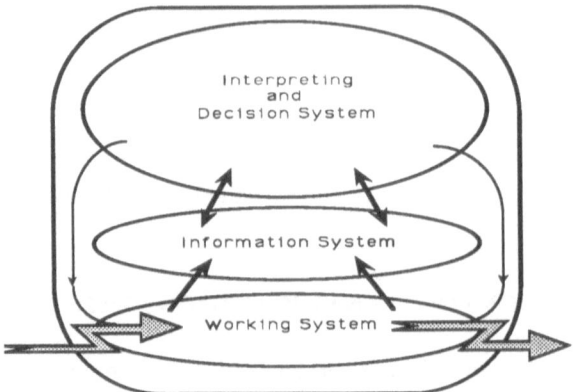

Figure 1: Overview of the organization of a complex system

For a robot operator, the input flow consists of a task to be executed according to a predefined plan. The output flow is a plan to be executed by lower level operators and/or informations about the plan which the robot operator executes. The operating system is the physical means of action, the information system is the interface which memorizes and interprets information and the decision system is in charge of designing a plan, using predefined rules, and making decisions.

II.2. Application to an "articulated gripper" operator

An articulated gripper, such as LRP's gripper [12], have the ability to grasp arbitrary objects and to impart arbitrary motions and forces to those objects to perform a wide range of manipulation tasks.

From a mechanical point of view, the gripper mechanism is characterized by having several open kinematic loops, which constitute fingers, terminated by friction contact joints. During manipulation, fingers and the grasped object form a multi-closed kinematic chain. If the hand is equipped with force sensors, it is possible to have an adaptative behavior of independent and coordinated motions of joints during grasping and manipulating subtasks. Sensory information can also indicate failures and successes.

In that case, the operator's input flow is the plan that be executed. One of them could be for instance :

> - *command the basic motion of each finger 'i' in order to bring the fingertips at the (X_i, Y_i, Z_i) coordinates.*
>
> - *try to make contact on each finger in the "Ni" direction.*
>
> - *exert a gripping effort of magnitude "Si".*
>
> - *move the grasped object in a cartesian direction, with a speed "v", until detection of a contact.*
>
> - *if no contact is made within the geometric reach of the object, abort the movement task but maintain the grasp.*

The output flow may include, among other things, reports on the execution of actions, physical execution of actions, or information to be sent to other effectors. The operating system executes basic functional actions, i.e. controlling the movement of an articulation, of a finger, or of all the fingers to produce a relative movement on the object.

The information system provides the system status variables. The sphere of decision consists of a set of production rules controlled by inference engine.

III. LEARNING TO INCREASE SPEED

III.1. Overview of the programming methodology

Basically, programming an " intelligent effector" means organizing the flow of informations between the different subsets of the complex system. Therefore, it is important to use a methodology. The methodology we selected allows to define an intelligent cycle of decision, as shown in Figure 2.

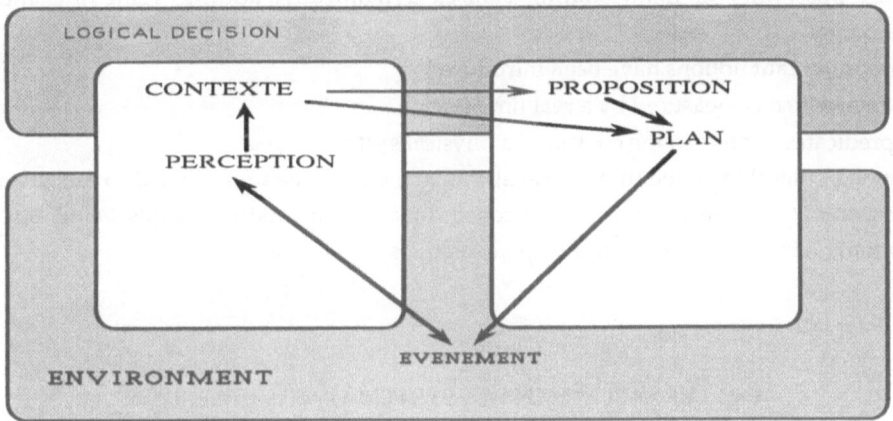

Figure 2 : Cycle of decision

In that case, the effector is first supplied with an initial time sequence of actions, or "ground plan" which is a non-determimism tree decomposition of states and transitions. The execution of this plan is monitored, and corrective patches can be made to the ground plan based on rules. Indeed, every action is conditioned by a set of constraints (pre-conditions) which are due to the existence of a context, and every execution of an action causes changes in that context (post-condition). Therefore, this type of methodology is perfectly suitable for an opportunist learning based on knowledge which has been previously learned through experimentation or reasoning. It allows to generate automaticaly the best decomposition of sub-goals for a given assembly goal.

III.2. A robotic oriented inference engine

The implementation of the above mentioned principles naturally requires the use of declarative programming in order to facilitate the introduction of heuristic rules and meta-knowledge. The "real time" constraint introduced by the robotic application requires, beside an efficient inference engine, the introduction of the time variable. For this purpose, we developed (using "C" language) a type 1 inference engine of the prolog kind, inspired by Van Caneghen's [13]. Since its size is limited (about 60 kbytes), it doesn't take up a lot of RAM storage and leaves plenty of space on standard robot control calculators for the rules base and the engines' execution stacks. The main characteristics of the engine are:
• a great efficiency, obtained through the prior encoding of rules,
• the possibility of freezing predicates,
• the possibility of manipulating clauses dynamically, adding, removing and decoding.
Two important notions have been introduced :
• time, which is measured by a real time clock,
• predicates relative to the control of a physical system.
Among other things, the first notion allows to measure the time needed to execute a task, to detect failures due to time constraints in unification attempts ("time-out" notion), to freeze predicates or postpone matings.

IV. EXPERIMENTAL SYSTEM OVERVIEW

IV.1 Configuration of the prototype

The articulated gripper we have built contains three fingers, each with three degrees of freedom. Fingertips are hemispherical surfaces made of elastomer, with a high coefficient of friction. The main characteristics of the structure are :
• non-redundant active joints,
• arbitrary motion of the grasped object,
• isostatic character of the mechanism,
• good maneuverability of the object (+-50mm,+-15°),
• basic finger geometry,
• high resolution in grasped active motion of the object.

Figure 3 : The LRP articulated gripper used in a left hand configuration.

In order to drive the nine joints of the device (six active joints and three passive joints) we minuaturied a sensing actuator for easier installation on joint axis. This integrated driving unit is composed of :
- • a direct current motor,
- • a harmonic drive gear with a high reduction ratio,
- • an absolute and an incremental optical encoder,
- • an elastic body which is collinear with the joint axis

The differential angular deformation of the elastic body, which is directly proportional to driving the torque transmitted at each joint, is detected by the two position sensor outputs.

IV.2. Realization of the object stiffness
The micro-manipulator structure can be used in an active or passive mode.
• active mode : the existence of electric drivers controlled in force and position leads us to realize an hybrid force-position control structure.
Local controllers provide the commands of actuators, from static and kinematic constraints specified in two controllable disjoint subspaces, one composed of 'n' wrench of constraints 'τs' describing the force system acting to the grasp object and

the other composed of the '(6-n)' twists of freedom 'τc' respecting reciprocity $(\tau c \ast \tau s)=0$. ($\ast$ denotes the reciprocal product of two screw quantities)

The expression for the motor torque to be applied to the ith active joint is the sum of :
- a contribution to the actuator torque from the displacement constraint τ_p, i
- a contribution to the force constraint τ_f, i

$$\tau, i = \tau_p, i + \tau_f, i$$

Both position and force use a P.D. feedback. Closed loop errors are directly servoed in joint space.

$$\tau_p, i = k_p, i \, (q_c, i - q_a, i) + k_v, i \, q_i$$
$$\tau_f, i = k_f, i \, [(q_m/N - q_s)_c, i - ((q_m/N - q_s)_a, i] + k_v, i \, q_i$$

where :
- q_a is the actual joint position,
- q_c is the command joint position,
- q_m is the actual motor position,
- N is the reducer ratio,
- k_p, k_v, and k_f are respectively the position, velocity and force gains,
- ka is the stiffness constant of the elastic element.

Joint positional incremental commands are obtained from the desired object motions in the gripper reference frame using the explicit linear relationship :

$$[dq] = M^{-1} [dx]$$

where [dq] is the generalized active joint displacement vector and [dx] is the six dimensional position/orientation object displacement vector in respect to the reference frame.

M^{-1} is the (6*6) "inverse jacobian" matrix whose rows are coordinates of the reciprocal complement of active joint screw axis expressed in a reference frame.

The static relation between the operational force vector [F] and the active joint torque/ force vector [T] is derived straightforwardly from the previous equations by using the principle of virtual work. The resulting equation is :

$$[T] = - \, {}^t M^{-1} ([F] - \, {}^t N[S])$$

[S] is a vector whose components are torques developed by passive joints, to apply the desired gripping forces to the object and N is the linear relationship for the passive joint displacement in terms of the twist representing the instantaneous motion of the grasped object.

• passive mode: After bringing into contact the parts to be mated, the structure's natural compliance is used to correct position errors.

If the interaction static wrench expressed at reference point P is $[F]_P$, the induced displacement vector of the object $[dx]_P$ is given by :

$$[dx]_P = (K\,{}^t\!M\,M)\,[F]_P, \qquad K = diag(k_1,....,k_6),$$

k_i being the stiffness constant of the elastic body place in the i-th active joint.

The above-mentioned control laws have been implemented on an integrated computer system. This structure involves three Motorola 68010-68881 single board computers and interface boards communicating a VME bus. See [14] for more details about the computer system architecture.

IV.3. Fine motion plan development

The assembly plan construction can be devided in two steps.

- 1) First, the user generates a sequence of elementary actions, or "ground plan", based mainly on a taxonomy of possible geometric configurations of the two parts and and transition actions between goal states, using a priori known part geometry.

- 2) From the deduced graph (in general the decomposition is not unique), an analysis is required in order to develop a set of rules which will control the triggering of actions (pre-conditions) and the identification of associate termination predicates of those actions (post-conditions).

During the execution of the plan, a partial redevelopment can take place using a patch plan, based on an identification of states and actions achieved through quantification and statistical analysis of the physical parameters.

The programming environment developed for this need is represented on figure 4.

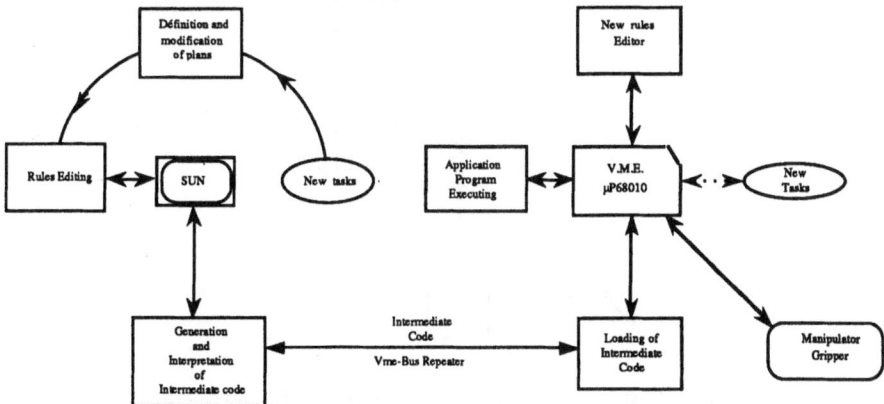

Figure 4 : Programming environment of the LRP manipulator-gripper

If we consider for instance a peg-in-a-hole task. The planning will be as follows :
• definition of a specific configuration (key state), which allows, starting from the initial state, to bring the two parts into contact so that the final assembly can always be made using hybrid motions.
For a cylindric assembly, the key state is reached when the peg touches the hole in two points.
From that state, the peg is inserted by means of two compliant motions :
• the axes of the two parts are aligned
• the insertion is achieved through an hybrid motion
Using the methodology we propose, we define a taxonomy of the states which are characterized by the measure of the pre-conditions. For each doublet of states, we define an action which allows the transition.
We will try to realize these actions in order to form the assembly plan.
The following elements are associated with each action:
- the pre-conditions to be realized to satisfy this transition,
- the post-conditions which have been realized, whose interpretation is a final state (or set of states),
- the exceptions that can occur.
With the two-hands assembly system we use, a possible deconposition of assembly actions is :
- displacement of the peg (part A) in the reference frame, specifying the position and the orientation of the frame R_A in relation to R_0 (the absolute coordinate system),
- displacement of the hole (part B) in the cartesian space. This displacement is specified in reference frame R_0,
- guarded displacement of A in order to make contacts,
- guarded displacement of B.

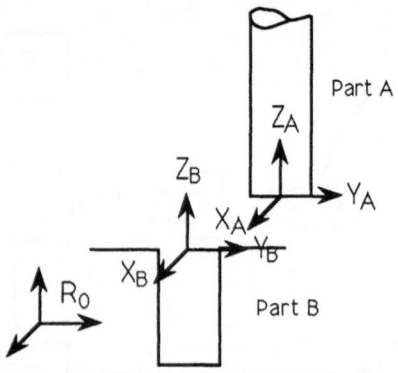

For a peg-in-a-hole task the initial plan is discarded by a linear graph .

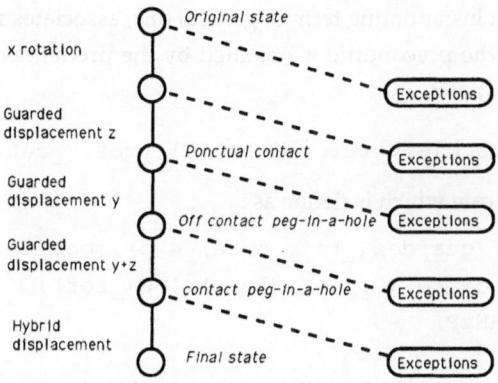

We give now, an example of actions. In the initial state, the two parts are not in contact. In the final state, we want them to be in contact. We use a guarded motion displacement action (G action), with a maximum distance and contact detection through variation of the constraints torque. In fact, a general action of guarded displacement along $\Delta L_m(x,y,z)$ is defined.

The preconditions are :
• the original configuration,
• physical parameters of the action :
       - variation of interaction static wrench $\Delta\tau_m$,
       - maximum displacements $\Delta L_m(x,y,z)$.

The postconditions are :
• normal termination : peg A is in contact with B ($\Delta\tau \geq \Delta\tau_m,\ \Delta L \geq \Delta L_m$). The database will be improve.
• raised exceptions :
       - absence of part B, defect of captors , which implies a $\Delta\tau < \Delta\tau_m$, and $\Delta L \leq \Delta L_m$
       - undervaluation of $.\Delta L_m$

In the first case, the actual solution is to keep out the plan and to warn the operator. In the other case, we can adjust ΔL_m , if it is possible (mechanical possibilities define a maximum ΔL_m and we set the same action with new preconditions. The operation in an "on-line" learning process which improve the numeric part of the knowledge base.

IV.4 Example of program.

The original plan is a list of actions. The planing program is a well-known plan (See for example Nilsson J. [15]).

For example, we describe here the G (guarded displacement) action rules.

Each G's action is an instantiating term of general one, associates with a specific data base, which define the preconditions obtained by the predefined predicate *Precond* (in this example _data =) .

```
G(_ref)  :- Precond(G, _ref, _data) G1(_ref, _data) ;
```

Lastly, G1 is a predicate which is define as :

```
G1(_ref,_data):-guarded(_ref,_data,_exp)!modifie(G,_ref,_exp) ;
G1(_ref,_):- modifie (G,_ref,dlmax) ! G(_ref);
G1(_ref,_ ):- HELP;
```

Guarded is a predicate which realizing the guarded displacement; it fails if the conditions (1) are not verified, otherwise the actual value of $D\!E$ is obtained in _exp. Modifie is a predefine predicate, which modifies the numeric knowledge base of the G action (with identification number _ref) .
If the action fails, we decide (if it is possible) to run again with a new value of $D\!E_m$. Ultimately, we decide to abort the plan.
In the same way, we can perform all guarded displacements and therefore the assembling plan.

VI. CONCLUSION

In this paper, we have described the realization of an "intelligent effector". A proposed methodology, based on the K.O.D. method allows an elaboration of an "on-line" learning plan. One of the advantages of this technique, is the possibility to adapt the plan to variations of behavior of robots and environment. An inference engine (type of Prolog) has been implemented for the control of an articulated gripper. The developed programming environment can be extended to other complex physical system.

ACKNOWLEDGEMENT

We would like to thank Francois Boudin for helping us to develop the hardware and software of the low-level controller of the LRP hand.

REFERENCES

[1] J.R. Kerr, "An analysis of multi-fingered hands", Ph. D. Stanford University, Dec. 84.

[2] Z. Ji, B. Roth, *"Direct computation of grasping forces for three-finger tip-prehension grasp"*, Journal of Mechanisms, Transmissions and Automation in Design, Vol 110, Dec 88.

[3] L. Romdhane, L. Duffy, *"Kinestatic Analysis of multi-fingered hands"*, Int Journal of Robotics Research, Vol 9, 90.

[4] R.P. Podhorodeski, A.A. Goldenberg, R.G. Fenton, *"Analytical bases for contact forces leading to internal forces for point friction contact grasps"*, Conf on Applied Mechanism and Robotics, Cincinnati, Nov. 90.

[5] Z.X. Li and J.F. Canny, S.S. Sastry, *"On motion planning for dexterous manipulation"*, Technical Report, N° UCB/ERL M89/12, University of California, Berkeley, 1989

[6] M. Erdmann, *"Using Back projection for Motion Planning with uncertainty"*, Int. Journal of Robotics Research, Vol 5, 86.

[7] T. Lozano-Perez, M.T. Mason, R.H Taylor, *"Automatic synthesis of fine-motion strategies for robots"*, Int Journal of Robotics Research, Vol 3, 84.

[8] C. Laugier, *"Traitement des incertitudes en programmation automatique des robots"*, Rapport de recherche n° 933. INRIA, Dec. 1988.

[9] R.A. Brooks. *"Symbolic Error Analysis and Robot Planning"*, Int. Journal of Robotics Research, Vol 4, 82.

[10] B. Dufay, J.C. Latombe, *"An approach to Auitomatic Robot Programming Based an Inductive Learning"*, Int. Journal of Robotics Research, Vol 3, 84

[11] J. Xiao, R. A. Voltz, *"Design and motion constraints of part mating planning in the presence of uncertainties"*, Proc IEEE Conf Robotics and Automation, 88.

[12] P. Bidaud and al, *"Application of a manipulator-gripper in an assembly cell"*, IEEE Conferences on Robotics and Automation, Raleigh 87.

[13] M. Van Caneghem, *"L'Anatomie de Prolog"*, Intereditions, Paris 86.

[14] P. Bidaud, D. Fontaine, *"Programming and control of a two-hand assembly system"*, Int. Journal of Robotics and Computer Integrated Manufacturing, Vol 6, 89.

[15] N. Nilsson, *"Principles of artificial intelligence"*, Tioga Publishing Company, Palo Atlo, California, 80.

Section 3: Grippers and Articulated Hands

Prehension is one of the basic functions that should eventually be achieved by a robot. Design of general purpose grippers, their control, and the integration with the robot arm system are three active directions of research to achieve manipulation.

Brock and Salisbury propose to implement a behavioural control scheme - as opposed to planning - on an arm/hand system in order to provide an adequate response time and adaptation in dynamic environments. The difficulty relies mainly in the lack of methods for analysis and design in such approaches. They formalize the problem in terms of behaviors associated to predefined effector actions, and a mechanism for selecting the appropriate behavior according to the sensory inputs and to memorized knowledge.

In their paper, Melchiorri and Vassura address the integration of the gripper with the arm system raising a very important issue: the hand should not be considered as a separate device attached to the arm and acting independently and in sequence with it. Instead, they argue that the mechanical design itself should take into account the structural integration of the whole system, including the hand, the arm and the sensors. The control strategy should also consider the system as a whole to exploit its redundancy for a more dexterous manipulation.

Gripper design can be simple and adapted to the manipulation of a few specific objects and therefore limited in its use, or sophisticated in order to be general purpose, but in this case usually heavy and expensive. Scherrer and Vischer propose a third approach: they report on the design of a general purpose gripper that is mechanically simple enough but is integrated with several sensors (force, tactile, optical, ...) and controlled by an adequate software. Generality is thus provided by its ability to place the fingers according to sensory data.

Manipulation cannot be achieved correctly in general without a good knowledge on the friction involved in the contact relationships between the fingers and the manipulated objects. Bicchi, Salisbury and Brock propose a procedure to measure the friction coefficients for rotation and translation using force-based sensors on the finger tips.

Implementation of Behavioral Control on a Robot Hand/Arm System

David L. Brock and J. Kenneth Salisbury

MIT Artificial Intelligence Laboratory

Cambridge, MA 02139

Abstract

Planning and behavior are two strategies which have characterized much of the recent research in robotic intelligent control. While planning algorithms provide provable convergence and analytic results, they are not generally robust and responsive in dynamic environments. Reactive behavior on the other hand, while designed for robust performance and rapid response, provide fewer analytic tools for design and evaluation. Our objective is to understand the mechanism of behavioral control and to develop analytic design tools, particularly in the area of robotic manipulation. In this paper we propose a general framework for reactive control and discuss some methods for analysis and design. We also analyze and implement a behavioral control scheme for the acquisition of a cylinder using a robot hand/arm system.

1 Introduction

The purpose of a robot is to perform a task. In a static environment planning and feedforward control can successfully perform many tasks. However, in a dynamic environment, successful execution is more difficult. In order to assure success, it is necessary to continually sense the environment and generate appropriate action based on the most current knowledge. Elaborate and long term plans would not be recommended since a sudden change in the environment may undo all subsequent action. The time frame of a particular action should therefore be consistent with the persistence of the relevant environmental states.

Rather than a detailed plan, a task could be partitioned into sequence of subtasks such that the results of each subtask are the prerequisites of the next. Conditions for execution or termination of each subtask would be continuously monitored to guarantee appropriate action at any given instant. Every environmental state as perceived through the sensors, must be an execution condition for at least one task in the sequence. Otherwise situations may arise for which the robot would have no response. This concept is similar to Universal Plans [Shoppers 1989], for which every environmental state corresponds to some node in a decision tree. The root of the decision tree is the goal, thus assuring every initial condition will eventually converge to the desired state.

The actions associated with each subtask may operate independently — each one monitoring conditions for its execution and overriding the operation of its predecessor. This is similar to the subsumption architecture which employs a hierarchy of small independent functions relating sensor input to motor output [Brooks 1989]. Instead of competing agents however, the proposed hierarchy would generate cooperative action directed toward the completion and maintenance of a specific goal.

Our objective is to first develop a control structure to achieve a task in the presence of both catastrophic and serendipitous changes in the environment. Second, actions associated with a task should be robust and achieve success for a class of situations rather than a particular instance. This also eases the burden of perception to those environmental states which distinguish classes. Third we seek to develop tools for the design and analysis of the control structure to ensure stability and task completion. Finally, it would be convenient for the control strategy to be decomposable into small independent elements allowing parallelization as well as incremental design and evaluation.

2 Background

Planning and behavior are two philosophies which have characterized much of the recent research in robotic intelligent control. Planning systems designed to achieve specific objectives have been developed, particularly in the areas of collision free motion, fine motion planning, grasping and part insertion [Brost 1986; Lozano-Perez 1976, 77, 83, 87, 89; Nguyen 1986]. Inspired in part by psychophysical research, biology, neural modeling and ethology [Gould 1982], behavioral robot research has attempted to develop stereotypical behavioral primitives [Bajcsy 1984; Grupen 1988; Iberall 1987; Jacobsen 1987; Stansfield 1988, 89, 90; Tomovic 1987] and architectures for the design of reactive strategies [Brooks 1985, 86, 88, 90; Connell 1987; Mataric 1990]. While planning systems provide analytic and provable results, they are generally poor in dynamic and unpredictable environments. Behavioral systems, on the other hand, are robust in changing surrounds, but generally lacking in analysis. Our objective is to understand the mechanisms of each control strategy and develop analytic tools, particularly in the area of robotic manipulation.

3 Theory

In this section we provide a theoretical framework, discuss the general algorithm and propose some analytic design tools. Basically we define a task as a region in state space and the achievement of a task as its inclusion of the knowledge space; that is, we define a task as a particular arrangement of the world which we know through sensing. To accomplish a task, we partition it into a sequence of subtasks or goal regions. The initial region in the sequence is automatically defined as the state space; thus every environmental state corresponds to some goal in the sequence. Each goal is partitioned further into disjoint subsets and an action, that is a trajectory or in the simplest case a velocity vector, is defined for each one. The objective is to design behavior, that is to derive a sequence of goals, partitions and actions to refine the knowledge space and accomplish the task.

3.1 Definitions

The *state space* S is the space of all parameters necessary to predict future events of the system. These include both the states of the robot and the mutable states of the environment.

The *knowledge space* $K \subset S$ represents the region of state space currently known to the system. Given perfect knowledge of the states, $K = \{s\}$ for some $s \in S$. However, more likely we will know some states to within some error, $\pi_i(K) = B_\epsilon(\pi_i(s))$, while others we will not know at all, $\pi_i(K) = \pi_i(S)$, where π_i is the projection function on the ith dimension. As a task is executed, we expect the knowledge space to become finer as interpretations of the current sensor values are incorporated into the previous knowledge of the system.

The *sensors* B represent the output of the physical sensors. For each $b \in B$ we associate an *interpretation region* of state space $I(b) \subset S$, representing knowledge gained through the sensor readings.

The *actuators* C are the affectors of the system. We define an actuator *command* $c \in C$ as a velocity in the continuous case or a displacement in the discrete. A *trajectory* t is a set of commands $t = \{c_\alpha\}$, where $\alpha \in [0,1]$ for the continuous or $\alpha \in \{0, \ldots, n\}$ for the discrete cases. Errors in the transmission of a command to the physical displacement of the system may introduce deviations from the actual and expected states, that is, $c \in B_\epsilon(c_\alpha)$.

We define a *task* as a *goal*, that is a region in state space $G \subset S$, or more generally as a sequence of goals

$$G_0 \supset G_1 \supset \cdots \supset G_n,$$

where $G_i \subset S$. As a convention we will define $G_0 = S$, so that every task is represented by a goal sequence and every point in the state space is in at least one goal.

We partition every goal G_i into a family of disjoint subsets R_{ij},

$$\bigcup_j R_{ij} = G_i \text{ and } \bigcap_j R_{ij} = \emptyset.$$

and for each region R_{ij} we associate an *action* a_{ij} which is a command $c \in C$ or a trajectory $t \in \{c_\alpha\}$. We know the system has achieved a particular goal if $K \subset G_i$ or is within a particular region if $K \subset R_{ij}$. If $K \subset G_n$ then the system has accomplished the task, otherwise we execute action a_{ij} of the associated region R_{ij} of the smallest goal region such that $K \subset G_i$.

The *forward projection* $F_{ij}(K) \subset S$ of an action a_{ij} is the set of all possible states after the execution of the action given both sensor and actuator errors. Also, we define $F_{ij\alpha}(K)$ as the forward projection after a single command of an action $c_\alpha \in a_{ij}$.

Finally we define *behavior* as the goals, regions and their associated actions $B = \{(R_{ij}, a_{ij})\}$. This definition includes aspects of the environment in the regions R_{ij}, stereotypical action a_{ij} and the relationship between them.

3.2 Framework

The idea is to generate appropriate action to accomplish a goal based on the most current knowledge of the world.

Suppose at certain intervals we acquire sensor readings $b_i \in B$ and determine its interpretation region $I_i = I(b_i) \subset S$. The evaluation of the current state is a classic problem in estimation. Here we will simply assume the knowledge state is the intersection of the previous knowledge state and the interpretation region,

$$K(i \mid i) = \begin{cases} I_i \cap K(i \mid i - 1) & \text{if } I_i \cap K(i \mid i - 1) \neq \emptyset \\ I_i & \text{otherwise,} \end{cases}$$

where $K(i \mid j)$ denotes the knowledge at i given interpretations $I_0, \ldots, I_j$.

Next we determine the smallest goal such that $G_j \supset K(i \mid i)$, find the associated region $R_{jk} \supset K(i \mid i)$ and execute the next command in the action a_{jk}.

If we require

$$K(i + 1 \mid i) = F_{jk\alpha}(K(i \mid i)) \subset R_{ij}$$

and

$$F_{ij}(K(i \mid i)) \subset G_{i+1},$$

and $K(i \mid i) \subset G_i$, then the system converges to the goal, even for finite many reorientations of the states occurring at random times during the execution of the task. This simply states that knowledge is not lost through the execution of an action and that the actions associated with a goal increase the knowledge so that it is included within the next goal.

3.3 Algorithm

The above discussion provides a framework for the analysis of behavior, but the algorithm for real time control is far more simple. At any given instance the robot only needs to know which action a_{ij} to execute. Actions are associated with regions within a particular goal and the achievement of a goal is known through the current knowledge state. If during the analysis we know the current knowledge states are equivalent to the current sensor interpretations, then we simply associate actions to current sensor values. However if the current knowledge and interpretations are not equivalent; that is the robot has acquired knowledge that it is no longer able to sense, then we must associate actions to both the current sensor values and some memory state. Thus the generic form of the algorithm is given in figure 1. The current sensor values and previous memory state are compared to those in each task, then a new state and action are issued. This form simply represents a distillation of the design and analysis.

3.4 Design

An important aspect of any proposed behavior based system is the development of design methodologies and analytic tools. These are necessary both to aid in the construction of behavior and to ensure stability and convergence. Since the system state determines the values of the sensors (within the limits of noise) and these determine the interpretation regions, then for each memory state we know the next action the system will execute. We can therefore describe the effects of the behavior as a vector or trajectory field on a space defined as the product of the state space and the memory states. We will call such a map a *behavior diagram*, since it completely describes the behavior of the system in a given

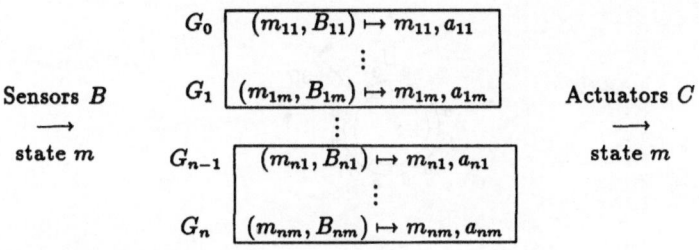

Figure 1: The computer matches the current sensor reading and previous memory state with those defined for each task and then issues a new state and actuator command. Actions issued by behavior further in the sequence override the actions of their predecessors.

environment. Although similar to potential or gradient fields used previously [Payton 1990, Arkin 1990], these diagrams extend to include both trajectories and memory state transitions. An important aspect of these maps is that a sudden environmental change can be represented as a sudden shift in position within the behavior diagram. To illustrate some of these concepts, we will consider a mobile robot under behavioral control, then employ the same techniques on a manipulator grasping problem.

3.5 Example

As an example, we will consider a mobile robot built by Maja Mataric, which employs the subsumption architecture through use of the behavioral language [Brooks 1990]. The cylindrical mobile robot translates and rotates independently and has has twelve sonar range sensors space evenly about its circumference, figure 2. Significant regions about the robot are described by thresholding sensor values or comparing their ratios. By considering both the robot and these critical regions we describe an extended object, figure 3. The position of this extended object within a particular environment describes an extended configuration space. This is similar to the extended configuration space proposed by [Villarreal 1991], which incorporates the effects of compliance during assembly. The present definition, however, also includes aspects of sensing and control as well as compliance. In figure 4 we show a projection of the extended configuration space for the robot encountering a wall. The vertical axis represents its displacement from the wall and horizontal axis its rotation. The control algorithm, as implemented by Mataric, assigns a displacement vector or state transition for each critical region. This is reflected in the behavior diagram, figure 5. The behavior of the robot is thus described as the motion of a point through this diagram and a sudden environmental change as a shift in this point.

Figure 5 shows the effects of the lower level "stroll" behavior. The length of the vectors define the translational or rotation displacements and the state transition regions define shifts to alternate diagrams. In this case, positions in the RECENT STOP region cause transitions to the right diagram. Many aspects of the system behavior become readily apparent, such as a collision region ①, limit cycle ②, and a stable region, line ③.

These diagrams also offer implications for behavioral design. The actions, that is

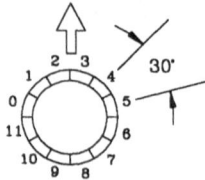

Figure 2: The cylindrical mobile robot translates and rotates independently and has twelve sonar range sensors space evenly about its circumference.

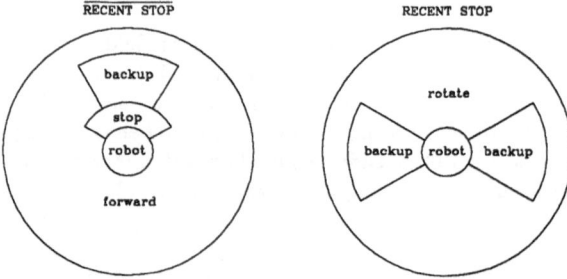

Figure 3: The robot and critical regions surrounding it represent an extended object which can be used to construct a configuration space map.

the displacement vectors or trajectories, must be compatible between adjoining regions to avoid chattering. Specifically, the components of the command vectors along the boundary normal in neighboring regions must be in the same direction. Design may also be possible by examining the extended configuration space and assigning vectors or trajectories to the disjoint regions.

In the subsumption architecture, state transitions are represented by triggered monostables, which are variables that when set remain true for a specific length of time. Since these variables represent knowledge states gained through sensing, their persistence for a fixed length of time may not be desirable, since newly acquired sensor values may negate their validity. The most recent sensor values should be compared with the previous knowledge state to yield a consistent representation.

4   Application

As an application of these concepts we will consider the task of grasping a cylinder with a robot hand/arm system, figure 6. The objective is to acquire and maintain a grasp in the presence of random changes in the position of the object. Furthermore we require the

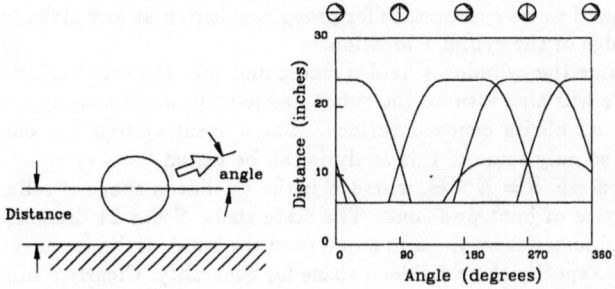

Figure 4: The extended configuration space is shown for the robot encountering a wall. The space is parameterized by the distance and inclination from the wall.

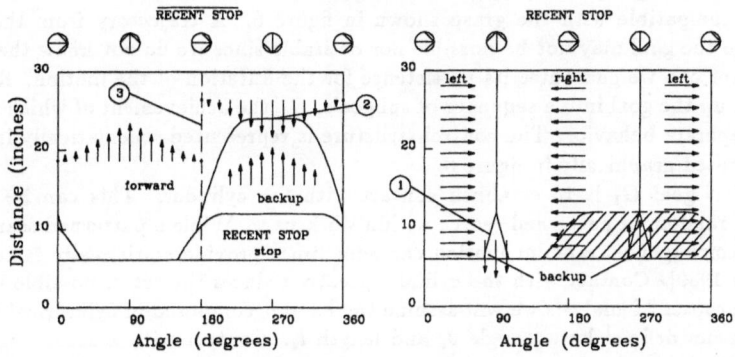

Figure 5: The diagrams illustrate the operation of the "stroll" behavior. The length of the vectors define the length of the displacement or angle relative to the wall. In the left diagram, positions in the RECENT STOP region cause a transition to the right diagram. A point remains in the right diagram for a fixed period of time after leaving the shaded region. This may cause aberrant behavior for moving obstacles, such as the case for region ① which causes the robot to collide with the wall. Also incompatible vectors at the interface of the boundaries cause chattering instabilities ② . On the other hand, line ③ represents a stable region, in which the robot neither moves toward nor away from the wall.

motion of the hand to be appropriate for grasp acquisition at any given instant, given its current knowledge of the cylinder location.

We will assume the cylinder is unobstructed and that the robot is initially unaware of its location. We will also assume the robot has joint position sensing and coarse tactile sensing located on all its exposed surface. The current system has only fingertip and palmar sensing so only some of this analysis can be tested.

The state space is $S = E \times \Theta$, where E is the Euclidean space of solid body positions and Θ is the space of joint positions. The state space S has 21 dimensions, 6 from the position of the object, 6 from the arm and 9 from the hand. Cylindrical symmetry reduces the order, but we specify the Euclidean space for generality. Clearly a full analysis of this space is difficult and is the primary motivation for subdividing the task into components.

The sensors B include joint position and tactile sensors and the interpretation regions I are the sets of possible joint angles and cylinder locations given both sensor error and resolution. For a cylinder in contact with a surface, we know its axis must pass through some point in the normal projection of that surface equal to its radius and the direction must be in the tangent space.

The actuator commands C are discrete displacements. In the current system, commands are issued every 0.028 seconds, which is the limit set by the PUMA controllers.

The task is to grasp the cylinder and the goal G is the set of joint angles and cylinder positions compatible with the grasp shown in figure 6. A trajectory from the current position to the goal may not be possible nor desirable since we do not know the cylinder location nor can we guarantee its persistence for the duration of the motion. Rather we will break up the goal into a sequence of subgoals, G_i, the achievement of which describes a set of separate behavior. The control structure is represented schematically in figure 7 and illustrated graphically in figure 8.

The first goal G_1 is to establish contact with the cylinder. This can be achieved through a random or patterned search within workspace. While a patterned search yields provable convergence, random motion can sometimes provide statistically faster results [Erdmann 1989]. Contact with the cylinder greatly reduces the set of possible locations. For the purposes of analysis we will assume the hand is composed of cylindrical links and tactile sensors defined by an angle θ_s and length l_s. Contact with a sensor implies the cylindrical axis passes through a section of a solid annulus A and its direction must be tangent to some point on the surface represented by a set D as shown in figure 9. The set A is defined by the maximum cylindrical radius r_{max}, the finger radius r_f and the radial and linear extent of the sensor θ_s, and l_s respectively, while the set D is simply defined by the angle θ_s. We will define the center x_c of A as the center of the minimum bounding sphere S. Hence $x_c = [0\, r_c\, l/2]^T$, where $r_c = (r_{max} + r_f)/2 \cos(\theta_s/2)$ and the radius of the bounding sphere is $r_b = \sqrt{r_c^2 + r_f^4 - 2r_c r_f \cos(\theta_s/2)}$. We also define a direction vector n_c from the center of the phalange to the point x_c.

Figure 10 shows the volume R in which the cylinder must reside. This set R is formed by moving the cylinder through all position while remaining in contact with the sensor. The set R thus defines a restriction on the space of collision free motion.

The second goal G_2 is to move the cylinder axis to within the forward projection of the palm. This is accomplished by moving the hand into the free space below the region R and orienting the palm with the normal n_c toward the point x_c. The vectors n_c and x_c can be pre-computed based on sensor arrangement and resolution. Stereotyped trajectories

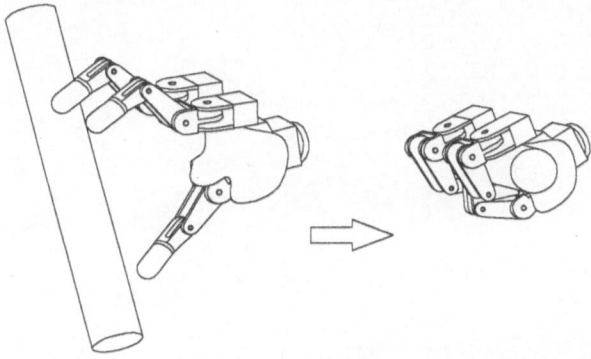

Figure 6: The objective is to acquire and maintain a grasp in the presence of sudden changes in the position the cylinder. The robot should also generating appropriate action based on its current knowledge of the cylinder's position. For the initial analysis we will assume the cylinder has a radius less than some maximum and that the hand interacts with it away for its endpoints.

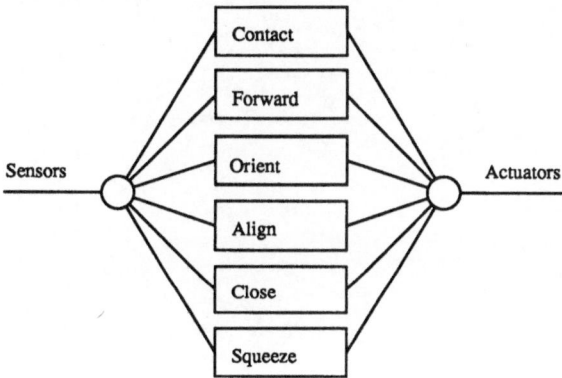

Figure 7: Independent behavior can operate concurrently to achieve a stable grasp of a cylinder. Output from action further down the sequence override that of its predecessors. The first behavior attempts to establish contact with the cylinder. The second positions the axis within the forward projection of the palm. The third orients the cylinder between the upper and lower fingers. The fourth aligns the cylinder with the principle axes of the hand and the final ones close the fingers and maintain the grasp.

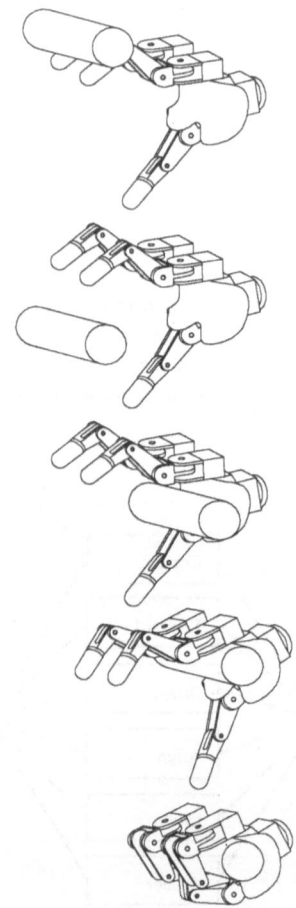

Figure 8: The first objective is to establish contact with the cylinder, the second to place the cylinder axis in the forward projection of the palm, the third to orient the cylinder between the fingers, the fourth to align with the principle axes of the hand and finally to close the fingers and grasp the cylinder.

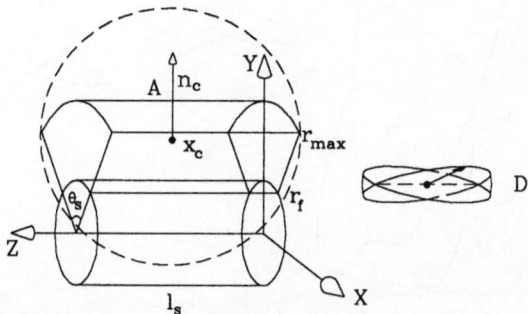

Figure 9: Contact with the sensor reduces the set of possible cylinder locations. The axis must pass through a point defined by a solid annular section A and the direction must be within a set D which is the tangent space of the sensor surface.

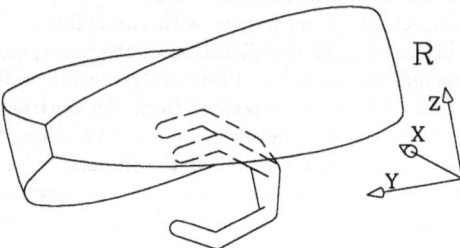

Figure 10: The locus of possible cylinder locations imposes a restriction on the free space movement of the hand.

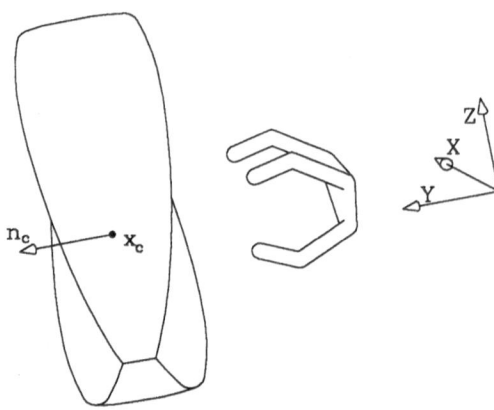

Figure 11: Contact with the outer contact sensor initiates the action which aligns the region of possible cylinder locations with the forward projection of the palm. The subsequent behavior simply uses the contact sensors on the fingers to center the cylinder between the upper and lower fingers.

may also be pre-computed within certain ranges of joint positions.

The third goal G_3 is to establish contact with the palm between the upper and lower fingers. The locus of possible cylinder locations, based on the alignment from the previous task is shown schematically in figure 11. The maximum inclination of the cylinder toward the hand is given by half the tactile radial resolution. The state space, at least locally, is four dimensional and can be parameterized by a distance y_a and x_a of a point through which the axis passes and an inclination θ_a and rotation ϕ_a as shown in figure 12. Figure 13 shows a representation of the four dimensional configuration space with an array of plots. Each plot shows the intersection of the fingers with the cylinder. The horizontal axis of each plot describes the distance y_a in the direction of the palm, and the vertical axis the rotation ϕ_a about the normal to the palm. Plots are given for various inclinations of the cylinder, $\theta_a = 70°$, $90°$ and $110°$, and deviations from the central axis of the hand $x_a = 0$ cm, 1 cm and 2 cm. Thus we can define a behavior to align the cylinder by simply rotating hand for contacts on the outer edges of the fingers and reversing direction and rotating for contacts between the upper two fingers. The distance of the reverse trajectory is defined by the worse case cylinder inclination which is defined by the sensor resolution $\theta_s/2$.

The behavior diagram shown in figure 14 illustrates the actions and state transitions defined for disjoint regions of the local state space. The goal G_3 is to contact the palm which is represented by the unobstructed region on the vertical axis annotated in the plot.

The fourth goal G_4 is the alignment of the cylinder with the palm. Since alignment with about the y-axis requires contact resolution unavailable with the current palm, we will consider only alignment about the z-axis. The palm sensor consists of a curved plate supported by mechanical springs, figure 15. Four infrared emitter/detector pairs measure

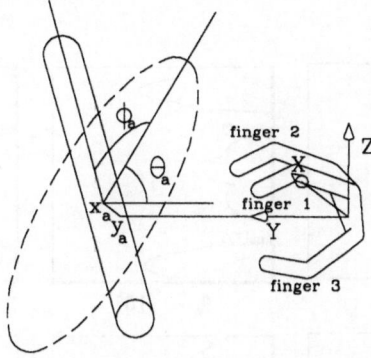

Figure 12: The space of possible cylinder locations can be parameterized by the distance from the center of the palm y_a, x_a and the relative angle between the hand and the cylinder θ_a and ϕ_a.

the displacement of the corners of the plate and thus deduce the location of contact [Brock 1990]. Figure 16 shows a simple planar model of palm. We parameterize the space of cylinder positions by y, the distance of the side of the cylinder to the center of the palm, and θ, the angle between the palm and cylindrical axes. The height of the opposing sides of the palmar plate were used to determine left, right, both or no contact; that is,

On	$h_1 < h_t$ and $h_2 < h_t$
Left	$h_2 - h_1 > h_d$
Right	$h_2 - h_1 < -h_d$
Off	otherwise.

and the corresponding extended configuration space boundaries are

Initial contact	$y \geq h_o$	$\theta = \pm \mathrm{atan}((y - h_o)/w)$
Bottom	$y \geq 0$	$\theta = \pm \mathrm{atan}(y/w)$
On	$y \leq h_t$	$\theta \leq \pm \mathrm{atan}((h_t - y)/l_s)$
Left/Right	$y \geq 0$	$\theta \leq \pm \mathrm{atan}((y - y_t)/w).$

Significant friction in the sliding support structure of the plate was found to cause simple rotate about undeformed spring thus yielding the simplified equation for the left/right contact. The alignment algorithm was simply

On	close fingers
Left	rotate about $\mathbf{r} = (r_x, r_y)$
Right	rotate about $\mathbf{r} = (-r_x, r_y)$
Off	move forward.

Figure 17 shows the behavior diagram and the experimental results for a judicious choice of rotation point $\mathbf{r} = (4.45 \text{ cm}, 0.13 \text{ cm})$, which is a point on the edge of the palmar plate. The angular and translational displacement were given by $\Delta \theta = 0.34°$ and $\Delta y = 0.26$ mm, respectively. Figure 18 shows a poor choice $\mathbf{r} = (-2.00 \text{ cm}, 0.0 \text{ cm})$ resulting in an incompatible vector field and chattering instability. Both strategies,

128

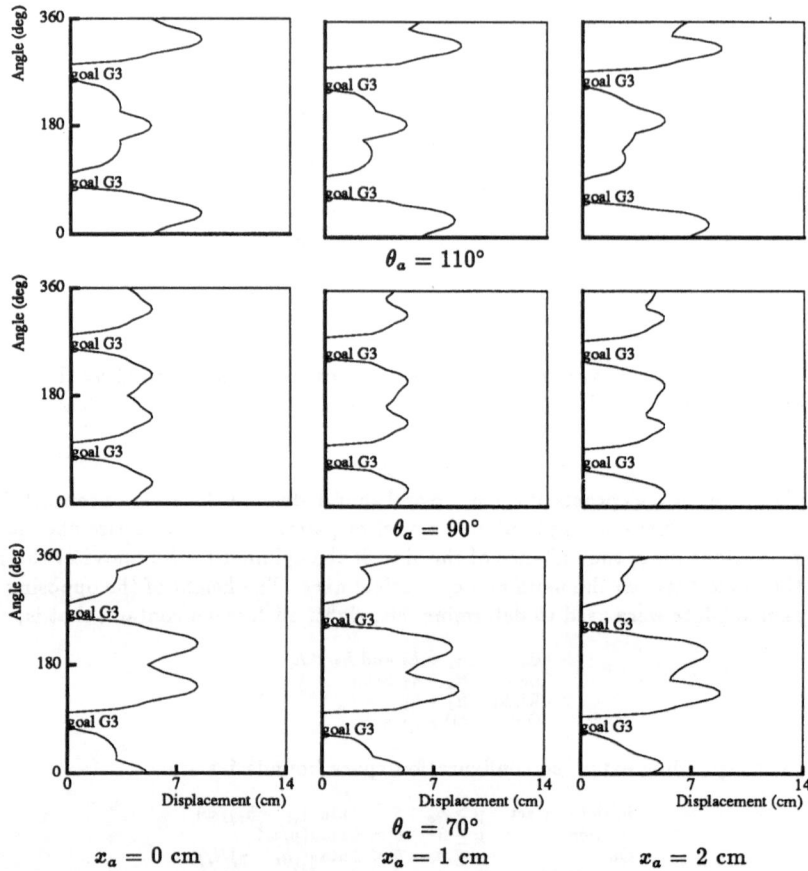

Figure 13: The slices of the four dimensional configuration space describing contacts between the hand and cylinder are shown above. The vertical axis of each plot represents the rotation ϕ_a of the cylinder about the normal to the palm and the horizontal axis the displacement y_a along that normal. The nine plots illustrate the configuration space for three inclinations of the cylinder $\theta_a = 70°$, $90°$ and $110°$ and three displacement from the center line $x_a = 0$ cm, 1 cm and 2 cm. The hand is positioned approximately as depicted in figure 12, with joint angles given by $\theta_1 = [40, 40, -10]$, $\theta_2 = [-40, 40, -10]$ and $\theta_3 = [0, -40, 10]$, where θ_i the position of finger i. In these plots, the radius of the cylinder is 2 cm and the radii of the cylindrical links are 1 cm. Also the link lengths and kinematics in this example are identical to the Salisbury Hand. The goal region G_3 for contact with the palm is also annotated in the graphs.

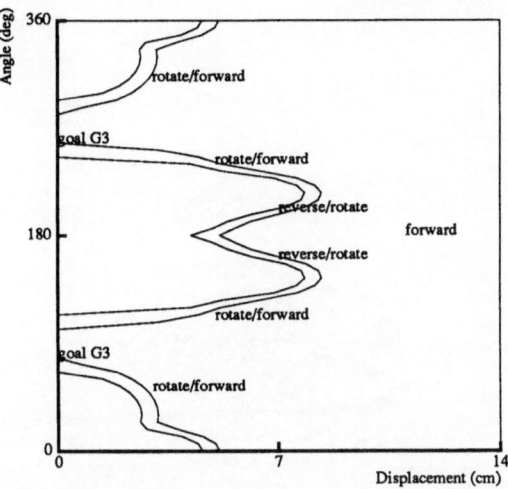

Figure 14: The behavior diagram illustrates state transition regions which initiate trajectories to align the cylinder. Contact between the upper two fingers necessitates reversing direction and rotating the hand.

however, converge to the goal, which is not the case for $r = (~0.0$ cm, 0.0 cm) and $\Delta\theta = 5°$, figure 19.

The final goal G_5 is achieved simply by closing the fingers. Force closure is achieved for cylindrical radii r such that

$$r \leq r_{max} = (l_f \sin\phi + h/2)/\cos\phi - r_f,$$

where ϕ is the friction cone angle, l_f the length of the finger, r_f the radius of the phalanges and h the separation between opposing fingers. For the Salisbury hand with urethane fingers $\phi = 25°$ [Bicchi 1991], $l_f = 7.9$ cm, $r_f = 1$ cm and $h = 8.12$ cm. Thus $r_{max} = 8.2$ cm which is significantly larger than the practical limit set by the radius of curvature of the palm, 3.2 cm. For an extensive analysis of force and form closure grasping by opposing fingers see [Chammas 1990].

The sequence of goals $G_0, \ldots, G_5$ and associated action describe a collection of stereotyped motion initiated by sensor events. This set of independent action effectively accomplishes a task in the presence of environmental disturbances.

5   Conclusion

The goal of this work is to develop a control structure which reacts rapidly and appropriately to sudden changes in the environment. Our objective is also to provide a set of analytic tools for the construction and evaluation of behavioral designs. Ideas such as

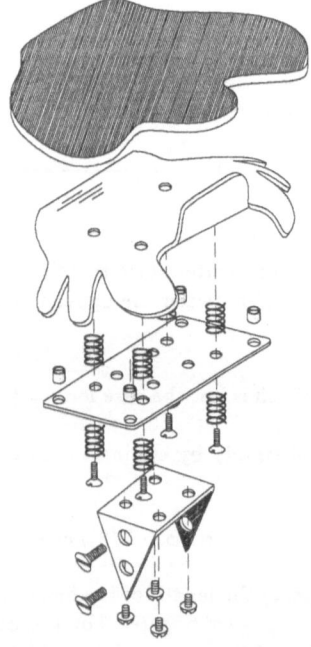

Figure 15: Optical range sensors detect the position of the palmar surface supported by simple mechanical springs.

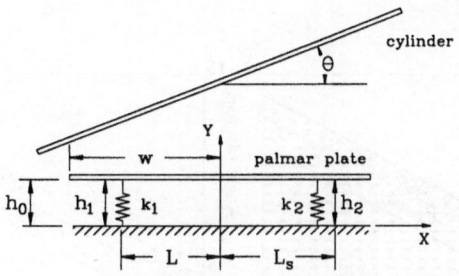

y	–	distance to side of cylinder
θ	–	angle between palm and cylinder
k_i	1355N/m	spring constant
w	4.20 cm	half width of palmar plate
l	2.90 cm	half width of springs
l_s	3.40 cm	half width of sensors
h_o	0.28 cm	rest height of palmar plate
h_t	0.12 cm	threshold
h_d	0.15 cm	difference threshold
y_t	0.12 cm	left/right threshold

Figure 16: Simple model of the palm sensor.

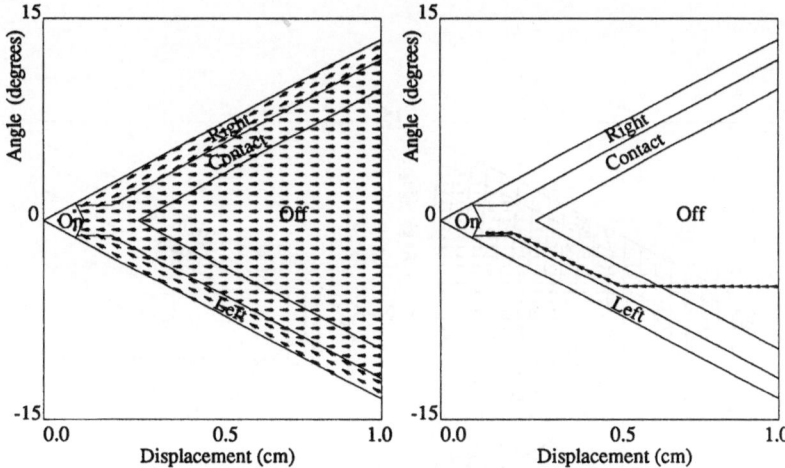

Figure 17: The behavior diagram combines the extended configuration space and associated actions. The horizontal axis represents the distance of the palm to side of cylinder and vertical axis the angle between them. Smooth convergence to the goal is demonstrated in both the behavior diagram to the left and the actual experimental trial on the right.

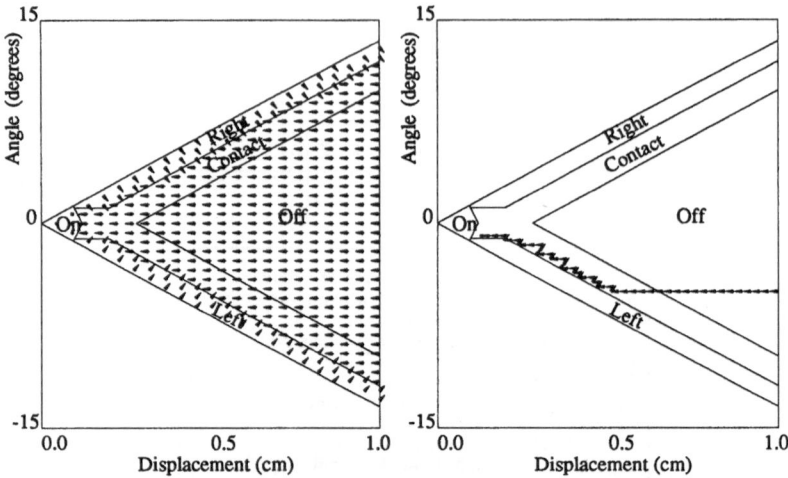

Figure 18: Chattering instabilities result from incompatible vector fields as shown theoretically in the diagram to the left and practically with the experimental data to the right.

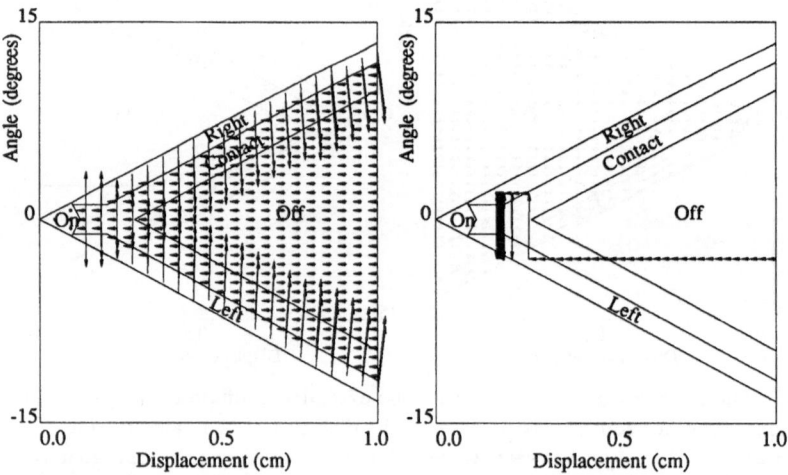

Figure 19: Limit cycles are possible as demonstrated both from the model and experiment.

extended objects, modified configuration spaces and behavioral diagrams are attempts at developing such tools.

6 ACKNOWLEDGMENTS

The authors would like to gratefully acknowledge the financial support from the following sources: NASA contract number NAG-9-319, Sandia National Laboratories under contract 75-2608, the Office of Naval Research University Research Initiative Program under Office of Naval Research contract N00014-86-K-0685, the Advanced Research Projects Agency of the Department of Defense under Office of Naval Research contract N00014-85-K-0124,

References

[1] Arkin, R.C., "Intergrating Behavioral, Perceptual, and World Knowledge in Reactive Navigation," *Designing Autonomous Agents*, ed. Maes, P., 1990.

[2] Bajcsy, Ruzena and Goldberg, Kenneth Y., "Active Touch and Robot Perception", *Cognition and Brain Theory*, Vol. 2, Summer 1984.

[3] Bicchi A., Salisbury J. K., and Brock, D. L., "Experimental Evaluation of Friction Characteristics with and Articulated Robotic Hand," *Second International Symposium On Experimental Robotics*, June 1991.

[4] Brock, D. L., "Contact Sensing Palm for the Salisbury Robot Hans," Sandia Report, June 1990.

[5] Brooks, R. A., "A Robust Layered Control System for a Mobile Robot," *MIT AI Memo* 864, MIT Artificial Intelligence Laboratory, September 1985.

[6] Brooks, R. A., "Achieving Artificial Intelligence Through Building Robots," *MIT AI Memo* 899, MIT Artificial Intelligence Laboratory, May 1986.

[7] Brooks, R. A., "A Robot that Walks: Emergent Behaviors from a Carefully Evolved Network," *Proceedings of the IEEE International Conference on Robotics and Automation*, Philadelphia, PN, 1988 .

[8] Brooks, R. A., "The Behavior Language; User's Guide," *MIT AI Memo* 1227, MIT Artificial Intelligence Laboratory, April 1990.

[9] Brost, R. C., "Automatic Grasp Planning in the Presence of Uncertainty," *Proceedings of the IEEE International Conference on Robotics and Automation*, San Fransico, CA, April 1986 .

[10] Chammas, C. Z., "Analysis and Implementation of Robust Grasping Behaviors," *MIT AI TR* 1237, MIT Artificial Intelligence Laboratory, May 1990.

[11] Connell, Jonathan, "Creature Design with the Subsumption Architecture," *Proceedings of IJCAI-87*, Milan, Italy, August 1987.

[12] Erdmann, M. A., "On Probabilistic Strategies for Robot Tasks," *MIT AI TR* 1155, MIT Artificial Intelligence Laboratory, August 1989.

[13] Gould, James L., Ethology - The Mechanisms and Evolution of Behavior, W.W. Norton and Company, New York, 1982.

[14] Grupen, Roderic A., "Behavior based control for autonomous robotic manipulation" University of Massachusetts, Dec. 13, 1988.

[15] Iberall, Thea, "The Nature of Human Prehension: Three Dextrous Hands in One", *Proceedings of the IEEE International Conference on Robotics and Automation,* Raleigh, NC, 1987 , p. 396-401.

[16] Jacobsen, Stephen C., et., al., "Behavior Based Design of Robot Effectors", Center for Engineering Design, Department of Mechanical Engineering, University of Utah, 1987.

[17] Lozano-Pérez, Tomás, "The Design of a Mechanical Assembly System," *MIT AI TR* 397, MIT Artificial Intelligence Laboratory, 1976.

[18] Lozano-Pérez, Tomás, "LAMA: A Language for Automatic Mechanical Assembly," *5th International Joint Conference on Artificial Intelligence,* MIT Cambridge, MA, August 1977, p. 710-716.

[19] Lozano-Pérez, Tomás, "Spatial Planning: A Configuration Space Approach," *IEEE Transactions on Computation,* Vol. C-32, No. 2, 1983, p. 108-120.

[20] Lozano-Pérez, Tomás, et. al, "Handey: A Task-Level Robot System," *Proceedings of the International Society of Robotics Research,* 1987, p. 123-130.

[21] Lozano-Pérez, Tomás, et. al., "Task-Level Planning of Pick-and-Place Robot Motions," *IEEE Computer,* March 1989, p. 21-29.

[22] Mataric, Maja J., "A Distributed Model for Mobile Robot Environment-Learning and Navigation," *MIT AI TR* 1228, MIT Artificial Intelligence Laboratory, May 1990.

[23] Nguyen, Van-Duc, "The Synthesis of Stable Force-closure Grasps," *MIT AI TR* 905, MIT Artificial Intelligence Laboratory, July 1986.

[24] Payton, D.W., "Internalized Plans: A representation for action resources," *Designing Autonomous Agents,* ed. Maes, P., 1990.

[25] Stansfield, S. A., "Reasoning About Grasping," AAAI-88.

[26] Stansfield, S. A., "Robotic Grasping of Unknown Objects: A Knowledge-Based Approach," Sandia Report SAND89 - 1087, June 1989.

[27] Stansfield, S. A., "Knowledge-based Robotic Grasping," *Proceedings of the IEEE International Conference on Robotics and Automation,* Cincinnati, OH, May 1990 , p. 1270-1275.

[28] Tomovic, R., Bekey G. and Karplus, W., "A Strategy for Grasp Synthesis with Multifingered Robot Hands," *Proceedings of the IEEE International Conference on Robotics and Automation,* Raleigh, NC, 1987 .

[29] Villarreal, A. and Asada, H., "A Geometric Representation of Distributed Compliance for the Assembly of Flexible Parts," *Proceedings of the IEEE International Conference on Robotics and Automation,* Sacramento, CA, April 1991 .

Mechanical and Control Issues
for Integration of an Arm-Hand Robotic System

Claudio Melchiorri[†] and Gabriele Vassura[‡]

† DEIS, Dipartimento di Elettronica Informatica e Sistemistica

‡ DIEM, Dipartimento di Ingegneria delle Costruzioni Meccaniche
University of Bologna
Via Risorgimento 2
40136 Bologna, Italy

Abstract

In this paper, after some considerations on what integration means for a dexterous manipulation system, the design features of the University of Bologna robotic hand system are presented. The prototype of the new hand-arm manipulator, currently at the set-up phase, is described in detail, focusing on the developed solutions with respect to sensorial equipment, hand-arm mechanical integration, and control algorithms for coordination of the arm and hand subsystems to obtain both structural and functional integration.

1   Introduction

The term integration, very used in defining modern automation, identifies one of the main goals of present research and development activity on robotic manipulation systems, i.e. to implement fully a system approach in the design and realization of robotic devices. This achievement implies the exploitation of results available from different research fields (such us sensors, actuators, mechanical design of hand and arms, control theory and computer science) by defining new purpose-oriented architectures of compatible and cooperating subsystems. The state of art exhibits many good implementations at subsystem level, (as articulated hands, sensors of every kind, fast and precise arms, smart actuators etc.), but it seems that in several fields a sufficient development of complete systems, suitable to host and usefully exploit the available technological solutions, has not been realized yet.

In this paper, the attention is focused on the design and realization of a particular robotic system for manipulation tasks: an integrated dexterous hand-arm system. In this context, we think that the "integration" has to be approached at to two basic different levels: the *structural integration* level and the *functional integration* one, each of them being an essential part of the development process.

The structural integration concerns design aspects related to the synthesis of the technological parts by which the expected functions will be accomplished. A major issue here is to join and harmonize parts that may have been separately conceived and developed, solving the problems of reciprocal compatibility by purposely developed systems configurations or even by part re-design. With respect to a robotic hand-arm system, main critical problems to be solved in order to achieve a satisfactory structural integration are: the definition of the arm kinematic structure, allowing the full exploitation of the hand dexterity, the compatibility of the arm and hand design with installation of complex, distributed multi-sensor systems, and the suitability of hand actuation and transmission to be housed within the robotic arm.

With respect to the first point, main critical issues to be considered are limitations in the working volume, in the achievable hand postures, and in arm's size and compatibility for operations in restricted areas.

As regards the second point, it's our opinion that more emphasis has to be given both to new sensor development and to finger structure design for achieving a real integration. In particular, from an analysis of the state of the art, it results that:

- the availability of sensors really suitable for phalange or fingertip mounting, including both transducing elements and conditioning electronics boards with the necessary wire bundles, is still limited, particularly if high density tactile sensoriality on double curvature pads with high degree of miniaturization is needed;

- the suitability of mechanical structures of fingers to host distributed sensoriality is poor, mainly because an unsatisfactory development of alternative configurations for tendon transmission imposes to adopt the bulky pulley-routing configuration, causing a lack of internal space for sensory devices.

With respect to the third point, i.e. the possibility of installing on the arm the actuation and transmission system of the hand, integration may be greatly helped by developing actuators more suitable to be housed within the robot arm and by abandoning the end-effector interchangeability concept, provided that a dexterous hand may operate as universal end-effector.

The structural integration is therefore achievable by a joined, coordinated design of the hand, the wrist and the arm, with a search for optimization that considers as targets not only the intrinsic performance of each single subsystem, but the general compatibility at first.

The functional integration can be defined with respect to two different aspects: the intrinsic capabilities of the single subsystems and the degree of cooperation with other subsystems that they can achieve in accomplishing a task.

At the single component or subsystem level, the first step of functional integration is to enlarge, if possible, its range of activity by adding new functions to those already assigned in order to fully exploit its potential to operation. An example is the application of the *whole-arm manipulation* concept, [1], which integrates the functionality of an intermediate link of a robotic arm with the capability of interacting with the environment. The link is not limited to be a simple carrier of the end-effector, but may be profitably used extending the intrinsic capabilities of the arm. A similar idea may be applied also at the hand level, giving to each phalange the capability of contacting the grasped object, thus greatly enlarging the manipulation functionality of the hand and realizing the *whole-hand manipulation* concept, see [2].

At the system level, functional integration means that the execution of a given task is accomplished by a simultaneous cooperation of subsystems, not simply by their alternation in executing different parts of the task. A typical example of non integration is how, for long time, robotic hands have been used in manipulation tasks. With a few recent exceptions, see [3], [4], [5], a robotic hand has been generally considered as a separate device, acting in sequence with the arm, and interfaced with it simply by on-off commands without any real coordination during the task execution. On the contrary, examples from nature show how the functional integration is essential to dexterity achievement: most of skilled manipulation tasks, such as writing, painting, sewing, etc., are accomplished by means of simultaneous coordinated motion of both the fingers and the wrist or the arm itself.

In the following, this paper presents the adopted solutions for the integration of the UB (University of Bologna) hand on an anthropomorphic robotic arm. In particular, in the first part of the paper the aspects related to the mechanical design for structural integration are illustrated, while in the second a kinematic control strategy developed in order to exploit the redundancy of the system to get cooperation of hand-arm in accomplishing manipulation tasks is presented as a contribution to functional integration. Some comments about the present and future planned activity and results conclude the paper.

2 Mechanical design for structural integration of the UB Hand

Since the current experimental and implementative activity is based on previous experiences in the realization of dexterous hands, in developing the presented new system a bottom-up approach has been followed. For this reason, at the moment the design activity has been mainly focused on the hand subsystem, using a commercially available anthropomorphic manipulator, a PUMA 560, as arm. As a matter of fact, we think that the version II UB hand, currently at the set-up phase, may be considered an intermediate step of evolution, a tool for progressive implementation and verification of integration issues rather than a definitive design.

Since the main goal is to achieve a good level of integration since now, even at the expense of some performance limitation mainly due to the adopted actuation concept and technology, the UB hand II has been conceived so as to be open to progressive implementation of new technological solutions as soon as they become available. Furthermore, some choices have been influenced by the will of reducing the overall complexity and cost, in order to make the system attractive for application.

2.1 Hand design for sensorial integration

Since one of the major expectations for a dexterous robotic hand in manipulation tasks is to have performances and capabilities similar to those of the human hand, size compatibility is one of the conditioning factors and, therefore, miniaturization issues must be pursued in the design phase. A purposely oriented design of the subsystems composing the finger must be developed in order to achieve reciprocal compatibility: the integration of sensors must be planned since the design phase, being very difficult to add them to a previous finger structure not designed *ad hoc*. In fact, a mechanical design, good for a low-sensorized hand, may result quite unacceptable when distributed sensor application is considered.

Since several subsystems must be integrated in a hand, many aspects are involved in the design phase: the mechanical articulated structure of joined links, the power transmission for joint actuation, the proprioceptive sensors, the exteroceptive sensors, the wiring connections, and, if required, the electronic circuits for local signal conditioning.

The configuration adopted for the sensorial system has remarkable influence on design. For example, when only fingertip manipulation is required, the installation of sensors occurs at the end of the kinematic chain, reducing the complexity both of the design and of the problem of housing the sensing devices and the related electrical wirings. Therefore, in this case no significant constraint is imposed to the finger structure design. When a distributed sensoriality is required, design problems are enhanced, both for increasing of wiring problems and for geometrical interference with phalange structure and internal transmission. If thin-layer epidermic distributed sensoriality can be arranged, making it possible to wrap the sensing equipment around the external surface of the phalange, problems are reduced. In fact, no structural differentiation must be provided on the links, because the sensing layer can be directly supported by the link structure. On the contrary, the differentiation becomes necessary whenever the intrinsic tactile (I.T.) sensing concept is adopted, [8], [9]. In this case, three mechanical elements are required for each finger phalange: an internal link belonging to the articulated structure (the skeleton), an external shell where contact occurs, and a connecting element acting as a strain transducer.

For the version II UB robotic hand, distributed I.T. sensing has been adopted. For each active link, an I.T. sensor has been envisaged, and therefore nine sensing modules are inserted in the three fingertips, five phalanges and in the palm. This sensorial equipment allows to fulfill whole hand manipulation requirements, i.e. to implement controlled manipulation procedures by contacting objects everywhere on the available links, in order to better approach human hand operating mode.

As to the kinematic architecture, the new design maintains the configuration adopted in the version I, which proved to be satisfactory. The full opposability of the thumb against the two upper fingers, the presence of a palm and the relative position of the fingers proved to be effective in achieving a great diversification of grasping modes.

The design of the new hand can be analyzed at three levels, namely the I.T. sensing modules, the finger organization, and the general configuration of the hand.

Each sensing module is basically made of a six-component force-torque miniaturized transducer and a small board with local conditioning electronics, realized by surface-

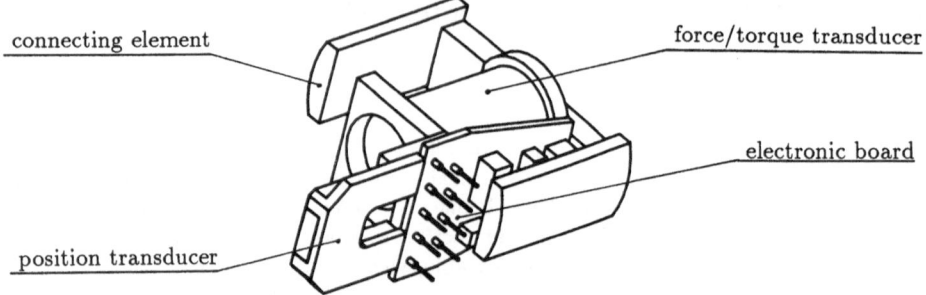

connecting element

force/torque transducer

electronic board

position transducer

Figure 1: The I.T. sensing module.

mounting hybrid circuit technology. The configuration adopted for transducer's body is the thin-walled cylindrical shell, which offers a good compromise between sensitivity and suitability to miniaturization.

Each sensor body is connected to the housing phalange frame by a prismatic profile. In the intermediate phalanges, the connection with the contact element is realized by means of a u-shaped element, see Fig. 1, while in the fingertips the sensor body is connected directly with the external shell. The semiconductor-type strain-gauges are placed according to optimization criteria [10], and permanent wiring is allowed by fixing the electronic board to the sensor body-frame. Each module provides further integration with proprioceptive Hall-effect based joint rotation sensor: global wiring is made at finger level by means of a flexible, purposely shaped printed circuit.

At finger level, a basic articulated structure is provided, made by CNC machined parts connected through ball bearing revolute pairs. The design emphasizes modularity and tends to permanent assembly solutions in order to increase reliability and reduce the number of parts. The design of the sensing module has been developed so as to be compatible with space requirements imposed by the pulley-routed internal tendon transmission. The general architecture of the finger structure is shown in Fig. 2, while the adopted shell configuration for fingertip and phalanges is illustrated in Fig. 3. The main characteristic is to have surfaces described by mathematical functions, in order to allow application of algorithms for determination of the contact centroid. The intermediate phalanges are covered with a cylindrical surface with elliptic cross-section, while the fingertip shells are revolution ellipsoids with the longitudinal axis inclined in the upward direction. A flat surface in the upper region enhances the approach capability to objects in presence of constraints, and is very useful in picking-up small objects on planar surfaces. The adopted configuration permits an easy removal of the external shells, so as to allow thorough accessibility and easy intervention on tendons and sensors. It also offers a free cross area all along each finger, between the inner structure and the outer shell, used to house wire routing.

Also at the hand level a modular structure has been adopted: three finger modules can

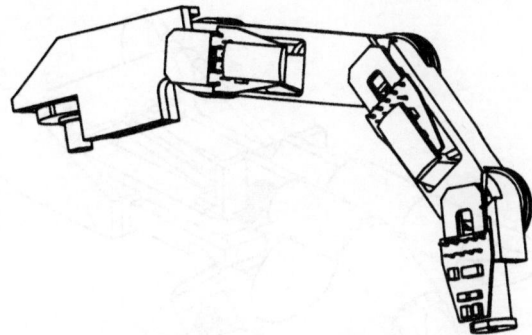

Figure 2: The finger structure.

be independently pre-assembled and then put together, each of them being comprehensive of the integrated subsystems of articulated skeleton, tendon transmission, sensory equipment. Two finger modules correspond to upper fingers, the third includes the thumb and the palm. Each module is attached to the hand frame, where a main electronic board hosts a DSP for sensor signal processing.

2.2   Hand design for hand-arm integration

A major critical problem in the mechanical integration of an articulated dexterous hand and a robot arm is the need of adopting remote actuation, due to the present unavailability of actuators compatible with direct-drive or within-hand installation. This means to properly group the motors of the hand into an actuator-box suitable to be hosted within the arm structure, to define a transmission scheme with acceptable performance, and to design wrist articulations that can be passed through by the transmission lines, with reduced coupling effects and limited overall complexity.

The most influential factor is the choice of the joint actuation scheme, that determines the required number of motors for each joint and the adoptable routing schemes for tendon-based power transmissions. As widely demonstrated, [11]-[14], joint actuation achieved by couples of agonistic-antagonistic tendons, individually driven by one motor each, can allow for best performance and can reduce problems due to routing-path variation caused by relative motion of the hand with respect to the arm. Unfortunately the bulk of the motor-box increases so that the installation on a robot arm should require for it a size not compatible with the real payload (the hand and the grasped object). Some other joint actuation schemes, with a reduced number of actuators (typically the one joint-one actuator scheme, or the n joints-n+1 actuators, adopted in the Salisbury's hand) can be more compatible with integration, but present more problems for compensating path length variations.

When minimization of the number of actuators is a primary design issue the one joint-one actuator scheme is the most logical choice. In this case, design for integration must

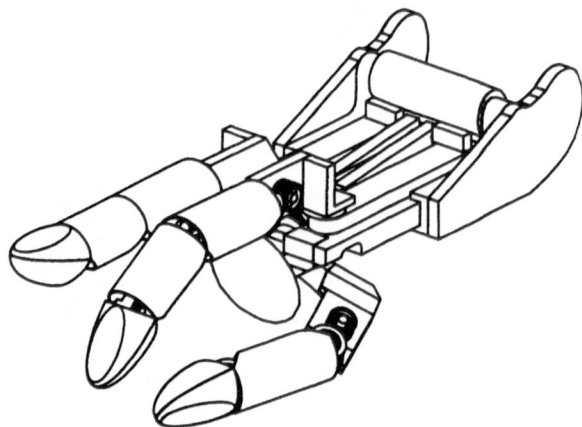

Figure 3: The UB hand II.

provide acceptable solutions to the problem of tendon preloading, keeping the routing length constant for any configuration of the wrist or providing for tension regulating mechanisms.

As to the tendon routing design, the choice of pulley or sheath routing greatly influences the performance of the transmission and its complexity. On the one hand pulley-routing gives best performance in terms of efficiency, but determines coupling effects in crossing intermediate joints and results in higher design complexity; on the other, sheath routing can avoid mechanical coupling and problems due to variation of routing length, but presents poor efficiency.

The configuration implemented for the UB hand has been defined according to the following issues:

- to adopt the minimal number of actuators;

- to cross the wrist without introducing coupling effects;

- to be suitable to future implementation of pre-tensioning variation for the whole tendon system;

- to get a mix between the efficiency of a pulley-routed transmission and the design simplicity of a sheath-routed one.

As already mentioned, the adopted solution is intended as a first step towards the future development of a system with a three degree-of-freedom wrist. In the present configuration, a bend d.o.f. in the wrist is realized by an articulation between the hand and the forearm, while a roll d.o.f. is obtained by a revolute pair between the forearm and a link connected to the robot arm. The actuator box is located on the forearm, so that only the bend articulation must be crossed by the tendons. Couples of preloaded antagonistic tendons, each driven by a single actuator, move the joints: the transmission is mixed as the tendons are routed with pulleys inside each finger, and with sheaths in their path from each finger to the hand-frame, to allow adduction-abduction displacement of the upper fingers and proximal rotation of the thumb, avoiding the more complex pulley-based design. This solution gives acceptable results in terms of mechanical performance, because the bending angles and their variations are small and the length of the sheathed stretch is short, thus reducing the effects of compliance and friction, [15].

From the hand to the actuator, the antagonistic tendons are guided parallel to each other and are routed across the wrist articulation by the same side of a floating-shaft pulley train. This makes the joint motion independent of the wrist motion, as the two antagonistic tendons are subject to the same path variation in winding or unwinding over the pulleys. A controlled variation of the pulley-shaft position results in compensation of tendon tension at any rotation angle of the wrist. A simple wrist-driven cam mechanism provides for constant tendon tension.

The proposed solution, see Fig. 3 and Fig. 4, though very simple, presents some attractive features, as the hand is mechanically decoupled from the wrist, the negative influence of sheaths is limited, the total number of pulleys is low and the design complexity is reduced. Moreover, since a tuning of tendon preload according to external load variation may be useful to enhance transmission performance, the pre-loading of all the tendons may be regulated by means of active variation of the floating shaft position with respect to the position determined by the cam. This is achievable by means of an additional actuator that can vary the length between the axes of the floating shaft and the follower.

2.3 Arm design for hand-arm integration

The integral re-design of the arm subsystem is surely the best way to provide for advanced solutions for dexterous manipulations.

However, in our case two main considerations were made. As already mentioned, the current activity follows a bottom-up approach, and therefore at the moment our interest is the design and realization of an "hand subsystem" suitable for the installation on an anthropomorphic arm. Moreover, since in general it is preferable to host the hand actuation box not too far from the hand itself, the re-design of only the terminal part of a commercial robot seemed to be a practical solution for our immediate purposes. Due to these reasons, a PUMA 560 robot has been chosen to host the hand system, with the replacement of the forearm and the wrist, as shown in Fig. 4.

3 A control technique for functional integration

As already mentioned in the introduction, functional integration is achieved by a well defined cooperation of the subsystems composing the device under consideration. This implies that during task executions all the parts of the hand-arm system must be coordinated in order to exploit the peculiarities of each component. Good examples of these "behaviors" may be found observing our actions during manipulations. For example, in pushing light objects we often use only the hand, resorting to an arm motion either when the fingers are close to a singular configuration (i.e. the object is close to the hand's workspace limit) or when we desire to maintain the fingers close to a desired configuration, suitable for manipulation tasks (such as writing or precise positioning of small objects). Even

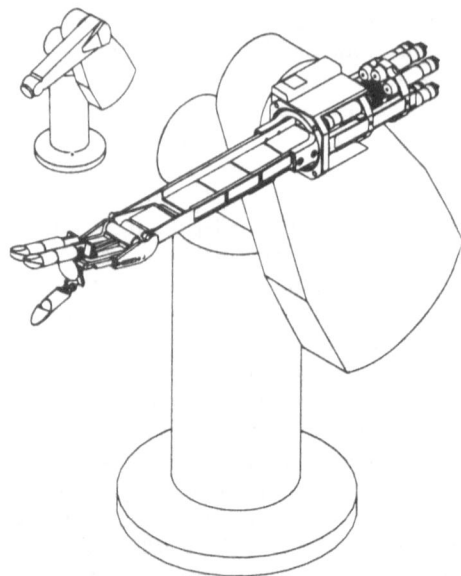

Figure 4: The UB Hand-Arm system.

if several important research project have been presented in the literature in the field of dexterous manipulation, poor consideration has been given to the the issue of coordinating the "activities" of the hand with those of the arm on which it is installed. Some examples may be found in [3], and [4], where optimality criteria have been used for the task planning of the arm and the hand motions, considering a stable and feasible grasp as a major goal, i.e. identifying the hand as a sort of master in the system and the arm as the slave. This choice reflects in a non homogeneous treatment of the two subsystems. A more natural approach would be to considers the hand-arm system as a single device, exploiting the intrinsic redundancy of the mechanism in order to obtain desired "behaviors" from the hand and from the arm.

Generally speaking, a robotic arm-hand system may be considered as a redundant manipulator. The arm, in general a 6 or 7 degree-of-freedom structure, has installed as end effector a parallel device which, normally, is a redundant device by itself. Even if a number of techniques have been proposed in the literature for the control of redundant systems, they can not be directly used in this context, since the device under consideration is characterized by some peculiarities that have to be considered in developing a control strategy.

The first characteristic is related to the type of redundancy: in this case a manipulator has a redundant parallel device, the hand, installed as end-effector, and therefore can not be considered a serial mechanical chain, as normally assumed for redundant systems.

A second peculiarity concerns the "behaviors" expected from this mechanism. Usually, different joint motions are involved, both in terms of amplitude and velocity, requiring different parts of the system to act in different ways. For example, in some tasks it may not be needed to move the arm, while in others the motion capabilities of the fingers may not suffice and a simultaneous well-coordinated motion of both the arm and the hand is required.

A third fact is that, intrinsically, a hand-arm system has to interact physically with the environment, and therefore also the force control problem must be considered in defining a control technique for this device.

In this section, an algorithm for the kinematic control of a dexterous hand-arm system is presented. The algorithm is based on the Jacobian transpose technique, a well know method for the solution of the inverse kinematic problem, see [19]-[22]. The basic formulation of this algorithm has been modified in order to account for the above mentioned characteristics of the considered robotic system. This technique has been developed in cooperation with Dr. J.K. Salisbury at the Artificial Intelligence Laboratory, MIT, where preliminary experiments on a hand-arm robotic system (a PUMA 560 with the Salisbury's hand) have been carried out, [5], [23].

3.1 Background

In manipulation tasks, it may be important not only to specify the object position/orientation, but also, once the object is grasped, to specify the relative displacements of the fingers. Therefore, in defining the problem, it may be convenient to consider an extended space instead of the usual 6 dimensional task space. Considering a hand with three fingers, the forward kinematics is expressed by

$$\mathbf{x(q)} = \left[\begin{array}{c} \mathbf{f(q)} \\ \mathbf{d(q)} \end{array} \right] \in \mathcal{R}^9$$

where $\mathbf{f(q)} \in \mathcal{R}^6$ are the forward kinematic equations relating the hand-arm joints values to the position/orientation of the object, and $\mathbf{d(q)} \in \mathcal{R}^3$ are the relative displacements of the fingertips, affected only by the joints of the hand. Assuming a grasp on a rigid object, and without slip at the contact points, the computation of $\mathbf{f(q)}$ and $\mathbf{d(q)}$ is quite straightforward.

Since the proposed algorithm is based on the Jacobian of the manipulator, we are interested in defining a differential relationship between the set of joint velocities and the set of object velocities and "internal velocities", i.e. the relative velocities of the fingertips. This is achieved by means of the grasp matrix $\mathbf{G}$, introduced in [16]. The grasp matrix relates forces and displacements of the object to the forces and displacements of the single fingers, i.e.:

$$\mathbf{F} = \mathbf{G}^T \mathcal{F} \tag{1}$$

$$\dot{\mathbf{x}}_f = \mathbf{G}^{-1} \dot{\mathbf{x}}_o \tag{2}$$

where:

$\mathbf{F} = \left[\begin{array}{ccc} \mathbf{F}_1^T & \mathbf{F}_2^T & \mathbf{F}_3^T \end{array} \right]^T$ are the forces applied at the fingertips;

$\mathcal{F} = \left[\begin{array}{ccccc} \mathbf{f}^T & \mathbf{t}^T & f_{12} & f_{23} & f_{31} \end{array} \right]^T$ are the 6 external and 3 internal forces acting on the object;

$\dot{\mathbf{x}}_f = \left[\begin{array}{ccc} \dot{\mathbf{x}}_{f1}^T & \dot{\mathbf{x}}_{f2}^T & \dot{\mathbf{x}}_{f3}^T \end{array} \right]^T$ are the 3 linear velocities of the fingertips;

$\dot{\mathbf{x}}_o = \begin{bmatrix} \mathbf{v}_o^T & \omega_o^T & \dot{d}_{12} & \dot{d}_{23} & \dot{d}_{31} \end{bmatrix}^T$ are the 6 velocities of the object and the relative velocities of the 3 contact points.

By means of the grasp matrix it is possible, see [23], to express the Jacobian of the hand-arm system as

$$\mathbf{J}_S = \begin{bmatrix} \mathbf{J}_P & \mathbf{J}_H \end{bmatrix} = \begin{bmatrix} \mathbf{J}_{Pf} & \mathbf{J}_{Hf} \\ 0 & \mathbf{J}_{Hd} \end{bmatrix}$$

where the subscripts P, H, f and d refer to the arm (PUMA), to the Hand, to the first six and to the last three components of the vector $\dot{\mathbf{x}}_o$ respectively. Therefore, the differential relationship between $\dot{\mathbf{x}}_o$ and $\dot{\mathbf{q}}_S$ is

$$\dot{\mathbf{x}}_o = \begin{bmatrix} \mathbf{J}_{Pf} & \mathbf{J}_{Hf} \\ 0 & \mathbf{J}_{Hd} \end{bmatrix} \begin{bmatrix} \dot{\mathbf{q}}_P \\ \dot{\mathbf{q}}_H \end{bmatrix} = \mathbf{J}_S \dot{\mathbf{q}}_S.$$

According to the basic Jacobian transpose algorithm, shown as a block diagram in Fig. 5, the joint velocities are computed as:

$$\dot{\mathbf{q}}_A = \lambda \, \mathbf{J}_S^T \mathbf{K}_E \mathbf{e} \tag{3}$$

and, therefore,

$$\begin{aligned} \dot{\mathbf{q}}_{PA} &= \lambda \mathbf{J}_{Pf}^T \mathbf{K}_{Ef} \mathbf{e}_f \\ \dot{\mathbf{q}}_{HA} &= \lambda (\mathbf{J}_{Hf}^T \mathbf{K}_{Ef} \mathbf{e}_f + \mathbf{J}_{Hd}^T \mathbf{K}_{Ed} \mathbf{e}_d) \end{aligned} \tag{4}$$

from which it is clear how the joints of the arm compensate only for errors in the position/orientation of the objects, while the joints of the fingers also affects the internal velocities. In Fig. 5, $\mathbf{x}_d(t)$ is the desired trajectory, $\mathbf{x}(t)$ is the trajectory of the manipulator, $\dot{\mathbf{q}}(t)$ and $\mathbf{q}(t)$ are the joint velocities and positions respectively, $\mathbf{J}$ is the Jacobian, $\mathbf{K}_E$ is a stiffness matrix, and $\lambda(> 0)$ is a gain affecting the convergence rate of the algorithm. By integration of eq. (4), one obtains the joint position trajectory $\mathbf{q}(t)$. A Lyapunov stability proof, see [19]-[23], shows that the obtained trajectory converges to the desired one $\mathbf{q}_d(t)$.

3.2   Considering the hand-arm system

In order to apply the Jacobian transpose technique of Fig. 5, some modifications are necessary to address the particularities of the hand-arm system. The first problem is that the method, in its basic form, does not make any distinction between the joints of the arm and of the hand. Therefore, the algorithm has to be modified to exploit the redundancy of the system in order to obtain some desired behaviors from the two subsystems according to the task being performed. The cases reported in the following refer to the attempt to try always to execute the desired manipulation task with only the hand, resorting to arm's movements when the fingers are close to singularities or when it is required to keep the hand in a desired configuration. Other strategies could be defined, leading to similar final conclusions.

In general, the modifications introduced in the basic algorithm must not interfere with the main task of the manipulator. This is accomplished if the modifications operate in the

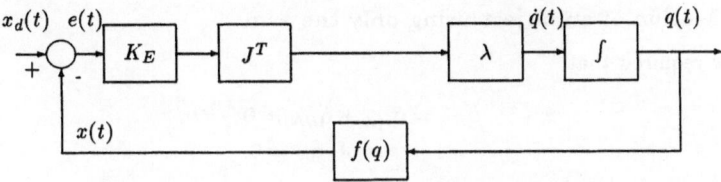

Figure 5: The basic Jacobian transpose scheme.

null space of the Jacobian $\mathbf{J}_S$, with the additional advantage that the stability proof of the algorithm still applies. Therefore, the final set of joint velocities $\dot{\mathbf{q}}_S$ may be computed as the sum of two terms:

$$\dot{\mathbf{q}}_S = \begin{bmatrix} \dot{\mathbf{q}}_{PS}^T & \dot{\mathbf{q}}_{HS}^T \end{bmatrix}^T = \dot{\mathbf{q}}_A + \dot{\mathbf{q}}_N = \lambda\, \mathbf{J}_S^T \mathbf{K}_E \mathbf{e} + \dot{\mathbf{q}}_N \tag{5}$$

where $\dot{\mathbf{q}}_A$ is the solution computed by the basic algorithm, and $\dot{\mathbf{q}}_N = \begin{bmatrix} \dot{\mathbf{q}}_{PN}^T & \dot{\mathbf{q}}_{HN}^T \end{bmatrix}^T \in Null(\mathbf{J}_S)$.

In the following, the term $\dot{\mathbf{q}}_N$ is computed considering three basic cases, i.e. the execution of a task involving only movements of the hand, only movements of the arm, and the execution of a task while reconfiguring the hand to a desired pose. Different behaviors of the hand-arm system may be obtained by a proper modulation of these three cases. Moreover, the introduction of the force feedback loop is presented.

3.2.1 Motion of an object using only the hand

If it is desired to move the grasped object by using only the hand, it follows that $\dot{\mathbf{q}}_{PS} = \mathbf{0}$, i.e., from (5),

$$\dot{\mathbf{q}}_{PS} = \dot{\mathbf{q}}_{PA} + \dot{\mathbf{q}}_{PN} = \mathbf{0}.$$

Since $\dot{\mathbf{q}}_N \in Null(\mathbf{J}_S)$, i.e. $\begin{bmatrix} \mathbf{J}_P & \mathbf{J}_H \end{bmatrix} \begin{bmatrix} \dot{\mathbf{q}}_{PN} \\ \dot{\mathbf{q}}_{HN} \end{bmatrix} = \mathbf{0}$, it follows:

$$\dot{\mathbf{q}}_{PN} = -\dot{\mathbf{q}}_{PA}$$
$$\dot{\mathbf{q}}_{HN} = \mathbf{J}_H^+ \mathbf{J}_P \dot{\mathbf{q}}_{PA}$$

or, in matrix form,

$$\dot{\mathbf{q}}_{N1} = \begin{bmatrix} -\mathbf{I}_p & \mathbf{0} \\ \mathbf{J}_H^+ \mathbf{J}_P & \mathbf{0} \end{bmatrix} \dot{\mathbf{q}}_A = \mathbf{P}_H \dot{\mathbf{q}}_A, \tag{6}$$

where $\mathbf{I}_p$ is the $n_p \mathrm{x} n_p$ identity matrix ($n_p = \dim(\dot{\mathbf{q}}_P)$). The matrix $\mathbf{P}_H$ in (6) generates, given the original solution $\dot{\mathbf{q}}_A$, a joint velocity term $\dot{\mathbf{q}}_{N1} \in Null(\mathbf{J}_S)$ which, added to $\dot{\mathbf{q}}_A$, satisfies the given constraint on the joint velocity vector ($\dot{\mathbf{q}}_{PS} = \mathbf{0}$). The matrix $\mathbf{P}_H$ may therefore be regarded as a non orthogonal projector of the given solution $\dot{\mathbf{q}}_A$ in $Null(\mathbf{J}_S)$.

3.2.2 Motion of an object using only the arm

It is now required that

$$\dot{q}_{HS} = \dot{q}_{HA} + \dot{q}_{HN} = 0$$
$$J_S \dot{q}_N = 0.$$

Hence, one computes

$$\dot{q}_{PN} = J_P^+ J_H \dot{q}_{HA}$$
$$\dot{q}_{HN} = -\dot{q}_{HA}$$

or

$$\dot{q}_{N2} = \begin{bmatrix} 0 & J_P^+ J_H \\ 0 & -I_h \end{bmatrix} \dot{q}_A = P_P \dot{q}_A, \tag{7}$$

where I_h is the $n_h \times n_h$ identity matrix, $(n_h = \dim(\dot{q}_H))$. The matrix P_P in (7) generates, given a solution $\dot{q}_A$, a joint velocity term $\dot{q}_N$ in $Null(J)$ which, added to $\dot{q}_A$, verifies the given constraints on the solution, $(\dot{q}_{HS} = 0)$. The matrix P_P, similarly to P_H in eq. (6), may be considered a projector of the given solution $\dot{q}_A$ in the null space of J_S.

3.2.3 Reconfiguring the hand to a desired pose

The goal now is to achieve the joint position vector q_H to a desired value q_{Hdes}, while the whole system is following a specified trajectory $x_{des}(t)$. A practical way is to generate a joint velocity term which compensates for the positional errors $(q_{Hdes} - q_H)$. In other words, the term $\dot{q}_N$ may now be computed from

$$0 = J_P \dot{q}_{PN} + J_H \dot{q}_{HN}$$
$$\dot{q}_{HN} = K(q_{Hdes} - q_H)$$

from which

$$\dot{q}_{PN} = -J_P^+ J_H K(q_{Hdes} - q_H)$$
$$\dot{q}_{HN} = K(q_{Hdes} - q_H)$$

or, in matrix form,

$$\dot{q}_{N3} = -P_P K(q_{Hdes} - q_H) \tag{8}$$

in which P_P is the same as in (7).

3.2.4 The force feedback loop

As previously mentioned, during task execution we need not only to generate suitable joint position trajectories, but also to control the forces applied to the environment. To this purpose, a further feedback loop is added to the scheme, leading to the final diagram shown in Fig. 6. In this scheme, F_d is a desired force, $e_F = (F_d - F)$ is the force error,

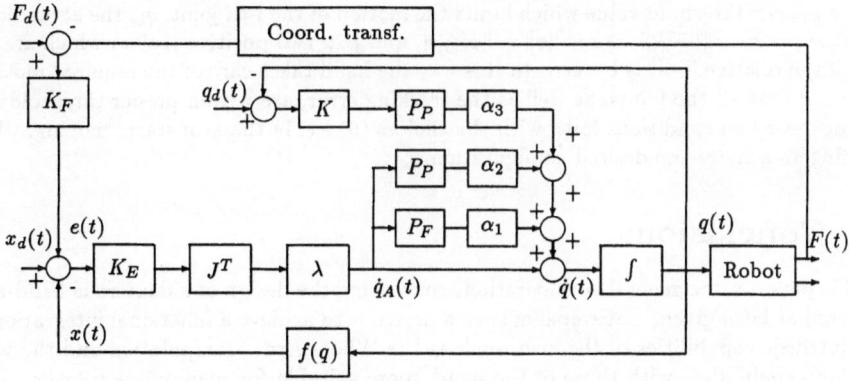

Figure 6: The modified scheme.

and $\mathbf{K}_F$ a compliance matrix which represents a model of the manipulator-environment interaction. Obviously, in this way the algorithm must be used in real time, and not simply as an off-line planner for joint trajectories.

In Fig. 6, the three terms $\dot{\mathbf{q}}_{Ni}$ of eqs. (6), (7), (8), are combined together. In the scheme, the three coefficients α_1, α_2, and α_3 ($0 \leq \alpha_i \leq 1$, i=1,2,3), are variable gains used to modulate the actions of the three terms $\dot{\mathbf{q}}_{N1}$, $\dot{\mathbf{q}}_{N2}$, and $\dot{\mathbf{q}}_{N3}$ on the original solution $\dot{\mathbf{q}}_A$.

3.2.5 Computation of the gains α_i

An interesting point is the definition of suitable strategies for the computation in real-time of the gains α_i. The final solution $\dot{\mathbf{q}}_S$ is

$$\dot{\mathbf{q}}_S = \dot{\mathbf{q}}_A + \alpha_1 \dot{\mathbf{q}}_{N1} + \alpha_2 \dot{\mathbf{q}}_{N2} + \alpha_3 \dot{\mathbf{q}}_{N3}.$$

By properly changing in real time the three factors α_i, it is possible to modulate the possible behaviors of the system. If $\alpha_i = 0$, the relative constraint term $\dot{\mathbf{q}}_{Ni}$ is neglected, while when $\alpha_i = 1$, the constraint is fully active.

If the desired behavior during manipulation is to use as much as possible the joints of the hand, resorting to arm's movements when the fingers are close to singularities or when it is required to keep the hand in a desired configuration, a possible initial choice is

$$\alpha_1 = 1, \ \alpha_2 = \alpha_3 = 0.$$

When, for example, the joints of the hand are close to a limit, or when the amplitude of the tracking error $\|\mathbf{e}\|$ is greater than a maximum allowed value $\|\mathbf{e}\|_{max}$, the values of the coefficients α_i must be modified. The adopted rules are

$$\alpha_1 = \begin{cases} 1 & \begin{aligned} &\text{if} - q_{Ti} < q_i < q_{Ti}; \\ &\text{and} \|\mathbf{e}\| < \|\mathbf{e}\|_{max} \end{aligned} \\ e^{-\mu \Delta q - \nu \Delta e} & \text{otherwise}; \end{cases} \tag{9}$$

$$\alpha_2 = (1 - \alpha_1) \tag{10}$$

$$\alpha_3 = (1 - \alpha_1) \tag{11}$$

where q_{Ti} is a threshold value which limits the motion of the i-th joint, q_{Li} the actual joint limit, $\Delta q = max \left\| \frac{q_i - q_{Ti}}{q_i - q_{Li}} \right\|$, $\Delta e = \|e\| - \|e_{max}\|$, and μ, ν two positive scalars which are set to 1 if the relative limit is broken. In this way the hand takes care of the required motion, provided that all the joints, as well as the tracking error, are within proper thresholds. If one of these two conditions fails, with the choices (9) - (11) the arm starts moving, while the fingers achieve the desired configuration q_d.

4   Conclusions

In this paper, some general considerations concerning the design of a dexterous hand-arm system has been given. Basic goal of such a design is to achieve a functional integration of the intrinsic capabilities of the arm, such as the Whole Arm Manipulation and the wide motion capabilities, with those of the hand, more suitable for manipulation tasks. This result is achieved with a proper structural integration of different components (and therefore is influential on the design of arm, hand, actuation system, and sensorial equipment), and with control strategies considering the whole system and all its capabilities.

Moreover, the currently adopted solutions regarding the mechanical design and the installation of the UB hand II on a manipulator have been illustrated and discussed, and a technique for the kinematic control of the device has been presented.

At the moment, the main components of the hand system have been realized, and the hand is being assembled. Its installation on a PUMA 560 is the major goal of the next laboratory activity. As regards the presented control strategy, some preliminary experiments have been carried out at the Artificial Intelligence Laboratory, MIT, on a manipulation system similar to the one discussed in this paper. The implemented algorithm consider only one finger of the Salisbury hand and the PUMA 560, i.e. a total of 9 degree-of-freedom. As soon as the UB hand-arm system will be available, the full algorithm will be implemented and verified.

Acknowledgments

The results presented in this paper are the result of the work of the team at the University of Bologna working at the UB hand project. In particular, the authors wish to thank S. Monti for the invaluable help in the assembly of the hand. Founding for this research has been provided by Progetto Finalizzato Robotica CNR, Contracts No. 89.00493.PF.67.3 and No. 90.00408.67. Claudio Melchiorri wishes to thank J.K. Salisbury for the stimulating discussions. Support for Claudio Melchiorri as Visiting Scientist at MIT has been provided by the NATO-CNR grant No. 215.23/15.

References

[1] Salisbury, J.K., "Whole Arm Manipulation", Proc. 4th Int. Symp. on Robotics Research, Santa Cruz, CA, R. Bolles and B. Roth Ed., MIT Press, Cambridge, MA, 1987.

[2] Vassura, G., Bicchi, A., "Whole Hand Manipulation: Design of an Articulated Hand Exploiting All Its Parts to Increase Dexterity", Proc. NATO Advanced Research

Workshop on Robots and Biological Systems, Il Ciocco, Tuscany, Italy June 26-30, 1989.

[3] Pollard, N., Lozano-Perez, T., "Grasp Stability and Feasibility for an Arm with an Articulated Hand", Proc. IEEE Int. Conf. on Robotics and Automation, Cincinnati, OH, 1990.

[4] Roberts, K.S., "Coordinating a Robot Arm and Multi-Finger Hand Using the Quaternion Representation", Proc. IEEE Int. Conf. on Robotics and Automation, Cincinnati, OH, 1990.

[5] Melchiorri, C., Salisbury, J.K., "An Algorithm for the Control of a Hand-Arm Robotic System", Proc. 5th Int. Conf. on Advanced Robotics, '91 ICAR, Pisa, Italy, June 19-22, 1991.

[6] Bologni, L., Caselli, S., Melchiorri, C., "Design Issues for the U.B. Robotic Hand" Proc. NATO Advanced Research Workshop on Robots with Redundancy, Salò, Italy June 27-July 1, 1988.

[7] Bonivento, C., Faldella, E., Vassura, G., "The U.B. Robotic Hand Project: Current State and Future Developments", Proc. 5th Int. Conf. on Advanced Robotics, '91 ICAR, Pisa, Italy, June 19-22, 1991.

[8] Brock, D.L., Chiu, S., "Environment Perception of an Articulated Robot Hand Using Contact Sensors", Proc ASME Winter Annual Meeting, Miami, FL, Nov. 1985.

[9] Bicchi, A., Dario, P., "Intrinsic Tactile Sensing for Artificial Hands", Robotics Research, R. Bolles and B. Roth Editors, MIT Press, 1987.

[10] Bicchi, A., "Strumenti e Metodi per il Controllo di Mani per Robot", Ph.D. Thesis, University of Bologna, 1989.

[11] Jacobsen, S.C., Wood, J.E., Knutti D.F., Biggers, K.B., "The UTAH-MIT Dexterous Hand: Work in Progress", Int. Jour. of Robotics Research, Vol. 3, No. 4, 1984.

[12] Jacobsen, S.C., Ko, H., Iversen, E.K., Davis, C.C., "Control Strategies for Tendon Driven Manipulators", IEEE Control System Magazine, Feb. 1990.

[13] Vossoughi, R., Donath, M., "Robot Finger Stiffness Control in the Presence of Mechanical Non-Linearities", ASME Trans., Vol. 110, Sept. 1988.

[14] Townsend, W.T., Salisbury, J.K., "The Efficiency Limits of Belt and Cable Drives", ASME Jour. of Mech., Transm., and Automation in Design, Vol. 110, Sept. 1988.

[15] Vassura, G., "On the Integration of an Articulated Dexterous Hand in a Robot Arm", Proc. 3rd. Int. Symp. on Robotics and Manufacturing, ISRAM'90, Vancouver, June 1990.

[16] Salisbury, J.K., Craig, J.J., "Articulated Hands: Force Control and Kinematic Issues", Int. Journal of Robotic Research, Vol. 1, No. 1, Spring 1982.

[17] Bonivento, C., Caselli, S., Faldella, E., Laschi, R., Melchiorri, C., Tonielli, A., "Control System Design of a Dexterous Hand for Industrial Robots", IFAC SYROCO'88, Karlsruhe, RFG, 1988.

[18] Kerr, J., Roth, B., "Analysis of Multifingered Hands", Int. Journal of Robotic Research, Vol. 4, No. 4, 1986.

[19] Balestrino, A., De Maria, G., Sciavicco, L., "Robust Control of Robotic Manipulators", Proc. 9th. IFAC World Congress, Budapest, Hungary, 1984.

[20] Wolovich, W.A., Elliott, H., "A Computational Technique for Inverse Kinematics", Proc. 23rd. Conf. on Decision and Control, Las Vegas, NV, Dec. 1984.

[21] Sciavicco, L., Siciliano, B., "A Solution Algorithm to the Inverse Kinematic Problem for Redundant Manipulator", IEEE Jour. of Robotics and Automation, Vol. 4, No. 4, Aug. 1988.

[22] Das, H., Slotine, J-J.E., Sheridan, T.B., "Inverse Kinematic Algorithms for Redundant Systems", Proc. IEEE Int. Conf. on Robotics and Automation, San Francisco, CA, 1988.

[23] Melchiorri, C., Salisbury, J.K., "Exploiting the redundancy of a hand/arm robotic system", A.I. Memo No. 1261, Artificial Intelligence Laboratory, MIT, Cambridge, MA, Oct. 1990.

Experimental Evaluation of Friction Characteristics With an Articulated Robotic Hand

Antonio Bicchi *
J. Kenneth Salisbury †
David L. Brock ‡

Massachusetts Institute of Technology
Artificial Intelligence Laboratory

Abstract

In this paper some applications of dexterous robotic hands to fine manipulation operations are discussed. Common to this kind of operations is the important role played by the frictional characteristics of the fingers and of manipulated objects. The paper discusses a procedure for measuring apparent coefficients of friction between the fingertips and manipulated objects. To take into account real contact phoenomena, the distinction is introduced between translational and rotational friction limits. Reported experiments rely on force-based (intrinsic) contact sensing devices, implemented in the phalanges of an articulated robot hand (Salisbury Hand). Data collected during these procedures can be subsequently used for tasks such as recognizing the superficial features of objects, controlling the internal grasp forces exerted by the hand on delicate objects, and following the contours of the surface of unknown objects.

1 Introduction

While robots seem to have saturated the easiest applications in the factory environment, a great potential remains for robots to be employed in unstructured surroundings and non-conventional tasks. In order for a robot to prove useful in such conditions, however, a higher level of dexterity in its operations than is currently available has still to be achieved. Dexterous articulated hands have been proposed since the early '80 by several researchers, e.g. Mason and Salisbury [1], Jacobsen et al. [2]. However, full exploitation of their capabilities has not yet been accomplished, mainly because of the lack of adequate contact sensing technology. Conventional, "skin-like" tactile sensing devices in fact, though intriguing for their similarity to the biological model, failed so far to provide an accurate, reliable feedback of the interactions between the robot hand and grasped objects. These devices also suffer from some other drawbacks, such as limited resolution and/or bandwidth, encumbrance, large number of connectors. A detailed report on the state-of-the-art for conventional tactile sensors can be found in [3].

Probably, the most severe limitation of skin-like tactile sensors is that frictional forces transmitted through contact can not be satisfactorily sensed. Pioneering work in designing skin-like

---

*Antonio Bicchi is currently a Research Associate on leave from the Department of Electric Systems and Automation (DSEA) and Centro "E. Piaggio," Università di Pisa, Italia.

†J. Kenneth Salisbury is currently a Research Scientist at the Artificial Intelligence Laboratory at Massachusetts Institute of Technology.

‡David Brock is currently a Ph.D. candidate in the Department of Mechanical Engineering at Massachusetts Institute of Technology.

tactile sensors with the capability of discerning tangential stresses, such as that presented by Hackwood et al. [4], and by Domenici et al. [5], has not till yet been applied to building practical devices. On the contrary, in some cases (e.g. for some piezoelectric film based sensors) friction forces produce aliasing in contact pressure measurements.

The fundamental role played by friction in the control of manipulation operations is apparent. Humans rely on friction forces in many of their grasping or manipulating operations: this is perhaps reflected in our uneasiness with slippery hands. Controlling the slippage of grasped objects, on the other hand, allows us to achieve high dexterity in object manipulation. In order to replicate at least part of the human manipulative capabilities, it is therefore necessary to be able to sense friction, and to control (to avoid, or, sometimes, to allow) slippage.

In the following, we will present some experiments intended to assess the feasibility of friction sensing with a non conventional contact sensing device, namely a force/torque based, or intrinsic, contact sensor. While the theory underlying this sensing principle is presented elsewhere [6], we show here that those sensors are well suited to deal with friction forces in fine manipulation operations, and can provide for a major advancement in the dexterity of robots.

2   Contact and Friction Model

An accurate model of the phenomenology related to contact is very difficult to describe. In fact, contact involves so many factors, both intrinsic to the touching bodies (such as their macro- and microscopic geometry and elastic properties), and extrinsic (lubrication, temperature etc.), that a comprehensive modelization is hopeless, except for very idealized cases. Research in the field of tribology has evidentiated the difficulty in providing a single model of friction valid over a broad range of circumstances. Armstrong [7] [8] provides a good reference and a vaste bibliography concerning tribologic problems relevant to robotic actuation and control. For our purposes here, however, so fine a physical description of contact phoenomena is not necessary. For instance, in many dextrous manipulation operations slippage is to be considered an undesirable loss of control to be avoided, so that we are primarily concerned with static friction limits. If a relative motion of the fingers on the object surface is desired (such is the case in the surface following task described by Balestrino and Bicchi [9], or in advanced dexterity operations as described by Brock [10]), it usually consists of a constant speed, quasi-static motion. In modeling the contact interaction between the end effector and the object surfaces, therefore, we will make some simplifying assumptions, such that an operative model of contact and friction results.

2.1   Contact Model and Contact Sensors

Consider the contact of two deformable bodies pressed against each other, as depicted in Fig. 1: we will refer to these bodies as to the *finger* and the *object*. The contact area will be in general composed of several non-connected small zones, distributed over the deformed surfaces in contact.

A force/torque-based (intrinsic) contact sensor can be realized in the fingers of the robot hand by employing a six-axis force sensor to measure the resultant force and torque exerted on the finger surface through contact [6]. Since intrinsic contact sensors are instrumental in the manipulative operations that we are going to describe, the concepts underlying their functioning will be briefly recalled.

A *contact centroid* on a contact surface is defined as a point c such that there exists a force p through that point, and a moment q normal to the surface at that point, which form a set of forces equivalent to the actual contact load. It can be shown that, under certain commonly verified assumptions, the contact centroid enjoys the property of being unique, and lies inside

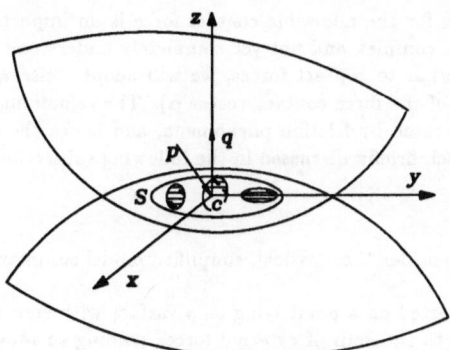

Figure 1: Soft-finger contact model

the convex hull S that encloses every contact zone. An intrinsic contact sensor is capable of providing the location of the contact centroid on the finger surface, along with the measurements of the components of the contact force and torque.

The above mentioned assumptions regard convexity and smoothness of the finger surface; it is also assumed that only compressive forces are exerted by contact (no adhesion). A further limitation applies to the allowable warping of the contact area over the contact surfaces; this bound can be quantitatively expressed in terms of the frictional properties of the surfaces and of their geometry. For a precise statement of these hypotheses, and for a proof of the contact centroid properties, the interested reader is referred to [6]. In the context of this paper, we will make a stronger assumption, i.e. that the contact area warping over the curved contacting surfaces is negligible altogether. This condition will be useful in modeling friction effects, and is satisfied in most practical circumstances of robotic manipulation.

Consider a cartesian reference frame with origin in the contact centroid c and z-axis aligned with the normal to the contacting surfaces. At each point $\mathbf{r} = (\mathbf{r}_x, \mathbf{r}_y)^T$ of the contact area, bodies mutually exert tractions $\mathbf{h}(\mathbf{r})$ whose normal component $\mathbf{h}_z$ is usually referred to as *pressure*, while $\mathbf{h}_x$ and $\mathbf{h}_y$ are friction components. With respect to that reference frame, the resultant contact force vector $\mathbf{p}$ is defined as

$$\mathbf{p} = \int_S \mathbf{h}(\mathbf{r}) \, dS$$

while the components of the resultant torque vector $\mathbf{q}$ are given, by definition of contact centroid, by

$$
\begin{aligned}
\mathbf{q}_x &= 0, \\
\mathbf{q}_y &= 0, \\
\mathbf{q}_z &= \int_S (\mathbf{h}_y \mathbf{r}_x - \mathbf{h}_x \mathbf{r}_y) \, dS.
\end{aligned}
$$

In order to control manipulation, we are interested in the physical constraints imposed on the components of $\mathbf{p}$ and $\mathbf{q}$. The pressures on the object, and their resultant $\mathbf{p}_z$, must be compressive and not larger than a limit value, which depends on the object and/or the finger strength. The limit value for the allowable contact force is an important factor in the choice of the grasp forces. Since complex and not yet completely understood mechanics are involved in the resistance of materials to contact forces, we will adopt conservative bounds on p_z (or possibly on the intensity of the force contact vector p). The remaining components of contact force and torque are generated by friction phenomena, and hence their limitations depend on the assumed friction model, briefly discussed in the following subsection.

2.2 Friction Model

In the following we will consider the classical, simplified model summarized as follows:

the friction force exerted on a point lying on a surface with zero relative velocity is equal and opposite to the sum of external forces tending to move the point. The maximum force that can be sustained in this way by friction is proportional (by a static coefficient of friction μ_s) to the normal force pressing the point on the surface. If the object is moving with respect to the surface, the friction force opposes the relative velocity with an intensity proportional (by a kinetic coefficient of friction μ_d, generally lower than μ_s) to the normal force.

This model of friction is expressed by the relationship

$$
\left. \begin{array}{l} h_x = f_x \\ h_y = f_y \end{array} \right\} \quad \begin{array}{ccc} v & = & 0 \\ \sqrt{h_x^2 + h_y^2} & \leq & \mu_s h_z \end{array}
$$

$$
\left. \begin{array}{l} h_x = \frac{-v_x}{\|v\|} h_z \mu_d \\ h_y = \frac{-v_y}{\|v\|} h_z \mu_d \end{array} \right\} \quad v \neq 0
$$

where $v = (v_x, v_y)^T$ is the velocity of the point relative to the surface in the contact plane (see Fig. 1), and f is the external solicitation on the contact. As it can be seen, possible dependencies of friction on slip rate or history are disregarded in this model, and a discontinuity in the friction force in corrispondence of a starting motion is admitted.

When a finite area of contact is considered, limitations on allowable friction forces are more involved. If we consider non-rigid surfaces, in particular, we have that some points in the contact area may move, while others do not. Therefore, it is possible to have mixed static and kinetic friction effects. Moreover, even in the assumption that the contact area is and remains in the contact plane before and immediately after slippage, the combined effect of soliciting forces f_x, f_y and torque t_z must be considered.

Consider first the case that a pure force f is applied to the bodies in contact at the contact centroid, and that the tangential component of f is gradually increased. At a certain instant (and in relation with the details of how f is transmitted to the contact interface), slippage will start at some contact points, where friction will abrubtly drop to its kinetic value. A redistribution of the contact load then follows, and more points will be lead to slip. If the tangential component of f is increased enough, the slip 'wavefront' will sweep the whole contact area, and the two surfaces will start sliding on each other (this phenomenon, which has similarities with classical plasticity models, gives rise to acoustic waves propagating in the bodies, whose detection could be used as a means for detecting slippage [11]). Since the contact centroid, for planar contacts, coincides with the center of friction of the contact area [1], the act of relative motion immediately following is a translation. The maximum intensity of the tangential component of the contact force f that can be absorbed by friction in this case obeys the limitation

$$\sqrt{f_x^2 + f_y^2} \le \int_S \mu(\mathbf{r})\, \mathbf{h}_z(\mathbf{r})\, dS. \tag{1}$$

where $\mu(\mathbf{r})$ equals either $\mu_s(\mathbf{r})$ or $\mu_d(\mathbf{r})$, depending upon the relative velocity $\mathbf{v}(\mathbf{r})$ at the contact area element dS. From an operative point of view this expression is impractical, and a more convenient expression can be obtained if we assume that an *equivalent* coefficient of friction μ_{eq} exists such that, at the instant when the last contact point starts slipping, we have

$$\sqrt{f_x^2 + f_y^2} = \mu_{eq} f_z \tag{2}$$

Such relationship can also be interpreted as if all points in the area of contact were starting to slip at the same instant: the local static coefficient of friction is substituted everywhere by a lower value, μ_{eq}, to take into account the apparent decrement of friction due to partial slip.

Consider now the case that a purely normal contact force is exerted between the finger and the object, and that the torsion t_z is gradually increased, until slippage occurs at every contact point. By symmetry reasons, the instantaneous center of rotation (COR) of the motion immediately following the first occurrence of sliding coincides with the contact centroid, so that we can write the upper bound of friction torque as

$$t_z \le \int_S \mu(\mathbf{r})\, \mathbf{h}_z(\mathbf{r})\, \|\mathbf{r}\|\, dS. \tag{3}$$

The evaluation of eq. 3 is not simple in general. Jameson [12] assumes that the results of Hertzian theory concerning the normal pressure distribution and the contact area shape hold even in the presence of friction forces, and gives an expression for a spherical fingertip and object:

$$t_z \le 0.59\, \mu_{eq}\, \alpha^{1/3}\, f_z^{4/3}, \tag{4}$$

where the equivalent friction coefficient μ_{eq} is introduced, as above, to take into account progressive slippage phoenomena, and α is a constant depending upon geometric and elastic properties of the contacting bodies [12].

From eq. 4 a general proportionality relationship may be hypothesized for generic non-conformal surfaces pressed against each other by a normal contact force f_z:

$$t_z \le \nu_{eq}\, f_z^{4/3}, \tag{5}$$

where ν_{eq} is a characteristic of the peculiar contact conditions. Note that the more-than-linear dependence of the maximum friction torque upon the contact force, may be interpreted as the result of the twofold effect of contact force, which increases local friction forces and enlarges the contact area (due to deformations of the bodies) as well.

If a torsion is applied at the contact along with tangential forces, the COR of slippage motion is not at the contact centroid, and therefore eq. (5) does not hold. The limits of friction resistance under combined shear and torsion loading have been studied by Jameson [12] and by Howe, Kao and Cutkosky [13], with reference to the case of contact between non-conformal (in particular, spherical) bodies. In this case, only the intensity of shear forces, $f_t = \sqrt{f_x^2 + f_y^2}$ is of interest, due to the symmetry of the problem. In the cited papers, it is shown that slippage occurs for combined loadings whose representative points in a shear force - torsion plane lie outside a convex curve (see Fig. 2). The authors also show that, given the limit values of friction under pure shear force (eq. (2)) and under pure torsion (eq. (5)), an ellipse approximates well the limit curve, while a linear interpolation provides a conservative approximation of the actual friction limit curve.

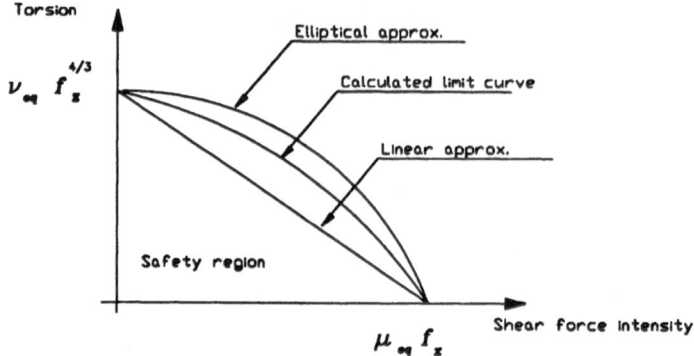

Figure 2: Sliding under combined torsion and shear loading for spherical surfaces (adapted from Howe, Kao and Cutkosky [12]

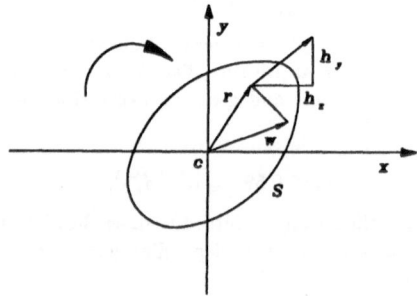

Figure 3: Sliding under combined torsion and shear loading for planar surfaces

The general case of contact between conformal surfaces (such as two portions of parallel planes, as occurs in grasping by parallel jaw grippers), is even more complex. With reference to Fig. 3, let the position of the COR in the contact centroid frame be the vector $\mathbf{w}$.

The maximum values that can be attained by friction forces and torque can be expressed in general as

$$|f_x| \leq \int_S \frac{\mathbf{r}_y - \mathbf{w}_y}{\|\mathbf{r} - \mathbf{w}\|} \mu(\mathbf{r}) \, \mathbf{h}_z(\mathbf{r}) \, dS, \tag{6}$$

$$|f_y| \leq \int_S \frac{\mathbf{r}_x - \mathbf{w}_x}{\|\mathbf{r} - \mathbf{w}\|} \mu(\mathbf{r}) \, \mathbf{h}_z(\mathbf{r}) \, dS, \tag{7}$$

$$|t_z| \leq \int_S \|\mathbf{r} - \mathbf{w}\| \mu(\mathbf{r}) \, \mathbf{h}_z(\mathbf{r}) \, dS + \mathbf{w}_x \mathbf{r}_y - \mathbf{w}_y \mathbf{r}_z. \tag{8}$$

In practice, the problem is posed in a different way: given the applied friction forces and torque (e.g. by means of intrinsic contact sensors: $\mathbf{p} = \mathbf{f}$, $\mathbf{q} = \mathbf{t}$), we wish to know how close the danger of slippage is. Since the COR position is not known a priori, we need to solve any two of the three equations above for r, and then obtain the bound on the third component of the frictional load. However, such solution requires both the knowledge of the contact area shape and of the pressure distribution over it, and can only be obtained numerically. References to this problem can be found in several authors' work, among which [1], [14], [13]; experimental results with an integrated (skin-like + intrinsic) tactile sensor have been described in [15].

3 Measurement of Friction Characteristics

As discussed above, it is rather difficult to predict, on theoretical grounds, the frictional characteristics of the contacts occurring in fine manipulation operations. On the other hand, such data are very important for the correct control of the robot hand. It is the authors' belief that, for most practical applications of robot hands, the evaluation of friction limits provided by the approximations described in Fig. 2 is sufficient to guarantee proper operation of the hand. To use those approximations (namely, the conservative linear approximation and the elliptical approximation), it is necessary to know the limit values of friction resistance in the cases of pure shear load and pure torsion. To this purpose, we devised a procedure for characterizing frictional effects by means of direct measurements in operative conditions, i.e. using exactly the same apparatus (hand, fingertips, objects) and in the same environmental conditions that are going to be met in the manipulative task.

3.1 Experimental Setup

The experiments described below have been carried out at the MIT AI Lab using contact sensors mounted on the distal phalanx of the fingers of the Salisbury articulated hand [1], fixed in turn to the wrist of a Unimation Puma manipulator. The low-level trajectory interpolation and control of the finger joints is performed by two 6-axis Unimation industrial controllers.

Part of the experiments have been carried out with a high-level control architecture consisting of a dedicated DEC-VAX computer and a Lisp machine (see [16]), running the hand programming environment OOLAH [17]. Later on, the high-level control system architecture has been modified and consists now of three processors Motorola 68030 on a VME bus, working in the parallel-processing, real-time environment provided by the VxWorks operating system.

We employed joint position sensors (encoders located at the actuator shafts). The force based contact sensors described in [18] have been used interchangeably with those described in [19]. The rest of the sensory equipment of the Salisbury hand (tendon tension sensors, palm contact sensor, and others) have not been employed in the experiments described below.

3.2 Translational Friction Characteristics

In Fig. 4 a generic object is grasped by means of the fingers of a robot hand. The fingers are supposed to be equipped with contact sensors, providing measurements of contact forces and torques. The position of the fingers relative to the grasped object is supposed to be such that the resulting grasp is stable enough to be robust with respect to small perturbations of the location of the fingers. The procedure to obtain static and kinetic coefficients of friction relative to translational slippage (due to pure shear) of the i-th finger on the grasped object surface is as follows:

- One of the fingers, e.g. the thumb, bearing the sensorized fingertip and opposing the other two fingers in the grasp, is imparted a time-increasing force in the contact plane, tending to move the finger along the object surface, while the normal component of the contact force is regulated at a constant value;

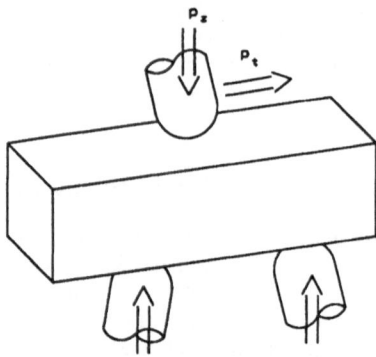

Figure 4: The robot hand fingertips and the object during the measurement of translational friction limits

- The starting motion of the finger is detected by joint position sensors, and the finger is stopped after a very short motion (1 mm);

- The force and torque transmitted by contact are measured by the fingertip sensor;

- The normal and tangential components of the contact force are computed by applying the intrinsic contact sensing algorithm to the force/torque data.

Finally, from observation of the time history of the friction ratio

$$R_f = \frac{\mathbf{p}_z}{\sqrt{\mathbf{p}_x^2 + \mathbf{p}_y^2}},$$

the desired information on friction coefficients can be derived. A typical plot of the (adimensional) values of R_f vs. time (in tenths of a second) in the experiment above described is reported in Fig. 5 (data are relative to an urethane-steel contact).

Observing this and many other experimental diagrams, a common pattern could be recognized by sight, consisting of two separate parts: initially, there is an almost linear increase of the friction ratio up to a maximum value, then a more or less abrupt change to a lower value. The latter is subsequently held without significant variations. Such pattern is in good agreement with the expectations based on the simplified classical model of friction discussed in section 2: the friction ratio increases during phase 1, until it reaches the (apparent) static coefficient of friction, after which R_f drops to a value corresponding to kinetic conditions.

The desired friction coefficients, μ_s and μ_d, might be obtained by considering the n samples of the friction ratio, $R_f(i), i = 1, n$ corresponding to instants $t = iT$ (T is the sampling period), simply as

$$\mu_s = \max_{i=1,n}\{R_f(i)\}, \qquad (9)$$

$$\mu_d = \frac{\sum_{i=s}^{n} R_f(i)}{n - s}. \qquad (10)$$

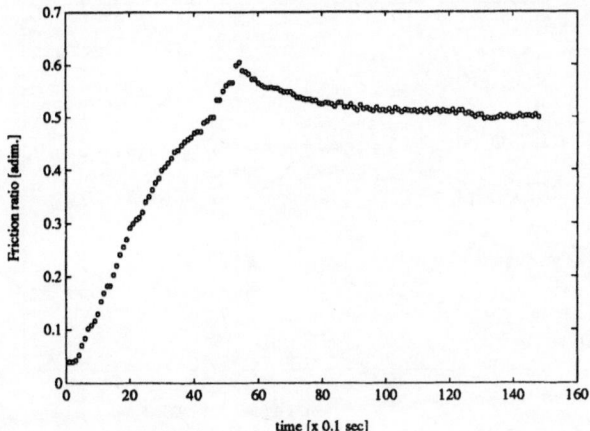

Figure 5: Friction ratio vs. time in a typical contact experiment between steel and urethane surfaces

where s is the sample corresponding to the maximum R_f, immediately preceding slippage. However, experimental tests repeatedly carried out on the same object in apparently similar conditions, showed that the estimates obtained through eq. (9) and eq. (10) are too sensitive to small perturbations of data, such as those due to sensor noise, vibrations of the robot structure, and in general to the poorly deterministic character of the complex phenomenon of friction.

To overcome this problem, a more robust algorithm has been devised, based on the idea of matching experimental data with the expected pattern of friction. The set of measurements $R_f(i)$ is split in two parts, preceding and following the kth sample. A least-squares linear approximation of the subset of data preceding the kth sample is evaluated as

$$\hat{R}_f(j) = a_k\, j + b_k, \tag{11}$$

$$a_k = \frac{k \sum_{j=1}^k j\, R_f(j) - \sum_{j=1}^k j\, \sum_{j=1}^k R_f(j)}{k \sum_{j=1}^k j^2 - (\sum_{j=1}^k j)^2}, \tag{12}$$

$$b_k = \frac{k \sum_{j=1}^k j^2 \sum_{j=1}^k R_f(j) - \sum_{j=1}^k j \sum_i^k j\, R_f(j)}{k \sum_{j=1}^k j^2 - (\sum_{j=1}^k j)^2}. \tag{13}$$

The remaining samples $R_f(i), k < i \leq n$ are approximated by a constant c_k equal to their mean value. The average quadratic errors corresponding to the approximations of the first and second data set are evaluated as

$$\sigma_1(k) = \frac{\sum_{j=1}^k (R_f(j) - a_k j - b_k)^2}{k}, \tag{14}$$

$$\sigma_2(k) = \frac{\sum_{j=k}^n (R_f(j) - c_k)^2}{k}, \tag{15}$$

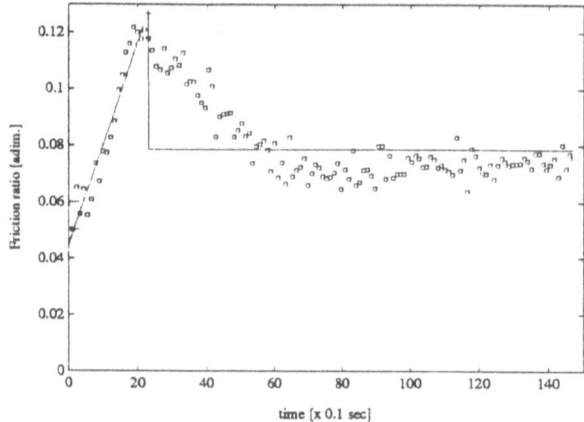

Figure 6: Friction ratio plot for an ABS plastic-steel pair

and their sum, $\sigma = \sigma 1 + \sigma 2$, is computed as k spans the time axis; the value $k = \hat{k}$ that minimizes $\sigma(k)$ is assumed to correspond with the beginning of slippage. Accordingly, the estimated friction coefficients are:

$$\mu_s = a_{\hat{k}}\hat{k} + b_{\hat{k}}, \qquad (16)$$
$$\mu_d = c_{\hat{k}}, \qquad (17)$$

while the sum of average quadratic errors $\sigma(\hat{k})$ is used as a measure of the dependability of these estimates.

The results of experimental tests carried out with three different pairs of finger and object cover materials are reported in Fig. 6, Fig. 7, and Fig. 8, respectively.

Superposed to the friction ratio measurements (small squares) are the lines corresponding to the approximating pattern above described. Particularly interesting is the plot of Fig. 8 (steel on a thickly painted wooden block), showing how the measured friction ratio bounces several times between the static and kinetic friction coefficient values before settling at the lower (kinetic) value. This phenomenon is due to stick-slip motion of the finger on the object surface (this irregular motion could be easily verified also by visual inspection).

It should be pointed out that the value of the estimates obtained through the method above is relative only to the peculiar operative conditions in which they have been obtained. This means that a comparison with friction coefficient measurements obtained by classical and more precise methods would not make much sense in this context, and are therefore avoided here. Therefore, reference to "apparent" friction coefficients is meant in the above discussion. The point in estimating friction characteristics with the methods above described is that their results are repeatable in similar operative conditions.

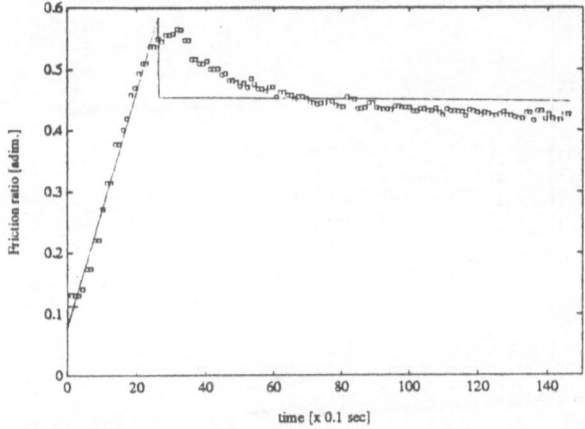

Figure 7: Friction ratio plot for an urethane-steel pair

3.3 Rotational Friction Characteristics

Another series of experiments have been carried out in order to evaluate the feasibility of a direct measurement of rotational friction (pure torsion) coefficients, ν_s and ν_d. The procedure followed in this case is similar to the one for translational friction data, differing only for the means of obtaining a rotational slippage between the object and the finger (see Fig. 9).

Recalling equation 5, we observe that in this case the ratio to be monitored during the tests is

$$R_s = \frac{q_z}{(p_z)^{\frac{4}{3}}}.$$

Three typical plots of R_s vs. time are reported in Fig. 10 a), b), and c). These plots correspond to the same finger and object cover materials, but different values of the normal component of the contact force, p_z, which was controlled to a reference value of $12N$, $9N$, and $6N$, respectively. The static and kinetic rotational coefficients of friction are derived from raw measurements using the pattern-matching algorithm above illustrated: Fig. 10 shows the piecewise linear approximation superimposed to experimental data. For the three reported cases, the estimates of ν_s and ν_d are $\nu_s = 0.813\,N^{-1/3}mm$, $\nu_d = 0.686N^{-1/3}mm$, $\nu_s = 0.814\,N^{-1/3}mm$, $\nu_d = 0.737N^{-1/3}mm$, $\nu_s = 0.812\,N^{-1/3}mm$, $\nu_d = 0.792N^{-1/3}mm$, respectively. It can be seen from these results that the relationship 5 is in good agreement with experimental data. It must also be noted that the scattering of data about the expected pattern in rotational friction measurements is higher than for translational slip: this is caused by the lower signal-to-noise ratio in the measurement of rotational friction coefficients, which are inherently small for non-conformal contact surfaces.

3.4 Object Identification by Friction Characteristics

In order to assess the repeatability of friction characteristics evaluation by the methods above described, they have been repeatedly applied to a large set of objects, following the phases below:

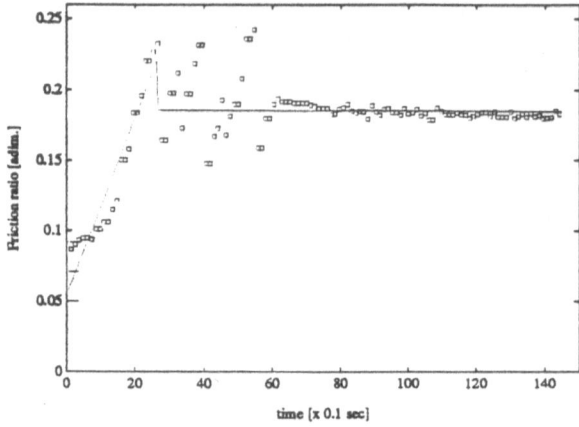

Figure 8: Friction ratio plot for a painted wood-steel pair

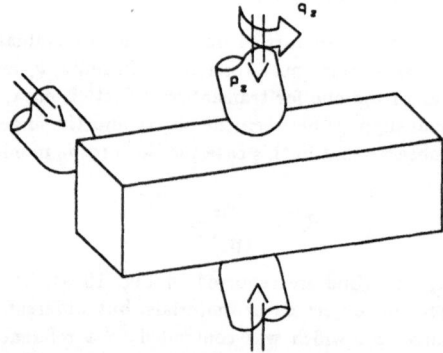

Figure 9: The experimental measurement of rotational friction limits

- objects of different shape and with different cover materials are collected in groups according to their superficial characteristics;

- the friction characteristics of an object out of each group are evaluated;

- the groups are merged into three classes, according to whether their kinetic translational friction coefficient is $\mu_d \leq 0.15$, $0.15 < \mu_d < 0.5$, $\mu_d \geq 0.5$;

- the objects are picked up by the robot hand at random, and are subject to friction measurement; based solely on friction data, the class of the objects are recognized.

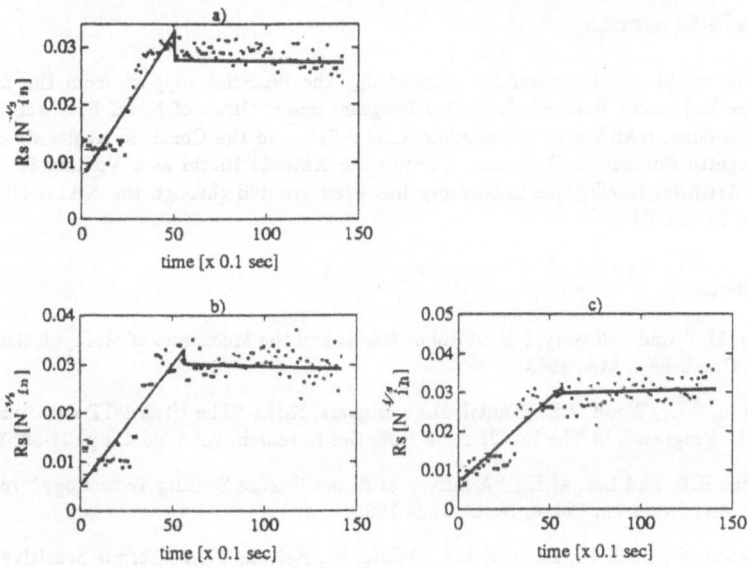

Figure 10: Rotational friction ratio plots for an urethane-steel pair. a) $p_z = 12\ N$; b) $p_z = 9\ N$; c) $p_z = 6\ N$.

The groups of objects in the first class gather small ABS boxes and Teflon objects of different shapes. In the second class are collected wooden blocks with thickly painted surface; the third is comprised of objects with high friction surfaces (such as rubber toys and hoses). The experiment has been repeated more than 50 times over a two-months period, with full success.

4   Conclusions

Friction forces play a fundamental role in dexterous manipulation. When planning or executing an operations involving contacts with the environment or with manipulated objects, the knowledge of the limits of friction forces is instrumental to control (or simply avoid) slippage. Theoretical study of these limits shows that not only the effects of pure shear forces (as most often is held in robotics literature), but also those of torsions applied at the contact must be taken into account. This important fact has been noted in previous works ([12], [13], [15]), indicating that a good approximation of the friction limit curve can be obtained starting from limits in the pure-shear and pure-torsion cases. In this paper, we presented results related to experimental measurements of such limits, that allow practical implementation of manipulative operations based on friction sensing, such as object recognition, adaptive grasping, controlled slippage, etc.

Acknowledgments

The authors would like to gratefully acknowledge the financial support from the following sources: the University Research Initiative Program under Office of Naval Research contract N00014-86-K-0685, NASA contract number NAG-9-319, and the Consiglio Nazionale delle Ricerche, Progetto Finalizzato Robotica. Support for Antonio Bicchi as a Visiting Scientist at the M.I.T Artificial Intelligence Laboratory has been granted through the NATO-CNR joint fellowship n.215.22/07.

References

[1] Mason, M.T. and Salisbury, J.K.: "Robot Hands and the Mechanics of Manipulation," MIT Press, Cambridge, MA, 1985.

[2] Jacobsen, S.C., Wood, J.E., Knutti, D.F., Biggers, K.B.: "The Utah-MIT Dextrous Hand: Work in Progress", in The Int. Jour. of Robotics Research, vol.3, no.4, pp. 21-50, 1984.

[3] Nicholls, H.R. and Lee, M.H.: "A Survey of Robot Tactile Sensing Technology," Int. Jour. of Robotics Research, Vol. 8, No .3, June 1989.

[4] Hackwood, S., Beni, G., Hornak, L.A., Wolfe, R., Nelson, T.J.: "Torque Sensitive Tactile Array for Robotics", in The Int. Jour. of Robotics Research vol.2, no. 2, 1983.

[5] Domenici, C., DeRossi, D., Bacci, A., Bennati, S.: "Shear Stress detection by a Piezoelectric Polymer Tactile Sensor", IEEE Trans. on Electr. Insulation, vol. EI 24, no.6, 1989.

[6] Bicchi, A., Salisbury, J.K., and Brock, D.L.: "Contact Sensing from Force Measurements", MIT AI Lab Memo no.1262, October 1990.

[7] Armstrong, B.: "Dynamics for Robot Control: Friction Modeling and ensuring excitation During Parameter Identification", Ph.D. Thesis, Electrical Engineering Dept., Stanford University, 1988.

[8] Armstrong-Helouvry, B.: "Stick-Slip Arising from Stribeck Friction", Proc. IEEE Int. Conf. on Robotics and Automation, 1990.

[9] Balestrino, A., and Bicchi, A.: "Adaptive Surface Following and Reconstruction Using Intrinsic Tactile Sensing," Proc. Int. Works. on Sensorial Integration for Industrial Robots, Zaragoza, Spain, 1989.

[10] Brock, D.L.: "Enhancing the dexterity of a Robot hand Using Controlled Slip", Proc. IEEE Int. conf. Robotics and Automation, 1988.

[11] Howe, R.D., and Cutkosky, M.R.: "Sensing Skin Acceleration for Slip and Texture Perception", Proc. IEEE Int. Conf. on Robotics and Automation, 1989.

[12] Jameson, J.W.: "Analythic Techniques for Automated Grasp", Doctoral Dissertation, Stanford University, 1985.

[13] Howe, R.D., Kao, I., and Cutkosky, M.R.: " The Sliding of Robot Fingers Under Combined Torsion and Shear Loading", Proc. IEEE Int. Conf. on Robotics and Automation, 1988.

[14] Goyal, S.: "Planar Sliding of a Rigid Body With Dry Friction: Limit Surfaces and Dynamics of motion", Ph.D. Thesis, Mech. Eng. Dept., Cornell University, Ithaca, 1989.

[15] Bicchi, A., Bergamasco, M., Dario, P., Fiorillo, A.S.: "Integrated Tactile Sensing for Gripper Fingers", Proc. Int. Conf. on Robot Vision and Sensory Control - RoViSeC '88, Zurich. I.F.S. Publications Ltd., 35-39 High Street, Kempston, Bedford MK 42 7BT, U.K., 1988.

[16] Salisbury, J.K., Brock, D.L., Chiu, S.L.: "Integrated Language, Sensing and Control for a Robot Hand", Proc. 3rd ISRR, Gouvieux, France, MIT Press, Cambridge MA, 1985.

[17] Chiu, S.L.: "Generating Compliant Motion of Objects with an Articulated Hand", M.S. Thesis, MIT AI Lab Technical Report 1029, 1985.

[18] Brock, D.L, and Chiu, S. "Environment Perceptions of an Articulated Robot Hand Using Contact Sensors," Proc. ASME Winter Annual Meeting, Miami, FL, 1985.

[19] Bicchi, A.: "On the Optimal Design of Multi-Axis Force Sensors", MIT AI Lab Memo no.1263, October 1990.

Intelligent Robot Gripper for General Purposes

H.-K. Scherrer and D. Vischer

Institute of Robotics, ETH,
Swiss Federal Institute of Technology,
CH-8092 Zürich, Switzerland

Abstract: In the past many different robot grippers have been developed to grasp one or a few specific objects. Those grippers are well suited for continuous work in structured environments and are thus employed for most of todays industrial applications. Some researchers, on the other hand, have focused their attention on sophisticated general purpose grippers with the dexterousness and kinematics similar to the human hand. This approach leads to mechanically refined but usually heavy and expensive grippers, which may be difficult to mount on a robot wrist. This paper presents another approach to the design and development of general purpose grippers. The basic idea is to integrate various task-dependent sensors with a mechanically still fairly simple gripper and combine it with an appropriate control software involving artificial-intelligence methods. This approach has led to the construction of the "COR-Gripper"* described in this paper. The gripper consists of three independent, parallel fingers and includes a force & torque sensor, tactile sensors as well as optical and ultrasonic proximity-sensors. It is currently mounted on a Puma 560, which is coupled to a vision system. The control-software is a combination of classical algorithms and AI-methods, such as neural networks. The latter is used to calculate the optimal finger-positions for grasping objects of arbitrary shape.

Keywords: robot, gripper, sensors, tactile sensors, neural networks, optimal grasping

1. Introduction

The work reported here is part of a larger research effort concerned with the development of a robot able to work in an unstructured environment cooperating with a human being. For monotonous and hard work it would often be desirable to use a robot to support the human operator in his task, i. e. to accomplish a cooperation of the robot with the operator. The robots contribution could be restricted to the laborious and repetitive work, whereas the human acts as a supervisor and handles exceptions. The title of this project currently running at ETH is "Cooperating Robot with Visual and Tactile Capabilities". The benchmark-test we have chosen for our project will be to remove dishes, such as glasses, cups and plates from the trays in the canteen of the ETH and put them into the dish-washer. Such tasks can only be achieved with the extensive use of sensor information. Expertise in many different research areas is required. Twelve Ph. D. candidates from four different institutes are involved in the project with various contributions, such as the development of a fast vision-system to analyze complex 3D-scenes as well as safety concepts and interface considerations for the cooperation between operator and robot.

The COR-Gripper described in this paper is a crucial component of the "Cooperating Robot". The gripper should be able to grasp various objects of different weights, shapes and materials. In the course of the benchmark-test, for instance, it should be able to handle heavy plates, breakable glasses and cups, soft paper napkins as well as comparably tiny

---

* COR: Cooperating Robot, see Introduction

objects, such as the spoons. Among those known objects we also expect a lot of unknown things, like left-overs, orange-juice bottles, milk containers or even purses, accidentally left on the trays.

The 3D-vision system of the "Cooperating Robot" transmits the information about the position and orientation of the objects to be grasped to the robot-controller. This information may not always be accurate and sometimes even faulty. The robot-software should account for this and use the sensors of the gripper to support the vision-system in its task and to avoid collisions, which may result in damage of expensive machinery, including the robot or the gripper itself, or even in injuries of the human operator.

2. CHOICE OF GRIPPER-KINEMATICS

In general, there are two classes of grasping: • form-stable grasping
                                                • force-stable grasping

In our approach we only use force-stable grasping, since the friction coefficient is hardly predictable for many of the considered objects and, since it is impossible to design a mechanically simple gripper for form-stable grasping, which allows to handle arbitrary-shaped objects.

The most common and simplest grippers consist of two parallel fingers or jaws with one degree of freedom. Those simple grippers are effective, where objects with two parallel surface-planes are grasped. If the jaws are not form adaptive, then the grasping stability may be low for round or spherical objects. To combine the advantages of this simple gripper designs with more general grasping abilities, we added a third finger to the two fingers of the parallel gripper. In Figure 1 we compare a parallel gripper, a three finger gripper and an artificial hand. It is suggested that the three-finger gripper is a good compromise between the sophisticated hand and the simple parallel gripper. Only for spherical objects the artificial hand has an obvious advantage over the three-finger gripper, because its form adaptive abilities lead to a better contact.

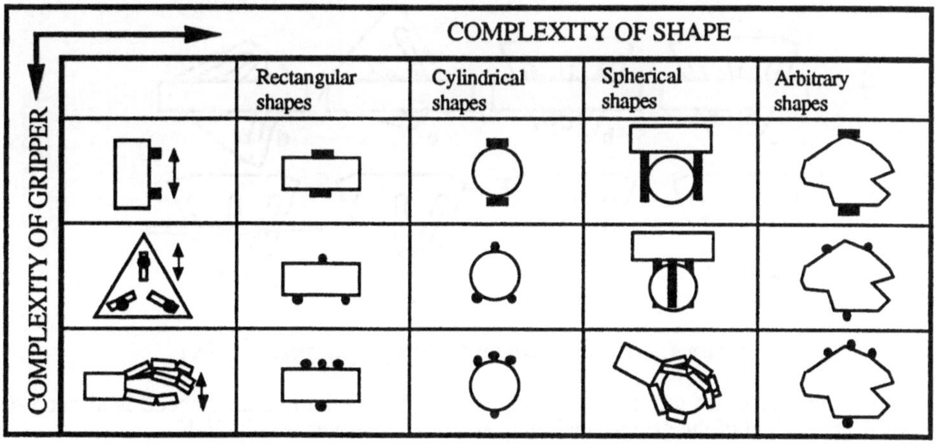

Figure 1: Comparison of parallel grippers, three-finger grippers and multi-fingered hands

3. MECHANICAL DESIGN

For our initial investigations we used a parallel gripper equipped with three parallel fingers. Besides being applicable to a restricted number of object-shapes this initial design was also limiting the actual working area of the robot. This problem arises for instance, when large, flat objects have to be grasped, such as plates. The plates can only be picked up from the side by moving the fingers underneath the rim, which is impossible with most parallel grippers.

In the final layout, the COR-Gripper has three fingers, which can be moved independently. This results in an optimal ratio between working area and outer dimensions of the gripper. For simplicity the prototype gripper has three identical finger modules, each consisting of a parallel guided finger, a ball screw, a belt drive and a DC-motor (Figure 2).

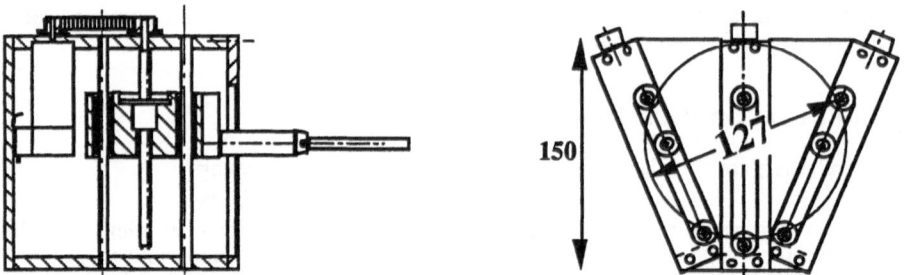

Figure 2: Linear module   *Figure 3: "Trident configuration"*

The three modules are arranged comparable to a trident with angles of 22.5 degrees between two modules (Figure 3). The arrangement of the finger modules allows for a maximum of possible triangle configurations of the three fingers within a certain area. The maximal diameter of tangible objects is about 120 mm but thin and elastic objects (Figure 4h) can also be handled. When the three fingers are positioned close to the lower margin of the parallel guide the gripper can also grasp flat objects such as plates. The fingers are absolutely symmetrical and it is possible to grasp concave objects like glasses or cups in three different ways, as shown in Figure 4e, f and g.

Figure 4 shows different configurations of the gripper. Usually there exist several solutions for grasping one specific object.

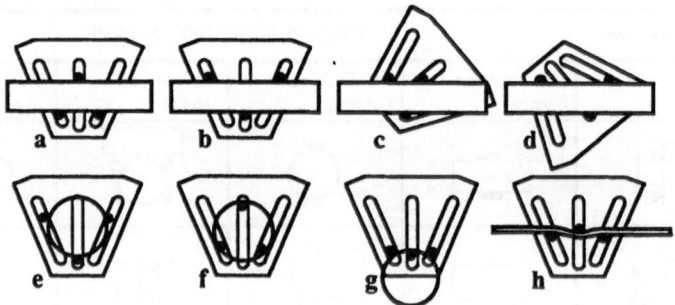

Figure 4: Multiple solutions for grasping a specific object

The maximal constant force of each finger is 50 N and the maximal speed is about 120 mm/s. The weight of the gripper including sensors and drives is only 24 N.

4. SENSORS

The various sensors for the prototype gripper are shown in Figure 5 and some of their features are listed in Table 1. We tried to use purchasable sensors, whenever this was possible. Most of the available sensors, however, proved to be too heavy or too spacious for our application. Thus, the strain-gauge sensors in the fingers had to be developed in our lab and both, ultrasonic and fibre-optic sensors will be modified for a future version of the gripper.

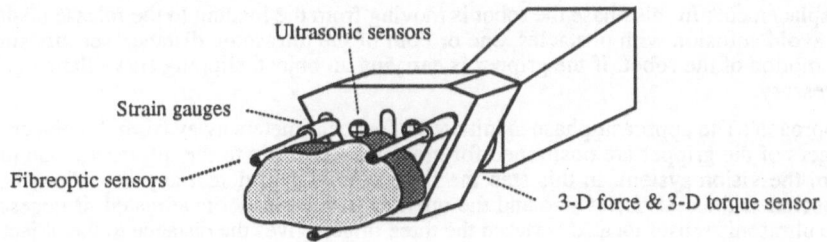

Ultrasonic sensors

Strain gauges

Fibreoptic sensors

3-D force & 3-D torque sensor

Figure 5: Sensors

Purpose	Sensor	Quantity	Range
• Contact-detection of gripper • Support of object recognition by Weight measurements	Force & torque sensor	3 forces 3 torques	- 66 ... + 66 N - 5.1 ... 5.1 Nm

• Contact-detection of finger
• Contact-force control
• Contact-position on the finger
• Slip surveillance
	Strain gauges
	(see Figure 6)
		12
		(4 bridges)
			Forces: - 50 ... 50 N
			          resolution: 0.25 N
			Positions: 0.. 60 mm
			          resolution: 1 mm

• Distance gripper to environment	Ultrasonic	2	80 ... 500 mm
• Distance fingertip to object	Fibreoptics	3	0 ... 30 mm
• Position of the fingers	Encoders	3	0 ... 92 mm resolution: 0.002 mm

Table 1: Sensors of the COR-Gripper

The design of the fingers, including the strain-gauge sensor, is presented in Figure 6. Each of the three fingers allow for the measurement of the two perpendicular contact-forces F_1 and F_2 acting on the finger and their corresponding points of attack, expressed by the positions s_1 and s_2.

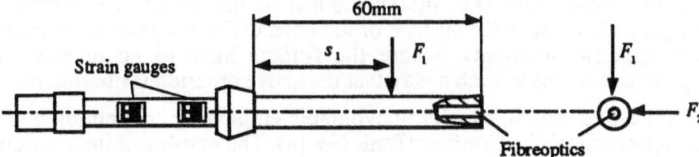

60mm

Strain gauges

s_1

F_1

F_1

F_2

Fibreoptics

Figure 6: Finger of the COR-Gripper with 4 strain-gauge bridges and fibreoptics

The two forces and two distances can be calculated from the measurements of four strain-gauge full-bridges, two for each direction. With this simple setup it is possible to get all the necessary information for slip detection and for a sophisticated finger-force control. All this sensor-information could also be acquired by a somewhat costly and complicated tactile touch-sensor. The strain-gauge sensor presented here is much cheaper to manufacture and easier to handle. Nevertheless it supplies almost as much information about the grasped object as a touch-sensor and provides quite accurate results.

Under ideal conditions the relation between the measured strains and the two forces & distances can easily be derived. Due to small imprecisions in the manufacturing and inaccurate positioning of the gauges, however, a cross talk in the range of 3% between the forces was observed. By means of a calibration, based on a least-square method, it was possible to eliminate most of the cross talk and strongly enhance the performance of the sensor.

According to the different tasks, we divide the possible operations of the robot into three categories:

- 'Displacement': In this phase the robot is moving from the loading to the release position. To avoid collision with obstacles, one or both of the ultrasonic distance-sensors survey the motion of the robot. If the gripper is carrying an object, slipping surveillance is also necessary.

- 'Approach': The approach-phase is initiated a few centimeters away from the object. The fingers of the gripper are positioned for grasping according to the information acquired from the vision system. In this step the fibre sensors will detect any deviations of the expected to the real object size and the opening of the gripper is adjusted, if necessary. The ultrasonic sensor located between the three fingers gives the distance to the object and may be used to decelerate the robot during the approach.

- 'Grasp': After a successful positioning, the gripper can be closed. For this step the force sensors in the fingers and the detection of the contact zones on the fingers are the most important sensor-informations. The 'grasping phase' ends with the withdrawal of the gripper. During withdrawal slippage surveillance is important.

5. OPTIMAL PLACEMENT OF THE FINGERS FOR GRASPING (NEURAL NET)

The computation of the optimal finger-positions and forces to grasp an object of arbitrary shape can be accomplished with different analytical or geometrical methods [2], [3]. Our method is based on a cascade of two neural networks [1]. Force-stable grasping of an object means that the sum of the finger-forces as well as the sum of torques, resulting from those forces, are zero. The friction force is not considered here, because it is unreliable and unpredictable in our environment. The equilibrium condition for forces and torques can be formulated as a quadratic minimization criterion:

$$A \; || \; sum \; of \; forces \; ||^2 + B \; || \; sum \; of \; moments \; ||^2 \qquad A,B > 0$$

Additional conditions accounting for the kinematics of the gripper and the convexity of the shape have been added.

The silhouette of the object to be grasped is supplied by the vision system. As shown in Figure 8, we divide the boundary of the silhouette into a number of segments. Their size is depending on the complexity (i.e. the curvature) of the shape. The finger-forces are assumed to be perpendicular to the surface of the object. The purpose of the neural net is to find the three specific segments, where the fingers have to be placed on, and the corresponding contact forces in such a way that the above criterion is minimized

To avoid an exhaustive search, usually unavoidable employing conventional methods, we decided for an approach with a Hopfield/Tank-Net [4]. The number of iterations required by the Hopfield net to reach a stable state, i.e. to find the minimum of the error function, is strongly depending on the initial state. We found that the convergence of net is much better, when this initial state was chosen in agreement with our (human) experience.

In order to imitate this factor of human experience, we preprocess the discretized vision data with a feedforward network [5] to find an appropriate initial condition for Hopfield net. Similar to a human, this feedforward network should also be able to gain experience through adequate learning. To train the feedforward net, we put together a huge collection of objects, where the optimal finger positions as well as the magnitude of the necessary forces are known or can be calculated in advance, such as triangles, squares and circles. A feedforward net has the property to serve as a storage for a certain amount of known objects. In addition to that, this type of net is also able to generalize a certain set of taught rules to some extent.

The teaching of the feedforward net is done offline. We do not use online-learning, since any unknown object would disturb the weights of the feedforward net. Once a solution is found for a previously unknown shape, we add it to the existing learn-set. During the next offline learning-phase the extended learn-set is used to train the feedforward net. By this method it is possible to improve the "experience" of the feedforward net, which consequently results in fast convergence for the Hopfield net.

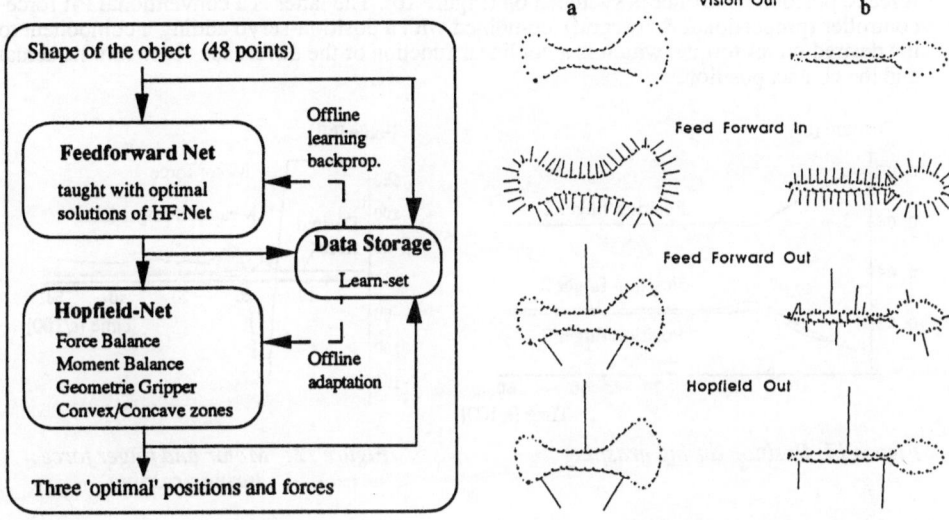

Figure 7: Cascade of neural networks

Figure 8: Examples of known & unknown objects

The taught shape, shown in Figure 8a, is 'recognized' by the feedforward net and its output is thus very similar to the final optimal solution provided by the Hopfield net after one single iteration. In the case, where the shape is unknown (Figure 8b), the feedforward net is able to detect similarities to known objects and derives a reasonable solution to serve as an input for the Hopfield net, which is than capable of finding the optimal solution after 150 iterations.

The cascade of the two neural networks proves to be effective to compute the optimal finger positions and forces for grasping known and unknown objects, leading to good accuracy and a comparable low computational effort. The method allows to grasp unknown objects on the first attempt with a high probability of success.

6. CONTROL CONCEPTS

The basic control concepts of the COR-Gripper are presented in Figure 9.

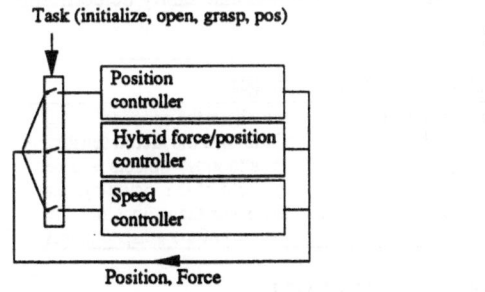

Figure 9: Basic control algorithms

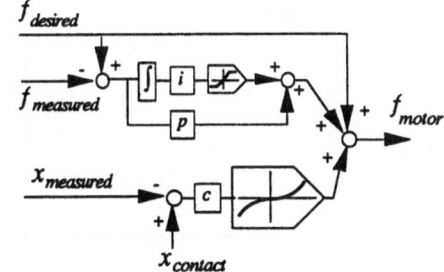

Figure 10: Hybrid force & position controller

At the lowest level three types of controllers for each of the three fingers are available. The appropriate algorithm is chosen according to the task and the sensor information. The finger encoders are initialized by means of the speed controllers. The position servos guide the fingers close to the object to be grasped. The last approach before contact is accomplished with a speed controller. After all three fingers have reached their contact-positions, a hybrid force & position controller is switched on (Figure 10). The latter is a conventional P-I force-controller (proportional & integral), combined with a position-servo adding a component to the desired motor torque, which is a nonlinear function of the differences between the actual and the contact position.

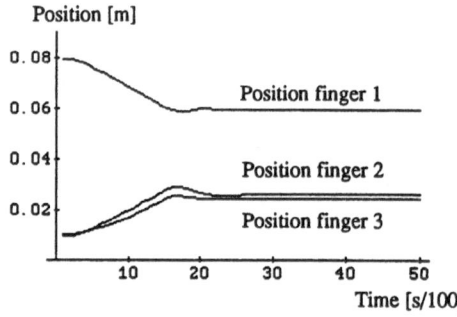

Figure 11: Position during grasping

Figure 11: Position during grasping   *Figure 12: Motor and finger force during grasping*

In the near future, we plan to replace the three independent single-input single-output controllers with a more sophisticated servo-mechanism, which will hopefully allow us to make the most of all the available sensor-information provided by the gripper.

7. SOFTWARE CONCEPT

The robot software consists of four different layers. The top layer is concerned with the task-planing and involves a high-level language, which includes commands such as "CloseGripper", "MoveRobot()" and "GetShapeFromVision". The next lower layer is organized according to the different system-components. We distinguish between the robot subsystem, the vision subsystem, the controller for the conveyor belts, which handles the flow of the trays, and the subsystem, which is in charge of detecting the human operator by means of a (second) vision-system. The robot subsystem contains the puma controller and the gripper controller. All the sensor and actuator-drivers are located in a further software-layer and the lowest layer includes all necessary communication-tools for the hardware-components, such as serial ports, AD/DA boards, encoders etc.

The gripper software provides nearly a dozen high-level commands for the user (Table 2). Those commands are similar to a robot programming language and can easily be combined to simple operations.

High Level Command	High Level Command	Purpose

`GripperOpen`
`GripperClose`
`GripperPos(h1,h2,h3)`
`GripperGrasp(f1,f2,f3,a1,a2,a3)`
	`GripperInit`
	`GripperCalib`
	`GripperSearch`
	
	`GripperClassify`
		Initializes the encoders
		Calibrates the tactile sensors
		Tries to grasp objects without
		the help of the vision-system
		Measures the weight and other
		features of the grasped object

Table 2: High-level commands for the gripper

The software for the gripper and the robot is implemented on a Motorola 68020 microprocessor-board, which communicates over a VME-bus with four input/output-bords and a M 68040, where the algorithms for the neural nets are executed.

8. CONCLUSIONS

An overview on the design and development of the COR-Gripper has been given. The gripper consists of three independent, parallel fingers and contains a force & torque sensor, tactile sensors for slip-detection as well as optical and ultrasonic proximity-sensors.

It has been demonstrated that a mechanically still fairly simple gripper is capable to handle a multitude of different objects. This can be achieved by combining the appropriate sensor-information with an "intelligent" software, employing classical algorithms and AI-methods, such as neural networks. As opposed to most multi-fingered hands, intended to imitate the dexterousness of the human hand, this gripper has the advantage of being quite compact and light. Therefore it may easily be mounted on a robot arm.

The cascade of the two neural networks proves to be effective to compute the optimal finger positions and forces for grasping known and unknown objects, leading to good accuracy and a comparable low computational effort. The method allows to grasp most unknown objects on the first attempt with a high probability of success.

Figure 11: COR-Gripper

Figure 12: Cooperating Robot

9. ACKNOWLEDGEMENTS

We are thankful to our colleagues from the Institute of Robotics, the staff of our electronics & mechanics workshop and all those students, who have contributed to this project.

10. REFERENCES

[1] Xu G., Scherrer H.-K., Application of neural networks on robot grippers, IJCNN Int Joint Conf. on Neural Networks, San Diego June 1990

[2] Ji Z., Roth B. Direct Computation of Grasping Forces for Three-Finger Tip-Prehension Grasps, Proc 20th ASME Mec Conf. Kissimmee, Florida, pp 125-138, Sept 1988

[3] Podhorodeski R., et a. , The Feasibility and Optimization of Grasp Contact Forces for Robotic Application, 10th Brazilian Congress of Mechanical Engineering, Coppe, 1989

[4] Hopfield J.J., Tank D.W., Neural computation of decisions in optimization problems, Biological Cybernetics, Springer Verlag Vol 52, pp 141-152, 1987

[5] Arnoczky B., Friedli T., Neuronales Netz für Robotergreifer, Diploma Thesis, Institute of Robotics, Feb. 91

Section 4: Robotic Systems and Task-Level Programming

How to organize the perception, decision and control software of a robotic system to achieve its tasks? How to define the task to the robot at an abstract level? Solving these issues paves the path to intelligent robots acting in the real world. The papers in this section address these problems in the contxt of different application fields.

The first paper by R. Taylor and several colleagues from IBM T. J. Watson Research Center and New York University Medical Center describes the architecture of a system that integrates several interfaces, sensing devices and manipulation aids. It assists a surgeon in planning and executing an optimal procedure for facio-cranial surgery, a precision demanding task. The system builds a 3D model of the patient's skull and extracts specific features. It compares it with a data base of normal models to help the surgeon producing a surgical plan. Then the systems also assists the surgeon during execution.

In the paper by G. Schrott, an environment for task-level programming in the context of flexible manufacturing is described. The system is organized into 5 levels integrating a rule-based system based on OPS5 for task planning, and the numerical processings necessary for sensing and control. A global object-oriented database includes the representations necessary for the user interface, task planning, coordination and control levels. The use of object-oriented data and knowledge is very adapted to such a hierarchical structure.

R. Chatila, R. Alami, B. Degallaix, V. Perebaskine, P. Gaborit and P. Moutarlier propose an architecture for the control system of an intervention mobile robot. Such robots are characterized by the fact that they have to accomplish tasks in remote, ill known environments. It is then necessary to endow the robot with the features necessary for autonomy. Interaction with the human operator should be as much as possible at a task level. The control structure of the robot is composed of a task interpretation and refinement part, and a set of execution functions. The task is planned and tele-programmed off-line, and the robot has the ability to execute it in different manners according to the environment conditions. An experimentation with indoor navigation is described.

Another aspect of remotely operated robots is presented in the paper by G. Hirzinger, G. Grunwald, B. Brunner and J. Heindl. The telerobotic system developed for the ROTEX experiment is an orbital system that integrates autonomous and teleoperated modes with shared control. The robot is equipped with a multisensory gripper. The teleoperation uses a 6 dof control ball and visual feedback from the robot's sensors and a stereovision system observing the scene. This is an important difference with the intervention robot problem in which the knowledge is provided to the human tele-operator only through the robot's sensors. In order to cope with time delays in the ROTEX experiment, a predictive display is used by the remote operator. One interesting feature of the system is the "tele-sensor-programming": the off-line teleprogramming phase produces the expected nominal sensory patterns that will be used as reference and refined during execution. At this phase, a stereo and laser system mounted on the gripper are used, for example to predict the motion of a floating object that is grasped in real time.

A Model-Based Optimal Planning and Execution System with Active Sensing and Passive Manipulation for Augmentation of Human Precision in Computer-Integrated Surgery

Russell H. Taylor (1), Court B. Cutting (2), Yong-yil Kim (1, 3), Alan D. Kalvin (1)
David Larose (1), Betsy Haddad (2), Deljou Khoramabadi (2), Marilyn Noz (2)
Robert Olyha (1), Nils Bruun (1), Dieter Grimm (1)

(1) IBM T. J. Watson Research Center; Yorktown Heights, New York 10598
Phone: (914)784-7796 Fax: (914)784-6282 Email: rht@watson.ibm.com
(2) Institute for Reconstructive Plastic Surgery; NYU Medical Center; New York, NY
10016
Phone: (212)263-5502 Email: cutting@mcirps3.med.nyu.edu
(3) Present address: Korea Institute of Science an Technology; Seoul, Korea 136-791

Abstract

Researchers at IBM and NYU Medical Center have recently begun development of a model-based system for optimal planning and augmented execution of precise osteotomies to correct craniofacial malformations. In these procedures, the facial bones are cut into several fragments and relocated to give the patient a more normal facial appearance. There is a significant synergy between better presurgical planning methods and the ability to execute the plans precisely and efficiently. The planning component of our system will transform CT images into a 3D geometric model of the patient's skull and assists the surgeon in planning an optimal procedure based on an analysis of the patient's anatomy compared to a database of normal anatomy. The surgical component will use realtime sensing to register the model-based surgical plan with the reality in the operating room. It will employ a variety of man-machine interface modalities (graphics, synthesized speech, etc.) together with passive manipulation aids to assist the surgeon in precise execution of his plan. This paper describes the overall system architecture, the proposed surgical procedure, implementation status, and some early experiments that we have performed.

Introduction and Background

Recent advances in medical imaging technology (CT, MRI, PET, etc.), coupled with advances in computer-based image processing and modelling capabilities have given physicians an unprecedented ability to model and visualize anatomical structures in live patients, and to use this information quantitatively in diagnosis and treatment planning.

The precision of image-based pre-surgical planning often greatly exceeds the precision of surgical execution. Typically, precise surgical execution has been limited to procedures (such as brain biopsies) for which a suitable stereotactic frame is available. The inconvenience and restricted applicability of these devices has led many researchers to explore the use of robotic devices to augment a surgeon's ability to perform geometrically precise tasks planned from computed tomography (CT) or other image data (e.g., [1], [2], [3], [4], [5]). The ultimate goal of this research is a partnership between a man (the surgeon) and machines (computers and robots) that seeks to exploit the capabilities of both to do a task *better* than either can do alone. Machines are very precise and untiring and can be equipped with any number of sensory feedback devices. Numerically controlled, robots can move a surgical instrument through an exactly defined trajectory with precisely controlled forces. On the other hand, the surgeon is very dexterous. He is also quite strong, fast, and is highly trained to exploit a variety of tactile, visual, and other cues. "Judgementally" controlled, he understands what is going on in the surgery and uses his dexterity, senses, and experience to execute the procedure. By nature, he wants to be in control of everything that goes on. However, he *must* rely on the machines to provide precision. How can he trust them not to harm the patient?

Augmentation with Simple Passive Aids

The most obvious way to prevent a robotic device from making an undesired motion is to make it incapable of moving of its own accord. Motor-less manipulators have been implemented, in which joint encoders are used to provide feedback to the surgeon on where his instruments are relative to his image-based surgical plan (e.g., [4], [6]). One important limitation of this approach is that it is often very difficult for a person to align a tool accurately in six degrees of freedom with only positional feedback. Passive manipulators, permitting free motion until locked, have been implemented for limb positioning, tissue retraction, instrument holding, and other applications in which accuracy is not important [7] [8]. In another case, Davies [9] implemented a passive manipulation aid with restricted degrees-of-freedom for prostate surgery, which was used clinically, after prototyping the necessary motions on an active robot.

Within the planning system, initial programs for image processing, model construction, and feature extractions have been completed. Substantial progress has been made on the interactive graphics and surgical simulation components. Independent work is underway in to construct the anatomical data base. Work on a full-blown surgical plan optimizer using point, line, and surface features is still in very early stages. However, an earlier plan optimizer based on point features is already in clinical use ([21], [15]) and could be used as an alternative.

Within the surgical system, we have developed interfaces to the Optotrak, an adequate pointing system, beacon mounting methods and means of tracking and displaying bone fragment motions relative to each other. A variety of PC-based graphics, voice, and tonal cues are available for feedback to the human, and we are in the process of debugging the service routines needed to provide a connection to the RS6000 for more sophisticated graphics and other online processing.

We have gone through several generations of the passive manipulator. The first version, shown in Figure 5, used a modified SCARA structure with a counterbalanced Z-arm for coarse (and fine Z) positioning, with a 5 DOF distal fine positioner. It had electric particle brakes with zero-backlash gearboxes for the four proximal joints and manual clamps distally. Our experience with this version was mixed. The distal fine positioning mechanism worked very well. However, the proximal (coarse) structure proved too compliant and the Z-arm had too much inertia to be effective as a fine positioning mechanism. Consequently, we modified the structure to provide a fine "Z" stage counterbalanced by a constant force spring and replaced the coarse positioning joints with a simpler and more rigid cartesian structure (Figure 6).

Several different versions of the end-of-arm tooling were developed. We eventually settled on an Elmed "Retract-Robot" (tm) surgical instrument holder [8]. Our initial implementation placed a Lord Corp. 6DOF force/torque sensor between the goniometer cradles and the instrument holder. In practice, we found that this arrangement did not leave sufficient clear working space, so we removed the force sensor. Subsequent versions will have a larger "standoff" distance, thus leaving more room for the force sensor and other tooling.

Experiments and Early Experience with the Surgical System

Pointing System

Our prototype pointer is shown in Figure 7, and has four LED beacons placed approximately at the corners of a 75mm by 150mm rectangle. The pointer tip is a needle approximately 90mm from the plane of the rectangle. We calibrate the pointer by placing it in a

Next, one or more "views" of the patient are obtained to measure the *relative* position of the beacons, which are all rigidly affixed to the skull. At this point, it is necessary to "register" the beacons to the CT coordinate system used to plan the procedure. The most straightforward technique for doing this is to use a calibrated pointer to identify known landmarks while simultaneously recording the positions of the beacons mounted to the patient's skull. These landmarks can either be anatomical features or fiducial markers rigidly affixed to the patient when he is CT'ed and replaced in the same place at the time of surgery.[3] Once this has been done, it is possible to compute the coordinates T_{ci} of each bone fragment B_i relative to the camera. The coordinates of fragment B_j relative to fragment B_i are, of course, $T_{ij} = T_{ci}^{-1} \bullet T_{cj}$

Next, the surgeon cuts free a bone fragment B_j. He now needs to manipulate this fragment so that it is in a desired pose $_{des}T_{ij}$ relative to another bone fragment B_i.[4] To do this, he places the center-of-rotation of the passive manipulator's goniometer cradle stages over the desired center-of-rotation of the bone fragment. He uses standard surgical tools to grasp the bone fragment firmly and uses the adjustable end-of-arm tooling clamp to grasp the instrument. At this point the bone fragment is rigidly affixed to the manipulation aid. He now uses the passive manipulation aid to realign the bone fragment so that $T_{cj} = T_{ci} \bullet {_{des}T_{ij}}$. In a typical alignment strategy, this will be done by first unlocking all "fine motion" degrees of freedom and manipulating the fragment into its approximate desired position, with the surgeon relying on his own tactile feedback and the force sensor information to verify that there is no undesired obstruction. Then, each degree-of-freedom will be successively brought into alignment and locked. Once the B_j has been aligned, standard surgical screw and plate methods are then used to affix it to B_i. Bone grafts are used as necessary to fill in any gaps between fragments.

This process is repeated until all bone fragments have been repositioned. Surgery then procedes normally.

Implementation Status

As of May 1991, many components of the total system architecture have been implemented, at least in preliminary form. However, system integration is just beginning. In particular, we have yet to integrate the planning system with the surgical system and more integration has still to be done integrating sub-components as well.

---

scheme is that the beacon carrier could be used to as a handle to assist in surgical manipulation. The disadvantage is that it is bulky.

[3] One possibility is to use a "mouth guard" customized to the patient's teeth to hold calibration phantoms and (possibly) LED beacons to eliminate the use of the pointer.

[4] For simplicity, this discussion will assume that B_i is rigidly affixed to the skull base, which is held fixed relative to the manipulator base during this part of the procedure. There are a number of options for relaxing this constraint.

such aids is that the surgeon provides all the motive force. Generally, the manipulation aid should interfere as little as possible with the surgeon's tactile "feel" for what is happening to the patient, while preserving the desired alignment once it is achieved. Six DOF manipulation aids with manually [8] and semi-automatically [7] actuated brakes have been developed for tissue retraction, instrument placement, and similar applications. One serious drawback of these systems is that they provide little assistance in actually achieving the desired alignment. Even without the additional inertia of a mechanical linkage, most people find it extremely difficult to achieve an accurate six degree-of-freedom alignment.

Our approach is to develop manipulation aids with computer controlled (or manually actuated) brakes to provide selective locking of *orthogonally decoupled* degrees-of-freedom resolved in a tool frame centered at a point reasonably far removed from the mechanism. This permits implementation of a variety of manipulation strategies in which the surgeon only needs to work on aligning a few (often, one) degrees-of-freedom at a time.

The basic structure of our present implementation is illustrated in Figure 4. There are three basic components. A 6 DOF *fine positioning system* consists of three counterbalanced linear stages carrying a conventional k_z axis and crossed goniometer cradle k_x and k_y axes with a rotation center about 150 mm from the mechanism. A standard 6 DOF surgical *adjustable tooling clamp* [8] permits a surgical instrument to be grasped at any desired position relative to the fine positioning stage. A three degree-of-freedom *coarse positioning system* is used to position the fine positioner's work volume at any desired location relative to the patient. One advantage of the coarse-fine structure is that it permits relatively large work volumes while limiting the inertia that the surgeon must cope with. The modularity is similarly very useful for experimentation.

Surgical Procedure

It is anticipated that surgery will procede normally up to the point where the surgeon is ready to perform the first planned osteotomy. At this point, the surgeon will affix at least three (and, usually, four or five) LED beacons to each (future) bone fragment. In our present planned implementation, the surgeon will affix each beacon by inserting a standard 1.5 mm surgical "K-wire" into the patient's bones, trimming the end at a convenient length, and then fitting a beacon carrier over the "stump." Depending on the location, the K-wires may be inserted percutaneously or beneath a skin flap but in a position where the beacons can be exposed to the camera. The beacon carriers are constructed so that the LED center is mounted coaxially with the K-wires and always "bottom out" against the end of the K-wires. This means that beacons may be removed and replaced for convenience during surgery, once all the initial locations have been measured.[2]

---

[2]   One alternative implementation might mount several beacons on a single rigid carrier that would be secured to the bones by standard orthopaedic bone screws or similar means. The advantage of this

The Sensing Subsystem provides information needed to register the reality on the operating table with the models from the surgical plan, for tracking motion of bone fragments, surgical instruments, etc. The principal geometric sensor in the present system is an Optotrak 3D digitizer manufactured by Northern Digital. This device uses three CCD linescan cameras to track active LED beacons. This system is fast and accurate, is much less readily confused by stray light than similar lateral-cell based devices, and (unlike electromagnetic field 6D sensors) is unaffected by metal in the operating theatre. The model in our laboratory is capable of producing 1000 3D positions/sec to an accuracy of about ±0.1 mm and of returning up to eight 6D positions with an additional overhead of about 10 ms beyond the 3D sampling time. Beacons are mounted on the patient for bone tracking (see below), on the manipulation aid for positional feedback, and on a pointer used as a designator by the surgeon. Other potential uses might include head tracking for a helmet-mounted display, tracking of additional surgical instruments, location of mouth or head-mounted CT fiducial markers on the patient, etc. A force sensor mounted on the passive manipulation aid (below) will be used to provide additional safety monitoring of manipulation forces. In the future, we anticipate incorporating a a number of additional sensing modalities, including normal computer vision, realtime radiography, and (possibly) redundant kinesthetic sensors in the manipulation aids to provide continuity when visual endpoint sensing is temporarily blocked.[1]

The Surgeon Interface uses a variety of modalities (graphics, synthesized voice, tonal cues, programmable impedence of manipulator joints, etc.) to provide online, realtime "advice" to the surgeon, based on the sensed relationship between the surgical plan and surgical execution. Eventually, we anticipate a quite sophisticated, "intelligent" system that uses its model of the task plan to automatically customize displays, select appropriate sensor tracking modes, and help interpret inputs from the surgeon. In this ultimate system, a helmet-mounted stereographic display might be used to project the surgical advice directly onto the surgeon's visual field, and the surgeon would use voice input to tell the system what he wants. Our initial plans are much more modest, and are intended to provide a framework for later growth. Initially, very simple realtime graphics and auditory cues will be provided for alignment. An online 3D model display will provide somewhat more detailed "snapshots" of bone fragment positions relative to the surgical plan. The surgeon will have a limited ability to modify the sequence interactively through standard menus, sterilizable computer input devices, and the pointing system. For instance, he could designate where, exactly, he has cut (or proposes to cut) an bone. The computer will be able to simulate this cutting action "online" and allow the surgeon to compare it with the cut proposed when the surgery was planned.

Passive Manipulation Aids are provided to assist the surgeon in precisely aligning bone fragments or in aligning his instruments relative to the patient. The defining charactistic of

---

[1] Preliminary experience with redundant joint encoders on one version of the passive manipulator has been mixed. Whether the extra design complexity is justified will necessarily depend on the particular application and implementation details.

representation ([19]) of each connected set of tissue classified as "bone" is constructed by a variation of Baker's "weaving wall" algorithm. In the third step (model simplification), coplanar faces are merged to reduce the size of the model to about 1/3 of the original number of faces. A rendering of a typical reconstruction can be seen in Figure 3. Although the reconstructed models are of very high quality, they are still characterized by very large data structures (300,000 faces for a typical skull). We are presently developing approximation methods that should very significantly reduce the model size for most applications.

The anatomical feature extractor identifies anatomical features from the models. These features include standard morphometric landmarks, ridge curves, and surface patches bounded by ridge curves and geodesics between landmarks [20]. The present implementation is semi-automatic. A technician "seeds" the search by identifying points on or near ridge curves, and the computer then locates and follows the ridge curves. We anticipate implementing a more automatic procedure some time in the future.

The surgical simulator permits a surgeon to interactively specify where he wishes to cut the bones apart and to manipulate the pieces graphically. It also permits him to display the bone fragment motions computed by the plan optimizer (described below) and to modify the plan as he chooses.

The anatomical data base summarizes anatomical feature information for "normal" individuals, and is being constructed in a parallel research activity at NYU.

The surgical plan optimizer uses information from the anatomical data base to compute optimal motions of each bone fragment to most closely approximate the corresponding anatomy of a "normal" individual of the same age, race, sex, and size as the patient. This component, like the anatomical data base, is based on the work of one of the authors (Cutting), Grayson, Bookstein, and McCarthy [21].

The surgical plan produced consists of the model data, the location and sequence of the cuts, the location of key anatomical features to be used in registering the patient to the model data. and the planned optimal motion of each bone fragment.

Surgical System

The surgical system assists the surgeon in carrying out his surgical plan. "Real time" (i.e., predictable latency) computation is provided by an enhanced PC/AT DOS system with a 33 MHz Intel 80386/387 processor. This machine supports all sensor and manipulator interfaces and provides a limited online graphics capability. It is connected via a local area network to an IBM RS6000 workstation which provides online 3D graphics and higher level functions.

Earlier work at NYU led to the development of optimal planning methods based on analysis of morphometric landmarks obtained from radiographs, together with an innovative surgical technique based on inter-occlusal splints to help with surgical execution ([14], [15]). Although helpful, these techniques are far from perfect. Better planning techniques, based on CT-derived 3D models of the bone surfaces, are needed. Even with the inter-occlusal splints, surgical execution is still significantly less precise than planning. Furthermore, the procedure is very time consuming, and the use of splints (necessary for accuracy) forces compromises in the surgical plan. In one recent case, for example, optimal positioning of the cheek bone would have required the soft palate to be stretched (temporarily) more than was possible [15]. There is significant synergy between better planning methods and better means of executing the plans that are developed.

Our joint research is in relatively early stages and is aimed at an in-vitro demonstration on plastic skull models, and does not directly address clinical qualification of in-vivo surgical devices. The goal is an integrated system based on 3D models derived from CT data [16]. These models will be used both for optimal planning and interactive presurgical simulation of the planned procedure and for realtime online "advice" to the surgeon during execution.

Subsequent sections will describe the overall system architecture, the proposed surgical procedure, implementation status, and some early experiments we have performed.

System Architecture

The overall architecture is illustrated in Figure 2 and may be broken down, roughly, into a presurgical modelling and planning subsystem, and a surgical subsystem.

Presurgical Planning System

The presurgical planning system uses models derived from CT images to assist the surgeon in planning precise surgical procedures. It runs on an IBM RS6000 workstation with advanced graphics hardware. The principal components of this system are discussed below.

The medical image database and display system supports archival, retrieval, low-level processing, and 2D display of CT, MRI, and other images. The system we are using (QSH) was developed by one of the authors (Noz) and Maguire at NYU Medical Center, where it is in clinical use [17].

The anatomical model builder transforms CT images into 3D solid models of the patient's anatomy [16] [18]. The process proceeds in three steps. In the first step (segmentation), each voxel in the CT data set is assigned a tissue classification label, based on an adaptive thresholding technique. In the second step (model reconstruction), a winged-edge boundary

Augmentation with a Stationary Robot: Stereotactic Surgery

In cases where only a single motion axis is required during the "in contact" phase of the surgery, the robot may be used essentially as a motorized stereotactic frame (e.g., [1], [2], [10]) A passive tool guide is placed at the desired position and orientation relative to the patient; brakes are applied; and robot power is turned off before any instrument touches the patient. The surgeon provides whatever motive force is needed for the surgical instruments themselves and relies on his own tactile senses for further feedback in performing the operation. This approach ameliorates, but does not entirely eliminate, the safety issues raised by the presence of an actively powered robot in close proximity to the patient and operating room personnel. Furthermore, maintaining accurate positioning is not always easy, since many robots tend to "sag" a bit when they are turned off or to "jump" when brakes are applied. Leaving power turned on and relying on the robot's servocontroller to maintain position introduces further safety exposures. Finally, the approach is limited to cases where a passive guide suffices. The surgeon cannot execute a complex pre-computed trajectory.

Augmentation with an Active Robot: Hip Replacement Surgery

Over the past several years, researchers at IBM and the University of California at Davis developed an image-directed robotic system to augment the performance of human surgeons in precise bone machining procedures in orthopaedic surgery, with cementless total hip replacement surgery as an initial application [5]. This application inherently requires computer controlled motion of the robot's end-effector while it is in contact with the patient. Thus, considerable attention had to be paid to safety checking mechanisms [11]. In-vitro experiments conducted with this system demonstrated an order-of-magnitude improvement in implant fit and placement accuracy, compared to standard manual preparation techniques ([12], [13]). A clinical trial on dogs needing hip replacement operations is underway, and the veterinary surgeon (Dr. H. A. Paul) has founded a startup venture (ISS, Inc.) to develop and market a version for use on humans.

Augmentation with Intelligent Passive Aids: Craniofacial Osteotomies

Researchers at IBM and NYU Medical Center have recently begun research on computer-integrated methods for optimal planning and augmented execution of precise osteotomies to correct cranio-facial malformations. In these procedures, the facial bones are cut into several fragments and relocated to give the patient a more normal facial appearance. Bone grafts are applied, together with metal plates to hold the fragments in place while the patient heals. Typical pre-operative and post-operative results may be seen in Figure 1. Although these results are dramatic, considerable improvement is often still possible.

number of different orientations with the tip at the same spot, c_{tip} relative to the camera and measuring the pose, F_{ptr}, of the pointer rigid body relative to the camera. Then c_{tip} and (more importantly) the displacement p_{tip} relative to F_{ptr} are found by linear regression on the relation

$$F_{ptr,i} \bullet p_{tip} - c_{tip} = 0$$

for multiple pointer poses $F_{ptr,i}$. Typical residual errors when this is done for 9 data points are on the order of 0.2 mm or less. When the calibrated pointer is then used to measure known dimensions on a steel rule placed at an arbitrary spatial pose, the values returned were verified to be consistently within 0.3 mm of the ruler values. An undetermined amount of this variation may be attributed in the difficulty of placing and holding the pointer tip accurately on a flexible ruler in free space. We are presently designing a more pencil-like pointer that should be rather more convenient to use.

Positioning with Active Brakes and Endpoint Sensing

We have experimented with various strategies for exploiting the computer settable brakes on the proximal joints of our initial manipulator. Using only endpoint sensing and a simple strategy of automatically setting brakes when each successive degree-of-freedom was aligned, and then iterating once, it was possible to place the center-of-rotation of the end effector to within about 0.5 cm of a desired target within a couple minutes. Providing a variable pitch auditory signal and/or computer graphics feedback speeded this process up considerably. However, the relatively large inertia and (more importantly) the structural compliance of the initial implementation made it difficult to improve significantly on the positioning without using the fine motion joints. The whole process felt clumsier than it needed to. This lead us to replace these joints with a more rigid cartesian structure (Figure 6), which we have yet to equip with active brakes.

In order to factor out the effects of manipulator structure on active braking, we have also experimented with the much simpler "direct drive" structure shown in Figure 8. Using this structure, we found that a simple strategy of setting the brake to 10 or 20 percent of its full torque when the manipulator is within 1 mm of its target, and setting full torque when it is within its final goal (0.2 mm of target) works quite well. With practice, a person can achieve and hold a desired alignment (in two degrees-of-freedom) in an average time of 40 seconds. The broader band prevents one from overshooting the mark, and then one can hunt around for the "detent" at the target. Performance and consistency is very significantly improved by the addition of a simple tonal indication of which error band the manipulator is in. Indeed, with tonal feedback and a single narrow locking band, we are consistently able to position the manipulator to ±0.15 mm in an average time of 13 seconds.

Positioning of Simulated Bone Fragments

We have also begun experiments to verify our ability to move and place bone fragments accurately using the method outlined in "Surgical Procedure," above. We mount two sets of four beacon carriers onto eight K-wires driven into a piece of simulated bony material, which is mounted on an XYZ stage (Figure 9). We point at simple landmarks on the simulated bone, measure the beacon positions, perform an "osteotomy," reposition one fragment, "graft" it with hot melt glue, and then point at the landmarks again to measure the result. Using this process, we are able to coarse-position the manipulator to within about 0.75 mm in about 1-2 minutes. After the fragment is cut free, we can consistently fine-position it in about 2-3 minutes so that $T_{c,moved}^{-1} \cdot T_{c,fixed}$, as measured by the beacons, is within 0.3 mm and 0.5° of the desired value. We believe that these values can be improved somewhat with experience and further development. Fragment positioning, as measured from the landmarks, is within about 0.8 mm and about 1.5° of the desired value, which is rather better than a surgeon can do unaided. A significant part of this larger error is believed to be due to the particular method chosen for pointing at a few hand-drawn landmarks on somewhat flexible foam "bones," combined with the rather poor ergonomics of our existing pointer design. Beyond this, there is also a certain circularity in using the Optotrak to measure the performance of a system that it is a part of, even though it has been independently calibrated. We are planning a more careful series of experiments using better (and more) landmarks for the initial registration and a coordinate measuring machine to verify accuracy of bone fragment motion after repositioning.[5]

Simulated Surgery on a Plastic Skull

We have also performed the following "bottom line" experiment, in which a number of markers were implanted in a plastic skull model (Figure 10 and Figure 11). The skull was located by pointing to three anatomical landmark points. A fragment was then cut free using an oscillating surgical saw and the skull base was moved to a new position and orientation. The fragment was then returned to its original position relative to the skull and secured with hot melt glue. Finally, the positions of the fragment and skull base were remeasured by pointing. The measurements were repeated three times and averaged. The results were similar to those in the previous experiment. As measured by the beacons, the relative displacement of the fragment relative to the skull base, was 0.4 mm and 0.4°. With further practice, we might reasonably expect to halve these values for in-vitro tests without significant changes to the apparatus. On the other hand, actual surgery may add additional factors limiting the ultimate accuracy that is obtained.

---

[5]   It should be noted that errors in determining the initial positions of beacons relative to the landmarks will be largely cancelled out in tracking *relative* bone fragment motion. For example, if the fragment-to-beacon misregistration is 1 mm and 1° and the fragment is relocated by 5° and 10 mm, the additional positioning error introduced will be about 0.26 mm.

The pointer measurements showed a displacement of 0.7 mm and 0.45°. These values are comparable to values of 0.4 mm and 0.7° obtained from a similar set of measurements taken before the skull was cut apart. A significant part of the measured displacement is doubtless due to the double use of a rather clumsy pointer and only a few landmarks. A more careful experiment is being planned. Nevertheless, these initial results are quite encouraging and (again) are rather better than a surgeon can do with existing techniques.

Future work

The primary goal for the immediate future is to complete integration of the entire system. Once this is accomplished, we anticipate a period of laboratory experimentation to help us determine (a) how accurately we can, in fact, plan and execute these procedures, and (b) what forms of man-machine interface are most useful in this environment.

At the same time, we are beginning to act on some of the lessons already learned. Experience with the setups used to construct the beacon holders showed the value of micrometer-style fine adjustments after the structure has been locked. We are therefore incorporating such adjustments into the manipulation aid. At the same time, we are enlarging the effective radius of the goniometer cradles to provide room for a force sensor and additional free workspace near the patient. This will permit extremely precise alignment of the tool endpoint. Since the surgeon will have little or no direct tactile feedback while using the micrometer screws, we anticipate carefully monitoring and displaying the force sensor values during this phase of the procedure, after the surgeon first tactilely verifies a free range of motion about the target pose.

At a future date, we also anticipate motorizing the micrometer fine adjustments. This will permit us to implement a number of "shared autonomy" strategies for instrument and bone fragment positioning. It will also provide an automatic readjustment capability to compensate for small perturbations in patient positioning, as well as an extremely precise micromachining capability that could be useful in a number of surgical applications. There are, of course, a number of crucial safety issues that must be solved before an active device is used in surgery [11]. Although this research is targeted at *in-vitro* feasibility demonstrations, and does not address qualification for *in-vivo* clinical trials, it is reasonable to expect that the passive version will be used first. In any case, the modularity of both the system architecture and passive manipulation system should permit ready experimentation and a natural evolution of capabilities.

Acknowledgements

We wish to thank Walter Carpini, Chris Hockey, Leon Kehl, and Michelle Melluish, of Northern Digital, Inc. for their friendly and helpful consultations on the Optotrak system and for their responsiveness in adapting their application programming interface to meet our needs. We also wish to thank Rudi Schmidt, of IBM Research Central Scientific Services, for innumerable "rush jobs" and for his assistance in design and fabrication of our beacon holders.

References

[1] Y. S. Kwoh, J. Hou, E. Jonckheere, and S. Hayati, "A Robot with Improved Absolute Positioning Accuracy for CT Guided Stereotactic Surgery," *IEEE Transactions on Biomedical Engineering*, pp. 153-161, February 1988.

[2] P. Cinquin, S. Lavallee, and J. Demongeot, "Computer Assisted Medical Interventions," *Proc. Second Workshop on Medical and Health Care Robots*, pp. 91-101, Newcastle-on-Tyne, Sept. 1989.

[3] S. Lavallee, "A New System for Computer Assisted Neurosurgery," *Proc. 11'th IEEE Engineering in Medicine and Biology Conf.*, pp. 926-927, Seattle, Nov 1989.

[4] Y. Kosugi, E. Watanabe, J. Goto, T. Watanabe, S. Yoshimoto, K. Takakura, and J. Ikebe, "An Articulated Neurosurgical Navigation System Using MRI and CT Images," *IEEE Transactions on Biomedical Engineering*, pp. 147-152, February 1988.

[5] Russell H. Taylor, Howard. A. Paul, Brent D. Mittelstadt, William Hanson, Peter Kazanzides, Joel F. Zuhars, Edward Glassman, Bela L. Musits, William L. Bargar, and William Williamson, "An Image-based Robotic System for Hip Replacement Surgery," *Journal of the Robotics Society of Japan*, pp. 111-116, October 1990.

[6] S. Lavallee, "Computer Assisted Interventionist Imaging: Application to the Vertebral Column Surgery," *Proc. 12'th IEEE Engineering in Medicine and Biology Conf.*, pp. 430-431, Phila., Nov. 1990.

[7] J. A. McEwen, C. R. Bussani, G. F. Auchinleck, and M. J. Breault, "Development and Initial Clinical Evaluation of Pre-Robotic and Robotic Retraction Systems for Surgery," *Proc. Second Workshop on Medical and Health Care Robots*, pp. 91-101, Newcastle-on-Tyne, Sept. 1989.

[8] Elmed Incorporated., Retract-Robot (tm), 1990.

[9] B. L. Davies, R. D. Hibberd, A. Timoney, and J. Wickham, "A Surgeon Robot for Prostatectomies," *Proc. Second Workshop on Medical and Health Care Robots*, pp. 91-101, Newcastle-on-Tyne, Sept. 1989.

[10] J. L. Garbini, R. G. Kaiura, J. A. Sidles, R. V. Larson, and F. A. Matson, "Robotic Instrumentation in Total Knee Arthroplasty," *Proc. 33rd Annual Meeting, Orthopaedic Research Society*, p. 413, San Francisco, January 1987.

[11] Russell H. Taylor, Howard. A. Paul, Peter Kazanzides, Brent D. Mittelstadt, William Hanson. Joel F. Zuhars, William Williamson, Bela L. Musits, Edward Glassman, and William L. Bargar, "Taming the Bull: Safety in a Precise Surgical Robot," *Proc. 1991 Conference on Advanced Robotics*, p. (Accepted), Pisa, Italy, June 1991.

[12] Russell H. Taylor, Howard. A. Paul, Brent D. Mittelstadt, William Hanson, Peter Kazanzides. Joel F. Zuhars, Edward Glassman, Bela L. Musits, William L. Bargar, and William Williamson, "An Image-based Robotic System for Precise Orthopaedic Surgery," *Proc. 12'th IEEE Medicine & Biology Conf.*, Phila., November 1990.

[13] B. D. Mittelstadt, **** *In Preparation* ****, PhD thesis, University of California at Davis. 1990.

[14] Court Cutting, MD, "Applications of Computer Graphics to the Evaluation and Treatment of Major Craniofacial Malformations," in J. Udupa and G. Herman, editor, *3D Imaging in Medicine*, pp. 163-189, Boca Raton: CRC Press, 1991.

[15] Court Cutting, MD, Barry Grayson, DDS, and Hie Chun Kim, "Precision Multi-Segment Bone Positioning Using Computer Aided Methods in Craniofacial Surgical Applications." *Proc. 12'th IEEE Medicine & Biology Conf.*, Phila., November 1990.

[16] A. D. Kalvin, C. B. Cutting, B. Haddad, and M. Noz, "Constructing topologically connected surfaces for the comprehensive analysis of 3-D medical structures," *Proc. SPIE Medical Imaging Conference V*, San Jose, February 1991.

[17] M. E. Noz and G. Q. Maguire, "QSH: A Minimal but Highly Portable Image Display and Processing Toolkit," *Computer Methods and Programs in Biomedicine*, pp. 229-240, Nov. 1988.

[18] A. D. Kalvin, *Segmentation and Surface-based Modelling of Objects in 3D Biomedical Images*, PhD thesis, New York University, New York, 1991.

[19] Bruce Baumgart, *Geometric Modeling for Computer Vision*, PhD thesis, Stanford University. 1974.

[20] F. L. Bookstein and C. B. Cutting, "A proposal for the apprehension of curving craniofacial form in three dimensions," in A. Burdi, K. Dryland-Vig, and K. Ribbens,, editor. *Craniofacial Morphogenesis and Dysmorphogenesis*, pp. 127-140, Ann Arbor: University of Michigan, 1988.

[21] Court Cutting, MD and Barry Grayson, DDS, "Three Dimensional Computer Aided Planning of Craniofacial Surgical Procedures," *Plastic and Reconstructive Surgery*, 1986.

Figures

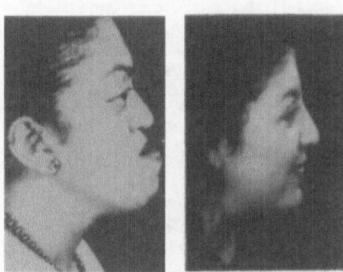

Figure 1. Preoperative and postoperative views. showing results of manual execution.

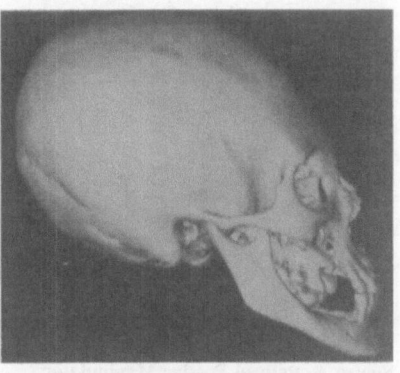

Figure 3. Rendering of typical skull model.

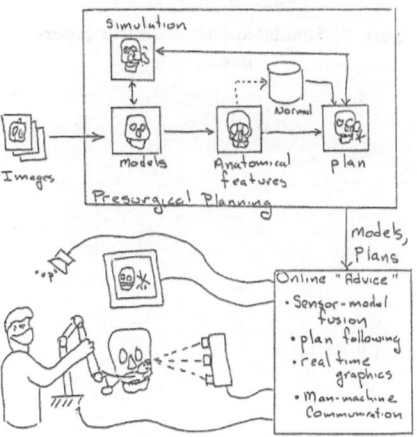

Figure 2. Overall Architecture

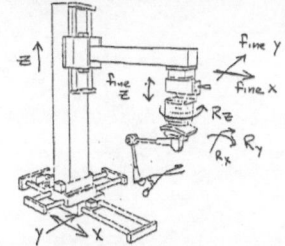

Figure 4. Passive Manipulator Structure

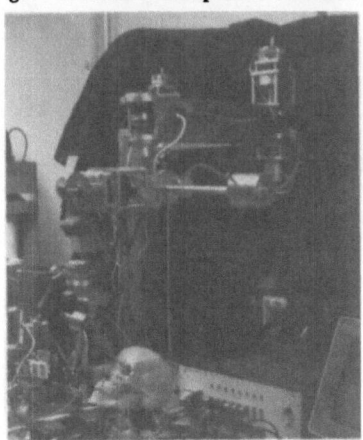

Figure 5. Initial Passive Manipulator

Figure 6. Refined Passive Manipulator

Figure 7. Pointer

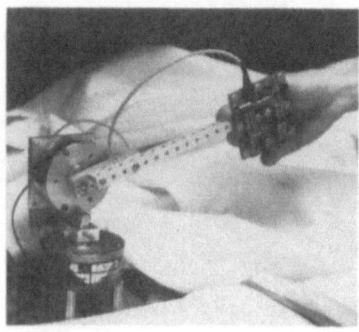

Figure 8. Computer braking experiment

Figure 9. Simulated bone fragment experiment

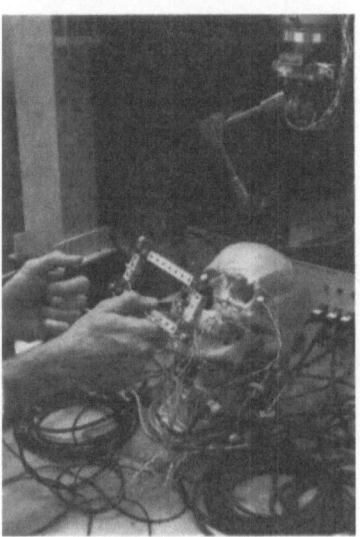

Figure 10. Locating landmarks on skull

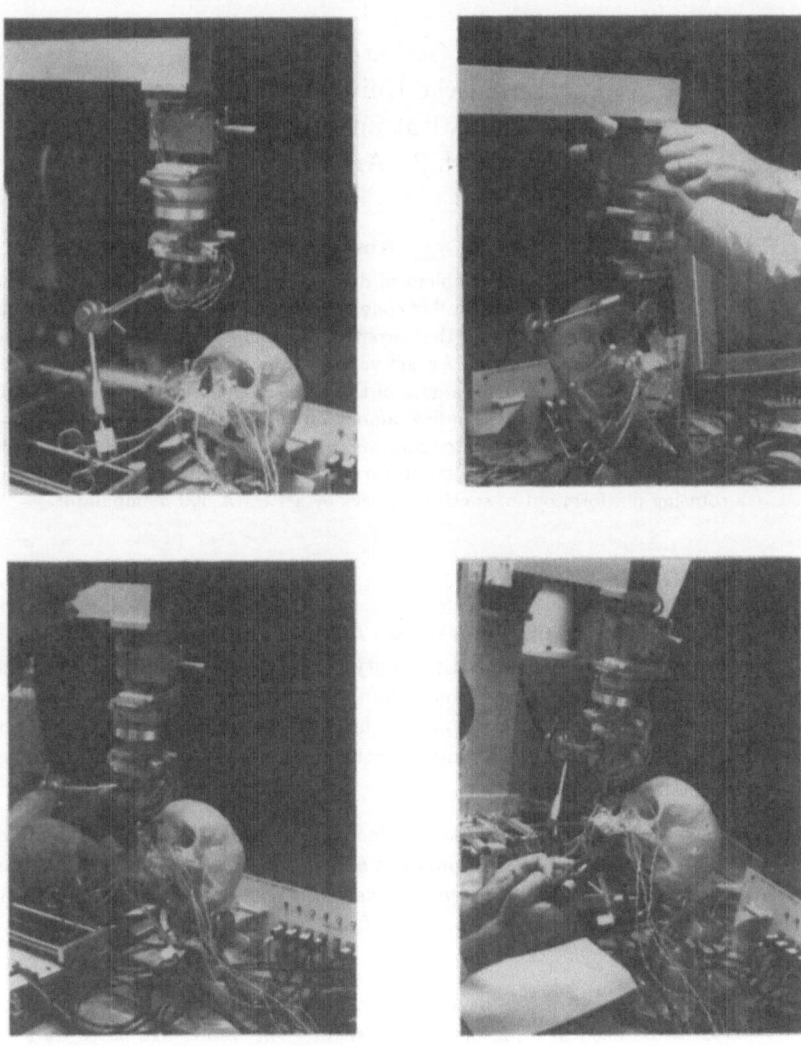

Figure 11. Simulated skull surgery

An Experimental Environment for Task-Level Programming of Robots

Gerhard Schrott

Technische Universität München

Institut für Informatik

D-8000 Munich 2 , Arcisstraße 21 , Germany

Abstract

This paper deals with the problem of decomposing high-level tasks of production control systems down to the level of code generation for manipulators and machines. Furthermore, the execution of the decomposed tasks, their synchronisation and the reaction on errors is treated. An active, object-oriented knowledge base serves as central location for the domain data of the experimental environment and as active component for the information flow among all autonomous units.

To demonstrate the implemented solution, a complex experimental environment has been prepared in our robotics laboratory. A wall with windows was to be built on a rotating platform out of specified blocks by a PUMA 560 manipulator.

1 Introduction

The joint research project SFB 331 (called *Information Processing in Autonomous Mobile Robot Systems*)[1] at the Technical University of Munich works with the aim of supporting autonomous and flexible automation within a shop-floor environment. Our investigations for this project are concentrated on task-level programming tools and distributed real-time knowledge-bases for flexible manufacturing. These areas contain the following sub-projects:

- The development of concepts for decomposing high-level tasks down to the level of code generation for manipulators and machines. This includes the definition of concepts and abstract representations for specifying the synchronization of robots and machines and the development of fast algorithms to solve the findpath problem for robots.

- The development of a knowledge-base shell to construct object-oriented knowledge bases to be used in a factory environment. The knowledge bases do not only contain data (like databases) but also manage procedural knowledge and, furthermore, provide the active distribution of knowledge to all processes connected.

---

[1]This work was partially supported by the Sonderforschungsbereich 331 (Informationsverarbeitung in autonomen, mobilen Handhabungssystemen)

- Hardware and software architectures for the integration of robot controllers, programmable controllers for DNC machine tools and sensor equipment.

The test bed for these investigations was constructed in our robotics laboratory using a PUMA 560, its sensor equipment and the appropriate distributed computer system. The aim was to demonstrate the capacity of the developed concepts and to show their real use in a factory environment. The model was defined so close to the hierarchical structure of a flexible manufacturing system that the results can easily be transferred to such systems.

Figure 1: The Experimental Environment

2   The Experimental Environment

The experiment as constructed in our robotics laboratory is shown in figure 1. A freely defined wall with "windows" is to be built on a rotating platform by the PUMA 560 manipulator. All the task needs to construct a wall is the shape of the wall (given by single-unit blocks) and the numbers of the different types of blocks available for that

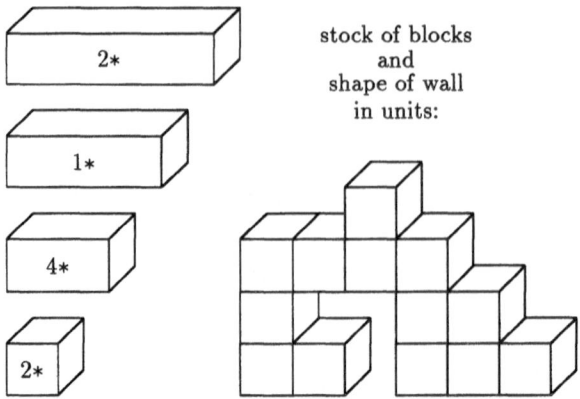

Figure 2: Specification of Building Task

task. Figure 2 shows the required inputs. From that definition, all activities to build the wall will be planned and started autonomously using the stored knowledge. If errors or problems occur while the task is being performed, the system recovers autonomously as far as possible.

Our model was implemented on a heterogeneous multi-computer system architecture. Figure 3 gives an overview of the processors used and how they are linked together. Mainly, the system consists of three MicroVAX stations and one rtVAX, all connected by DECnet.

The main computer is a VAXstation 3100 running under the VMS operating system. It contains the graphical user interface and all processes of the application, i.e. the wall planner, the task-transforming layers and the path finder.

The VMS MicroVAX is the host system for the rtVAX and also for the PUMA's controlling unit, where it can download executable files. Furthermore, this MicroVAX contains the modular robot programming environment MRL [5], which allows the programmer to develop very large VAL II applications in a module-oriented high-level language with structured data types.

The global real-time knowledge base [7] is situated on a DECStation 3100 running the ULTRIX operating system. The current version of the knowledge base, however, is also available on a VMS-VAXstation. The graphics screen of the knowledge base station is used to show on-line displays of the current status of the wall, and to monitor the exchange of orders, tasks and receipts between the different planning layers.

The rtVAX controls the processors for the electrical grippers and the force-torque sensor and it supervises the rotating platform. Additionally, the rtVAX provides the Cartesian real-time interface (CRI) to the PUMA 560's controlling unit. The CRI makes it possible to send single robot commands (VAL II) to the PUMA 560, and get receipts after their execution. Additionally on-line trajectory modifications for sensor feedback are implemented.

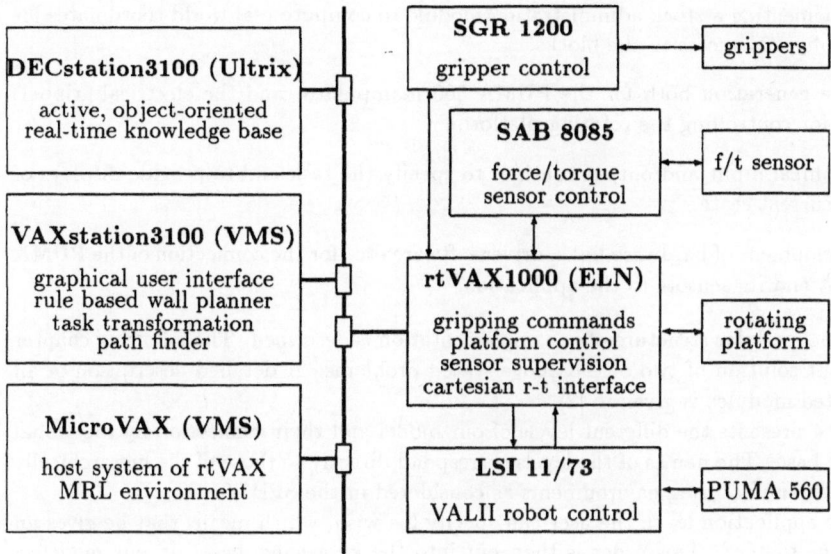

Figure 3: The Hardware Architecture

3   Task-Level Programming

Task-level programming is situated on a level higher than programming in a robot programming language and is usually synonymous to intelligent robot systems.

On this level it is not programmed by a sequence of robot move instructions how a task is to be solved but by the specification of the task itself, e.g. in our experiment *build specified wall*. This specification is analyzed and transformed into a set of primitive actions directly executable by the manipulators and machine tools; transposed to our modell, the PUMA 560 is controlled via the Cartesian Real-time Interface (CRI), whereas, the rotating disk, the electrical gripper and the force/torque sensor are supervised by parallel processes in the rtVAX.

To realize task-level programming for our intended application, the following problems had to be solved:

- Modelling the domain in a knowledge base.

- Implementing a planner to build the wall by means of the available blocks.

- Planning the synchronization between the manipulator and the turning platform, because a block can only be put on the platform when it is halted.

- Planning collision-free paths between stock and platform, and planning the regrasping and re-orientating of those blocks which have to be positioned vertically.

- Error detection on machine level and recovery mechanisms

- Implementing a stock administration module to compute real-world coordinates for the physical access to the blocks

- Code generation both for the PUMA 560 manipulator and the electrical gripper, and for controlling the rotating platform.

- Graphical input and output routines to specify the task and to provide displays of the current state.

- Development of hardware links, driver software etc. for the connection of the PUMA robot and its sensors to the application.

First the modular structure of our implementation is described. The following chapter explains our solution of two of the above stated problems. A detailed description of all implemented modules is given in [2].

Figure 4 presents the different levels of our model and their connection by the global knowledge base. The names of the levels correspond directly to those of the hierarchically structured manufacturing environments as considered in the SFB 331.

On the application level, the user can specify his wish, which means that he gives an order to the system. This order is then put into the knowledge base. In our model, a comfortable graphics environment allows the user to define the intended shape of the wall by just clicking into a grid window with the mouse device. An additional menu requires the number of each sort of block to specify the stock for the current order.

The global production control gets the order and starts constructing the partlist. The output of the rule-based planner, implemented in OPS5 is a set of quintuples $< a, s, x, z, d >$ or a message that the wall cannot be built with the given stock. In the first case, each quintuple determines that the action a "put" is to be done with a block of type s (i.e. of length 1,2,3 or 4) at the coordinates x, z with the direction d (horizontal or vertical). x and z are not real-world coordinates but only relative positions in the x, z shape of the specified wall. The set of quintuples is transformed to a sequence which defines the order of performing the pick-and-place commands by the robot.

Having got this sequence of quintuples, the production control enters them into the knowledge base one by one as objects of the class *task*. The next *task* is only put into the knowledge base if a positive receipt of the task coordination level has been received for the previous one. In case of a negative acknowledgement, this means that the block was not positioned correctly, the production control tries to recover and continue building the wall completely.

Our model deals with errors of the kind "lost block" and "grip error". Both mean that a pick-and-place task could not be performed correctly. If the local production control of our model gets such a receipt, it starts retrying the task with another block. If this is not possible, it tries to replan the rest of the specified wall with the blocks still available at that moment. Replanning can also result in actions to take blocks back from the partly finished wall and to reuse it in another ordering. If replanning is successful, the local production control continues entering tasks into the knowledge base.

Each time the task coordination level receives a new task, it has to decompose it into primitive actions. Such primitives are, for example, gripping a block, planning a

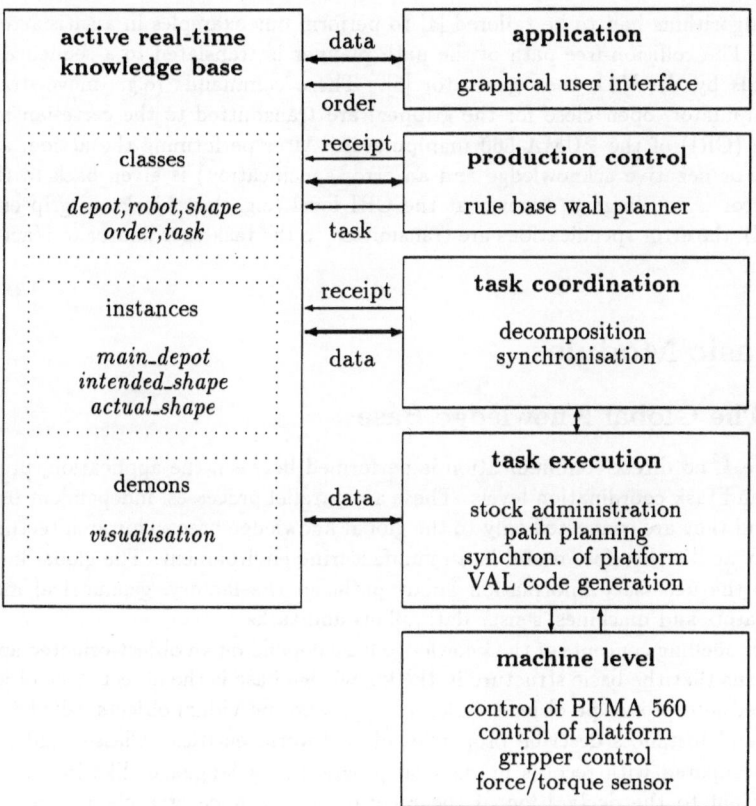

Figure 4: The Levels of Task Decomposition

path, transporting the block, stopping the platform, and others. The primitives are then used by the different modules of the task execution level. It is the task coordinator's responsibility to synchronize these modules and to call them at the right time [3]. When all primitives are completely executed, the task coordinator puts an "ok" receipt into the knowledge base, which then will automatically be distributed to the production control.

The modules of the task execution level have to perform the following activities: a synchronizer is responsible for sending stop-and-go signals to the rotating platform. The task transformer must ensure that blocks are put onto the stationary platform only. It also receives the actual angle of the platform and later provides it to the path planner. A stock administrator computes the real-world coordinates of blocks in the stock. Up to that level, the tasks only specify the use of a block of a certain kind. The stock administrator then decides which specific block will be used for the current task.

The path planner computes a collision-free path for transporting a block from its stock position to the platform, and for inserting the block into the wall. As the complexity of the findpath problem is polynomial in the number of degrees of freedom of a manipulator, special algorithms had to be tailored [4] to perform our examples in a satisfactory limit of time. The collision-free path of the path planner is translated to a sequence of VAL commands by the VAL code generator [5]. These commands (e.g. move-straight for the manipulator, open/close for the gripper) are transmitted to the cartesian real-time interface (CRI) of the PUMA 560 manipulator. After performing the action, a receipt (positive or negative acknowledge and an error specification) is given back to the CRI. If the error is too large to recover at the CRI level (e.g. lost blocks or grip errors are detected), the error specifications are transmitted to the task coordinator to react on that situation.

4 Basic Modules

4.1 The Global Knowledge Base

In Figure 4, no direct communication is performed between the application, production control and task coordination levels. These are parallel processes, independent from each other, and they are connected only to the global knowledge base. This architecture yields flexibility and is therefore useful in a manufacturing environment. The global knowledge provides the necessary information about paths in the factory, geometrical models of manipulators and machines, sensor data, plans and tasks.

The modelling concepts of the knowledge base depend on an object-oriented approach. This means that the basic structure in the knowledge base is the object. The objects can be divided into prototype objects called *classes* and individual objects called *instances*. Classes and instances describe properties of real-world entities. Classes and instances can be compared with records in classical programming languages like Pascal. A class is equivalent to the declaration of the record. An instance of a class correlates with the corresponding set of values for the items of the record of the class. The items are synonymous to the so-called *attributes* of a knowledge base class. An attribute stores information about one property of a real-world entity. Attributes are classified into *class, instance* and *administration attributes*.

Class attributes describe properties the values of which are equal in all instances of this class. Instance attributes express properties which may be different in every instance of the class. Administration attributes contain information on which instances are an element of the class and on the inheritance of attributes. A class inherits all the attributes of its superclasses, which means that the attributes defined in the superclasses are added to the one defined in the considered class. In the same way, the subclasses of a class inherit all attributes of that class.

Furthermore, a demon concept is integrated, i.e. attached procedures which are activated automatically by data-accessing events [6]. It is used to store procedural knowledge and also to send messages actively to connected processes. Therefore, the knowledge base can distribute important data immediately after they have been inserted. The connected processes need no longer poll interesting data (e.g. status flags, sensor information) from the knowledge base but get it automatically as soon as it is available. This mechanism was shown to be very useful in our model for exchanging tasks and receipts between the different control levels. In this sense, the knowledge base works similar to an active black-board, onto which all agents can put their messages but need not worry about how to address them to the right destination.

Our concrete knowledge base contains a model of the domain of the whole experimental application, i.e. there are classes (and instances) to describe orders, tasks, actions, blocks, the stock, a robot, positions and so on. These data can be accessed on all software levels of our implementation (see figure 4). As the knowledge base works as an autonomous process, an inter-process communication, based on DECnet and shared memory, provides the interface to distributed processes on different computers.

4.2 Knowledge-Based Task Planning

A knowledge-based task planner can be defined as a function

$$f : K^* \times T \to P^*$$

where K is the set of states of the knowledge base, T the set of tasks and P the set of primitive actions [3]. K^* expresses that the changes in the knowledge base are immediately combined into the act of planning and keep the planning running. The function f is (for a specific task) given by a set of rules which we call behavior pattern, as it describes the behavior of an autonomous unit when it solves a given task. The concept offers the advantage that the behavior patterns can be executed by a standard interpreter - we use OPS5. The single behavior rules have a uniform structure

$$\{<s> \text{ (state ^phase p ^current c)}\}$$
$$(\text{condition}_1)$$
$$\cdots$$
$$(\text{condition}_n)$$
$$\to$$
$$(\text{action}_1)$$
$$\cdots$$
$$(\text{action}_m)$$

They consist of a condition and an action part. If all conditions of the condition part are satisfied, the rule is ready to fire; that means, the actions of the action part are executed. The first condition is especially significant because it introduces a state concept into the behavior patterns.

The task planner is divided into three parts. These are

- the Task Interpreter (TI)

- the Interpreter of the Behavior Patterns (IBP)

- the primitive actions of the application

The Task Interpreter (TI) establishes the task-oriented interface for the application level. The actual interpretation of the tasks is done by looking for the appropriate behavior patterns in the knowledge base. These behavior patterns are passed on to the IBP, which performs the actual transformation of the task into a sequence of primitive actions. Furthermore, the TI handles the finish and error messages of the IBP. The implementation of the TI as a separate process has the advantage that the TI can always receive new tasks even if planning is active. By this means it is possible for the TI to do tasks of high priority (for example abortion of a running task planning) immediately after they appear.

The Interpreter of the Behavior Patterns (IBP) is the essential part of the task planner. It provides the asynchronous fact entry in the OPS5 working memory, a watch mechanism for knowledge base elements and a finite automaton-like model with states and transitions (called state concept) for behavior patterns. Furthermore, it is easily possible to set up timeouts to restrict the period of occurring wait states. The domain-dependent part provides the error and exception handling which is given by fixed behavior patterns built into the IBP.

The state concept takes into account that the rules of the behavior patterns fire sequentially in most cases in a relatively strict order. However, the primitive actions as results of a fired rule are carried out concurrently. Mostly, calling a primitive action means in fact sending a message to a separate process where it actually is performed. The primitive actions then run parallel to the IBP process on the same processor (e.g. the path-finding algorithm for the robot effector [4]) or on a separate processor (e.g. the control of the robot). When a primitive action is finished, this is indicated by writing an entry into the knowledge base. Nevertheless, there is also the possibility of waiting for its end immediately after a call for a primitive action.

The following example of a behavior pattern is taken from our experiment:

```
Begin_BP:
; behavior pattern to get a block from a depot
; parameters:
; p1: type = length of block (B1,B2,B3,B4)
; p2: source (MAIN_DEPOT,FITUP_DEPOT)
; p3: destination (MAIN_DEPOT,FITUP_DEPOT,[WALL,<x>,<z>,d>])

{ <s> (state  ^phase GET  ^current START) }
  (parameter  ^p1  <object>)
  (parameter  ^p2  <source>)
  (parameter  ^p3  <dest>)
-->
  remove_fact (<s>)
  bind (<ok>, true)
  <ok> := find_path (<ok>, <object>, <source>, <dest>)
  <ok> := generate_commands (<ok>)
  send_commands_to_VAL (<ok>, GET)
  make_fact (state  ^phase PUT  ^current START)

END_BP.
```

The above behavior rule can fire only if the IBP is actually in a phase GET at the START point. It will fire if the three parameter facts are also present. The state fact is then removed and functions of the task execution modules are called:

- **find_path** starts the path planner to compute an obstacle-free path (i.e. a sequence of intermediate points) from the source to the destination point. The path is stored in the global knowledge base.

- **generate_commands** calls the code generator, which transforms the sequence of intermediate points (read from the knowledge base) to a sequence of MOVES-like commands for the PUMA 560 and inserts them into the knowledge base.

- **send_commands_to_VAL** causes the VAL driver to pass the commands to the manipulator. This function does not wait until all commands are transmitted to VAL. It is finished after the VAL driver has accepted its job. The VAL driver, however, asynchronously sends receipts for every command to the IBP. If one of them is negative, the error-handling rules of the IBP will react.

The computed results of the different functions are not inserted as facts into the rule interpreter but put into the knowledge base. There they can be accessed by every interested function. The last actions of the behavior pattern above determines the new state of the IBP.

5 Conclusion

Task-level programming of autonomous robot systems requires solutions in different areas of software engineering. Methods of artificial intelligence are integrated for rule based transformation of tasks and for knowledge representation. Numerical problems have to be solved for motion planning and sensor data processing. Our experiment has prooved our ideas and solutions to these complex problems.

The future work will concentrate on distributed real-time knowledge bases and will improve the task transformation facilities. Sensor feedback loops including image processing , complex motion and grip planning will be integrated to achieve a sophisticated task-level programming robot system.

Finally, I would like to thank my colleagues Klaus Fischer and Johann Schweiger for their contributions to this paper.

References

[1] Bocionek S., Fischer K.: *Task-Oriented Programming with Cooperating Rule-Based Modules*. To appear in: Int. Journal for Engineering Applications of Art. Intelligence, Pineridge Press Periodicals, 1989

[2] Bocionek S.: *Task-Level Programming of Manipulators: A Case Study*. TU Munich, Institut für Informatik, Report TUM I9001, January 1990, 52 pages.

[3] Fischer K.: *Knowledge-Based Task Planning for Autonomous Robot Systems*. Proc. of the Int. Conf. on Intelligent Autonomous Systems, Amsterdam (Netherlands), December 1989 , p. 761-771.

[4] Glavina B.: *Solving Findpath by Combination of Goal- Directed and Randomized Search.* In: IEEE Int. Conf. on Robotics and Automation, Cincinatti (Ohio), 1990 , p. 1718-1723.

[5] Hagg E., Fischer K.: *Off-line Programming Environment for Robotic Applications.* Proc. of the INCOM '89, Madrid, Sept. 1989

[6] Meyfarth R.: *Demon Concepts in CIM Knowledge Bases.* In: IEEE Int. Conf. on Robotics and Automation, Cincinatti (Ohio) 1990, p. 902-907

[7] Schweiger J., Siegert H.-J.: *An objectoriented Knowledge Base for CIM-Environments.* Proc. of the International Workshop 'Information Processing in Autonomous Mobile Robots', München (Germany), Springer, March 1991 , p. 231-245.

An Architecture for Task Interpretation and Execution Control for Intervention Robots: Preliminary Experiments

Raja Chatila   Rachid Alami   Bernard Degallaix
Victor Pérébaskine
Paul Gaborit   Philippe Moutarlier
Laboratoire d'Automatique et d'Analyse des Systèmes (LAAS-CNRS)
7, Ave. du Colonel Roche 31077 Toulouse cedex - France

Abstract

This paper deals with the software system design of robots that can replace or assist humans in intervention tasks in hazardous environments. We consider missions that include navigation from a site to another, execution of specific tasks, and transmission of data. Communication difficulties with a remote ground station (delays and bandwidth), as well as time constraints on the tasks make it necessary to endow such robots with a high level of autonomy in the execution of a well specified mission. However, it is today unrealistic to build completely autonomous (and useful) robots. This paper presents an approach to answer these requirements; it describes an on-board robot system architecture that enables the robot to interpret its mission according to the context, and to control and adapt in real time the execution of its actions.

1 Introduction

Intervention robots are machines that have to perform tasks in difficult environments which are often remote or of difficult or dangerous access. In addition, according to the application context, specific constraints, related to the interaction with the human operator, to the task itself, and to the environment have to be taken into account:

- Communication constraints: time delays (for example one way light travel time between Mars and Earth is between 5 and 20 minutes); impossibility or loss of communication (a rover on Mars may communicate with Earth only at certain periods of the day because of visibility); low bandwidth.

- Task constraints: duration (e.g., motion only during daytime in planet exploration); non-repetitiveness: a human operator should be able to assign a variety of tasks to the robot, and to modify them according to returned information.

- Environment constraints: the environment is <u>partially</u> known and with <u>little</u> accuracy before the intervention, and may remain so for the human operator.

The question addressed in this paper is: "What robot architecture is adapted to cope with these requirements and constraints?".

These constraints forbid, in many applications, the use of classical teleoperation to control intervention robots, as well as telerobotics-like [15] approaches, wherein the human operator is still tightly in the control loop even if the robot has the capacity to execute some tasks automatically [9].

On the other hand, research on autonomous mobile robots is not (yet) able to produce an intelligent agent capable of accomplishing completely by itself a general objective such as "Collect *interesting* rock samples", non withstanding the fact that the specification of this objective itself cannot be really done beforehand (what is an *interesting* rock?).

In another stream of thinking, *collective intelligence* is considered to be more easy to accomplish by building small robots which have complete autonomy but limited behavior, and wherein *"the basic components that make up the system are designed in such a way that the desired functionality emerges as a global side-effect of their massively parallel, simple behavior"* [7]. Of course, the problem is now "how can global functionality emerge from local behavior"[5]. The shortcoming of such a system, even if successfull as defined, is that it is *not* programmable, i.e., it is necessary to fully specify the task at robot design stage.

In the *"task level"* teleprogramming approach presented in this paper, the robot is provided with *executable* missions that will be carried out autonomously. These missions (plans) are refined into actions according to specified execution modalities and to the actual execution context [13, 2]. The robot control structure includes the systems necessary for task interpretation and autonomous execution. The system is aimed to be generic and applicable to several domain instances of intervention robots: planetary rovers [6], underwater vehicles [11], disaster reaction [13], etc.

Section 2 presents the functional architecture of the overall system and and section 3 presents the operation of the on-board supervision and task refinement and execution control system.

In order to illustrate the approach we will consider a Mars Rover as an example of intervention robots. This case study is motivated by the french national project VAP[1][6].

Once safely on Mars, the rover is expected to be operational for about one year and to carry out eight hours per day activities (during daylight). Missions will have three main aspects:

- experiments within a worksite, such as sample collection, drilling, or various measurements,

- autonomous navigation between two worksites, within time and space (route) constraints given with the mission. Navigation includes natural landmark recognition, inertial localization, environment modelling for motion planning and execution on rough terrain ...

- worksite modelling and data transmission to Earth.

---

[1]VAP: Autonomous Planetary Vehicle, a project of the french national space agency CNES.

2 A Functional Architecture for Task-Level Teleprogramming and Action Execution Control

The proposed functional architecture (Figure 1) involves two distinct systems: the *Operator Station* and the on-board *Robot Control System* (RCS).

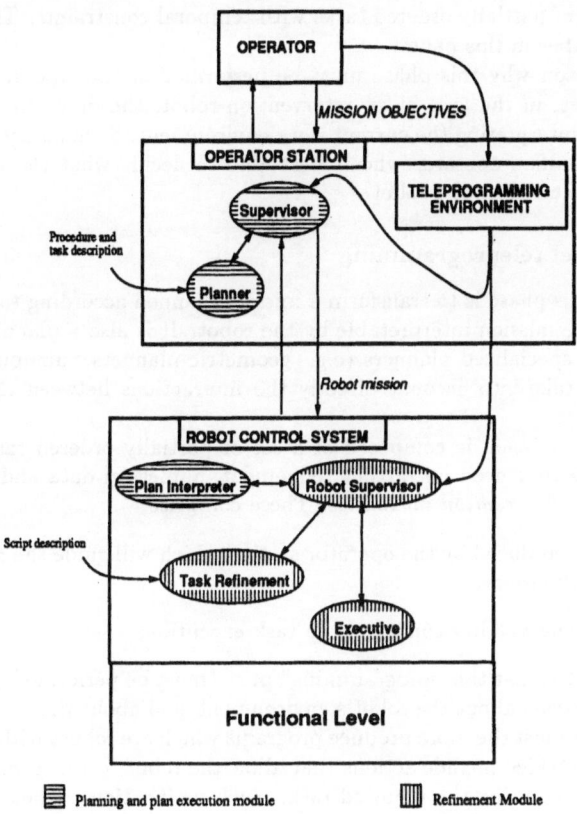

Figure 1: Global Functional Architecture

2.1 The Operator Station

The Operator Station includes the necessary functions to allow a human operator to build an *executable mission*, i.e., a mission that can be interpreted by the RCS, and to supervise its execution by the robot.

Its main components are a Mission Planner which determines the tasks to be achieved and their ordering and a Task-level Teleprogramming environment which provides information that will allow the Robot Control System to refine and execute the tasks.

2.1.1 Mission Planning

The mission planning phase is based on general action planning techniques, including temporal reasoning, since it is necessary to take into account time constraints in robots that have to act in the real world.

We have developed a temporal planning system called *IxTeT* (Indexed Time Table) [12] which can reason on symbolic and numeric temporal relations between time instants. It produces a set of partially ordered tasks with temporal constraints. This issue will not be developed further in this paper.

The main reason why this phase must be performed at the Operator Station stems from the fact that, in the case of an intervention robot, the determination of the goal itself is based on interpreting the current work environment. Such an activity is naturally performed by a human operator who is best able to decide what the robot should do, given the data acquired by the robot.

2.1.2 Task-level teleprogramming

The purpose of this phase is to transform a mission planned according to some objectives into an executable mission interpretable by the robot. It is also a planning phase. However, it relies on specialized planners (e.g. geometric planners, manipulation planners) that are able to take into account directly the interactions between the robot and its environment.

An executable mission is composed of a set of partially ordered tasks (the *mission planq*, procedures that need no further refinement, numerical data and task-dependant data structures, and *execution modalities*. These comprise:

- information produced by the operator station which will guide the refinement of the mission by the robot;

- constraints and validity conditions on task execution;

One key aspect is that this "programming" phase must be performed using partial and inaccurate information about the robot's environment, and about the consequences of the robot's actions. It must therefore produce programs which are robust with respect to these uncertainties, and which include actions that allow the robot to verify without ambiguity that it is indeed executing the required task. Such verification cannot be performed in terms of absolute values (which cannot be precisely known at programming time), but in terms of relations between the robot and identifiable features in its environment.

This means that the resulting program must rely on *sensor-based actions* (e.g. feature tracking) to allow the robot to permanently adapt its behavior and take appropriate actions when it detects any discrepancy between the planned state and the actual state of the world. Note that these requirements may drastically influence the way in which a goal is refined into tasks. For example, the need for robust execution of a motion between two points may entail a choice of trajectory which is not necessarily optimal in length or duration, but which allows the robot to track some feature.

Finally, depending upon the difficulty of the task and the nature of the environment, the teleprogramming phase can sometimes rely on the capacity of the RCS to successfully interpret high-level commands. But in some other situations, the task must be deeply detailed by the operator station, which normally requires previous acquisition and transmission of large amounts of data.

In the case of the planetary rover, the tele-programming of navigation tasks will be limited to the determination of itineraries and natural landmarks based on low-resolution images obtained from the orbiter. Further task refinement is performed autonomously by the RCS. Other tasks, like sample collection, must be programmed at the ground station in more details using "in situ" information acquired by the rover.

2.2 The Robot Control System

Once the executable mission is prepared at the Operator Station, it is sent to the on-board system called the *Robot Control System*.

Because the robot is working in a remote and a priori little known environment, significant differences between the planned (expected) state of the world and its true state are bound to appear. Therefore the robot is *fully autonomous* at "task level". It receives tasks that it transforms into sequences of actions using its own interpretation and planning capacities, and executes these actions while being reactive to asynchronous events and environment conditions. However, it may decide (by itself according to the context, or because it was programmed explicitly to do so) to contact the Operator Station when necessary (and possible) in order to solve some difficulties. Conversely, the Operator Station may also supervise the execution if possible, and if necessary interfere with action execution.

Thus, the robot is endowed with the capacities of (i) interpreting the mission according to the actual situation, and (ii) autonomous execution of its tasks. The on-board **Robot Control System** (figure 1) is composed of two levels: a supervision and planning level, and a functional level.

2.2.1 The Supervisory Level

The supervision level comprises:

- The **Robot Supervisor**, which manages the overall robot system and interacts with the operator station. The supervisor makes use of two resources:

  - the **Plan Interpreter and monitor** that is in charge of requesting (to the supervisor) the execution of the different tasks in the plan, of verifying that their execution satisfies the different constraints included in the plan, and of maintaining the necessary world state description.

  - The **Task Refinement System** which refines tasks into actions, taking into account the execution modalities specified with the plan, and the actual situation of the robot.

- The **Executive**, which is in charge of managing and controlling the robot's fucntional modules in order to execute actions.

2.2.2 The Functional Level

In order to control the robot's actions, we think that it is useful to define its basic functions in a systematic and formal way so that they can be controlled according to their specific features, while being easy to combine, modify or redesign [2]. We therefore defined in [14] and extended in [10] *robot modules* and introduced *primitive function types* within them [1] [8].

Essentially, a module embeds primitive robot functions which share common data or resources. An internal control task called the "module manager" is responsible for receiving requests to perform these functions from the robot controller, and for otherwise managing the module. Each function being well defined, its activation or termination must respect certain conditions that the module manager verifies.

A module may read data exported by other modules, and may output its own processing results to exported data structures (EDS). At a given time, a module can be executing several functions. All of the functions of each module are pre-defined at the system design stage.

The robot primitive functions fall into four different *types* according to their functioning mode.

Servers. Functions executed upon request. The processing result is exported in a pre-defined data structure to be accessed by the requesting module, or sent directly (message).

Filters. Functions started upon request (or systematically) and then run continuously at a given frequency until stopped. Their results are output in a data structure that is updated at the mentioned frequency.

Servo-processes. These functions implement a closed-loop between a perception function (related to processing sensory data) and an "action" function (related to robot effectors).

Monitors. These functions are used to detect a given situation or event. They verify, at a given frequency or at reception of an asynchronous signal, the occurrence of an event and react by generating themselves a signal (sent to the supervisory level or another pre-specified destination).

Both filters and servo-processes may be stopped, but cannot stop by themselves, unless they detect an abnormal condition; in this case, an appropriate signal is sent to the supervisory level of robot control system.

3 Plan Interpretation, Task Refinement and Execution Control

3.1 Introduction

After the planning and programming phase, an *executable mission* is produced composed of tasks, and possibly for some parts, of more low-level actions produced by the programming phase. The mission plan has to be tranformed by the robot control system into

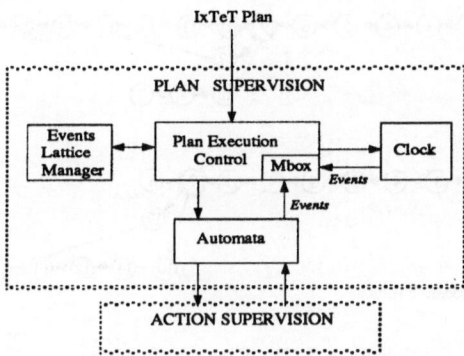

Figure 2: The Plan Interpreter

executable primitive actions, selected by taking into account the state of the environment and of the robot as the mission is executed. Mission execution will be globally managed by the robot supervisor, that makes use of the plan interpreter for scheduling the tasks, of the task "refiner" for precising their execution, of the robot executive that manages the robot functional modules for actually executing them.

3.2 Mission Interpretation

Plans, as produced by IxTeT, embed a set of partially ordered tasks, together with temporal constraints such as minimal and maximal expected durations, and synchronisation with expected external events or absolute dates.

The plan interpreter is in charge of monitoring the execution of the tasks involved in the plan. It has therefore to verify that the tasks produce indeed the expected effects, and must react in case of discrepancy. Some of the tasks need to be further refined according to the actual situation. They will be sent by the supervisor to the task refinement system.

3.2.1 Plan Execution Monitoring

Instants in the time lattice correspond to different event types: beginning or end of task execution, intermediate events produced by tasks during their execution, expected external events (which occur independently from robot actions). Besides, numerical bounds for dates and durations may be attached to some time-points or intervals.

The plan execution control process interacts with a clock by requiring the sending of messages at given absolute dates and with the robot supervisor by requiring task execution or cancellation (fig 2).

Messages received from the clock authorize the system to state the occurence of dated expected events and to monitor the task minimal and maximal expected durations.

Messages received from the robot supervisor are 'filtered' by the associated automata and transformed into events which correspond either to instants planned in the time lattice (in case of nominal execution) or to unexpected instants otherwise.

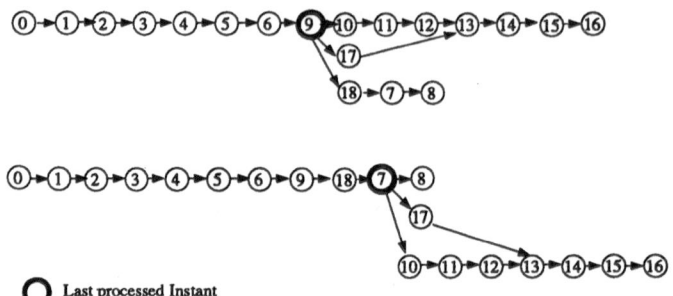

Figure 3: Plan interpretation

The plan interpreter starts from a given instant in the time lattice and 'executes' the time lattice by performing the following actions:

- At any moment, it considers only the time instants whose predecessors have been processed and whose planned occurence date is compatible with the current time.

- It requires the execution of a new task when it reaches the instant which corresponds to its beginning.

- While processing incoming events, the interpreter verifies that they correspond to planned instants and that they satisfy the planned ordering and numerical time constraints. If it is the case, the absolute date corresponding to the occurence of the events is considered, inducing a progressive 'linearisation' of the time lattice (see figure 3).

- In the current implementation, the only 'standard' reaction performed by the plan interpreter in case of non-nominal situations is to stop the current tasks, and update the world state according to events incoming from task execution until all tasks are stopped. This is performed by inserting new time instants in the time lattice corresponding to the transitions as defined by the automata.

Figure 3 shows the plan produced at two different stages of interpretation. The first time lattice represents a situation wherein the rover is navigating from *site-2* to *site-3* (indeed, instant 9 is in the past while instant 10 is still in the future). It has previously acquired a panoramic view of *site-2* (between instants 5 and 6), however it has to wait for instant 18 (visibility with the orbiter) in order to send the data (between instants 7 and 8).

In the second lattice, instant 18 has occured; the rover is performing two tasks in parallel (navigating and sending data to the orbiter).

3.2.2 Using Automata To Monitor Task Execution

In order to allow the plan interpreter to monitor task execution and to act on them while they are executed, each task is modelled by a finite state automaton (FSA).

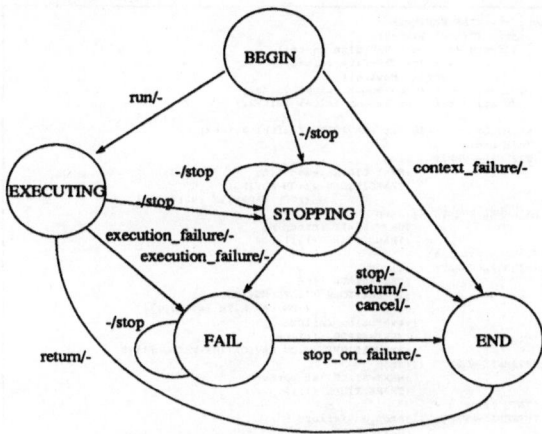

Figure 4: A Generic Automaton for task execution. Events are represented by a pair internal/external

The FSA associated to a task not only models the nominal task execution (as defined in the task description manipulated by the planner) but also non-nominal situations (exceptions), as well as the different actions which should be taken then by the plan interpreter.

In the finite state automata we use, state transitions can be caused by 'internal events' (i.e. events generated by the task execution process and transmitted to the plan interpreter) or by 'external events' (i.e. events due to an action of the plan interpreter on the task while it is executed).

Figure 4 shows a generic automaton for action execution.

Typical internal events are: RUN (beginning of execution), RETURN (nominal end of execution), STOP_ON_FAILURE (end due to a failure), STOPPED (end of execution caused by an external STOP request).

A Typical external event is STOP (corresponding to a stop request issued by the plan interpreter). Other external events are related to the task and force a state transition in the execution.

For each task, an automaton class is provided. An example is given in figure 5. Note that each automaton state change produces transitions in the world state as it is maintained by IxTeT. The example in figure 5 shows a decription of automaton associated to the task *TASK-NAVIGATE*.

Whenever a task execution is requested by the plan interpreter, an automaton of the associated class is instanciated and initialized. It will then represent the execution of the task as it is viewed by the plan interpreter.

```
(ixtet:Task TASK-NAVIGATE
     :args (?site1 ?site2)
     :effects ((:is-met VAP-Site ?site1)
               (:meets VAP-Site ?site2)
               (:equal Moving))
     :duration-min (min-length ?site1 ?site2)
     :duration-max (max-length ?site1 ?site2))

(DEFINE-AUTOMATE-CLASS TASK-NAVIGATE (?site1 ?site2)
  (:STATE begin
   : INTERNAL-EVENTS ((:run
                        :NEXT-STATE :executing
                        :TRANSITIONS (((:ON MOVING)
                                       (:OFF VAP-SITE ?site1))))
    :EXTERNAL-EVENTS ((:stop
                        :NEXT-STATE :stopping
                        :TRANSITIONS ()))))
  (:STATE executing
   : INTERNAL-EVENTS ((:return
                        :NEXT-STATE :end
                        :TRANSITIONS (((:OFF MOVING)
                                       (:ON VAP-SITE ?site2)))
                       (:execution_failure
                        :NEXT-STATE :failure
                        :TRANSITIONS (((:ON NAVIGATION-FAILURE))))
    :EXTERNAL-EVENTS ((:stop
                        :NEXT-STATE :stopping
                        :TRANSITIONS ()))))
  (:STATE failure
   : INTERNAL-EVENTS ((:stop_on_failure
                        :NEXT-STATE :end
                        :TRANSITIONS (((:OFF MOVING)
                                       (:ON VAP-SITE CURRENT)
                                       (:ON VALID-PATH CURRENT ?site1)
                                       (:OFF VALID-PATH ?site1 ?site2))))
    :EXTERNAL-EVENTS ((:stop
                        :NEXT-STATE :stopping
                        :TRANSITIONS ()))))
  ...)
```

Figure 5: Automaton description

3.3 Task Refinement

Task Refinement transforms a task into specific actions that are adapted to the actual context. As an example, a motion task may be executed in different ways:

- a displacement using only dead-reckoning systems to guide the movement.

- a closed-loop motion using a perceptual feature of the environment (landmark tracking, edge following of a large object, rim following, etc.).

These two modes correspond to the execution of different *scripts*, associated with the task "move". Script selection is based on testing conditions using the acquired data. Scripts have variables as arguments, that are instanciated at execution time. The execution of a script is similar to the execution of a *program*.

3.4 Task Execution

Action execution are represented by *activities*. The execution of a script corresponds to a global activity. An activity is thus equivalent to the execution of a program, and is analogous to the notion of process in an computer operating system. A simple activity is the execution of a function by a module. An activity may cause the emission of requests to other modules, starting *children activities*. The module in the parent activity is then a client of the module in the child activity. An activity may be the parent of many children, but may be the child of only one other activity. A set of rules and mechanisms were developed to create and manage activities. Two basic mechanisms are activity creation and message transmission between two activities.

At a given moment, the set of activities represent the functions being executed in the robot system. The activity structure is a tree with a parent-child relationship. The tree evolves while the robot is executing. An activity communicates with its (single) parent activity via "up-signals" and with its (eventual) child activities via "down-signals". The activity hierarchy is not predefined and depends on the current task.

Regardless of the specific processing it performs, an activity must be able to react to signals sent by its parent or its children. In particular, it must be able to react to asynchronous signals within a pre-defined bounded time delay.

An activity is also represented by a finite automaton. Its state changes are dependent on external or internal signals. Control flow between a mother and child activities is implemented as typed messages that cause a state change. Specifc mechanisms permit the propagation of a state change along the activity tree.

Important features of the notion of activity are:

- Activities provide for the management of a hierarchy of actions without imposing a fixed number of layers.

- All activities, regardless of level, may be treated in the same way.

- No assumptions are made about the nature of the inter-connections between activities: the activity hierarchy is *not* pre-defined. We might require that a "high level" activity starts and manages a "low level" activity at any level. This knowledge must however exist in the modules.

- All mechanisms for managing activities (starting, terminating, etc.) and the communication between them do not depend upon any programming language constructs; in this sense, they resemble part of an operating system.

- The concept of activity permits reactivity at all levels: at any moment, each activity in the tree structure is able to respond to asynchronous events.

Figure 6 represents an example of activity tree at one stage of the execution of a *go-to(location)* task. The "root" box represent the mother activity of this task, within the executive. The "monitor-11" box is the mother activity of a main child-activity (monitor-12) and of an associated monitoring activity (timer-200: a time-out for this task). When the corresponding event occurs, the main activity is stopped. "Monitor-12" is a monitoring activity also: the traveled distance should not exceed a given maximum (here 60 meters). Sequence in the "sequence-6" box means that the following children are to be executed in sequence (this defines a sub-activity block). "go-to-xy" and "exec-traj" are the two sequential activities. The first is the "go to" coordinate location, and the second is the trajectory execution phase within it. Other possible phases are "build-model" (environment modelling) and "build-traj" (trajectory planning). The last box "move-5" is the primitive action move being executed.

Figure 7 shows the currently configuration of the implemented modules.

Figure 8 shows the execution of the *go-to(location)* task in a laboratory experiment, where the robot discovers its environment and builds a model of it while it navigates to reach the assigned location.

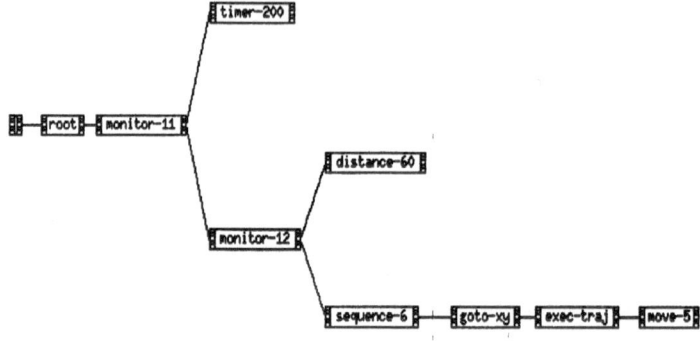

Figure 6: Activity tree during execution of a GOTO task

4   Conclusion

We presented in this paper a global approach to task planning and execution for intervention robots. Such robots are characterized by the fact that they have to be able to autonomously execute their actions in a partially and poorly known environment in order to accomplish missions and tasks specified and programmed by a human user. The robot have to interpret the tasks according to the context and its evolution, and to achieve autonomous execution.

The main interest of the plan interpretation as it is proposed in this paper is to maintain and update a complete history of the plan execution not only in nominal cases but also when a failure occurs. This is performed using automata which model tasks execution and their interaction with the plan interpreter. However other representations may be used. Indeed, we are working on an extension using on a rule-based system which should allow not only to model task execution but also combined effects due to the simultaneous execution of several tasks as well as domain dependent knowledge allowing to infer new transitions from starting from a set of observed events.

While it is still necessary to further deepen some aspects, the experimental results show that this task-level teleprogramming approach is sound and applicable.

References

[1] R. Alami, R. Chatila, M. Devy, and M. Vaisset, System architecture and processes for robot control. Technical report, Laboratoire d'Automatique et d'Analyse des Systèmes (C.N.R.S.), Toulouse (France), June 1990.

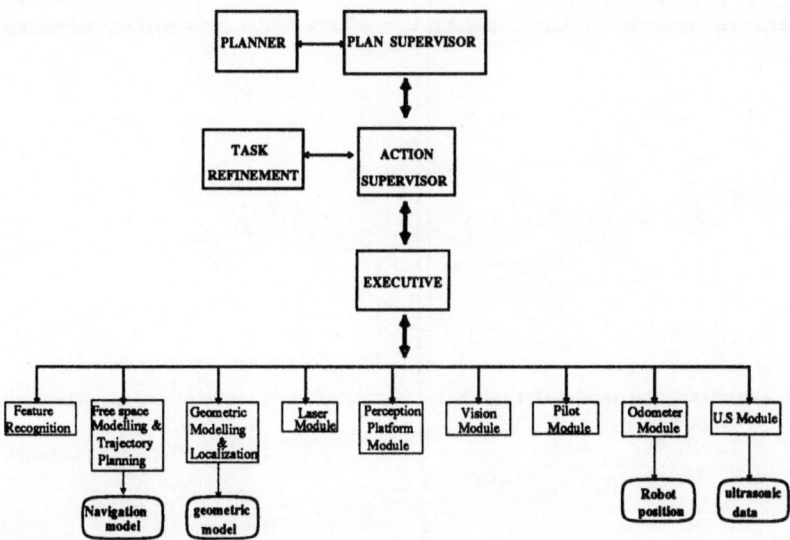

Figure 7: Implemented architecture

[2] R. Alami, R. Chatila, and P. Freedman. Task level programming for intervention robots. In *IARP 1st Workshop on Mobile Robots for Subsea Environments, Monterey, California (USA)*, pages 119–136, October 1990.

[3] Amine Mounir Alaoui. Raisonnement temporel pour la planification et la reconnaissance de situations. Thèse de l'Université Paul Sabatier, Toulouse (France), Laboratoire d'Automatique et d'Analyse des Systèmes (C.N.R.S.), October 1990.

[4] J. F. Allen. Towards a general theory of action and time. *Artificial Intelligence*, 23:123–154, 1984.

[5] C. M. Angle and R. A. Brooks. Small planetary rovers. In *IEEE International Workshop on Intelligent Robots and Systems (IROS '90), Tsuchiura (Japan)*, pages 383–388, July 1990.

[6] L. Boissier and G. Giralt. Autonomous planetary rover (v.a.p.). In *IARP Workshop on Robotics in Space*, Pisa, Italy, June 1991.

[7] R. A. Brooks, P. Maes, and G. Moore. Lunar base construction robots. In *IEEE International Workshop on Intelligent Robots and Systems (IROS '90), Tsuchiura (Japan)*, pages 389–392, July 1990.

[8] R. Ferraz De Camargo, R. Chatila, and R. Alami. A distributed evolvable control architecture for mobile robots. In *'91 International Conference on Advanced Robotics (ICAR),Pisa (Italy)*, pages 1646–1649, 1991.

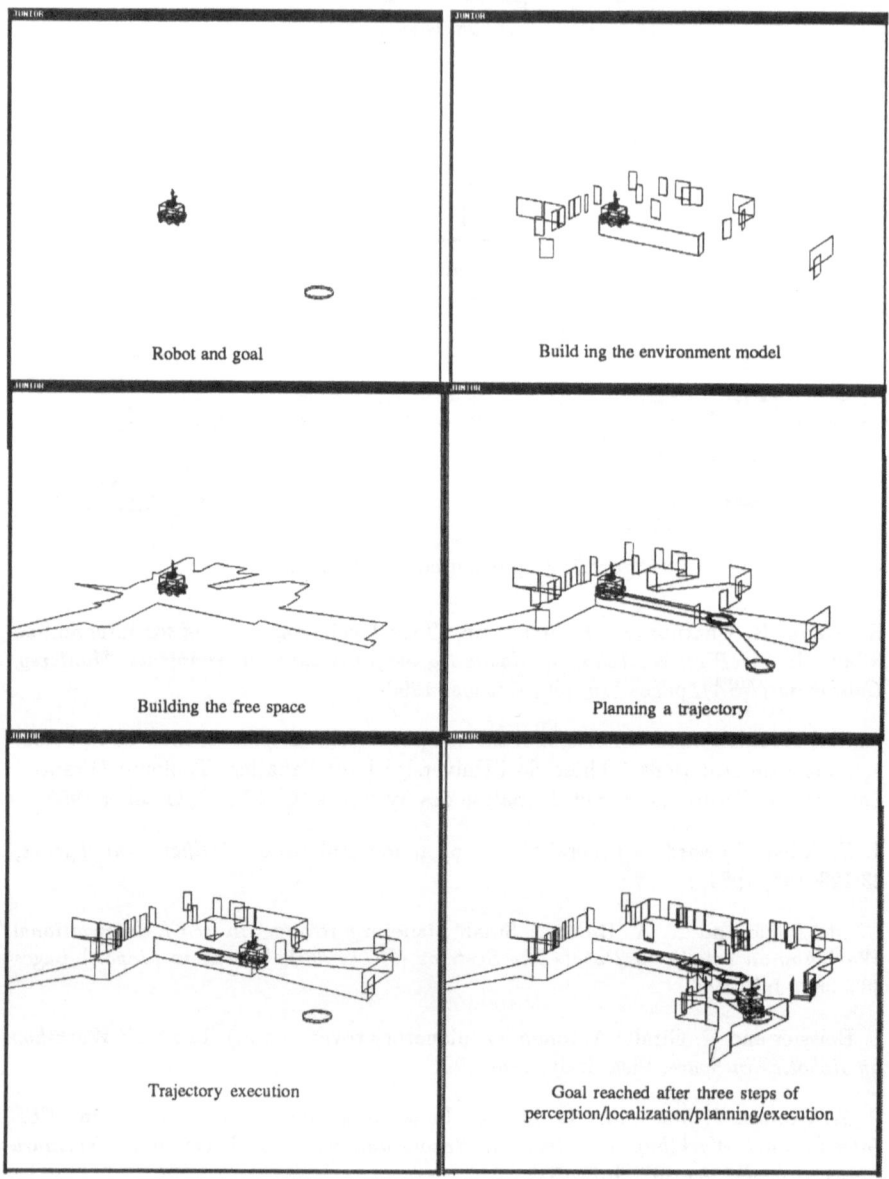

Figure 8: Experimental Execution of a *goto(location)* task.

[9] R. Chatila, R. Alami, and G. Giralt. Task-level programmable intervention autonomous robots. In *Mechatronics and Robotics I, P. A. MacConaill, P. Drews and K.-H. Robrock Eds, IOS Press*, pages 77 – 87, 1991.

[10] R. Chatila and R. Ferraz De Camargo. Open architecture design and intertask/intermodule communication for an autonomous mobile robot. In *IEEE International Workshop On Intelligent Robots and Systems, Tsuchiura, Japan*, July 1990.

[11] L. Floury and R. Gable. The wireline reentry in deep ocean dsdp/odp holes. Technical Report DITI/ICA-91/172-LF/DB, IFREMER - Centre de Brest, July 1991.

[12] M. Ghallab, R. Alami, and R. Chatila. Dealing with Time in Planning and Execution Monitoring. In R. Bolles, editor, *Robotics Research: The Fourth International Symposium*. MIT Press, Mass., 1988.

[13] R. Laurette, A. de Saint Vincent, R. Alami, R. Chatila, and V. Pérébaskine. Supervision and control of the amr intervention robot. In *'91 International Conference on Advanced Robotics (ICAR),Pisa (Italy)*, pages 1057–1062, June 1991.

[14] F. R. Noreils, A. Khoumsi, G. Bauzil, and R. Chatila. Reactive processes for mobile robot control. In *International Conference on Advanced Robotics (ICAR)*, 1989.

[15] T. Sheridan. Telerobotics. In *IEEE Int. Conf. on Robotics and Automation (Workshop on Integration of AI and Robotic Systems*, 1989.

A Sensor-based Telerobotic System for the Space Robot Experiment ROTEX

G. Hirzinger, G. Grunwald, B. Brunner, J. Heindl

D L R

(German Aerospace Research Establishment)

Oberpfaffenhofen, D-8031 Wessling

Abstract

The space robot technology experiment ROTEX to fly with the next spacelab mission D2 in 1993 provides a sensor-controlled robot which is supposed to work in an autonomous mode, teleoperated by astronauts, and teleoperated from ground. The robot's key features are its multisensory gripper and the local ("shared") sensory feedback schemes. The corresponding man-machine interface concepts using a 6 dof control ball and visual feedback to the human operator are explained. Sensory simulation on ground using advanced stereo graphics is supposed to predict the sensor based path refinement on board, while realtime fusion of stereo images and laser range information helps to predict the motion of floating objects to be grasped. The telesensorprogramming concept is explained as well as the learning schemes involved.

INTRODUCTION

Automation and robotics (A&R) will become one of the most attractive areas in space technology, it will allow for experiment-handling, material processing, assembly and servicing with a very limited amount of highly expensive manned missions, and the expectation of an extensive technology transfer from space to earth seems to be much more justified than in many other areas.

This is one of the main reasons why several activities towards space robotics have started in a number of countries, one of the largest projects being NASA's Flight Telerobotic Servicer (FTS), which is supposed to help already in the space station's construction, with a first experimental flight test scheduled for 93. For the European Space Agency ESA and especially for Germany a major contribution to the space station will be the subsystem "man tended free flyer" (MTFF), a laboratory which will be visited by astronauts only once per half year, but work automatically the rest of the time. In addition to study activities with respect to robot operation in such a laboratory a robot technology experiment ROTEX will fly with the next spacelab-mission D2 in early 93. ROTEX is kind of a starting shot for an active European participation in space automation and robotics.

ROTEX OVERALL CONFIGURATION

The main features are as follows (fig. 1):

- A small, six-axis robot (working space ~ 1 m) flies inside a space-lab rack (fig. 2). Its gripper is provided with a number of sensors, especially two 6-axis force-torque wrist sensors, tactile arrays, grasping force control, an array of 9 laser-range finders and a tiny pair of stereo cameras to provide a stereo image out of the gripper; in addition a fixed pair of cameras will provide a stereo image of the robot's working area.

- In order to demonstrate servicing prototype capabilities three basic tasks are envisioned:

  a) assembling a mechanical grid structure
  b) connecting/disconnecting an electrical plug (ORU-exchange using a "bajonet closure")
  c) grasping a floating object (fig. 3)

  Local sensory feedback from the multisensory gripper is a key issue in the telerobotic concepts used, so we will address the gripper's components in more detail.

- The envisioned operational modes are

  - automatic
    (preprogramming on ground
    (reprogramming from ground)
  - teleoperation on board (astronauts using stereo-TV-monitor)
  - teleoperation from ground (using predictive computer graphics)
  - tele-sensor programming
    (learning by showing in a completely simulated world including sensory perception with sensorbased execution later on-board).

- The main goals of the experiment are:

  - To verify joint control (including friction models) under zero gravity as well as μg-motion planning concepts based on the requirement that the robot's accelerations while moving must not disturb any μg-experiments nearby.
  - To demonstrate and verify the use of DLR's sensorbased 6 dof-handcontrollers ("control balls") under zero gravity.
  - To demonstrate the combination of a complex, multisensory robot system with powerful man-machine-interfaces (as are 3D-stereo-computergraphics, control- ball, stereo imaging, voice input-output), that allow teleoperation from ground, too.

SENSORY COMPONENTS

The gripper sensors belong to the new generation of DLR robot sensors based on a multisensory concept with all analog preprocessing and digital computations performed inside each sensor or at least in the wrist in a completely modular way (fig. 4 and 5). Using a high speed serial bus only two signal wires are coming out of the gripper (carrying all sensory information), augmented by two 20 kHz-power supply wires, from which the sensors derive their DC-power supply voltages via tiny transformers themselves.

This same concept with a "busmaster" board exchanging all relevant information with the robot control system via a nearly delay-less dualport-memory is used also (see fig. 5) to connect all joints of a new light-weight astronaut training manipulator (see below). Up to now the serial bus speed has been 375 kBaud, but will now be raised to 4 MBaud.

In the gripper 15 sensors are provided, in particular (fig. 4 and fig. 7):

  a) an array of 9 laser range finders based on triangulation, one "big" sensor (i.e. half the size of a match box) for the wider range of 3-35 cm (see fig. 6), and smaller ones in each finger for lower ranges of 0-3 cm. The "long range" finder will be provided with a scanning mechanism to aid the visual determination of position/ orientation of the floating object to be grasped.

b) A tactile array of 4 x 8 sensing elements (conductive rubber "taxels") in each finger. The dimensions of the tactile area is 32 x 16 mm. The binary state of each taxel is serially transmitted through analog multiplexers without additional wiring.

c) A "stiff" 6 axis force–torque sensor based on strain–gauge measurements and a compliant optical one. Originally it seemed necessary to make a final decision between these two principles, but as indicated in fig. 4 and fig. 7 they finally were combined into a ring–shaped system around the gripper drive, the instrumented compliance being lockable and unlockable electrically. Shaping these sensors as rings around the gripper drive shows up different advantages:

- it does not prolong the axial wrist length
- it brings the measurement system nearer to the origin of forces–torques and yields a better ratio of torque range to force range than achievable with a compact form.

The optically measuring instrumented compliance was e.g. described in /10/.

The stiff, strain–gauged force–torque sensor is a completely new design no longer based on spokes or bars but membranes. It performs automatic temperature compensation based on the temperature characteristic as stored during the calibration process and continues operating reliably with reduced accuracy if one of the strain–gauges is damaged. The ring-shaped form of this new sensor containing all electronics and preprocessing may be seen in fig. 8.

d) A pair of tiny stereo CCD-cameras, the CCD's plus optics plus minimum electronics taking a volume smaller than a match–box, too.

e) An electrical gripper drive, the motor of which is treated like a sensor with respect to the data bus and the 20 kHz power supply connections. The design criteria for this drive are outlined in the next chapter.

With more than 1000 tiny SMD electronic (fig. 7) and several hundred mechanical components the ROTEX gripper is one of the most complex multisensory endeffectors that have been built so far.The gripper was not space qualifiable on component level, so it had to undergo vibration, temperature, off-gasing and EMC-tests as a whole.

Sensory devices are also involved in the man-machine interfaces for astronauts and on-ground operators in form of the sensor or control balls used as 6 dof-hand controllers (fig. 9). The compliant 6 axis force-torque sensor principle as used in the wrist is - in a more compact form - integrated into a plastic hollow ball. The device does not only serve as a 6-dimensional joystick (for real and graphically simulated robot hands as well as complex graphic sceneries) but at the same time issues forces and torques as exerted by the human operator - an important feature in our telerobotic concept.

ACTUATORS

Space technology may become a major development drive for advanced light weight robots. For ROTEX two design problems were of crucial interest:

a) to arrive at an electrical gripper drive that allows fine positioning and reasonable grasp force feedforward control with grasping forces up $\approx$ 300 N (the gripper without sensors weighing $\approx$ 10 N).

b) to arrive at revolute joint drives for a 1g-compatible training manipulator with very high reduction but extremely compact construction and integrated torque measurement and control.

Rotational–Translational Gearing

In the gripper the problem is to transform the motor's high–speed rotational motion into a fairly slow axial motion to move the fingers (fig. 4). For this type of transmission a new, extremely low friction mechanical spindle concept has been designed (for details see /10/).

What we have gained with this motor–gearing combination is a small prismatic drive (applicable also in a robot joint), which used as gripper drive allows to exert grasp forces up to 300 N with a gripper weight of 5 N and a grasp speed of about 15 cm/sec; without measuring and feeding back grasp forces we arrived at a feedforward grasp force control resolution of $\approx$ 1-2 N ($<$ 1 % of max force) with high repeatability. Reduction rate referring to the finger rotation is $\approx$ 1 : 1000.

Rotational–Rotational Gearing

The "phase-shift" ideas as outlined in the last chapter have meanwhile been transferred to pure rotational gearings, too (for details see e.g. /10/). In an advanced version these gearings will imply inductive torque sensing and feedback control integrated in the joints. Presently a light-weight 1:1 replica of the ROTEX arm is close to be finished, where the new joint design is integrated into a carbon fibre grid construction (see /10/). As the ROTEX flight model cannot sustain itself in 1 g environment, this light-weight robot design aiming at a 10 : 1 weight reduction will be used for the astronaut training on ground and is thought to be a first step towards a real space light-weight robot.

THE TELEROBOTIC CONCEPTS

a) Shared Control

The fine motion planning concept of ROTEX is a "shared" control concept based on local sensory feedback (e.g. /3/). Its main features - briefly outlined - are (fig. 10 and 11):

Rudimentary commands Δx (the former deviations between master and slave arm) are derived either from the sensor ball's forces-torques f (using a sort of artificial stiffness relation $\Delta x = S_x^{-1}f$) or from a path generator connecting preprogrammed points (Δx being the difference between the path generator's, i.e. "master's", position and the commanded "slave" robot position x_{com}). Due to the above-mentioned artificial stiffness relation these commands are interpreted in a dual way, i.e. in case of free robot motion they are interpreted as pure translational/rotational commands; however if the robot senses contact with the environment, they are projected into the mutually orthogonal "sensor-controlled" (index f) and "position-controlled" (index p) subspaces, following the constraint frame concept of Mason /2/. These subspaces are generated by the robot autonomously using a priori information about the relevant phase of the task and actual sensory information: in a future stage the robot is supposed to discern the different task phases (e.g. in a peg-in-hole or assembly task) automatically. Of course the component Δx_p projected into the position controlled subspace is used to feed the position controller; the component f_f projected into the sensor-controlled subspace is either compared with the sensed forces f_{sens} to feed (via the robot's cartesian stiffness S_R) the orthogonally complementary force control law, (which in fact is another position controller yielding a velocity $\dot{x}_f$), or it is neglected and replaced by some nominal force vector f_{nom} to be kept constant e.g. in case of contour following. We prefer to talk about sensor-controlled instead of force-controlled subspaces, as non-tactile (e.g. distance) information may be interpreted as pseudo-force information easily, the more as we are using the robot's positional interface anyway. However we omit details as treated in /3/ concerning transformations between the different cartesian frames (e.g. hand system, inertial system etc.). Note that in case of telemanipulation via a human operator although we are not feeding the forces back to the human arm (as does "bilateral" force control with the well-known stability problems in case of delays), the operator is sure that the robot is fully under his control and he easily may lock up doors, assemble parts or plug in connectors in a kind of "shared control". In other words, the human operator (via stereovision and 3D-graphics) is enclosed in the feedback loop on a very high level but low band-width, while the low level sensory loops are closed on-board at the robot directly with high bandwidth. Thus we try to prepare a supervisory control technique (/5/) that will allow to shift more and more autonomy to the robot while always offering real-time human interference. One of the main issues here is that in case of non-negligible time delays stereo-visual feedback is replaced by predictive stereo graphics, leaving the basic structures untouched, independent on whether astronauts or on ground personnel are operating the robot in a space application.

b) Predictive control and Tele-sensor-programming

When teleoperating a robot in a spacecraft from ground or from another spacecraft so far away that a relay satellite is necessary for communication, the delay times are the crucial problem. Predictive computer graphics seems to be the only way to overcome this problem. Fig. 12 is to outline that the human operator at the remote workstation handles the control ball by looking at a "predicted" graphics model of the robot. The control commands issued to this instantaneously reacting robot simulator are sent to the remote robot as well using the time-delayed transmission links. Now the ground-station computer and simulation system contains a model of the uplink and downlink delay lines as well as a model of the actual states of the real robot and its environment (especially moving objects).

The most crucial problems lie

- in the derivation of "output data" (e.g. positions/orientations) of moving objects from say stereo images and range finders (we treat this as a special preprocessing problem discussed in /6/). Fig. 12 is especially well suited to understand the grasping of floating objects, as with the simple dynamics of a floating object the estimator is just a so called "extended Kalman filter". An observer solution for a linearized model with large time delays has been given in /7/.

- in a precise simulation of sensory perception and feedback in case of a static environment; this implies the use of a knowledge-based world model.

Let us focus on the second topic in more detail. Teleoperation from ground in case of an assembly operation implies realistic real-time simulation of robot environment and sensory perception thus that the predictive graphics motion contains sensor-based fine path generation in an on-line closed loop scheme, too (fig. 13). Another very challenging mode is characterized by sensor-based task learning using graphics simulation. Hereby the robot is graphically guided through the task (off-line on ground), storing not only the relevant hand frames but also the corresponding nominal sensory patterns (graphically simulated) for later (e.g. associative) recall and reference (fig. 10 and 14) in the on-board execution phase, after these data sets have been sent up. Thus we might talk of storing "situations" (fig. 10). Basically this technique may be characterized as "sensorbased teaching by showing"; let it call **tele-sensor-programming**; indeed it is a form of off-line-programming which tries to overcome the well-known problems of conventional approaches, especially the fact that simulated and real world are not identical. But instead of e.g. calibrating the robot (which is only half the story) **tele-sensor-programming** provides the real robot with simulated sensory data that refer to **relative** positions between gripper and environment thus compensating for any kind of inaccuracies in the absolute positions of robot and real world.

In both modes realistic simulation of the robot's environment and especially the sensory information presumably perceived by the real robot is of crucial importance.

Two levels of world modelling are involved:

i) coarse modelling on object level. In addition to approximating each workcell (including robot) part by a convex polytope for fast object detection on distance sensor level, an alternative approach partitions the space occupied by the objects on a polyhedron level, similar to the cell-tree representation in /8/. We have implemented an efficient index structure on convex closed polyhedron with the advantage of a fast yet very precise search for actually interesting objects. Application of this object representation is the computation of object intersection with the rays of the laserscanner as well as collision detection. Of course a fast restriction of the interesting regions to a minimal number of polyhedrons is desired. These interesting regions (leaves of the cell tree) are mapped into the local databases on which the realtime laser distance simulation runs.

The leaves of the cell-tree characterize the interesting regions. Each leaf wears a local data base on which the sensorsimulation runs. As the cell-tree is specified only for convex polyhedrons, the knots passed during object selection consist of the convex hulls of the objects involved.

In the leaves (local databases) of the cell-tree, i.e. after object selection, a relation is evaluated which assigns the real areas of a e.g. concave original object to the areas of the convex hull. This area relation is generated during construction of the geometric worldmodel. Intersection and interference checks preceded by a corresponding object selection are thus realizable in an efficient way for convex as well as for concave objects.

For force simulations the interesting regions are determined in a similar way based on the motion direction of the tool center point and a potential overlapping of colliding parts.

ii) fine modelling on surface level (local data base level): After determining the colliding objects a fine modelling on surface level of the object representation e.g. for precise laser range simulations is important. An exact boundary representation model is available via the geometry modeller CATIA. After modification and extension of the given data basis different local databases are accessible which refer to the single objects only. For the range finder simulations we assume that the optical features of the object surfaces are negligible and that it is sufficient to work on the geometric models of the workcell parts.

In case of force-torque simulation, i.e. detection of real collision of e.g. two polyhedrons, calculation of the exact force-torque values is based on the gripper's motion direction, surface orientation and stiffness of sensor compliance (following the approach given in /9/). Compliance of environment as well as friction is not taken in account so far. Due to low motion speeds we even neglect the robot's dynamics and base our computations on virtual interference of sensor and colliding object. Motion and force constraint equations are derived from the interference data and the simulated forces/torques are computed from these constraint equations and the known sensor characteristics. Our main problems presently are to find <u>realtime</u> solutions for virtual interference of bodies and derivation of the constraint equations.

There are errors expected in the sensor simulation and therefore not only the gross commands of the TM command device (control ball) are running into the simulation system, but also the real sensor data coming back from space (including the real robot position) are used to improve the robot's and its sensor's models (fig. 13). Fig. 15 outlines the different computer and graphics systems making up the telerobotic ground station and the information and data flow between them. A VME-bus-based shared memory concept has been implemented to connect the systems with high speed (6 MB/sec). Note that in addition to the shaded stereographics SGI system a RASTER TECH wireframe 3D display is of great help showing robot joint loadings, all sensory data as vectors and time history recordings, but also the predicted and executed robot paths.

c) Learning and self-improvement

In fig. 14 an associative mapping for relating stored (simulated) and actual sensory patterns is indicated. Due to computational on-board restrictions in ROTEX only analytical techniques are applied at this stage (e.g. calculating relative distance and orientation out of the range finder arrays). However for potential successor flights (socalled "COLUMBUS space station precursor flights") a more general training concept including stereovision is in preparation, where a training (using associative mappings or neural nets) is used to relate a nominal sensory perception (and its corresponding relative position/orientation) to the actual sensory perception (and the corresponding - may be - nonnominal relative position/orientation), see fig. 16.

The associative mappings mentioned above have been tested successfully in self-improving schemes for feedback controllers (/11/) that will be applied in ROTEX, too. Due to uncertainties in friction modelling under zero gravity the ROTEX joint controllers contain "friction model state observers" that allow for some adaptation of control parameters in the joints. But in addition the outer sensory feedback loop parameters may be updated from ground after using the down-link informations about the on-board sensory data and generated paths.

In total the learning scheme consists of 3 stages. The lowest, the controller, is trained on the basis of examples for optimal performance. The teacher's information, i.e. the nominal control commands as computed by an optimization procedure depending on the controller's inputs are provided by the second stage. This second stage considers control differences as generated with an a priori chosen controller and corrects the issued control commands in a way so that the control differences disappear. Only a crude plant information is needed for this; it is taken from the plant model, i.e. from the third stage.

Up to now among the three stages modelling, optimization and controller improvement the model and the controller have been realized by trainable associative memories; they were originally derived from the CMAC (cerebellar model articulation controller) of J. ALBUS /12/, but meanwhile they are better characterized as a tabular knowledge base, and they are now to be replaced by neural nets.

CONCLUSION

The ROTEX proposal as a first step of Germany's engagement in space robotics aims at the demonstration of a fairly complex system with a multisensory robot on board and human telerobotic inferference that makes use of sensor-based 6 dof handcontrollers, new concepts for predictive 3D-computergraphics and stereo display. For us teleoperation from ground is a very challenging technique that forces us to move even stronger into on-board autonomy. The control structures are thus that the human operator may step more and more towards supervisory control without changing the control loop structures. However we feel that for a number of years remotely operating robots will show up limited intelligence only, so that human "anytime" interference will remain important for a long time. Advanced real-time stereo graphics seems to become one of the most powerful man-machine interfaces in the attempts to overcome large transmission delay times, using appropriate estimation schemes in connection with modelling and real-time update of robot environment and sensory perception. The telesensor-programming concepts involved might help to make off-line-programming much more effective also in purely terrestrial applications.

REFERENCES

/1/ S.Lee, G. Bekey, A.K. Bejczy
"Computer control of space-borne teleoperators with sensory feedback", Proceedings IEEE Conference on Robotics and Automation, S. 205-214, St. Louis, Missouri, 25-28 March 1985.

/2/ M.T. Mason
"Compliance and force control for computer controlled manipulators", IEEE Trans. on Systems, Man and Cybernetics, Vol SMC-11, No. 6 (1981, 418-432).

/3/ G. Hirzinger, K. Landzettel
"Sensory feedback structures for robots with supervised learning". Proceedings IEEE Conference, Int. Conference on Robotics and Automation, S. 627-635, St. Louis, Missouri, March 1985.

/4/ G. Hirzinger, J. Heindl
" Sensor programming, a new way for teaching a robot paths and forces torques simultaneously". 3rd Int. Conference on Robot Vision and Sensory Controls, Cambridge, Massachusetts/USA, Nov. 7-10. 1983.

/5/ T.B. Sheridan
"Human supervisory control of robot systems.
Proceedings IEEE Conference, Int. Conference on Robotics and Automation, San Francisco, April 7-10, 1986.

/6/ B.C. Vemuri, G. Skofteland
"Motion estimation from multi-sensor data for tele-robotics", IEEE Workshop on Intelligent Motion Control, Istanbul, August 20-22, 1990.

/7/ G. Hirzinger, J. Heindl, K. Landzettel
"Predictive and knowledge-based telerobotic control concepts". IEEE Conference on Robotics and Automation, Scottsdale, Arizona, May 14-19, 1989.

/8/ O. Guenther
"Efficient structures for geometric data management", Lecture notes in Computer Science, Berlin, Heidelberg, New York, Springer 1988.

/9/ P. Simkens, J. De Schutter, H. Van Brussel
"Force/Torque Sensor Emulation in a Compliant Motion Simulation System",
Proceedings of CSME Mechanical Engineering Forum, Toronto, 1990.

/10/ J. Dietrich, G. Hirzinger, B. Gombert, J. Schott
"On a Unified Concept for a New Generation of Light-Weight-Robots", Proceedings of the Conference ISER, Int. Symposium on Experimental Robotics, Montreal, Canada, June 1989.

/11/ F. Lange
"A Learning Concept for Improving Robot Force Control", IFAC Symposium on Robot Control, Karlsruhe, Oct. 1988.

/12/ J. S. Albus
"A New Approach to Manipulator Control: The Cerebellar Model Articulation Controller (CMAC)", Transactions of the ASME, Journal of Dynamic Systems, Measurement and Control, pp. 221-227, Sept. 1975.

/13/ S. Hayati, S.T. Venkataraman
"Design and Implementation of a Robot Control System with Traded and Shared Control Capability", Proceedings IEEE Conference Robotics and Automation, Scottsdale, 1989.

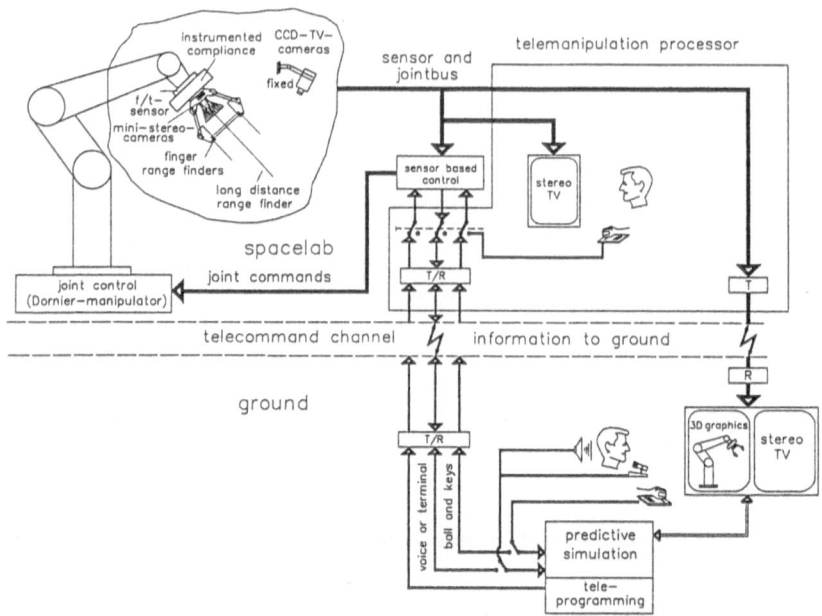

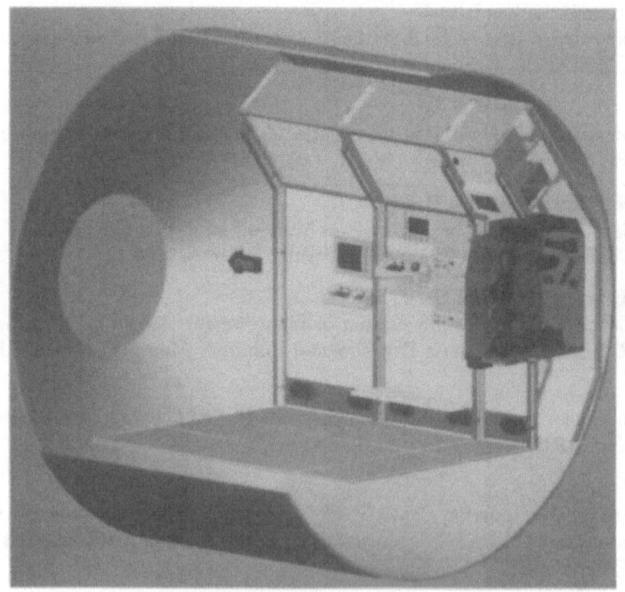

Fig.1 Schematic representation of ROTEX (upper part) and integration in spacelab (picture)

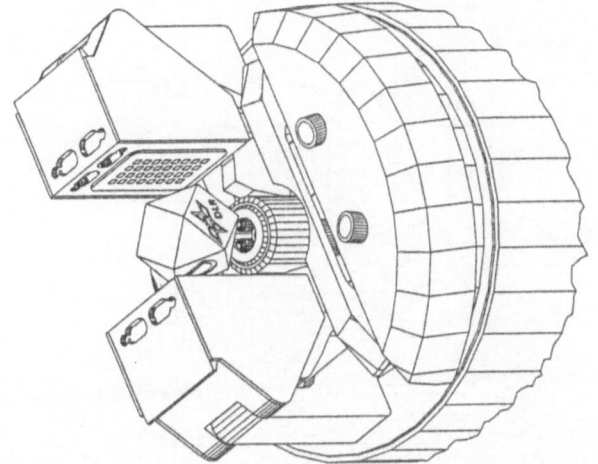

Fig.3 Grasping a floating object

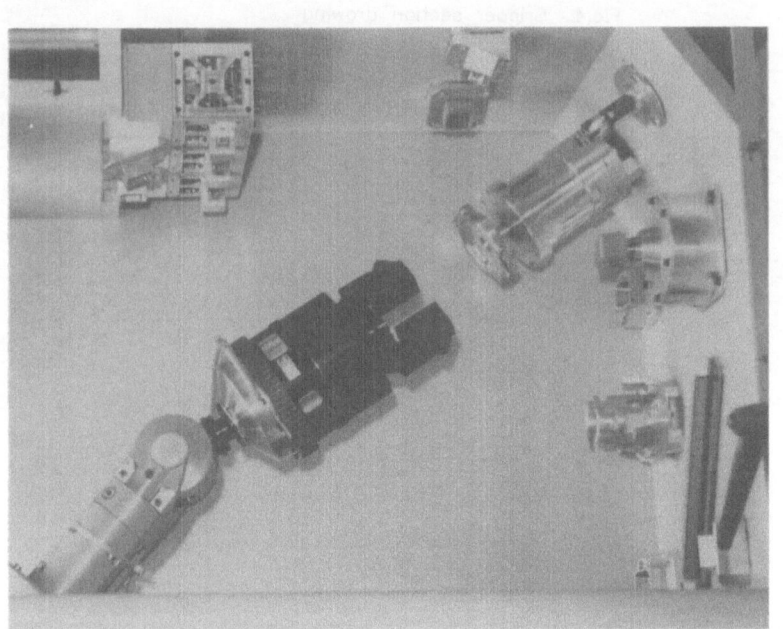

Fig.2 ROTEX robot and experiment set–up in
DLR laboratory

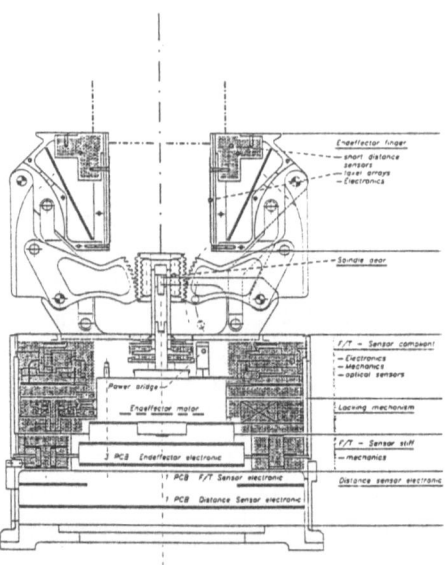

Fig.4 Gripper section drawing

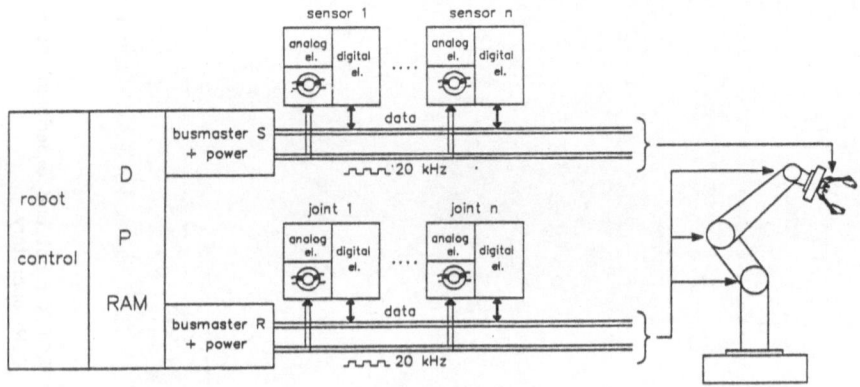

Fig.5 Information and power transfer in the ROTEX multisensory gripper and in the joints developed for a new light—weight robot

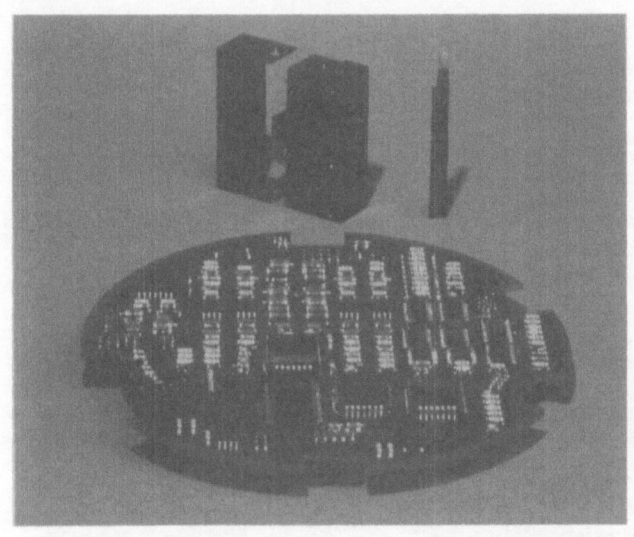

Fig.6 Medium range sensor and wrist electronic
board for all range finders

Fig.7 The multisensory ROTEX gripper integrates
more than 1000 tiny SMD-components

Fig.8   Ring—shape structure of a new membrane
based strain—gauge sensor with all
preprocessing integrated

Fig.9   On—ground telecommand system with control
ball and gripper position / force control joysticks

fine path generation

gross path generation

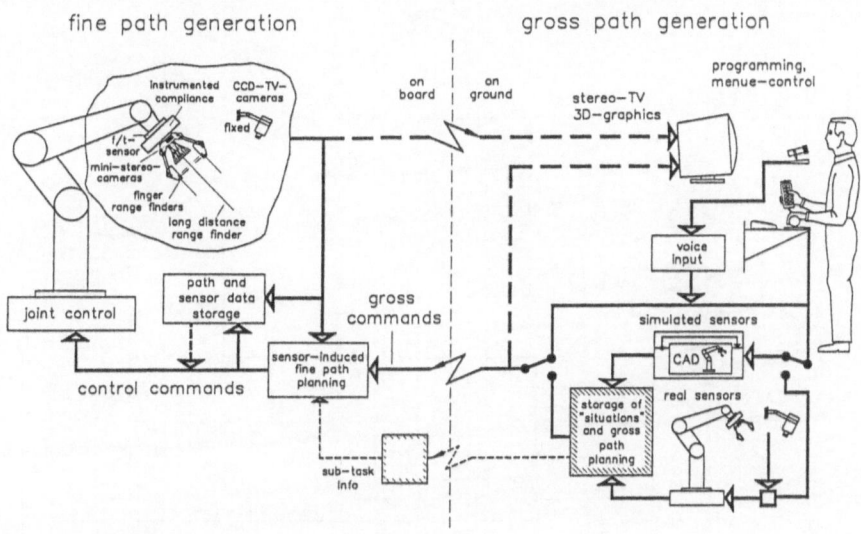

Fig.10 Overall loop structures for the sensor—based telerobotic concept

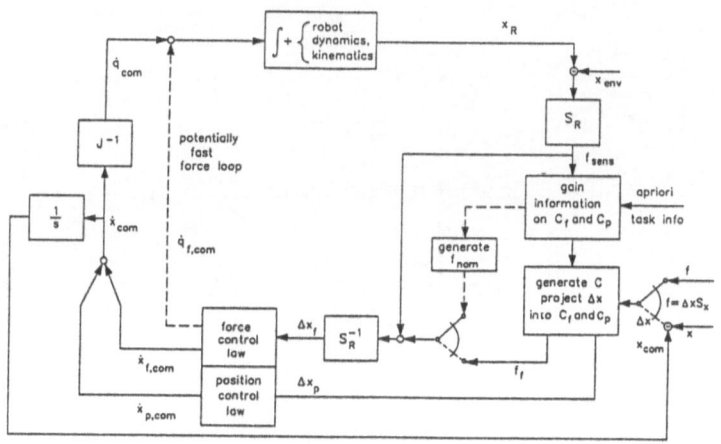

Fig.11 A local closed loop concept with automatic generation of force and position controlled directions and artificial robot stiffness

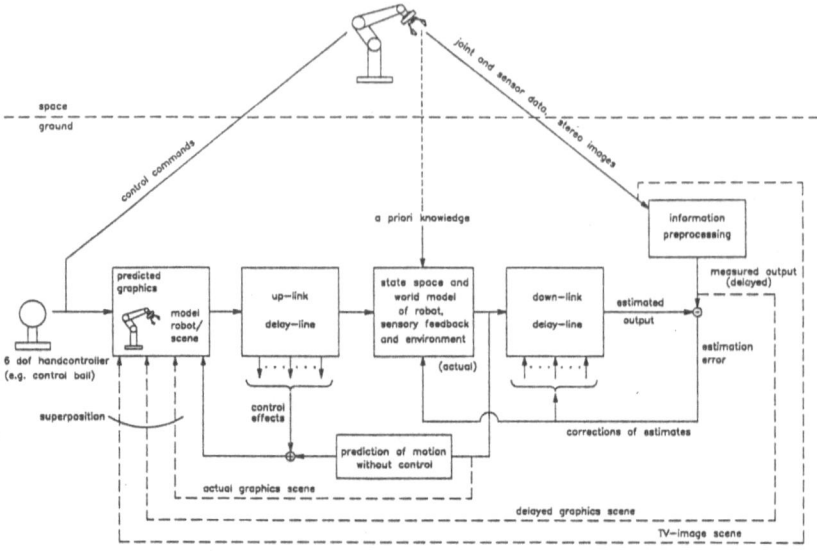

Fig.12   Block structure of predictive estimation scheme

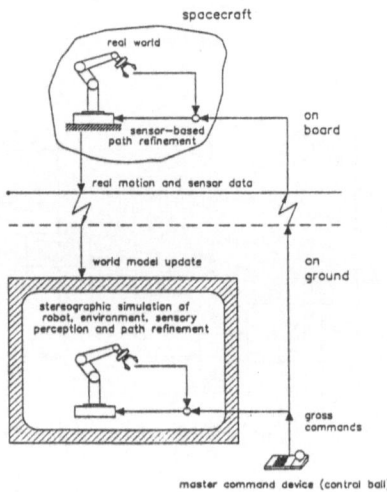

Fig.13   Presimulation of sensory perception and path
refinement in case of teleoperation from ground

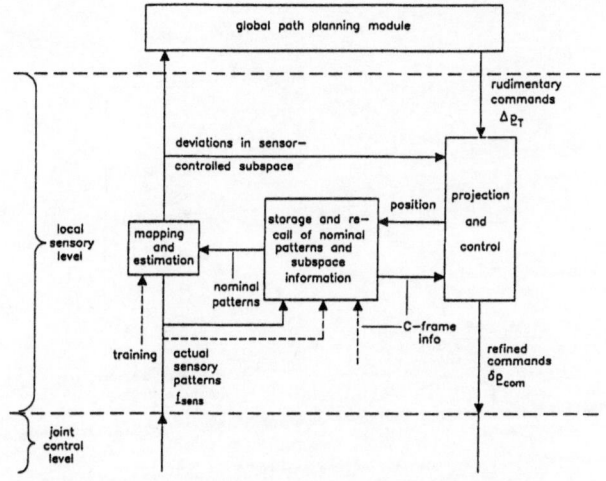

Fig.14 Storage and recall of nominal sensory patterns
— a key issue in the telesensorprogramming concept

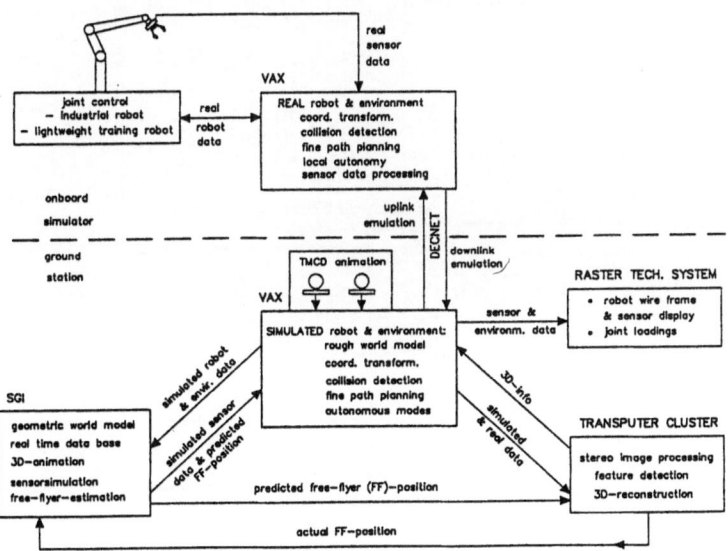

Fig.15 ROTEX groundstation configuration

238

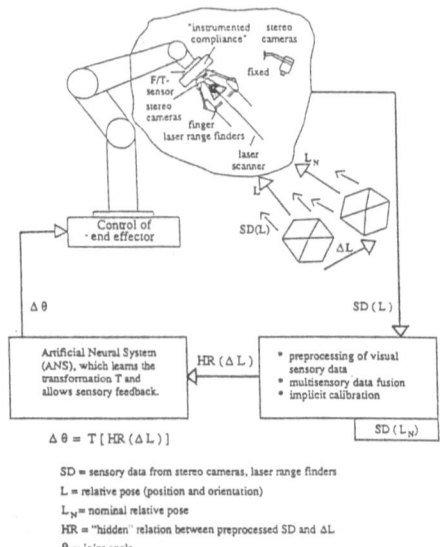

$$\Delta\theta = T[HR(\Delta L)]$$

SD = sensory data from stereo cameras, laser range finders
L = relative pose (position and orientation)
L_N = nominal relative pose
HR = "hidden" relation between preprocessed SD and ΔL
θ = joint angle

Fig.16   A learning scheme for relating differences
between nominal and nonnominal multisensory
patterns to position/orientation errors relative
to an object

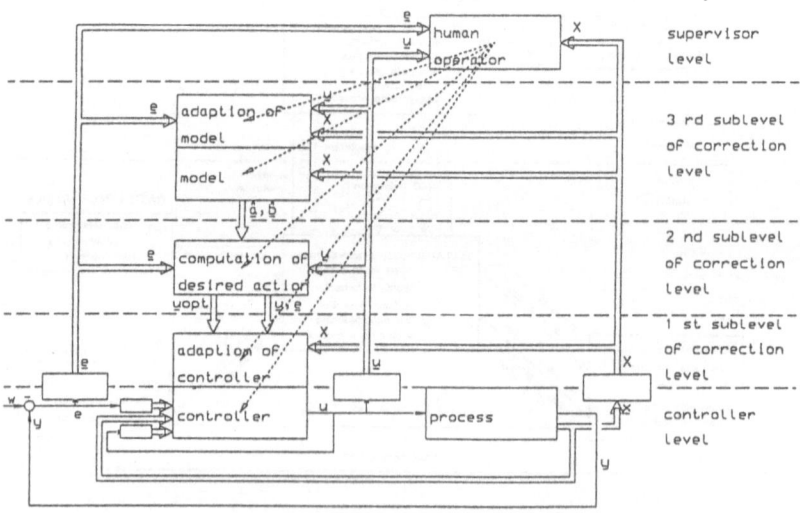

Fig.17   The on—ground scheme for improving the ROTEX
local sensory feedback loops during the flight

Section 5: Manipulation Planning and Control

Motion planning is inherently the basis of what we call robot intelligence or autonomy. What a new generation of robots needs is the capability to perfom global plans in reasonably short time although off-line, but then to adapt these paths dynamically via real time control, e.g. via sensory feedback. At present there is an enormous gap in bandwith between these two phases of autonomous robot motion.

The first paper of this section by Quinlan and Khatib addresses the problem of closing this gap by designing a real-time path planner based on a bitmap representation of the configuration space, aiming at a response time of a fraction of a second for an arbitrary path (after an initial precomputation time of e.g. half a minute). As an appropriate complement to this kind of preplanning the "elastic band" approach is presented, aiming at the on-line deformation of the original path due to sensory information in a non-static environment.

Sensor-based, i.e. force-controlled manipulation of deformable objects by a two-arm robot is described in the paper of Delebarre, Degoulange, Dauchez and Bouffard-Vercelli. The use of hybrid force control as usually proposed not only for one-arm but also two-arm robots in case of rigid objects to be handled serves as a basis. But the authors then emphasize the treatment of internal force control for solving the task of deforming an object and keeping this deformation for a class of assembly problems. Manipulation of a spring with two PUMA's is demonstrated as an experiment.

Planning safe trajectories in an environment which is partly static, partly changing due to moving objects like other robots is the focus of the paper by Bellier, Laugier, Mazer and Troccaz. They combine two path planners, a local one that takes the manipulator away from the constrained initial and final configurations and a global one which is responsible for generating a safe trajectory for the first three degrees of freedom.

Whenever a robot tries to do things equivalently or better than humans in nonindustrial "all-day life" tasks, this normally attracts a lot of attention. Ping-pong playing belongs to this category as well as juggling a ball, presented in the paper of Rizzi and Koditschek. This paper, focussing on spatial juggling using stereo vision, is a theoretically and experimentally remarkable extension and foundation of earlier work restricted to planar juggling. A key issue for these kind of problems is the motion of a mirror algorithm, the generalization of which to spatial juggling is outlined in the paper. A detailed overall description of the new juggling machine is given as well.

Towards Real-Time Execution of Motion Tasks

Sean Quinlan and Oussama Khatib

Robotics Laboratory
Computer Science Department
Stanford University

Abstract

The goal of the research described in this paper is to build robotic systems that can execute motion tasks in real time. We present two ideas towards this goal: a real-time path planner for static three-dimensional configuration spaces and a new framework for the integration of planning and control. The planner is based on a new class of cells, *slippery cells*, incorporated in an approximate cell decomposition algorithm. The *elastic band* concept is proposed to provide an effective link between planning and execution. With elastic bands, a path is treated as a flexible entity. The initial configuration of the elastic band is the path provided by the planner. The shape of the elastic dynamically evolves during execution, using sensory data about the environment.

1 Introduction

The goal of the research described in this paper is to build robotic systems that can execute motion tasks in real time. The desired system would fulfill commands to move to a desired position while avoiding obstacles in the environment. Our additional goal of achieving real-time performance relates to the responsiveness of the system; the time to execute the task is of importance and is both a motivating force in the search for new ideas and a means of comparing alternative solutions.

To execute a motion task a robot must combine the ability to plan motions and to execute them. In addition, the robot has sensors that enable it to monitor the effect of its actions and to perceive the environment. One approach to building robot systems is illustrated in Figure 1. In this approach, the robot uses its sensors to build a world model that is passed to a path planner. A task from the user is specified to the path planner which generates a trajectory for the robot to follow. A motion controller is used to track the path using feedback from sensory information of position and velocity. Three observations can be made about the above scheme:

First, with this approach, the responsiveness of the robot is totally dependent on the speed of the path planner. Complete solutions to path planning are computational complex. In recent years, much progress has been made in developing practical algorithms, which

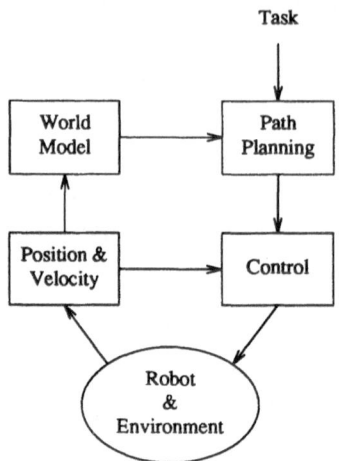

Figure 1: One scheme for a robot system

are efficient (Latombe 1991, Barraquand and Latombe 1991, Lengyel et. al. 1990). For typical three-dimensional problems, current implementations can generate a path in several seconds.

Second, it is a non-trivial problem to connect a path planner and a control system. A common approach is the use of trajectories generated by time parameterizing the path produced by the planner. A dynamic model of the robot can be used to insure that the generated trajectory is feasible or, better yet, optimal (Bobrow, Dubowsky, and Gibson 1983). Unfortunately, path planners are often designed to find any feasible path, with little attention to its suitability for execution. Planners may try to produce kinematically short paths but rarely do they use any knowledge of the robot dynamics. The result is that the path may be undesirable from a control point of view. For example, the path may have abrupt changes in direction or maintain little clearance from obstacles, thus requiring the robot to move slowly. Optimally short paths suffer substantially from this problem.

Third, there is an underlying assumption that the motion plan will be valid for the time it takes the robot to execute it. The validity of this assumption depends on the degree to which the world model represents the environment. At one extreme, there is a complete model of a static environment and thus a path can be used blindly. At the other extreme, there is no model of the changing environment and thus the concept of a preplanned path is meaningless. In reality, the situation is somewhere in between: there is some a priori knowledge that the planner can use and, during execution, the robot uses sensors to detect unexpected obstacles and uncertainties in the environment.

In the scheme shown in Figure 1, sensory information is used to build a world model; if changes in the environment are detected during execution then the world model will change. Under this scheme, if the world model is changed during execution, the obvious action is to halt the robot and plan a new path. In such situations, the iteration through the world modeler and path planner can be considered as a feedback loop. This loop, even with fast path planners, is expected to be the bottleneck in building responsive systems.

The first part of this paper describes a real-time path planner for static three-dimensional configuration spaces. This planner is based on a variation of the approximate cell decomposition method. The method uses a new class of cells, called *slippery cells*, that have a general shape and can typically describe the free space with fewer cells than other approximate cell decompositions. The result is a corresponding reduction in the required search time.

The second part of this paper describes a new framework to deal with the second and third observations above. In this framework, an intermediate level based on the *elastic band* concept is created between the planner and controller. With the elastic band, a path is treated as a flexible entity. Modifications to the path are made incrementally, using an artificial potential field (Khatib 1986). These modifications use knowledge about the obstacles in the environment, the dynamics of the robot, and sensory data.

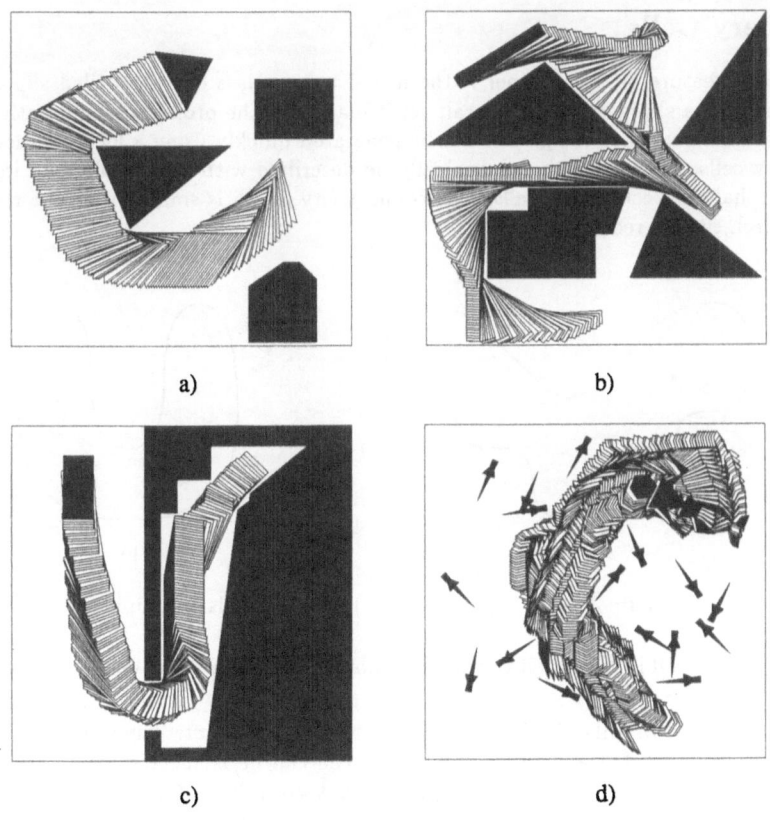

a)   b)

c)   d)

Figure 2: Some examples of paths generated by the planner

2 A Real-Time Planner

This section describes a path planner that operates in real time for static three-dimensional configuration spaces. Problems such as those shown in Figure 2 were solved by the planner in about one tenth of a second on a DECstation 5000.

The planner is based on the approximate cell decomposition method (Brooks and Lozano-Pérez 1985, Zhu and Latombe 1991). The robot's free space is decomposed into simple regions, called *cells*, such that a path between any two configurations in a cell can easily be generated. A connectivity graph of the cells is constructed and searched; the result is a sequence of cells from which a path can be computed. The planner is approximate in that the union of the cells may be a subset of the free space. The free space is represented to a resolution; the higher the resolution, the more accurate the approximation becomes. The effect of this approximation is that the planner may fail to find a path when one exists.

Slippery Cells

The novel feature of this planner is the use of a new class of cells, called *slippery cells*. The shape of a slippery cell is general, yet it maintains the property that a path between any two configurations in the cell can be generated quickly using a local method. Using slippery cells, the free space can typically be described with fewer cells than implementations that use rectangular cells. The connectivity graph is smaller with the result that the search time is reduced.

a)   b)

Figure 3: a) A slippery cell. b) A cell that is not slippery

<u>Definition</u> A cell is slippery if the outward normal from any point on the boundary does not intersect the cell.

Figure 3 shows two cells; the first is slippery, the second is not. The important property of slippery cells is that any two points in the cell can be connected by a path using an algorithm that is local. Although the proof is not given here, the intuition is as follows. To get from point A to point B, move in a straight line towards B until either B or the cell boundary is reached. In the case where the boundary is reached first, slide along the boundary by projecting the desired motion towards B into the plane tangential to the boundary. If the motion vector towards B points away from the boundary then stop sliding and move directly towards B. As the outward normal does not intersect the cell, the vector of desired motion towards B will always contain a component tangential to the boundary. It impossible to become "stuck" while sliding on the surface; the cell is slippery. It can also be shown that the motion monotonically decreases the distance to B and thus will reach its goal. Figure 4 gives an example of a path within a slippery cell.

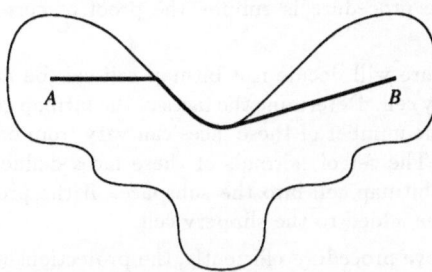

Figure 4: A path in a slippery cell

Free Space Decomposition

To use slippery cells in a planner, an algorithm is needed to decompose the free space into a set of slippery cells. As with other approximate cell decompositions, the set of cells should be disjoint and represent the free space as closely as possible. Instead of approaching this problem directly, we use an intermediate representation based on bitmaps. The bitmap configuration space, which is approximate, is then decomposed into slippery cells.

A bitmap representation is constructed by dividing the configuration space into a grid of equal sized rectangular cells. Each cells is labeled *free* or *forbidden*. Free cells are contained entirely in the free space, while forbidden cells intersect a configuration space obstacle. Such a representation can be stored in a computer as an array of bits of the appropriate dimension, hence the name bitmap. A bitmap representation is an approximate cell decomposition of the configuration space and is accurate to the resolution of the grid. For example, the three dimensional configuration spaces used to solve the problems in Figure 2 were represented as a $128 \times 128 \times 128$ grid of cells using a $128 \times 128 \times 128$ array of bits.

This paper does not address the issue of how to compute a bitmap configuration space for a robot in an environment but several approaches have been proposed (Branicky and Newman 1990; Lengyel et. al. 1990). One scheme constructs the bitmap from an analytical description of the configuration space by rendering the obstacles in an operation similar to drawing a graphical image. For the remainder of this paper we will assume that a bitmap configuration space has been generated for the environment.

To describe the bitmap configuration space as a set of disjoint slippery cells, each free bitmap cell is labeled with an integer indicating the slippery cell to which it belongs. The idea behind the labeling algorithm is the following. Find a free bitmap cell that has not been labeled as part of a slippery cell. This cell becomes the seed of a new slippery cell. The seed is grown by adding neighboring unlabeled bitmap cells that are in the free space, while preserving the slippery cell property. When no more bitmap cells can be added, the process is repeated to generate another slippery cell. If no unlabeled free bitmap cells are found, then the algorithm is complete.

The abstruse aspect of the algorithm is determining whether an unlabeled free bitmap cell may be added to a slippery cell. A simple, but computationally expensive solution is to check all bitmap cells that intersect outward normals of the bitmap cell in question. Surprisingly, we have found a procedure that determines this condition in a single memory access. Although the procedure is simple, the proof of correctness is involved and is omitted here.

The following procedure will decide if a bitmap cell can be added to a slippery cell to produce a new slippery cell. Determine the faces of the bitmap cell which will be connected to the slippery cell; the number of these faces can vary from one to the dimension of the configuration space. The set of normals of these faces define a subspace. Project the slippery cell and the bitmap cell into the subspace. If the projections are disjoint then the bitmap cell may be added to the slippery cell.

To implement the above procedure efficiently, the projections of the slippery cell into all possible subspaces are maintained while the slippery cell is grown. When a bitmap cell is added to the slippery cell, the projections are also updated. To determine whether a particular bitmap cell may be added, a single memory access into the appropriate projection is all that is needed.

To give a more concrete description of the entire algorithm, the pseudo code for the two dimensional case is given below. The input to the algorithm is a two dimensional array, C, describing the configuration space as a set of free and forbidden bitmap cells. The output is the same array, with each free bitmap cell labeled by the slippery cell to which it belongs. The following convention is used for the elements of C: negative one is a forbidden cell, zero is a free cell, and $i > 0$ is part of the ith slippery cell. Two one dimensional arrays, X and Y, record the projection of the current slippery cell onto the x-axis and y-axis respectively. A list, l, queues the bitmap cells to be examined.

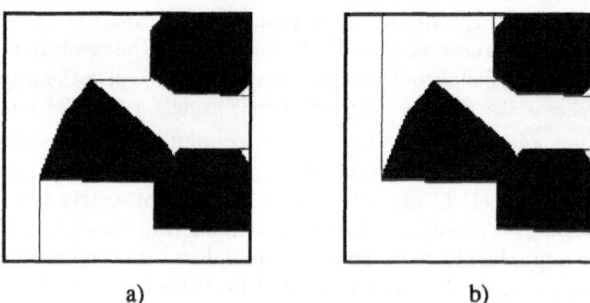

a)   b)

Figure 5: Two decomposition of a two dimensional configuration space into slippery cells. The dark region represent the obstacles and the lines denote the boundaries of the cells.

```
procedure slippery;
begin
        cell ← 0;
        for each element q in C do begin
                if C[q] = 0 then begin
                        cell ← cell + 1;
                        initialize l to q;
                        while l is not empty do begin
                                p ← first element of l;
                                if check(p) then begin
                                        C[p] ← cell;
                                        X[pₓ] ← cell;
                                        Y[pᵧ] ← cell;
                                        append the four neighbors of p to l;
                                end
                        end
                end
        end
end;

function check(q) : Boolean;
begin
        /*
         * Determine the axes along which the bitmap cell will connect
         * to the slippery cell
         */
        xaxis ← C[q+(1,0)] = cell or C[q+(-1,0)] = cell;
        yaxis ← C[q+(0,1)] = cell or C[q+(0,-1)] = cell;

        /* Check the appropriate projection */
        if xaxis and yaxis then
                return C[q] ≠ cell;
        if xaxis then
                return X[qₓ] ≠ cell;
        if yaxis then
                return Y[qᵧ] ≠ cell;

        /* The bitmap cell must be a seed and is thus a slippery cell */
        return true;
end;
```

The decomposition into slippery cells produced by the above algorithm is neither unique
nor optimal. Figure 5 gives an example of two decompositions produced for a two dimensional configuration space. The different decompositions are the result of scanning
the cells of the bitmap configuration space in a different order and thus different bitmap
cells are selected as the seeds for slippery cells. Although a given decomposition is not
optimal, it can be shown that each slippery cell is maximal under the constraints of the
configuration space obstacles and the previously grown slippery cells; for a given slippery
cell, there is no larger slippery cell that completely covers the other cell but does not
intersect the bitmap cell that are forbidden or already labeled as part of another slippery
cell.

The algorithm is efficient and generates a small number of cells. The time to examine
a single bitmap cell is $O(1)$ and each cell is examined at most $2d$ times, where d is

Table 1: Number of cells for the examples in Figure 2.

Example	Free Bitmap Cells	Slippery Cells	Octree Method
a	1238839	500	50179
b	628991	134	22781
c	154521	408	45405
d	114097	1010	31171

the dimension of the configuration space. For a bitmap with n cells, the total time to generate the slippery cells is $O(dn)$. The examples shown in Figure 2 were generated with a $128 \times 128 \times 128$ bitmap and the computation of the slippery cells took about twenty seconds on a DECstation 5000. For these examples, Table 1 compares the number of slippery cells with the much larger number of cells produced by an Octree decomposition at the same resolution (Faverjon 1984).

The Path Planner

After generating a set of slippery cells, a connectivity graph is constructed. The nodes of the graph represent the slippery cells while the arcs represent the adjacent cells. The graph can be constructed by scanning the array in which the bitmap cells are label. The scanning is an operation of order $O(dn)$ and, for the examples in Figure 2, takes about five seconds on a DECstation 5000. Table 2 gives the number of arcs in the connectivity graphs and the average number of connections per cell. As can be seen, the number of connections is low and appears to be a result of the typical size distribution of slippery cells. There tends to be a small number of large cells that describe the vast majority of the free space. Around these large cells lie many tiny cells that describe the complex surfaces of the configuration space obstacles. These tiny cells are often only connected to a single cell, namely their neighboring large cell.

Table 2: Number of in the connectivity graphs

Example	Number of Arcs	Average connectivity
a	1059	2.12
b	211	1.57
c	651	1.60
d	1673	1.66

A path between a start and goal configuration is computed in three steps. First, a sequence of adjacent slippery cells is found such that the start and goal configurations are enclosed in the first and last cell respectively. Second, for each pair of adjacent cells in the sequence, a via point is found that lies on the boundary of both cells. Third, a path is computed by concatenating segments of path between consecutive via points. These steps are the same as most other cell decomposition methods.

A sequence of cells is created by searching the connectivity graph. Ideally, the sequence should contain the optimal path and, in the context of path planning, the optimal path is typically the shortest path. However, finding such a sequence is a difficult problem; instead, in this planner, we use a heuristic intended to generate a "good" sequence. In particular, we generate the sequence that has the least number of cells by using a breadth first search. Many other heuristics are possible, for example minimizing the volume of the cells in the sequence, and we intend to investigate their relative merits.

The time to search a graph is order $O(e)$ where e is the number of arcs in the graph. For the examples in Figure 2, the connectivity graphs are small and the search time is less than one hundredth of a second.

The selection of via points also influences the length of the resulting path. The simplest solution is to have a single, predetermined via point for each pair of adjacent slippery cells. However, as some slippery cells are large, the fixed via points could lead to long, unnecessary detours. The current implementation determines the via points using a greedy search. From the start configuration, the closest point is found on the boundary between the current cell and the next cell in the sequence. This point becomes the first via point. The second via point is found in the same manner, except that the first via point is now the position from which the distance is calculated. This is repeated until the last cell is reached. As with any greedy search, the result is not optimal but produces reasonable results. The implementation of the greedy search requires many candidate boundary points to be examined. For the examples in Figure 2, the selection of via points takes about one tenth of a second.

Finally, the via points are connected to form a path. Path segments are found between the start configuration and the first via point, the final via point the goal, and each pair of consecutive via points. Each path segment lies entirely within a slippery cell and can be generated using the algorithm, described previously, for finding such paths. The path segments are then concatenated to form the complete path. The time to generate the path segments is proportional to the length of the path. For the examples in Figure 2, the path was constructed in about one hundredth of a second.

For a static, three-dimensional configuration space that is represented to a reasonable resolution, the planner described here can be considered to operate in real time. If the environment in which a robot moves does not change, then the configuration space will be static. The formation of a set of slippery cells and the generation of a connectivity map are then precomputation steps. When a task is specified, a path can be generated by searching the connectivity graph, determining via points, and concatenating path segments. Additional paths can be generated by repeating these steps without recomputing a decomposition of the configuration space. For a three dimensional configuration space represented to a resolution of $128 \times 128 \times 128$, the total time to generate a path is a fraction of a second.

The major limitation of this planner is the large memory it requires. For a three-dimensional configuration space, represented to a resolution of $128 \times 128 \times 128$, about four megabytes of memory are needed. At the same resolution, a six-dimensional configuration space requires sixteen terabytes of memory. The memory requirements restrict the planner to three or four dimensional configuration spaces.

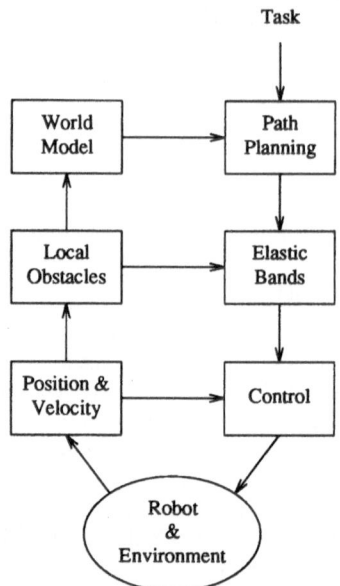

Figure 6: Proposed structure for a robotics system

3 Elastic Bands

The elastic band concept is aimed at providing an effective link between planning and execution. With the elastic band, a path is treated as a flexible entity. The initial configuration of the elastic band is the path provided by the planner. The shape of the elastic dynamically evolves during execution, using sensory data about the environment. This provide a tight connection between the robot and its environment while preserving the global knowledge needed to solve general tasks.

In this framework, the robot system has the structure shown in Figure 6. The system is a three-level hierarchy, with each level forming a closed loop with the environment. Global path planning at the highest level with the longest time cycle, the elastic band in the middle to handle local sensory data about the environment, and a controller at the bottom that closes the loop with the robots actuators and associated sensors.

The elastic band is modified by subjecting it to an artificial field of forces. These forces are designed to affect the path's length, smoothness, and obstacle clearance. The changes in the shape of the elastic band are incremental. These changes are restricted to maintain the elastic in the free space and thus it remains a globally free path.

For example, the application of an internal contraction force could be used to remove any slack and reduce the length of the elastic band. Since the elastic band is to remain in the free space, the elastic will shorten until it lies on the boundary of the configuration space obstacles. Figure 7-a shows a path obtained from a planner in a two-dimensional configuration space. This path is used as the initial configuration for an elastic band. Figure 7-b shows the elastic band after the application of an internal contraction force.

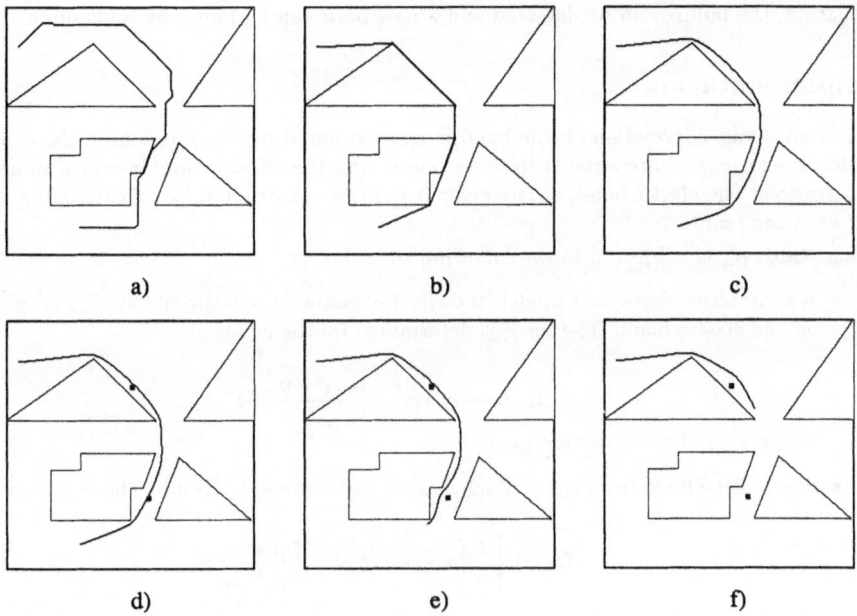

Figure 7: a) A path generated by a planner. b) Applying an internal contraction force. c) Applying both an internal contraction force and an external repulsion force. d) Some unknown obstacles on the path. e) and f) Avoiding the unknown obstacles during execution.

The internal contraction forces produce paths that are locally optimal in length. However, these paths lie on the boundaries of the configuration space obstacles and may exhibit discontinuities in direction. Obstacle clearance and first order continuity of the elastic are required for a the trajectory tracking level. One method to achieve both properties is to apply a repulsive field of forces. Each point on the elastic band is repelled from the closest point on the configuration space obstacles' boundaries. For the above example, the application of such a force field is shown in Figure 7-c.

As the robot moves along the elastic band, sensory information is used to deform the shape of the elastic while maintaining a path in the free space to the goal. This mechanism allows the robot to deal with unexpected obstacles. When such obstacles are detected, repulsive forces are applied to deform the elastic and to provide collision avoidance. Figure 7-d shows two unknown obstacles along the elastic. Figures 7-e and 7-f show the result of moving along such an elastic. As the robot approached the obstacles, the sensors detect them and the elastic is deformed.

Obviously, if the changes in the environment are large, the elastic band could fail to produce a feasible path even if one globally exists. This problem is typical of local collision avoidance methods and is the primary reason that path planning is needed. In such a situation, the failure can be detected and a new path can be found by replanning.

Implementation

The continuous curve of an elastic band is approximated by set of points in the configuration space. Forces are applied to these points and the effect is simulated by numerical integration. The elastic band, at any given instant, is constructed by concatenating paths between the points.

Each point, p_i, is subjected to the following three forces.

- A contraction force that models the effect of connecting a spring between each point on the elastic band. The force is determined by the equation

$$\mathbf{f}_c = -k_c \left(\mathbf{p}_i - \frac{\mathbf{p}_{i+1} + \mathbf{p}_{i-1}}{2} \right).$$

where k_c is the contraction gain.

- A repulsive force from the obstacles in the environment given by the equation

$$\mathbf{f}_r = \begin{cases} k_r (\frac{1}{\rho} - \frac{1}{\rho_0}) \frac{1}{\rho^2} \frac{\partial \rho}{\partial \mathbf{p}} & \text{if } \rho \le \rho_0 ; \\ 0 & \text{if } \rho > \rho_0; \end{cases}$$

where k_r is the repulsive gain, $\rho(\mathbf{p})$ is the distance at the point $\mathbf{p}$ to the nearest obstacle, and ρ_0 is the maximum distance at which the repulsive is applied. $\frac{\partial \rho}{\partial \mathbf{x}}$ denotes the partial derivative vector of the distance from $\mathbf{p}$ to the obstacle

$$\frac{\partial \rho}{\partial \mathbf{p}} = [\frac{\partial \rho}{\partial x} \ \frac{\partial \rho}{\partial y} \ \frac{\partial \rho}{\partial z}]^T.$$

If $\mathbf{q}$ is the position of the closest obstacle, the distance ρ is given by $\rho = |\ \mathbf{p} - \mathbf{q}\ |$ and

$$\frac{\partial \rho}{\partial \mathbf{p}} = \frac{\mathbf{p} - \mathbf{q}}{\rho}$$

- A velocity damping force given by

$$\mathbf{f}_v = -k_v \dot{\mathbf{p}}_i$$

where k_v is a velocity damping gain. The damping force decreases the settling time for the elastic band.

To ensure that the elastic band remain in the free space, a bitmap representation of the configuration space is used[1] The points are restricted to the free bitmap cells and each

---

[1]See the previous section for details of bitmap configuration spaces

point must maintain a distance from the two neighboring points that is less than the width of a bitmap cell. These restrictions enable a continuous curve that is enclosed in the free space to be constructed between neighboring points.

For the static portion of the environment, the distance ρ and its derivatives are efficiently computed using a lookup table. Sensory data is used to deal with the dynamic part of the environment and unexpected obstacles. In the current implementation, the shapes of unexpected obstacles are restricted to hyper-spheres.

The total time to compute the force for a point is $O(u)$, where u is the number of unexpected obstacles. For a small number of unexpected obstacles, our current implementation can perform about one hundred thousand point evaluations a second. A typical path can be described with less than 1000 points thus we achieve a servo-rate of about one hundred hertz.

The major limitation of the current implementation is the memory required for a bitmap representation of the configuration space. With current technology, such a representation is feasible only for configuration spaces of dimensions of three or four. We are currently investigating techniques that do not require an explicit representation of the entire configuration space.

4   Conclusion

Planning is expected to be the bottleneck in building systems that can execute motion tasks in real time. This paper describes a real-time path planner for static three-dimensional configuration spaces. The planner achieves this speed by precomputing a compact description of the configuration space using slippery cells. The shape of a slippery cell is general, yet it maintains the property that a path between any two configurations in the cell can be generated using a local method. With slippery cells, the free space is generally described with fewer cells than implementations that use rectangular cells. The connectivity graph is smaller with the result that the search time is reduced. After decomposition of the free space, the planner can compute paths between start and goal configurations in a fraction of a second on a typical workstation.

Elastic bands constitutes an effective framework for dealing with real-time collision-free motion control for a robot operating in an evolving environment. A planner provides an initial path that is a solution to the problem of moving a robot between a start and goal configuration. Incremental adjustments to the path are made while maintaining a global path in the free space. These modifications are based on sensory data about the environment and desired criteria concerning the path, such as length, smoothness, and obstacle clearance. Implemented as a real-time servo-loop, an elastic band provides many of the benefits of reactive systems without sacrificing global planning.

5   Acknowledgments

This material is based upon work supported under a National Science Foundation Graduate Fellowship and DARPA contract (DAAA21-89-C0002). Many thanks to Jim Bobrow,

Jean-Claude Latombe, Ross Quinlan, and Mark Yim for comments and suggestions on the work described here.

6 References

Barraquand, J. and Latombe, J. C. 1991. "Robot Motion Planning: A Distributed Representation Approach," *The International Journal of Robotics Research*, MIT Press, 10(6).

Bobrow, J., Dubowsky, S., and Gibson, J. 1983. "On the Optimal Control of Robotic Manipulator with Actuator Constraints," *Proc. American Control Conference*, San Francisco, pp. 782-787.

Branicky, M. S. and Newman, W. S. 1990. "Rapid Computation of Configuration Space Obstacles," *Proc. IEEE International Conference on Robotics and Automation*, Cincinatti.

Brooks, R. A., and Lozano-Pérez, T. 1985. "A Subdivision Algorithm in Configuration Space for Find-Path with Rotation," *IEEE Transactions on Systems, Man, and Cybernetics*, SMC-15 (2), pp. 244-233

Faverjon, A. B. 1984. "Obstacle Avoidance using an Octree in the Configuration Space of a Manipulator," *Proc. IEEE International Conference on Robotics and Automation*, Atlanta. pp. 504-512

Khatib, O. "Real-Time Obstacle Avoidance for Manipulators and Mobile Robots," *International Journal of Robotic Research*, vol. 5, no. 1, Spring 1986, pp. 90-98.

Latombe, J. C. 1991. *Robot Motion Planning*, Kluwer Academic Publishers, Boston.

Lengyel, J. Reichert, M., Donald, B. R., and Greenberg, D. P. 1990. "Real-Time Robot Motion Planning Using Rasterizing Computer Graphics Hardware," *Proc. SIGGraph*, Dallas.

Zhu, D. and Latombe, J. C. 1991. "New Heuristic Algorithms for Efficient Hierarchical Path Planning," *IEEE Transactions on Robotics and Automation*, Vol. 7, No. 1. pp. 9-20.

FORCE CONTROL OF A TWO-ARM ROBOT MANIPULATING
A DEFORMABLE OBJECT

Xavier Delebarre **, Eric Dégoulange*, Pierre Dauchez*, Yann Bouffard-Vercelli*

* LAMM - URA CNRS 1480 - UM II	** Dassault Aviation - 78 Quai Dassault
34095 Montpellier Cedex 5 - France	92214 Saint Cloud - France

ABSTRACT

A hybrid symmetric control scheme initially developed for the manipulation of rigid objects with two arms has been extended to the case of deformable objects. We briefly present this theoretical approach and give more details about the implemented control scheme. This implementation has been done on a multiprocessor architecture, which controls a robot composed of two Puma's 560. Experiments such as the manipulation of rigid objects with internal force control, or the assembly of two objects, with each being held by one arm have been carried out. However, this paper emphasizes the manipulation of deformable objects, i.e. their force controlled deformation and their transport while deformed. Several quantitative results are given, in the simple case of a one-dimensional spring.

1. INTRODUCTION

Many tasks involving the cooperation of two arms require some kind of force control [1, 2, 3, 4]. For instance, in unstructured environments such as space or ocean, two cooperating arms can be used to assemble two objects, with each being held by one arm. In this case, it is very important to be able to control contact forces between the two objects, in order to facilitate the insertion. Another important application is the assembly of an object held by two arms, on a fixed device. This requires controlling internal and external forces [5]: external force control is used during the assembly phase; in addition, if the object is rigid, internal force control is used to prevent any mechanical constraint within the object during its displacement; if the object is flexible and has to be deformed before assembly (for example, a shock absorber on a car), internal force control is used for deforming the object and for maintaining this deformation. During the deformation phase, control of the relative position of the arms is not efficient because it does not allow the system to reject some disturbances such as the slippage of the object within the grippers: if this situation occurs, although the final relative posture of the arms is the desired one, the final shape of the object is not the desired one. On the opposite, this desired shape will be obtained by controlling the internal forces produced by the deformation.

In order to reduce the implementation burden, it would be nice to find a unified approach for describing these various tasks. Since the actual controlled quantities (components of position and force) depend on the task, a hybrid control with selection matrices seems attractive. The solution with which we started has been proposed earlier by Uchiyama and Dauchez, for manipulating rigid objects with two arms [5]. Section 2 presents very briefly the main results of this approach and explains how we can use it for deformable objects. Section 3 describes our experimental setup and the control scheme we have actually implemented. Experimental results are shown in Section 4 and finally some conclusions are given in Section 5.

2. FORCE CONTROL OF A TWO-ARM ROBOT

2.1. Block diagram

If we want to hybrid control a two-arm robot manipulating a rigid object, we must first define the position and force controlled vectors. In the general case, these two vectors should have 12 components (for instance, if we have two six-degree-of-freedom arms). Intuitively, the absolute motion of the object ($\mathbf{p}_a$) and the relative motion of the arms ($\mathbf{p}_r$) can constitute the position vector $\mathbf{z} = [\mathbf{p}_a{}^T \ \mathbf{p}_r{}^T]^T$. Correspondingly, the external forces on the object ($\mathbf{f}_a$) and the internal forces in the object ($\mathbf{f}_r$) are defined to constitute the force vector $\mathbf{h} = [\mathbf{f}_a{}^T \ \mathbf{f}_r{}^T]^T$. $\mathbf{z}$ and $\mathbf{h}$ can then be used in a hybrid scheme (**Figure 1**), similar to that proposed in [6] for single-arm robots.

Subscript r represents reference values and subscript c represents current values. The vector $\boldsymbol{\xi}$ represents the joint angles of the two arms. The vector $\boldsymbol{\kappa}$ represents the forces and moments measured by the force sensors at the wrists of the arms. $\mathbf{e}_z$ and $\mathbf{e}_h$ are the command vectors to correct respectively position error and force error. Their sum $\mathbf{e}_r$ is the command vector to the actuators. $\mathbf{K}_z$ and $\mathbf{K}_h$ convert respectively the velocity and torque commands in jointspace into motor commands. These quantities are also such that the vectors e_z and e_h are of the same nature. This induces quite complex expressions of K_z and K_h, depending in general upon the dynamic parameters of the robot. G_z and G_h represent the position and force control laws. To find a stable control law for force control is a difficult problem which is not addressed in this paper [7]. J is the jacobian matrix corresponding to the kinematic transformation from ξ to z. B_a is a matrix which transforms the orientation error into a rotation vector [8]. S is a diagonal matrix which selects position or force control for such or such component. When the two arms firmly hold a rigid object, the actual controlled quantities are p_a and f_r (the first 6 components of z are controlled in position, the last 6 in force).

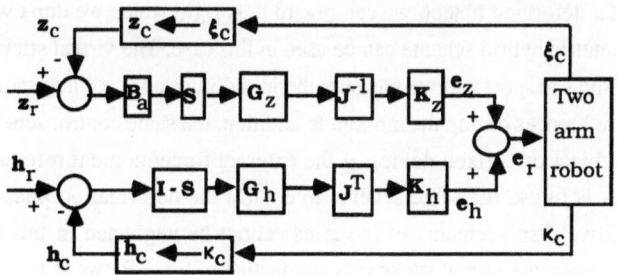

Figure 1: Hybrid control scheme for a two-arm robot

2.2. Definitions of the controlled vectors

Uchiyama and Dauchez have proposed mathematical definitions of z and h, based upon the definition of two "virtual sticks". These sticks are imaginary rigid bodies which join two arms hand-points O_{h1} and O_{h2} to the origin O_a of a frame Σ_a attached to the object. Therefore, the tips of the sticks coincide, unless the grippers slip on the object. Anyhow, the discrepancy between these tips is assumed to stay small. Then, z and h are derived by using position and force calculations at the tips of the sticks. The two arms are used for these calculations and therefore play similar roles. Consequently, the control scheme of **Figure 1** is called a symmetric hybrid control scheme.

z is defined by : $\quad z = [p_a^T \; p_r^T]^T = [1/2 \,(p_{b1}+p_{b2})^T \; (p_{b1}- p_{b2})^T]^T$

where p_{bi} represents the position and the orientation with respect to a fixed frame Σ_o of a frame Σ_{bi} attached to the tip of virtual stick i.

h is defined by : $\quad h = [f_a^T \; f_r^T]^T = [(f_{b1}+ f_{b2})^T \; 1/2 \,(f_{b1}- f_{b2})^T]^T$

where f_{bi} represents the force and moment at the tip of virtual stick i, due to the force and moment exerted on the object by arm i. f_{bi} can then be measured by the force sensor i.

2.3. Manipulation of a deformable object

To transport a deformed object, we can regard it as rigid, since we don't want its form to change. The symmetric hybrid scheme can be used in this case. The virtual sticks are defined at the beginning of the transport and we measure the initial internal force and moment which are used as constant references during the motion. In addition, the same control scheme can be used to assemble the object on a fixed device, if the external force/moment references have been defined. Our idea is to use the same scheme to control the deformation phase of the object, although the relative displacements of the arms cannot be neglected in this case. The only solution is then to make the virtual sticks vary in function of time. Between two close moments, the object does not deform much and consequently the frames Σ_{b1} and Σ_{b2} stay close to each other. If, after the elementary displacement, we redefine two virtual sticks, the tips of which coincide, we can use the same method again for the next short time interval, and so on. After each elementary motion, we can calculate the relative position and orientation p_r of these sticks, and cancel this value for the next time interval. To do so, we can use the time evolution of vector p_a (which represents the object trajectory) to redefine the virtual sticks: they still join O_{h1} and O_{h2} to O_a, which varies in space according to the reference given for p_a. The initialization of the process can be easily done in defining the virtual sticks and therefore the frame Σ_a, before any deformation. The point O_a is attached to the object. Σ_a cannot be physically attached to the object, because the deformation of the object would generally produce the deformation of Σ_a, which would not be othonormal anymore. Consequently, the "trajectory" defined by the reference p_a corresponds to:

- a displacement of the object, as far as the translational components of $\mathbf{p_a}$ are concerned, since they represent the absolute position of the point O_a attached to the object;

- a displacement of the arms, as far as the rotational components of $\mathbf{p_a}$ are concerned, since they represent the orientation of the frame Σ_a, which is not attached to the object, but which is the average of the frames Σ_{b1} and Σ_{b2}. This displacement of the arms is a global displacement which produces a rotation of the object, without producing any additional deformation of this object; indeed, the external and internal states of the object belong to orthogonal subspaces [8].

Remark: since the trajectory of the frame Σ_a can be controlled, the deformation task can be achieved while moving the object. In particular, if we specify $\mathbf{p_a}$ = constant, Σ_a becomes the frame about which the whole deformation occurs. Since this frame can be initialized in various ways, the object can be deformed with various absolute displacements of each arm. Of course we always obtain the same final shape of the object.

3. IMPLEMENTED CONTROL SCHEME

3.1. Experimental setup

In order to test the validity of our method, we have installed an experimental setup built around two Puma's 560. Each arm is equipped with a six-axis analog force sensor (built by a french company called AICO). The original Unimate controllers of these arms cannot be used for our purpose because first, they are not suitable for force control and second, our symmetric hybrid scheme requires a single controller (since the system is seen as a single robot with two arms and not as two robots).We have replaced them by a multiprocessor controller based on a VME bus built by AICO. More precisely, this controller currently includes:

- 5 processing boards (68020's with coprocessor, 16 MHz, 1Mb RAM);
- 1 memory board (1 Mb for storing global data);
- 2 counter boards (8 channels, for the 12 optical encoders signals);
- 2 A/D boards (16 channels, for the 12 potentiometers signals and for the 12 force sensors signals);
- 2 D/A boards (8 channels, for sending the commands to the 12 motors, via the amplifiers

of the Unimate controllers).

This architecture is programmed by a PC connected to it via a serial link (**Figure 2**).

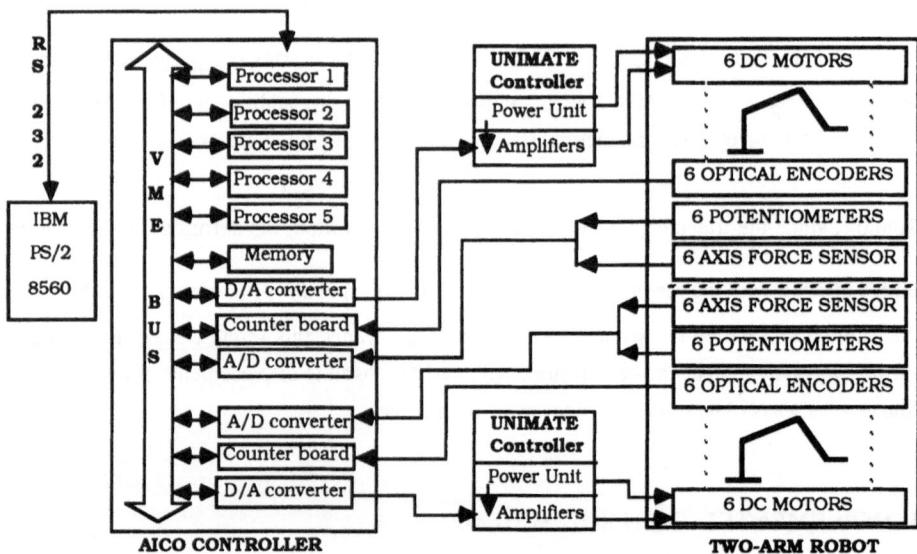

Figure 2: Connections between some components of the experimental setup

3.2. Modified block diagram

We started our implementation with the position loop of **Figure 1.** This loop shows a control law G_z in cartesian space. This solution offers the possibility to "stiffen" some particular cartesian directions and seems therefore quite interesting. However, after G_z, we used the pseudo-inverse J^+ of the jacobian J, instead of the inverse J^{-1}, in order to avoid any computation problem in case of singularity; this solution is such that the two arms "feel" the same gains. But our two PUMA's do not show the same stiffness for the same gains; therefore it was very difficult to tune the gains in the cartesian space. We then decided to add a PID controller in the joint space, after the pseudo-inverse, and to only use a proportional gain for G_z. Thanks to this PID in joint space, the behavior of the robot got much nicer.

As mentioned in section 2.1, $\mathbf{K}_z$ and $\mathbf{K}_h$ depend upon the dynamics of the system. Their calculations are not compatible with the computation power of our controller. We therefore decided to neglect the dynamic effects and to simplify the control diagram by introducing an elastic model in the force control loop. This model is represented by a proportional gain which transforms errors on the desired force/moment vector into position and orientation increments.

The calculation of $\mathbf{J}^+$ can be done numerically, but since $\mathbf{J}$ has a high dimension (for example (12x12)), this calculation is quite long (80 ms in our case). However, the computation time of a pseudo-inverse drastically decreases when the dimension of the matrix decreases (in our case, 13ms for a (6x6) matrix). Therefore, it is interesting, in order to get $\mathbf{J}^+$, to be able to compute the pseudo-inverses of smaller dimension matrices. In our case, it can be shown [8] that $\mathbf{J} = \mathbf{B}\,\mathbf{J}_b$, where $\mathbf{B}$ is a regular (12 x 12) matrix and where:

$$\mathbf{J}_b = \begin{bmatrix} \mathbf{J}_1 & 0 \\ 0 & \mathbf{J}_2 \end{bmatrix}$$

In this particular case, we cannot write $\mathbf{J}^+ = \mathbf{J}_b{}^+ \mathbf{B}^{-1}$. Nevertheless, we can show that this formula allows us to compute one solution for the joint velocities, in function of the workspace velocities (which is the only thing we need). We finally get the control diagram of **Figure 3**, which involves the inversion of two (6x6) matrices. The jointspace PID runs at 3 ms, while the sampling period of the overall system is 51 ms.

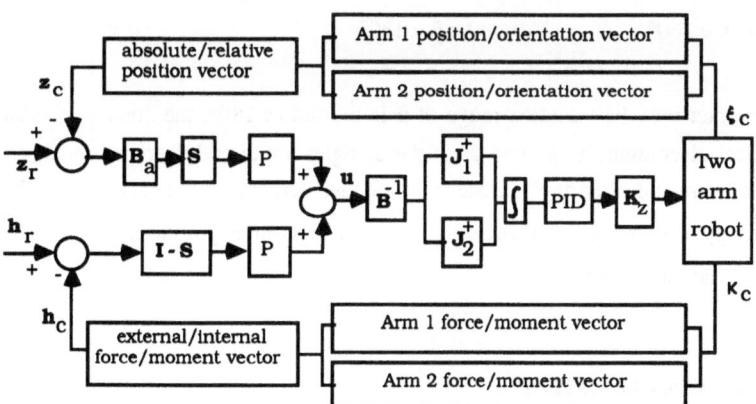

Figure 3: Modified control diagram

3.3. Representation of the orientations

The translational components of **z** are described in cartesian coordinates, which is a very usual choice. The decriptions for the rotational components can be of various kinds, including Euler angles, Roll-Pitch-Yaw angles, homogeneous transformations, ... Actually, it turns out that none of these representations is efficient for describing all the tasks which can be performed, especially during a learning phase with a teach pendant. As a matter of fact, some representations such as Euler angles may not show a possible direct rotation about a certain axis. Therefore, it would be interesting to be able to change the representation of the orientations, depending upon the task or even the sub-task which has to be described.

This modification could be done in the servo-loop by introducing a conversion block in the feedback chain and a subsequent adaptation in the direct chain. The programming burden would then increase very much. In addition, the gains of the various control laws would have to be re-tuned for each representation. It is much easier to servo the position/orientation vector **z** by choosing a fixed representation and to introduce a conversion block outside the servo loop, i.e. at the trajectory generation level. In other words, the user specifies the desired trajectory (absolute and relative) with the most convenient parameters, and a programmable transformation block possibly converts these parameters into those used for servoing (constituting z_r in **Figure 3**). As far as we are concerned, the homogeneous matrices have been chosen for representing **z**. The main reason for this choice is due to the problem encountered with angle representations. If we take the classical ZXZ Euler angles, they are defined in the following domains:

$$0 < \alpha < 360° \qquad\qquad 0 < \beta < 180° \qquad\qquad 0 < \gamma < 360°$$

Besides the fact that the variation range of β is limited to 180°, the other two angles show mathematical discontinuities around 0°. If these angles are directly servoed in **Figure 3**, the differences between desired angles close to 0° and their actual values can be close to 360°. This difference appears in the error vector $z_r - z_c$ and of course produces a non-desired and dangerous behavior of the robot.

4. EXPERIMENTAL SETUP

4.1. Qualitative experiments

With the control scheme of **Figure 3**, we have performed several experiments with our PUMA's, including:

- the manipulation of rigid objects firmly held, with control of internal and/or external forces;

- the manipulation of objects non-firmly held; more precisely, three boxes are transported together by compressing them between the two end effectors. One component of internal force is then controlled to avoid dropping the boxes; of course, external force control can still be selected;

- a cylindrical assembly in space, regarding the contact forces as internal forces. This approach was also used, for demonstration purpose, to light a match held by one arm on a match box held by the other arm. For these two experiments, we also implemented the approach phase of the arms, based upon a relative description of the task [9].

In addition, we have tried our method to manipulate a one-dimensional spring (deformable object); we have obtained several quantitative results which are presented in the next section.

4.2. Manipulation of a spring

Two experiments are presented. The first one deals with the deformation of a spring, using internal force control and the second one concerns the transport phase of this deformed object with different control modes. Before any manipulation, we have to define the virtual sticks. We have chosen to define them such that Σ_{b1}, Σ_{b2} and Σ_a are located in the middle of the spring, as can be shown in **Figure 4**. With this particular definition, the two arms contribute similarly to the motions. To measure the internal force induced in the spring by the two arms during all the manipulations, we have put a HITACHI force sensor as a spy between the two AICO force sensors which are used in the force control loop of our hybrid scheme (**Figure 4**).

4.2.1. Stretching the spring

The task consists in stretching the spring of 1 cm. It would be possible to do it by only using relative position control of each arm, but if the spring slips a little bit in the grippers, the final length would not be reached. If the know the spring stiffness (10 Ncm^{-1} in our case), it is much better to use force control to prevent such a disturbance.

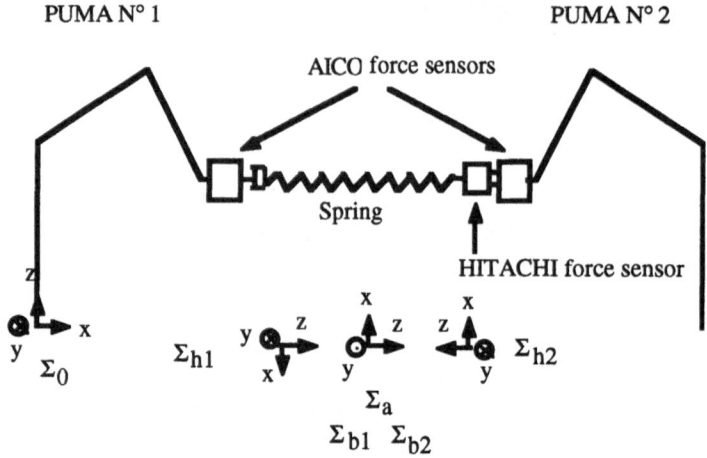

Figure 4: Definitions of the various frames

The initial length of each virtual stick is 13 cm. The corresponding initial internal force in the spring is -22.75 N (with our definition of the internal forces) and the task consists in adding -10 N to this force to reach the correct final length of the spring. Before using a fifth order polynomial to change the reference value from -22.75 N to -32.75 N, we have to define the selection matrix such that control of this internal force is activated. During the deformation, we have used the algorithm presented in paragraph 2.3 to change the length of the virtual sticks. **Figure 5-curve a** presents the reference value for the internal force and **curves b, c,** and **d** the response of the system with force loop gains tuned respectively to 20, 5, and 2 times higher than the real stiffness of the spring. For **case (b)**, the time response is really too slow and in **case (d)** we can observe an overshoot. Nevertheless, for **cases (c)** and **(d)**, no steady-state error is observed. By using the algorithm to change the virtual sticks during the deformation, the length of each stick has increased from 13 cm to 13.5 cm, which gives a total stretching of 1 cm and therefore proves that no slippage in the grippers has occured.

4.2.2. Carrying the deformed spring

In this second experiment, the two-arm robot has to carry the deformed spring without modifying its shape. To do that, several solutions can be thought of. The first one is to position control each arm separately; the second one is to use our hybrid control scheme in full position control; the third one is to use again this scheme with internal force control.

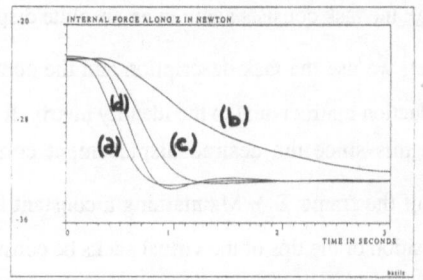

Figure 5: Internal force in deforming a spring

<u>Separate position controls</u>: for each arm, the task consists in an absolute displacement of 20 cm along the z axis of Σ_o. The initial length of the spring is 27 cm and has to remain constant. **Figure 6-curve a** shows this initial length, while **curve b** represents the real length computed via position measurements on the arms. The maximum error induced in the spring's length by the motions is 0.2 cm, as it is confirmed by the spy force sensor (**Figure 7-curve b**). According to **Figure 6-curve b**, the final length of the spring is correct; however, **Figure 7-curve b** shows that this is not true, since the final force is different from the initial one. This is due to a bad knowledge of the geometric parameters of the arms, which induces errors in computing the positions and orientations of each arm.

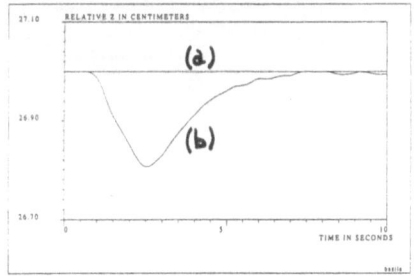

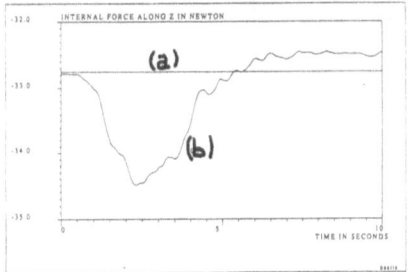

Figure 6: Spring length   Figure 7: Internal force

<u>Symmetric position control</u>: the task consists again in an absolute displacement of 20 cm along the z axis of Σ_o. This time, we use the task description and the control scheme presented in sections 2 and 3, with a selection matrix equal to the identity matrix. It is much more convenient to describe the task like this since the desired displacement corresponds to an absolute translation of the object (of the frame Σ_a). Maintaining a constant length of the spring only implies that the relative position of the tips of the virtual sticks be constant (and null).

Figure 8-curve a presents the reference value for the absolute displacement of the object, whereas **curves b** and **c** are the responses of the system. The proportional gain G_z used for **curve c** is 4 times higher than the one used for **curve b**. We can then observe a much better dynamic behavior. **Figure 9** presents the influence of this gain on the relative position along the spring direction. This gain does not have much influence on the error magnitude but does have one on the duration of this error.

This is confirmed by the information given by the spy force sensor (**Figure 10**). Here again, we can notice a steady-state error which confirms the bad knowledge of the geometric parameters of the arms.

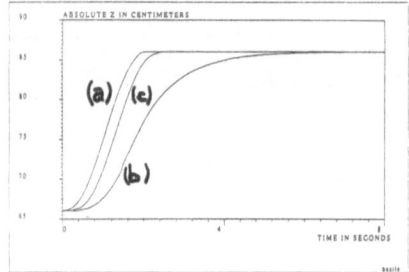

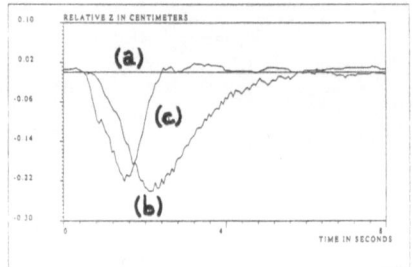

Figure 8: Absolute translation   Figure 9: Relative motion

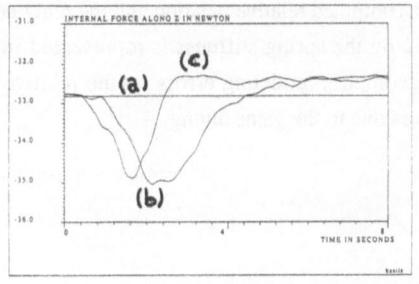

Figure 10: Internal force for symmetric position control

<u>Symmetric force control</u>: in order to reject the errors due to the bad knowledge of the geometric parameters, we can use our symmetric hybrid scheme, with control of internal force along the spring direction. **Figure 11** presents the force response of the system with gains tuned respectively to 20 (**curve b**), 5 (**curve c**) and 2 (**curve d**) times higher than the real spring's stiffness. For a well adapted gain (**curve d**), the maximum error is 0.7 N which corresponds to 0.07 cm. Now, there is no more steady-state error for the spring's length, although the computed (but not controlled) relative position of the sticks does not show it (**Figure 12**).

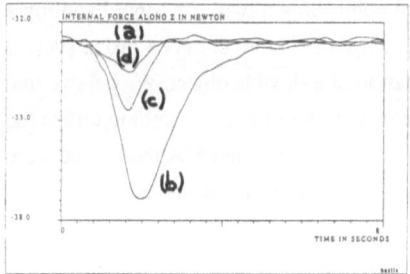

Figure 11: Internal force   Figure 12: Relative motion

The difference between this computed relative position and the real one obtained by dividing the internal force of **Figure 11** by the spring stiffness is represented in **Figure 13**. This shows the only influence of the geometric modeling errors on the relative positioning of the arms, independently from the errors due to the gains tuning.

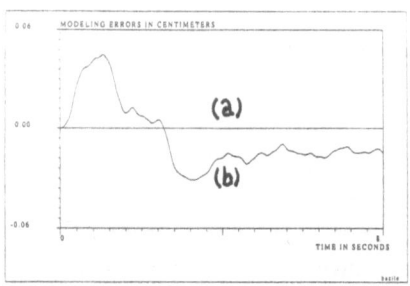

Figure 13: Influence of the modeling error on the relative position

5. CONCLUSION

Starting from an idea initially developed for force controlling a two-arm robot handling a rigid object, we have proposed a solution which can be used for more complex tasks, such as assembly in space of two objects or controlled deformation of a flexible object. We believe that the main advantage of such a unified approach is to allow the user of a real system to utilize the same control principle and similar programs for many different experiments. Although we were able to realize several experiments, we know that a lot of work still has to be done.

On a theoretical viewpoint, some research directions we have identified concern:

- the on-line modification of the selection matrix which would not affect the system stability (especially between two phases of the same task);
- the relationships between the shape of a complex deformed object and the corresponding internal forces;
- the use of a dynamic hybrid control to replace the elastic model currently implemented.

On a practical viewpoint, we plan to make many measurements in order to analyze the performances of our system. In particular, we are interested in comparing the results given by a pure position controller, a master/slave controller (these two controllers are also implemented on our system) and our hybrid controller, as far as speed, precision and percentage of success of the task are concerned (for various tasks). In parallel, we intend to modify our controller architecture in order to reduce the sampling period: we are currently investigating a solution based upon transputer boards.

6. REFERENCES

[1] E. Nakano, S. Ozaki, T. Ishida, I. Kato, "Cooperational Control of the Antropomorphous Manipulator MELARM", Proc. 4th ISIR, Tokyo, November 1974, pp. 251-260.

[2] H. Bruhm, K. Neusser, "An Active Compliance Scheme for Robots with Cooperating Arms", Proc. 3rd ICAR, Versailles, October 1987, pp. 469-480.

[3] S. Hayati, K. Tso, T. Lee, "Generalized Master/Slave Coordination and Control for a Dual Arm Robotic System", Proc. 2nd Int. Symp. on Robotics and Manufacturing, Albuquerque, November 1988, pp. 421-430.

[4] X. Yun, "Nonlinear Feedback Control of Two Manipulators in Presence of Environmental Constraints", Proc. IEEE Int. Conf. on Robotics and Automation, Scottsdale, May 1989, pp. 1252-1257.

[5] M. Uchiyama, P. Dauchez, "A Symmetric Hybrid Position/Force Control Scheme for the Coordination of Two Robots", Proc. IEEE Int. Conf. on Robotics and Automation, Philadelphia, April 1988, pp. 350-356.

[6] M.H. Raibert, J.J. Craig, "Hybrid Position/Force Control of Manipulators", Trans. ASME, Journal of Dynamic Systems, Measurement, and Control, Vol. 103, N° 2, 1981, pp. 126-133.

[7] T. Yabuta, A. Chona, "Stability of Force Control Servomechanism for Manipulator, Proc. 9th IASTED Int. Symp. on Robotics and Automation, Santa Barbara, May 1987, pp. 44-48.

[8] P. Dauchez, "Task Description for the Symmetric Hybrid Control of a Two-Arm Robot Manipulator", Technical Report # 90011, LAMM, April 1990.

[9] P. Dauchez, R. Zapata, "Coordinated Control ot Two Cooperative Manipulators : The Use of a Kinematic Model", Proc. 15th ISIR, Tokyo, September 1985, pp. 641-648.

A practical system for planning safe trajectories for manipulator robots

C. Bellier, C. Laugier, E. Mazer and J. Troccaz*
IRIMAG-IMAG/LIFIA
46, Avenue Félix Viallet, 38031 Grenoble cedex - FRANCE

Abstract

This paper deals with the problem of planning safe trajectories for a manipulator robot moving in an environment including both static obstacles and objects (other robots, conveyors ...) whose positions may change during the interval of time separating two planning steps. In order to solve this problem for a general manipulator having six degrees of freedom, we have developed and implemented a method combining two different path planners: a local planner whose purpose is to take the manipulator away from the constrained initial and final configurations, and a global planner which is in charge of generating a safe trajectory for the first three degrees of freedom of the arm. Both the local and the global planners are based on a fast iterative algorithm for computing the distance between pairs of objects. These planners have been implemented within a robot programming system including a solid modeler dedicated to robotics applications and a connection to a robot controller.

1 Introduction

Planning safe trajectories for a manipulator robot operating within an assembly workcell requires to make use of an efficient algorithm having the main following characteristics: (1) the system operates in a known and structured environment, (2) the position and the orientation of some objects (other robots, conveyors ...) may change independently of the robot considered, (3) the planning algorithm must be able to deal with up to six degrees of freedom, and (4) the model of the task (world states and the associated configuration space models) have to be updated as fast as possible after each planning step. Since solving this problem for a robot operating in a full dynamic world is too complex (according to the current state of the art), we have studied a more restricted subproblem: how to plan safe trajectories for a six axes manipulator robot moving in an environment including both static obstacles and objects whose positions may changed during the interval of time separating two planning steps ?

In order to solve this problem, we have developed and implemented a method combining two different path planners: a local planner whose purpose is to take the manipulator

---

*On leave at TIMB, Département d'Informatique Médicale, Faculté de Médecine, Domaine de la Merci, 38700 La Tronche

away from the constrained initial and final configurations, and a global planner which is in charge of generating a safe trajectory for the first three degrees of freedom of the arm (excluding wrist rotations). Both the local and the global planners are based on a fast iterative algorithm for computing the distance between pairs of objects. The local path planner has been implemented using the method described in [4]. The global path planner and the connection between the two planners are described in the rest of the paper. These planners have been implemented within a robot programming system (called ACT) including a solid modeler dedicated to robotics applications and a connection to a robot controller [1].

A large number of theoretical studies and of practical algorithms developed for the purpose of solving various instances of the general path planning problem can be found in the contemporary litterature (see [7] and [6]). But very little studies have given rise to the implementation of practical systems capable of controlling a true robot. As previously mentionned, the method described in this paper has been developed as a component of a robot programming system. It has been devised for the purpose of providing the user with interactive software tools for programming manipulation tasks involving several sequences of collision-free motions. Reasonable computation times —typically, from a few seconds to a few minutes when dealing with intricate situations— have been obtained by combining several well chosen techniques: modelling the robot world using hierarchical models [4], computing distances between pairs of objects (link-obstacle) using a fast iterative algorithm [2], computing ranges of valid configurations using a discretized representation [7] [6], and constructing an approximate model of the configuration space by combining several 2^n-trees [3] [5].

The section 2 presents the outline of our approach along with the used methods and models. The section 3 describes the method which has been devised for the purpose of computing in an efficient way the ranges of legal configurations to associate to each robot's link. The experimental results obtained by applying the method to a SCARA manipulator having four degrees of freedom and to a general six axes revolute robot are presented and discussed in the section 4.

2 Outline of our approach

2.1 Preliminary definitions and notations

The purpose of this subsection is to briefly introduce the concept of configuration space along with the used notations. Let A be an articulated system made up of p $(p \geq 1)$ rigid elements L_i (called links) moving in the cartesian space $W = R^3$. A *configuration* q of A is a minimal set of parameters —i.e. a vector in a n-dimensional space, with $n \leq p$— which completely specifies the position and orientation in W of each link L_i of A. n represents the number of degrees of freedom (d.o.f) of A; n is not equal to p (i.e. $n < p$) if A includes some coupled joints. Let $A(q)$ be the location in W of A when A is in the configuration q. The set of the possible values for the configuration vector q is the *configuration space* C_A associated to A, and the sets of configurations q which generate a collision between $A(q)$ and an obstacle in W are called the C-obstacles. Let I_i be the range of values that can be associated to the i-th joint of the robot; C_A is generally characterized by the cartesian product $I_1 \times I_2 \times \cdots \times I_n$. The associated *free-space* is the difference between C_A and the set of C-obstacles.

In the sequel, we will denote by $sweep(L_{i+1}, L_{i+2}, \cdots, L_n)$ the swept volume produced by executing a full movement for the links $L_{i+1}, L_{i+2}, \cdots, L_n$ (i.e. for all $q_{i+1} \in I_{i+1}, q_{i+2} \in I_{i+2}, \cdots, q_n \in I_n$); by the same time, the links $L_1, L_2, \cdots, L_i$ are supposed to be locked at a given location characterized by a partial configuration vector $(q_1, q_2, \cdots, q_i)$. We will denote by $A(q_1, q_2, \cdots, q_{i-1})$ the location in W of the links $L_1, L_2, \cdots, L_{i-1}$ when the configuration parameters $q_1, q_2, \cdots, q_{i-1}$ are fixed to a set of given values —in other words, the reference point of the link L_i is set to a given location characterized by the vector $(q_1, q_2, \cdots, q_{i-1})-$.

2.2   Overview of the path planners

The local path planner is directly related to the planner described in [4]. It makes use of a local approximation of the configuration space, constructed at each planning step using a set of tangent hyperplanes. One feature of this planner is its ability to cope with a large number of degrees of freedom. Another important feature of this planner (relatively to our problem) is its ability to process incompletely specified goals: for instance, a trajectory for the end effector can be only specified in terms of the trajectory of a reference point, leaving in this way the orientation of the end effector unspecified. This property is very useful for connecting the local planner with the global planner.

The global path planner makes use of an approximate representation of the robot configuration space. This representation is built using a recursive algorithm operating on an adapted discretization of the joint space (see [7] and [6]). As we will see further, the configuration space model is obtained by combining several 2^n-trees respectively representing the set of C-obstacles which is associated to the static obstacles and several C-obstacles characterizing the obstacles which can move independently of the robot. This computation is done using fast boolean operators [5]. In our approach, each 2^n-tree is constructed by recursively computing the ranges of legal configurations associated to each link of the arm. This is done using an iterative algorithm based on the distance computation between pairs of objects of the type "link-obstacle" (see section 3). Thanks to this algorithm, the system is capable of handling objects made up of complex geometric shapes (polyhedra, cylinders, cones, polytopes [1], $\cdots$). Finally, collision-free trajectories are generated using a classical A^* algorithm operating upon a graph representing the connectivity of the free-space —a node in this graph represents the free cell (i.e. a connected set of safe configurations) which is associated to a particular node of the previous 2^n-tree—.

This approach can be theoretically applied for any type of robot, but the intrinsic algorithmic complexity of the path planning problem makes it unpractical when $n > 4$ (n is the number of d.o.f of the robot). This is why the global planner is only used for planning the motions of the first three links of the arm (i.e. wrist positioning), assuming for that purpose that the next links are temporarily locked at some fixed locations. Then, safe paths involving wrist re-orientations have to be planned by combining global and local solutions (see section 4).

---

[1] a polytope is defined as the convex hull of a set of vertices

2.3 Constructing the configuration space model

An important step of the global path planning process is to construct an explicit representation of the robot configuration space C_{robot}. As previously mentionned, our system makes use of the method described in [7] and [6] for constructing an approximate model of C_{robot}. This method consists in recursively computing the ranges of configuration parameters which generate a collision with an obstacle in W, using a hierarchical model of the world and an adapted discretization of the joint space. In the current version of the system, the shape approximation level to consider and the discretization parameter associated to each d.o.f of the robot are specified by the user. The implemented method operates recursively among the n d.o.f of the robot, using the following algorithm to deal with the i-th link L_i:

$\textbf{Config}(q_1, q_2, \cdots, q_{i-1}, L_i)$

1. /* Set L_i to its initial location */
   - Compute $A(q_1, q_2, \cdots, q_{i-1})$ using the previously computed configuration parameters $q_1, q_2, \cdots, q_{i-1}$.
   - Determine the initial configuration parameters q_i (see section 3).

2. Compute the collision ranges $C_i = Range(q_i, L_i)$;
   /* The computed values $C_i \in$ C-obstacles */

3. If $(i < n)$ then
   /* L_i is not the last link of the arm */
   Compute the potential collisions $C_i^* = Range(q_i, sweep(L_{i+1}, L_{i+2}, \cdots, L_n))$;
   For each q_i^j in $(Discretization(C_i^* - C_i))$, Return $(Config(q_1, q_2, \cdots, q_{i-1}, q_i^j, L_{i+1}))$.

The ranges of q_i's values generating true collisions (by considering the solid components of L_i) and potential collisions (by considering the swept volumes) are computed using the function "Range" which is described in the section 3. The components which are considered for this computation are extracted from the hierarchical world model according to the chosen shape approximation parameter.

Remark: This algorithm can also be applied for robots having coupled joints, assuming for that purpose that closed kinematics chains, like parallelograms, are represented by open kinematics chains after having arbitrarily removed one of the coupled joints. In this case, all the links associated to a coupled joint have to be considered together at step 2 of the algorithm.

2.4 Constructing and searching the free-space model

As previously mentionned, our system makes use of a 2^n-tree representation for characterizing the part of C_{robot} which does not include any C-obstacle. Such a tree-like structure includes leaves representing either sets of safe configurations or C-obstacles; it also includes intermediate nodes defining mixed sets of configurations (figure 1 illustrates). Since the global planner is only used for planning the movements of the first three links of the arm (see section 2.2), we will make use of a 2^3-tree representation (an octree) for modeling the associated configuration space.

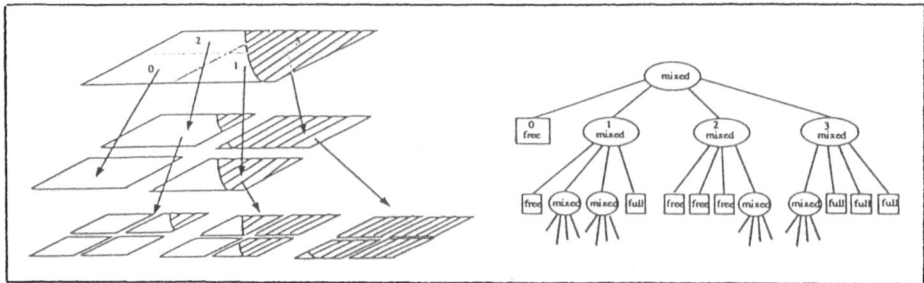

Figure 1: Representing a two dimensional configuration space using a 2^2-tree (a quadtree).

The octree representation has already been used in [3] for modeling a three-dimensional configuration space. It has been shown in [5] that such a tree-like structure can be easily constructed, modified and searched using simple and fast recursive algorithms (several instances of these algorithms exist depending on the dimension of the space considered). For instance, combining several octrees can be done using some fast boolean operators. This property is very useful for solving our problem, because it permits us to quickly update the configuration space model at each planning step. This is why, we have defined C_{robot} as the combination of several octrees : an octree representing the set of C-obstacles which is associated to the set of static obstacles, and several octrees representing the C-obstacles which are associated to the mobile obstacles (other robots, conveyors, ...). Then, simple updating operations (adding, deleting or changing the position/orientation of an object) can be executed by only modifying the involved octrees (before combining them together).

Searching for a safe-path in the previous representation requires to make use of an explicit characterization of the connectivity of the free-space. Since two adjacent regions in C_{robot} are generally represented by nodes located in different places in the tree structure (see figure 2), it is necessary to construct a *connectivity graph* linking together the nodes representing adjacent free regions of C_{robot}. This graph is called the free-space model C_{free}. It is built using a straightforward algorithm consisting in recursively determining the free neighbours of the current node (there are six possible searching directions for an octree) [5].

Then, searching for a safe-path in the C_{free} consists in searching the connectivity graph. This is done using a classical A^* algorithm and a cost function based on a particular distance in C_{robot} [6]. In the current implementation of the system, the considered distance is defined as follows: $d(q, q') = \max_{i=1,n} |q_i - q_i'|$.

Let q_{init}, q_{end} and $(N_1, N_2, \cdots, N_m)$ be respectively the initial configuration, the goal configuration and an ordered set of nodes representing the free-path generated using the previous A^* algorithm. The last step of the trajectory planning algorithm is to select in $(N_1, N_2, \cdots, N_m)$ a "good" connected set of configurations (i.e. as "short" and

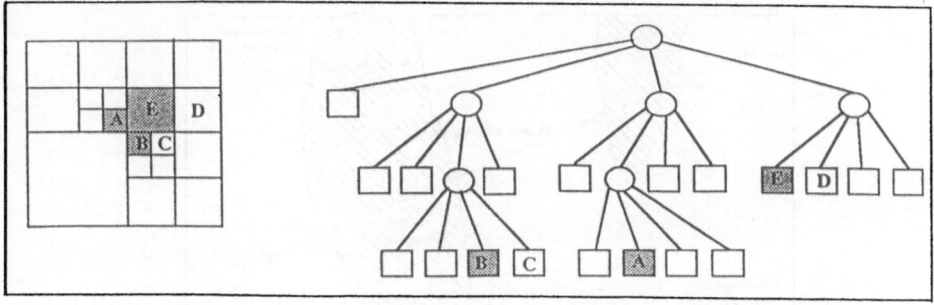

Figure 2: The regions A, B and E in C_{robot} are adjacent, but the associated nodes in the tree structure are located in very different places.

as "smooth" as possible) starting at q_{init} and endind at q_{end}. Since any node N_i is a parallelepiped in C_{robot}[2], any pair of configurations in N_i can be connected by a straight line in N_i. Consequently, the problem to solve may be converted into a more simple one consisting in determining a set of suitable configurations on the boundaries shared by the consecutive pairs of cells of the computed path. In our system, the new configuration q^i selected at the step i (for i=2,...,m) of the algorithm verifies the following property: $q^i \in N_{i-1} \cap N_i$ and $d(q^{i-1}, q^i)$ is minimal.

When no safe path exists at the current level of representation (q_{init} and q_{end} belong to two non connected components of C_{free}), the system provides the user with a simple failure diagnosis. The purpose of this diagnosis generation is to help the robot programming process: for instance, the user may decide to modify some discretization parameters at a given planning step; it also may decide to change the current location of a particular obstacle which is suspected to prevent the system to find a solution. Since a complete failure diagnosis is difficult to obtain, we have decided to only produce the list of the obstacles which generate non connected components of C_{free} and which are relevant according to the searched trajectory. A good heuristic is to analyze the C-obstacles which are located along the shortest trajectory connecting q_{init} to q_{end} (see figure 3).

3 Computing ranges of legal configurations

3.1 Fast distance computation and associated models

Both the local and the global planners make use of Euclidian distance computations for constructing the required configuration space models (see section 2.2). Since such computations are performed at each planning step for a large number of pairs of elementary objects (components of the robot links and of the obstacles), it is of a major importance to reduce the associated computation time. This is why we have implemented a fast iterative algorithm which takes full advantage of the hierarchical structure of the world

---

[2] N_i is a component of the octree representation

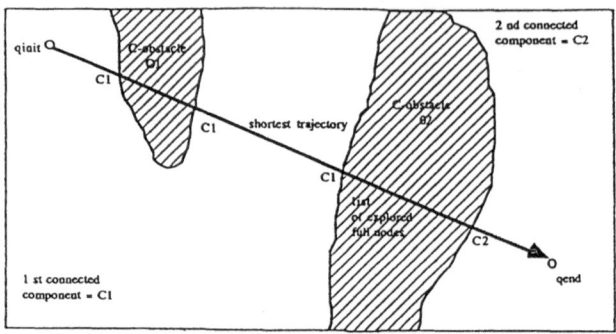

Figure 3: Identification of the obstacles which generate non connected components of C_{free} along the shortest trajectory.

model. This algorithm is based on the method proposed in [2] and in [4] for computing the distance between pairs of convex shapes and polytopes. It has been implemented within the ACT system [1] in such a way that complex geometric shapes can be handled by the path planners (i.e. objects made up of polyhedra, cylinders, cones, spheres, polytopes, ...).

Since the distance computation algorithm operates upon convex shapes and polytopes, non convex objects have to be decomposed into convex components and polynomial surfaces have to be approximated using simple polytopes (a planar polytope is associated to each patch of the discretized surface). This preprocessing phase is automatically performed by the system. It is completed by a modeling process whose purpose is to generate a hierarchy of enclosing volumes, by grouping together the previous components in an appropriate manner (in order to minimize the lost volume). This is obviously very useful for improving the distance determination procedure: this way, an obstacle which is far from the robot can be quickly processed using a rough resolution. As we will see in the sections 3.2 and 3.3, two different algorithms (aimed at respectively processing prismatic and revolute joints) have been developed for computing the ranges of legal configurations to associate to a given robot joint. Both algorithms make use of the previous distance computation procedure. They operate upon several hierarchical world models: one associated to the set of static obstacles, one associated to each mobile obstacle, and one associated to each robot link.

3.2 Computing collision ranges : the case of a prismatic joint

Let A be a convex component of the considered robot link, B be a convex component belonging to an obstacle, and $\vec{u}$ be the unit vector associated to the joint axis. In the sequel, we will refer by $P(X, \vec{u})$ the plane defined as follows: $P(X, \vec{u})$ is tangent to the convex shape X, $\vec{u}$ is the normal vector associated to $P(X, \vec{u})$, and $\vec{u}$ is oriented towards the exterior of X. Such a plane can be easily obtained using our distance computation procedure (by determining the closest points on X from a plane $P(\vec{u})$ located on the boundary of W).

Thanks to the convexity property, any range of configurations of A which generate a collision between A and B is an interval of the type $[q_{sup}, q_{inf}]$. The configuration q_{sup} (resp. q_{inf}) can be easily determined by "moving" A along the $-\vec{u}$ (resp. $+\vec{u}$) direction, starting from a point which generates no collision between A and B. A straightforward way to determine this starting point is to superpose two particular planes: the relation "A above B" can be obtained by placing the plane $P(A, -\vec{u})$ onto the plane $P(B, \vec{u})$; the relation "A under B" can be characterized in the same way using the planes $P(A, \vec{u})$ and $P(B, -\vec{u})$. Then, the following algorithm can be used to compute q_{sup} (a similar algorithm exists for q_{inf}):

1. Determine q such that the relation "$A(q)$ above B" holds;
   $d = d(A(q), B)$

2. While $(d \geq \epsilon)$ do
   $q \leftarrow Translate(A(q), -\vec{u}, d)$; $d^* = d(A(q), B)$;
   If $(d^* > d)$ then Return (no collision) else $d = d^*$

3. Return (q)

This algorithm is applied for each pair (A, B) belonging to the involved pairs of objects (link-obstacle). The resulting collision range (which is possibly made up of several non connected intervals) is defined as the union of the collision ranges associated to the processed pairs (A, B).

3.3   Computing collision ranges : the case of a revolute joint

Computing the collision ranges associated to a revolute joint can be done using a similar method. The main difficulty in this case comes from the fact that the iterative movements applied to A have as consequence to move each point of A along an arc of circle whose lenght depends on both the rotation angle and the gyration radius (see below).

Since no simple geometric property can be used for defining an initial collision-free configuration of A, we have decided to heuristically select this configuration when no obvious solution exists. The applied heuristic consists in exploring the discretized orientation domain of A (as shown in the figure 4), until a collision-free configuration has been found.

Let $[q_{sup}, q_{inf}]$ be the collision range associated to the pair (A, B). The following algorithm can be used to compute q_{sup} (a similar algorithm exists for q_{inf}):

1. If $Slice(P(A, \vec{u}), P(A, -\vec{u})) \cap Slice(P(B, \vec{u}), P(B, -\vec{u})) = \emptyset$ then Return (no collision)

2. Determine q such that $d(A(q), B) > \epsilon$;
   $d = d(A(q), B)$; $\theta = 0$;

3. While $(d \geq \epsilon)$ do
 $\theta^* = Angle(A, B, d); \theta = \theta + \theta^*;$
 If $(\theta \geq 2\pi)$ then Return (no collision); /* A can rotate freely */
 $q \leftarrow Rotate(A(q), \vec{u}, +\theta^*); d = d(A(q), B);$

4. Return (q)

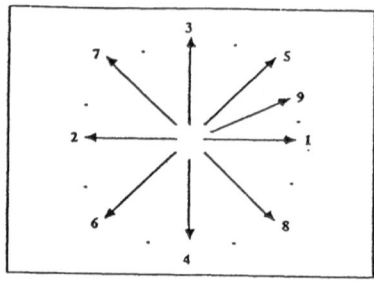

Figure 4: Searching for an initial collision-free orientation of A

The efficiency of this algorithm clearly depends on the characteristics of the function "Angle" whose purpose is to compute the next safe rotation to apply. A straightforward way for performing this computation is to make use of the relation $\theta = d/\rho$, where ρ is the maximum gyration radius of A. But, this approach generates a slow convergence when $\rho \gg d$. A more appropriate method for computing θ in this case, consists in only taking into consideration the gyration radius of some relevant points of A (namely, the points which will reach B first). Such points are referred as the "control points". In the current version of the system, they are defined as follows: $T_1 = P(A, \vec{u}) \cap P(A, \vec{n}) \cap P(A, \vec{n} \wedge \vec{u})$ and $T_2 = P(A, -\vec{u}) \cap P(A, \vec{n}) \cap P(A, \vec{n} \wedge \vec{u})$ where $\vec{n}$ is the vector associated to the distance $d(A, B)$ and oriented towards the exterior of A (see figure 5). Then, a simple algorithm can be used for computing θ:

$$\rho_1 = d(T_1, \vec{u}); \rho_2 = d(T_2, \vec{u}); \rho = max(\rho_1, \rho_2);$$

If $d(\vec{u}, P(B, -\vec{n})) \geq \rho$ then $\theta = \pi/2$ else $\theta = d/\rho;$

Finally, the resulting collision range is obtained by computing the union of the collision ranges which have been obtained for each pair (A, B).

4 Implementation and experiments

As previously mentionned, the method described in this paper has been implemented within a robot programming system (called ACT) including a solid modeler dedicated to robotics applications and a connection to a robot controller [1]. Both the ACT system and our path planner have been implemented in the C language on a Silicon Graphics workstation. The ACT system is currently marketed by the ALEPH Technologies company as a tool for developing industrial applications. It can also be used as a software basis to develop new robot programming techniques in an academic environment. Several experiments have been successfully processed using a workcell environment containing a SCARA robot having four d.o.f and a six axes revolute robot (see figure 6a).

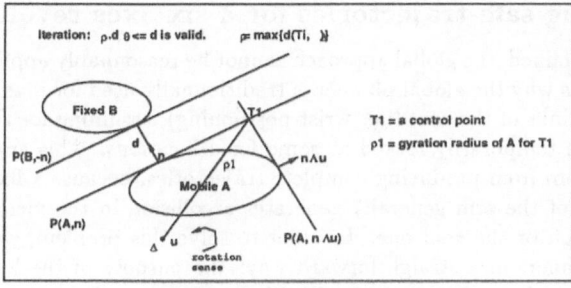

Figure 5: Determining the control points

4.1 Planning safe trajectories for a SCARA robot

Our method has been first applied for planning safe trajectories for the SCARA robot. In this case, each component of the other robot is considered as a mobile obstacle (i.e. as an obstacle whose position may change during the time separating two planning steps). Since the related planning process involves four d.o.f, we have decided to only make use of the global method for solving this problem. Let C_{robot} be the configuration space associated to the first three joints θ_1, θ_2 and z of the SCARA robot. Then, three models of C_{robot} have to be constructed for planning a full safe trajectory: one with the wrist joint θ_3 locked in the initial orientation, one with θ_3 locked in the goal orientation, and one using the volume swept by the wrist and the payload when moving from the initial orientation to the goal one. This approach, consisting in generating a safe trajectory made up of three parts, works fairly well when the robot workspace contains enough free space for executing the required re-orientations of the wrist. Processing the example shown in the figure 6a has required about 50s of elapsed time for constructing and displaying the initial configuration space models (figure 6b illustrates); afterwards, only a few seconds of elapsed time (from 1s to 5s) are required for generating a safe path in this environment and for displaying the obtained solution.

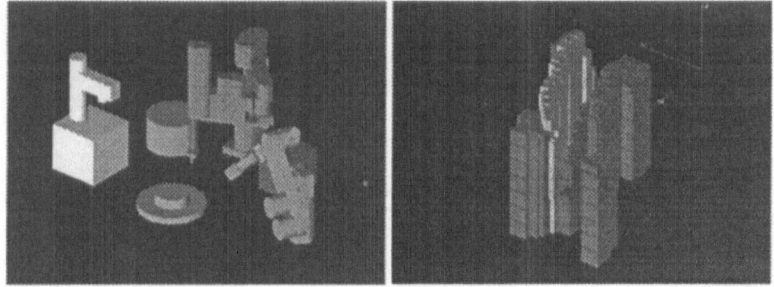

Figure 6: (a) A workcell containing a Scara robot and a six axes revolute robot. (b) The configuration space model associated to the first three d.o.f of the Scara robot

4.2 Planning safe trajectories for a six axes revolute robot

As previously explained, the global approach cannot be reasonably applied for more than three d.o.f. This is why the global planner is traditionnally used for planning the motions of the first three links of the arm (i.e. wrist positioning), assuming for that purpose that the next links are temporarily locked at some fixed locations. This approach obviously prevents the system from producing complete trajectories, because a fixed configuration of the last joints of the arm generally generates a collision in the vicinity of either the initial configuration or the goal one. In order to solve this problem, we have combined our two path planners in a straightforward way: the purpose of the local planner is to take the robot away from the constrained initial and goal configurations, and the global planner is in charge of generating a safe trajectory for moving the arm from the new initial configuration to the new goal one. Using this approach, the wrist re-orientations may be performed in two ways: either the local planner generates a solution which includes the required wrist rotations (this is generally possible for small re-orientations), or the global planner searches for an appropriate free-area (using the swept volume) in which the wrist reconfiguration can take place. In our implementation, the new initial and goal configurations, computed by the local planner, belong to an area where the wrist re-orientation can take place. For that purpose, the joints for $n > 3$ are set in an enclosing volume (an enclosing sphere for a six axes revolute robot), and the local planner is in charge of taking this volume away from the constrained initial and goal configurations.

This method has been applied for planning safe trajectories for several types of robots (including a robot having coupling joints), and for several simple and complex environments. As an example, the configuration space models associated to a revolute robot operating within a workspace containing a car made up of 1000 faces, has been constructed in approximatively 5 minutes of elapsed time. After having executed this initial operation, the system was able to generate a complete safe trajectory in 12s (let us remember that the initial configuration space model is computed at the initialization step, and that all the workspace modifications are then processed using partials computations).

Acknowledgements

The work presented in this paper has been partly supported by the DRET military agency (Direction des Recherches et Etudes Techniques) and by ALEPH Technologies (an industrial company). We would like to thank J.M. Lefevre for his contribution to this project.

References

[1] J. Troccaz E. Mazer and al. Act: a robot programming environment. In *IEEE Int. Conf. on Robotics and Automation*, April 1991. Sacramento, California.

[2] D.W. Johnson E.G. Gilbert and S.S. Keerthi. A fast procedure for computing the distance between objects. *IEEE Journbal of Robotics and Automation*, 4:193–203, 1988.

[3] B. Faverjon. Obstacle avoidance using an octree in the configuration space of a manipulator. In *IEEE Robotics and Automation*, March 1984. Atlanta.

[4] B. Faverjon and P. Tournassoud. A local based approach for path planning manipulators with a high number of degrees of freedom. In *IEEE Int. Conf. on Robotics and Automation*, March 1987. Raleigh.

[5] G. Garcia. *Contribution à la modélisation d'objets et à la détection de collisions en robotique à l'aide d'arbres octaux.* PhD thesis, Ecole Nationale Supérieure de Mécanique de Nantes, September 1989. In French.

[6] C. Laugier. Planning robot motions in the sharp system. In *CAD-Based Programming for Sensory Robots, Edited by Bahram Ravani, NATO ASI Seri F*, volume 50, pages 151–188, 1988. Springer Verlag.

[7] T. Lozano-Pérez. A simple motion-planning algorithm for general robot manipulators. In *IEEE Int. Jour. on Robotics and Automation*, June 1987.

Preliminary Experiments in Spatial Robot Juggling

A. A. Rizzi and D. E. Koditschek *

Center for Systems Science, Yale University
New Haven, CT 06520-1968

Abstract

In a continuing program of research in robotic control of intermittent dynamical tasks, we have constructed a three degree of freedom robot capable of "juggling" a ball falling freely in the earth's gravitational field. This work is a direct extension of that previously reported in [7, 3, 5, 4]. The present paper offers a comprehensive description of the new experimental apparatus and a brief account of the more general kinematic, dynamical, and computational understanding of the previous work that underlie the operation of this new machine.

1 Introduction

In our continuing research on dynamically dexterous robots we have recently completed the construction of a second generation juggling machine. Its forebear, a mechanically trivial system that used a single motor to rotate a bar parallel to a near-vertical frictionless plane was capable of juggling one or two pucks sensed by a grid of wires into a specified stable periodic motion through repeated batting [7, 3, 6]. In this second generation machine, a three degree of freedom direct drive arm (Figure 1) relies on a field rate stereo vision system to bat an artificially illuminated ping-pong ball into a specified periodic vertical motion. Despite the considerably greater kinematic, dynamical, and computational complexity of the new machine, its principle of operation represents a straightforward generalization of the ideas introduced in the previous planar study. Moreover, its empirical performance reveals strong robust stability properties similar to those predicted and empirically demonstrated in the original machine. The arm will successfully bring a wide diversity of initial conditions to the specified periodic veritical motion through repeated batting. Recovery from significant perturbations introduced by unmodeled external forces applied during the ball's free flight is quite reliable. We typically log thousands and thousands of successive impacts before a random imperfection in the wooden paddle drives the ball out of the robot's workspace.

The work presented here represents the first application of the controllers developed in [3] to a multi-axis robot, and demonstrates the capabilities of the Büghler arm and the Cyclops vision system. Both of these systems have been developed at the Yale University Robotics Laboratory to facilitate our investigations into robot control of intermittent dynamical tasks. Thus, the present paper takes on the strong aspect of displaying the fruits of previous research. We offer a comprehensive description of the components of the new apparatus in Section 2. Section 3

---

*This work has been supported in part by the Superior Electric Corporation, SGS Thomson-INMOS Corporation and the National Science Foundation under a Presidential Young Investigator Award held by the second author.

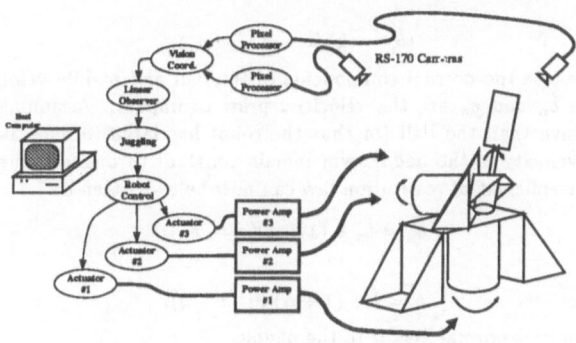

Figure 1: The Yale Spatial Juggling System

briefly reviews the notion of a mirror algorithm, the sole repository of all "juggling intelligence" in our system, and displays its generalization to the present kinematics. Section 4 provides a system level "tour" describing the manner in which the machine's physical and computational architecture is coordinated to realize the desired task. The paper concludes with a brief outline of our near-term future research directions.

2 Juggling Apparatus

This section describes the constituent pieces of our juggling machine. The system, pictured in Figure 1, consists of three major components: an environment (the ball); the robot; and an environmental sensor (the vision system). We now describe in fairly specific terms the hardware underlying each component and propose a (necessarily simplified) mathematical model in each case that describes its properties in isolation.

2.1 Environment: Striking a Ball in Flight

The two properties of the ball relevant to juggling are its flight dynamics (behavior while away from the paddle), and its impact dynamics (how it interacts with the paddle/robot). For simplicity we have chosen to model the ball's flight dynamics as a point mass under the influence of gravity. This gives rise to the flight model

$$\ddot{b} = \tilde{a}, \tag{1}$$

where $b \in B = \mathbb{R}^3$, and $\tilde{a} = (0, 0, -\gamma)^T$ is the acceleration vector experienced by the ball due to gravity.

Suppose a ball with trajectory $b(t)$ collides with the paddle in robot configuration $q \in Q$ at some point, p on the paddle which has a linear velocity v. Letting $\mathcal{T} \triangleq B \times Q$ denote the total configuration space of the problem, we seek a description of how the ball's phase, $(b, \dot{b}) \in TB$ is changed by the robot's phase, $(q, \dot{q}) \in TQ$ at an impact.

As in [7, 6] we will assume that the components of the ball's velocity tangent to the paddle at instant of contact are unchanged, while the normal component is governed by the simplistic (but standard [14]) coefficient of restitution law. For some $\alpha \in [0, 1]$ this impact model can be expressed as

$$(\dot{b}'_n - v'_n) = -\alpha(\dot{b}_n - v_n), \tag{2}$$

where $\dot{b}'_n$ and v'_n denote the normal components of the ball and paddle velocities immediately after impact, while $\dot{b}_n$ and v_n are the velocities prior to impact. Assuming that the paddle is much more massive than the ball (or that the robot has large torques at its disposal), we conclude that the velocity of the paddle will remain constant throughout the impact ($v' = v$). It follows that the coefficient of restitution law can now be re-written as

$$\dot{b}'_n = \dot{b}_n + (1 + \alpha)(v_n - \dot{b}_n). \tag{3}$$

and, hence,

$$\dot{b}' = \dot{b} + (1 + \alpha)nn^{\mathrm{T}}(v - \dot{b}), \tag{4}$$

where n denotes the unit normal vector to the paddle.

2.2 Robot Kinematics: An Almost Spherical Arm

At the heart of the juggling system resides a three degree of freedom robot — the Bühgler Arm[1] — equipped at its end effector with a paddle. The revolute joints give rise to the familiar difficulties in computing and analyzing the robot's inverse kinematics. Moreover, as in our earlier work, the presence of revolute kinematics introduces a strongly nonlinear component to the "environmental control system", an abstract discrete dynamical system with respect to which we find it effective to encode the juggling task.

The robot kinematics relevant to the task of batting a ball relates the machine's configuration to the normal vector at a point in its paddle. In order to represent this formally we parametrize the paddle's surface geometry. Let $\bar{p}$ represent (in homogeneous coordinates) a planar transformation taking points in the unit box, $\mathcal{S} \triangleq [0,1] \times [0,1]$ diffeomorphically onto the paddle's (finite) surface area expressed with respect to the gripper frame, $\mathcal{F}_g$. Associated with each point on the paddle's surface, $\bar{p}(s)$ is the unit normal, $\bar{n}(s)$, again, the homogeneous coordinate representation of the vector with respect to $\mathcal{F}_g$. The paddle's "Gauss map" [15] is now parametrized as [2]

$$N : \mathcal{S} \to \mathcal{N}(3) : s \mapsto [\bar{n}(s), \bar{p}(s)]; \qquad \mathcal{N}(3) = \mathbb{R}^3 \times S^2. \tag{5}$$

Denote by $H(q)$ the robot's forward kinematic map taking a configuration, $q \in \mathcal{Q}$, to the homogeneous matrix representation of the gripper frame with respect to the base. The world frame representation of any paddle normal at a point is thus specified by the extended forward kinematic map,

$$G : \tilde{\mathcal{Q}} \to \mathcal{N}(3) : (q, s) \mapsto [n(q, s), p(q, s)] = H(q)N(s); \qquad \tilde{\mathcal{Q}} = \mathcal{Q} \times \mathcal{S}. \tag{6}$$

At the cost of a little more notation, it will prove helpful to define the projections,

$$\pi_{\mathcal{Q}}(q, s) = q; \qquad \pi_{\mathcal{S}}(q, s) = s.$$

The linear velocity of the hit point due the robot's motion may now be written explicitly as

$$v = \sum_{i=1}^{dim\mathcal{Q}} \dot{q}_i D_{q_i} H(q)p(s) = D_q p\,\dot{q} = Dp\,\Pi_{\mathcal{Q}}\dot{q}; \qquad \Pi_{\mathcal{Q}} \triangleq [D\pi_{\mathcal{Q}}]^{\mathrm{T}} = \begin{bmatrix} I_{dim\mathcal{Q}\times dim\mathcal{Q}} \\ 0_{dim\mathcal{S}\times dim\mathcal{Q}} \end{bmatrix}, \tag{7}$$

---

[1] Pronounced *byōog'–ler*.

[2] The appearance of n in (4) suggests that it is really a force vector, thus we will define the possible normal vectors at a prescribed spatial point, $\mathcal{N}(3) \triangleq \mathbb{R}^3 \times S^2$ as lying in the space dual to the infinitesimal velocities at the point, $\mathcal{N}(3) \subset T^*\mathcal{B}$.

Additionally lying in the total configuration space is the contact submanifold, $\mathcal{C}$, — the set of ball/robot configurations where the ball is in contact with the paddle — given by

$$\mathcal{C} \triangleq \{(b,q) \in \mathcal{T} : \exists s \in \mathcal{S}, b = p(q,s)\}.$$

which is evidently the only place that the normal appearing in (4) becomes relevant. Since $\bar{p}$ is one-to-one by assumption there is a map $s_c : \mathcal{C} \rightarrow \mathcal{S}$ such that

$$b = p(q, s_c(b,q)). \tag{8}$$

Combining (7), and (8) we may now rewrite the impact event (4) in terms of a "collision map" $\mathbf{c} : T\mathcal{C} \rightarrow T\mathcal{B}$, as

$$\begin{aligned} \dot{b}' &= \dot{b} + \mathbf{c}(b,\dot{b},q,\dot{q}) \\ \mathbf{c}(b,\dot{b},q,\dot{q}) &\triangleq -(1+\alpha)n(q,s_c(b,q))n^{\mathrm{T}}(q,s_c(b,q))\left(\dot{b} - Dp\,\Pi_{\mathcal{Q}}\dot{q}\right). \end{aligned} \tag{9}$$

Choosing a gripper frame, $\mathcal{F}_g$, for the Bühgler Arm depicted in Figure 1 located at the base of the of the paddle (the point of intersection of the second and third joints) whose x-axis is aligned with the paddle's normal and whose z-axis is directed along the paddle's major axis, we have

$$N(s) = \left[\begin{bmatrix} 1 \\ 0 \\ 0 \\ 0 \end{bmatrix}, \begin{bmatrix} d_g \\ s_1 \\ s_2 \\ 1 \end{bmatrix} \right],$$

and we will artificially set $s_1 = 0, s_2 = s \in [\underline{s}, \bar{s}]$ for reasons to be made clear below. The frame transformation, $H(q)$, is developed in [13], and yields a forward kinematic map of the form

$$G(q,s) = [n(q,s), p(q,s)] \tag{10}$$

$$n(q,s) = \begin{bmatrix} \cos(q_1)\cos(q_2)\cos(q_3) - \sin(q_1)\sin(q_3) \\ \cos(q_2)\cos(q_3)\sin(q_1) + \cos(q_1)\sin(q_3) \\ -\cos(q_3)\sin(q_2) \\ 0 \end{bmatrix}$$

$$p(q,s) = \begin{bmatrix} -(\sin(q_1)d_2) + (\cos(q_1)\cos(q_2)\cos(q_3) - \sin(q_1)\sin(q_3))\,d_g + \cos(q_1)\sin(q_2)s_2 \\ \cos(q_1)d_2 + (\cos(q_2)\cos(q_3)\sin(q_1) + \cos(q_1)\sin(q_3))\,d_g + \sin(q_1)\sin(q_2)s_2 \\ -(\cos(q_3)\sin(q_2)d_g) + \cos(q_2)s_2 \\ 1 \end{bmatrix}.$$

Analysis of the jacobian of p shows that it is rank three away from the surface defined by

$$\delta_p(q,s) \triangleq (s_2^2 + \cos^2(q_2)\cos^2(q_3))(\sin^2(q_2) + \cos^2(q_3)) = 0,$$

thus away from $\delta_p = 0$ we can define $Dp^{\dagger}$, the right inverse of Dp, and the workspace is now given by $\mathcal{W} \triangleq p(\tilde{\mathcal{Q}} - \mathcal{H})$ where

$$\mathcal{H} \triangleq \left\{\tilde{q} \in \tilde{\mathcal{Q}} : \delta_p(\tilde{q}) = 0\right\}$$

Finally the inverse kinematic image of a point $b \in \mathcal{W}$ may be readily computed as

$$p^{-1}(b) = \begin{bmatrix} \mathrm{ArcTan2}\,(-b_2, -b_1) + \mathrm{ArcSin}\left(\frac{\sin(q_3)d_2}{\sqrt{b_1^2 + b_2^2}}\right) \\ -\frac{\pi}{2} + \mathrm{ArcTan2}\left(b_3, \sqrt{b_1^2 + b_2^2 - \sin(q_3)d_2^2}\right) - \mathrm{ArcSin}\left(\frac{\cos(q_3)d_g}{\sqrt{b^T b - \sin^2(q_3)d_2^2}}\right) \\ q_3 \\ 0 \\ \sqrt{b^T b - \sin^2(q_3)d_2^2 - \cos^2(q_3)d_g^2} \end{bmatrix} \; ; \; q_3 \in S^1, \tag{11}$$

with the freely chosen parameter, q_3, describing the one dimensional set of robot configurations capable of reaching the point b. Having simplified the kinematics via the artificial joint constraint, $s_1 \equiv 0$, the paddle contact map may simply be read off the inverse kinematics function,

$$s_c(b,q) = \pi_S \circ p^{-1}(b) = \begin{bmatrix} 0 \\ \sqrt{b^T b - \sin^2(q_3)d_2^2 - \cos^2(q_3)d_g^2} \end{bmatrix}.$$

2.3 Sensors: A Field Rate Stereo Vision System

Two RS-170 CCD television cameras with 1/2000sec. electronic shutters constitute the "eyes" of the juggling system. In order to make this task tractable we have simplified the environment the vision system must interpret. The "world" as seen by the cameras contains only one white ball against a black background. The CYCLOPS vision system, described in Section 2.4, allows the straightforward integration of these cameras into the larger system.

Following Andersson's experience in real-time visual servoing [1] we employ the result of a first order moment computation applied to a small window of a threshold-sampled (that is, binary valued) image of each camera's output. Thresholding, of course, necessitates a visually structured environment, and we presently illuminate white ping-pong balls with halogen lamps while putting black matte cloth cowling on the robot, floor, and curtaining off any background scene.

2.3.1 Triangulation

In order to simplify the construction of a trangulator for this vision system, we have employed a simple projective camera model. Let $\mathcal{F}_c$ be a frame of reference whose origin is at the focal point and whose z-axis is directed toward the image plane of this camera. Let $^c p = [p_x, p_y, p_z, 1]^T$ denote the homogeneous representation with respect to this frame of some spatial point. Then the camera, with focal length f, transforms this quantity as

$$^c u = f \begin{bmatrix} p_x/p_z \\ p_y/p_z \\ 1 \\ 0 \end{bmatrix} \triangleq \mathbf{p}_f(\,^c p\,). \tag{12}$$

Here, $^c u \in \mathbb{R}^4$ is the homogeneous vector, with respect to $\mathcal{F}_c$, joining the oragin of $\mathcal{F}_c$ to the image plane coordinates of $^c p$. Thus, for a camera whose position and orientation relative to the base frame, $\mathcal{F}_0$ are described by the homogeneous matrix $^0 H_c$, the projection of a point, $^0 p$ is

$$u = \mathbf{p}_f(\,^c H_0 \,^0 p\,).$$

Given two such cameras separated in space, whose frames of reference with respect to $\mathcal{F}_0$ are represented by $^0 H_l$ and $^0 H_r$, it is straightforward to derive a *triangulation function*, $\mathbf{p}^\dagger$, capable of reconstructing the spatial location of a point, given its projection in both images. In particular if projection onto the *right* and *left* image planes is given by

$$^r u_r = \mathbf{p}_{f_r}(\,^r H_0 \,^0 p\,),$$

and

$$^l u_l = \mathbf{p}_{f_l}(\,^l H_0 \,^0 p\,)$$

respectively, a (by no means unique) triangulation function is given by

$$\mathbf{p}^\dagger(\,^r u_r , \,^l u_l\,) \triangleq \frac{1}{2}\left(\,^0 H_r\,(o + t_r \,^r u_r) + \,^0 H_l\,(o + t_l \,^l u_l)\,\right), \tag{13}$$

where

$$\begin{bmatrix} t_r \\ t_l \end{bmatrix} \triangleq \left(C^T C \right)^{-1} C^T (o - {}^r H_0 \, {}^0 H_l \, o)$$

and

$$o \triangleq [0 \quad 0 \quad 0 \quad 1]^T ; \; C \triangleq [-{}^r u_r \mid {}^r u_l]; \; {}^r u_l = {}^r H_0 \, {}^0 H_l \, {}^l u_l .$$

This amounts to finding the midpoint of the biperpendicular line segment joining the two lines defined by ${}^r u_r$ and ${}^r u_l$. Note that there is considerable freedom in the definition of $\mathbf{p}^\dagger$ since it maps a four dimensional space (the two image plane vectors) onto a space of dimension three (ball position).

Finally it is worth noting that although the implementation of a triangulation system of this type is simple, the measurement of the parameters required for its construction is quite difficult. A short description of the automatic method of calibration we have chosen to use in the juggling system can be found in Appendix A.

2.3.2 Signal Processing

In practice it is necessary to associate a signal processing system with the sensor to facilitate interpretation of the data. For the vision system in use here, sensor interpretation consists of estimating the ball's position and velocity, correcting for the latency of the vision system, and improving the data rate out of the sensor system – the 60 Hz of the vision system is far below the bandwidth of the robot control system.

Given reports of the ball's position from the triangulator it is straightforward to build a linear observer for the full state — positions and velocities — since the linear dynamical system defined by (1) is observable. In point of fact, it is not the ball's position, b_n, which is input to the observer, but the result of a series of computations applied to the cameras' image planes, and this "detail" comprises the chief source of difficulty in building cartesian sensors of this nature.

2.4   Controller: A Flexibly Reconfigurable Computational Network

All of the growing number of experimental projects within the the Yale University Robotics Laboratory are controlled by widely various sized networks of Transputers produced by the INMOS division of SGS-Thomson. Pricing and availability of both hardware and software tools make this a natural choice as the building block for what we have come to think of as a computational "patch panel." The recourse to parallel computation considerably boosts the processing power per unit cost that we can bring to bear on any laboratory application. At the same time the serial communication links have facilitated quick network development and modification.

The choice of the INMOS product line represents a strategy which standardizes and places the burden of parallelism — inter-processor communications support, software, and development environment — around a commercial product, while customizing the computational "identity" of particular nodes by recourse to special purpose hardware. We provide here a brief sketch of the XP/DCS family of boards, a line of I/O and memory customized Transputer nodes developed within the Yale Robotics Lab and presently employed in all our control experiments. The backbone of this system is the XP/DCS CPU, providing a transputer and bus extender. By coupling an XP/DCS to an IO/MOD a computational node can be customized for interfacing to moderate bandwidth hardware. Similarly joining up to eight XP/DCSs to individual CYCLOPS frame memory boards, then ganging these together under a single video digitizer forms a programmable field rate monocular vision system.

The **XP/DCS processor** The XP/DCS (produced by Evergreen Designs) was designed in conjunction with the Yale Robotics Laboratory in 1987 [9] in order to meet both the computational and I/O requirements presented by robotic tasks. The board is based on the INMOS T800 processor, a 32 bit scalar processor capable of 10 MIPS and 1.5 MFLOP (sustained) with four bidirectional 20MHz DMA driven communication links and 4 Kbytes of internal (1 cycle) RAM. The processor is augmented with an additional 1-4 Mbytes of dynamic RAM (3 cycle), and an I/O connector which presents the T800's bus to a daughter board.

IO/MOD The IO/MOD (also produced by Evergreen Designs) allows an XP/DCS to "communicate" with custom hardware in a simple fashion. In order to properly implement the ideal "processing path panel" it is essential that the integration of new sensors and actuators be simple and fast. The IO/MOD augments an XP/DCS by providing a 32 bit latched bidirectional data bus, six 4 bit wide digital output ports, and eight digital input signals, all of which are mapped directly into the memory space of the T800.

CYCLOPS Vision System Much like the IO/MOD the CYCLOPS system has been designed to augment a set of XP/DCS boards for a particular sensing task — vision. In actuality there are three major components to the vision system [8]:

Digitizer: Digitizes an incoming RS-170 video signal and outputs it in digital form over a pixel bus.

Filter: A filter board capable of performing real-time 2D convolution on an image may be placed on the pixel bus.

Frame Memory: In much the same fashion as the IO/MOD the CYCLOPS Memory Board augments an XP/DCS with 128 Kbytes of video memory. By associating up to eight memory boards with a pixel bus it becomes easy to construct a real-time parallel processing vision system.

3 Juggling Algorithm

This section offers a brief review of how the juggling analysis and control methodology originally introduced for the planar system [7] may be extended in a straightforward manner to the present apparatus. After introducing the "environmental control system," an abstract dynamical system formed by composing the free flight and impact models, it becomes possible to encode an elementary dexterous task, the "vertical one juggle," as an equilibrium state — a fixed point. A simple computation reveals that every achievable vertical one juggle can be made a fixed point, and conversely, the only fixed points of the environmental control system are those that encode a vertical one juggle. Leapfrogging the intermediate linearized analysis of our planar work [3], we then immediately follow with a description of a continuous robot reference trajectory generation strategy, the "mirror law," whose implementation gives rise to the juggling behavior.

3.1 Task Encoding

Denote by $\mathcal{V}$ the robot's choices of impact normal velocity for each workspace location. Suppose that the robot strikes the ball in state $w_j = (b_j, \dot{b}_j)$ at time s with a velocity at normal $v_j = (q, \dot{q}) \in \mathcal{V}$ and allows the ball to fly freely until time $s + t_j$. According to (9) derived in the previous section, composition with time of flight yields the "environmental control system"

$$w_{j+1} = f(w_j, v_j, t_j) \triangleq A_{t_j} w_j + a_{t_j} + \begin{bmatrix} t_j \mathbf{c}(w_j, v_j) \\ \mathbf{c}(w_j, v_j) \end{bmatrix}, \tag{14}$$

that we will now be concerned with as a controlled system defined by the dynamics

$$f : T\mathcal{B} \times \mathcal{V} \times \mathbb{R} \to T\mathcal{B},$$

with control inputs in $\mathcal{V} \times \mathbb{R}$ (v_j and t_j).

Probably the simplest systematic behavior of this system imaginable (beyond the ball at rest on the paddle), is a periodic vertical motion of the ball. In particular, we want to be able to specify an arbitrary "apex" point, and from arbitrary initial conditions, force the ball to attain a periodic trajectory which passes through that apex point. This corresponds exactly to the choice of a fixed point, w^*, in (14), of the form

$$w^* = \begin{bmatrix} b^* \\ \dot{b}^* \end{bmatrix}; \quad b^* \in \mathbb{R}^3; \quad \dot{b}^* = \begin{bmatrix} 0 \\ 0 \\ \nu \end{bmatrix}; \quad \nu \in \mathbb{R}^3, \tag{15}$$

denoting a ball state-at-impact occurring at a specified location, with a velocity which implies purely vertical motion and whose magnitude is sufficient to bring it to a pre-specified height during free flight. Denote this four degree of freedom set of vertical one-juggles by the symbol $\mathcal{J}$.

The question remains as to which tasks in $\mathcal{J}$ can be achieved by the robot's actions. In particular we wish to determine which elements of $\mathcal{J}$ can be made fixed points of (14). Analysis of the fixed point conditions imposes the following requirements on w^*:

$$\dot{b}^* = \frac{1}{2}\lambda \tilde{a} \tag{16}$$

and for some $(q, \dot{q}) \in T\mathcal{Q}$ and $\lambda \in \mathbb{R}^+$,

$$p(q, s_c(b^*, q)) = b^* \text{ and } \mathbf{c}(b^*, \dot{b}^*, q, \dot{q}) = -\lambda \tilde{a}. \tag{17}$$

Every element of $\mathcal{J}$ satisfies (16), since this simply enforces that the task be a vertical one-juggle. For the Bühgler Arm (17) necessitates that n be aligned with $\tilde{a}$ so as not to impart some horizontal velocity on the ball. From (10) it is clear that this will only be the case when $q \in \mathcal{Q}^*$, where

$$\mathcal{Q}^* \stackrel{\triangle}{=} \{q \in \mathcal{Q} : \cos(q_3)\sin(q_2) = -1\}.$$

Thus, we can conclude that only those elements of $\mathcal{J}$ satisfying the condition $b^* \in p(\mathcal{Q}^*)$ will be fixable. In particular, $\mathcal{Q}^*$ corresponds to the paddle being positioned parallel to the floor, and thus $p(\mathcal{Q}^*)$ is an annulus above the floor, as is intuitively expected.

This simple analysis now furnishes the means of tuning the spatial locus and height of the desired vertical one juggle. The fixed-point input satisfying these conditions, u^*, is given by

$$u^* = \begin{bmatrix} \frac{2}{\gamma}\|\dot{b}^*\| \\ \left(\frac{\alpha-1}{\alpha+1}\right)Dp^{\dagger}\dot{b}^* \end{bmatrix}.$$

3.2 Controlling the Vertical One-Juggle via a Mirror Law

Say that the abstract feedback law for (14), $g : \mathcal{W} \to \mathcal{V} \times \mathbb{R}$, is a verticle one-juggle strategy if it induces a closed loop system,

$$f_g(w) = f(w, g(w)), \tag{18}$$

for which $w^* \in \mathcal{J}$ is asymptotically stable fixed point. For our original planar machine [3] it was shown that the linearization of the analogous system to (14) was controllable around every vertical one juggle task. A similar analysis has not yet been completed for the Bühgler Arm, although a similar result is expected. Experiments with the planar system revealed that the linearized perspective was inadequate: the domain of attraction resulting from locally stabilizing linear state feedback was smaller than the resolution of the robot's sensors [3].

Instead, in [7] a rather different juggling strategy was proposed that implicitly realized an effective discrete feedback policy, g, by requiring the robot to track a distorted reflection of the ball's continuous trajectory. This policy, the "mirror law," may be represented as a map $m : T\mathcal{B} \to \mathcal{Q}$, so that the robot's reference trajectory is determined by

$$q(t) = m(w(t)).$$

For a one degree of freedom environment it is not hard to show that this policy results in a (essentially) globally asymptotically stable fixed point [5]. For a two degree of freedom environment, we have shown that local asymptotic stability results [3]. The spatial analysis is in progress.

The juggling algorithm used in the present work is a direct extension of this "mirror" law to the spatial juggling problem. In particular begin by using (11) to define the the joint space position of the ball

$$\begin{bmatrix} \phi_b \\ \theta_b \\ \psi_b \\ s_b \end{bmatrix} \stackrel{\Delta}{=} p^{-1}(b). \tag{19}$$

We now seek to express formulaically a robot strategy that causes the paddle to respond to the motions of the ball in four ways:

(i) $q_{d1} = \phi_b$ causes the paddle tracks under the ball at all times.

(ii) The paddle "mirrors" the vertical motion of the ball through the action of θ_b on q_{d2} as expressed by the original planar mirror law [7].

(iii) Radial motion of the ball causes the paddle to raise and lower, resulting in the normal being adjusted to correct for radial deviation in the ball position.

(iv) Lateral motion of the ball causes the paddle to roll, again adjusting the normal so as to correct for lateral position errors.

To this end, define the ball's *vertical energy* and *radial distance* as

$$\eta \stackrel{\Delta}{=} \gamma b_z + \frac{1}{2}\dot{b}_z^2 \quad \text{and,} \quad \rho_b \stackrel{\Delta}{=} \sin(\theta_b)s_b \tag{20}$$

respectively. The complete mirror law combines these two measures with a set point description ($\bar{\eta}$, $\bar{\rho}$, and $\bar{\phi}$) to form the function

$$q_d = m(w) \stackrel{\Delta}{=} \begin{bmatrix} \overbrace{\phi_b}^{(i)} \\ \underbrace{-\frac{\pi}{2} - (\kappa_0 + \kappa_1(\eta - \bar{\eta}))\left(\theta_b + \frac{\pi}{2}\right)}_{(ii)} + \underbrace{\kappa_{00}(\rho_b - \bar{\rho}_b) + \kappa_{01}\dot{\rho}_b}_{(iii)} \\ \underbrace{\kappa_{10}(\phi_b - \bar{\phi}_b) + \kappa_{11}\dot{\phi}_b}_{(iv)} \end{bmatrix}. \tag{21}$$

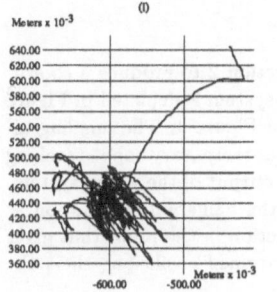 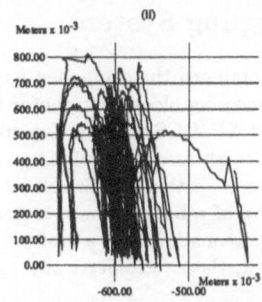 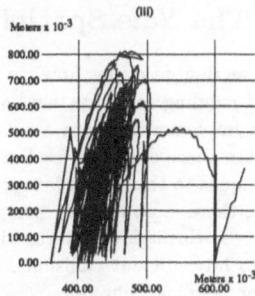

Figure 2: One-Juggle ball trajectory: (i) X-Y projection (ii) X-Z projection and (iii) Y-Z projection.

For implementation, the on-line reference trajectory formed by passing the ball's state trajectory, $w(t)$, through this transformation must be passed to the robot tracking controller. As described in Section 4.4, the high performance inverse dynamics tracking schemes that we presently employ require the availability of a target velocity and acceleration profile as well. By design $m(w)$ is differentiable and the time derivatives of w are known – at least away from an impact event. Thus, denoting by

$$F(w) \triangleq \left[\begin{array}{c} w_2 \\ \tilde{a} \end{array} \right]$$

the spatial vector field corresponding to the ball's continuous free flight dynamics (1), we know that $q_d(t) = m(w(t))$ implies

$$\dot{q}_d = Dm\, F$$

and

$$\ddot{q}_d = Dm\, DF\, F + [F \otimes I]^\mathsf{T} D^2 m\, F$$

In practice, these terms are computed symbolically from (21) and F.

We have succeeded in implementing the one-juggle task as defined above on the Bühgler arm. The overall performance of the constituent pieces of the system – vision module, juggling algorithm, and robot controller – have each been outstanding, allowing for performance that is gratifyingly similar to the impressive robustness and reliability of the planar juggling system. We typically record thousands of impacts (hours of juggling) before random system imperfections (electrical noise, paddle inconsistencies) result in failure. Figure 2 shows the three projections of the ball's trajectory for a typical run. As can be seen the system is capable of containing the ball within roughly 15cm of the target position above the floor and 10cm of the target height of 60cm.

It is worth noting that in the x-z and y-z projections there is evidently spurious acceleration of the ball in both the x and y directions. Tracing this phenomenon through the system confirmed an earlier suspicion; our assumption that gravity is exactly aligned with the axis of rotation of the base motor is indeed erroneous. Correction of this calibration error requires the addition of trivial workspace cues (a plumb bob) to allow the direction of the gravitational force to be calibrated along with the remainder of the system. This correction is now in progress.

4 The Yale Spatial Juggling System

This section describes how we have combined the components of Section 2 to produce a coordinated juggling robot system. An engineering block diagram for this system is depicted in Figure 3. Its implementation in a network of XP/DCS nodes is depicted in Figure 4. The juggling algorithm these diagrams realize is a straightforward application of contemporary robot tracking techniques to the mirror law presented in Section 3 as driven by the output of the vision system. Thus, there is no explicit pre-planning of robot motions. Instead, the ball's natural motion as perceived through the stereo vision system stimulates a "reflex" reaction in the robot that gives rise to intermittent collisions. In turn, these "geometrically programmed" collisions elicit the proper juggling behavior.

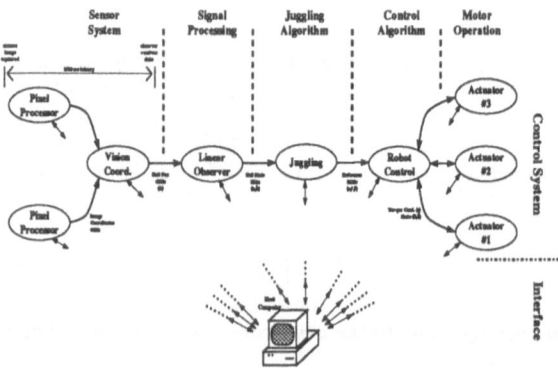

Figure 3: The Juggling System Controller Block Diagram.

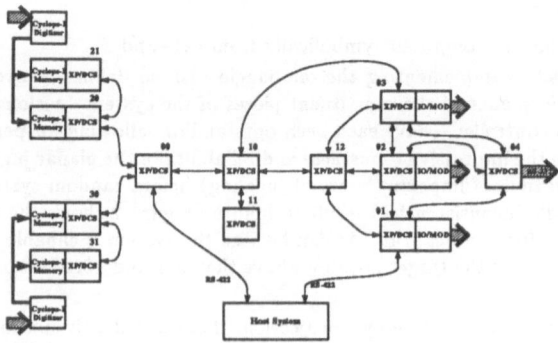

Figure 4: The Juggling System Network Diagram.

4.1 Vision: Environmental Sensing

The vision system must provide sufficient information regarding the state of the environment. In our present implementation we have so structured the visual environment as to make the task conceptually trivial and computationally tractable. In particular, the vision system need only extract the three dimensional position of a single white ball against a black background.

To perform even this apparently trivial task in real time we require two CYCLOPS vision systems — one for each "eye" — introducing a total of four nodes. In both Cyclops systems two memory boards, each with an associated XP/DCS processor, are attached to the digitizer. Computational limitations currently allow the system to process "windows" of less than 3000 pixels out of an image of 131, 072 pixels (a 256 × 512 image).

Figure 5 depicts the flow of events on the five processors used for vision processing during an image cycle. The cycle begins with a field being delivered to one memory board associated with each camera (two of processors 20, 21, 30, 31). Note that these two events are guaranteed to occur simultaneously through the use of synchronized cameras. After the images have been deposited in the memory boards the centroid of the ball is estimated by calculating the first order moments over a window centered around the position of the ball, as determined by the most recent field (depicted by arrows in Figure 5). Upon completion, the image coordinate locations are passed to the neighboring pixel processors, for use in determining window location for the next field, and up to the triangulation process which is located on processor 00 of Figure 4. Once the low-level pixel processors have determined the ball location in a pair of images, stereo triangulation introduced in Section 2.3 locates the position of the ball in space with respect to a fixed reference coordinate system.

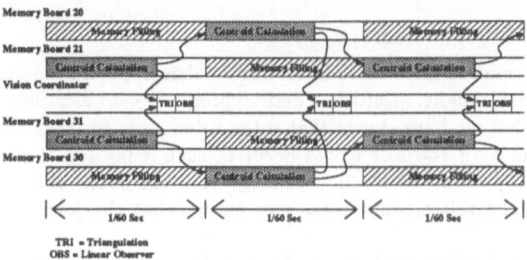

Figure 5: Timing for CYCLOPS vision system

4.2 Signal Processing

The signal processing block must "interpret" the environment and present it in a fashion that is acceptable for use by the remainder of the system. In this simple context, "interpretation" means producing good estimates of the ball's position and velocity at the current time. This is accomplished by connecting the output of the triangulator to a standard linear observer.

The timing diagram in Figure 5 shows that the vision block adds an unavoidable 1/30 sec. delay between the time an image is captured and the time a spatial position measurement has been formed. The ball's flight model presented in Section 2.1 is a sufficiently simple dynamical system that its future can be predicted with reasonable accuracy and, accordingly, a current estimate of the state is formed by integrating that delayed state estimate forward in time one field interval.

The data must now be passed to the interpolator. The task here involves stepping up the unacceptably slow data rate of the vision block (60 Hz): the time constant of the actuators is near 200 Hz. This interpolation stage uses the flight model of the ball, integrating the current state estimate of the ball forward over small time steps allows the data rate to be boosted from 60 Hz to 1 kHz.

This sequence of calculations is carried out on processor 00, (the coincidence with the triangulation process is incidental). The implementation of these signal processing functions is divided into two independent event driven processes. The first of these runs the observer and predictor, which are synchronized with the triangulation system and thereby with the cameras. Thus the sampling rate for the observer is set by the field rate of the cameras. The second process increases the effective data rate by scheduling itself at 1 msec intervals and updates its estimates of the ball state at each interval.

4.3 Juggling

The execution of the jugling algorithms developed in [7, 4] and Section 3 are implemented in this segment of the network. The evaluation of (21) is again carried out on processor 00, where both the high-level vision and signal processing are performed. The implementation consists of a single process which evaluates q_d, $\dot{q}_d$, and $\ddot{q}_d$ whenever new state information is received – yet another example of how we use naturally occuring events within the system to initiate computation. Since the input of this process is connected to the output of the interpolator the reference trajectory fed to the controller will be updated at the same rate as the output of the interpolator (1 kHz).

4.4 Robot Control

The geometric transformation introduced in Section 3, when applied to the joint space coordinate representation of the ball's flight, results in a desired profile of joint locations over time, $q_d(t)$. For the planar juggling robot we have shown that if the robot were to track exactly this "reference signal," then collisions with the ball would occur in such a fashion that the desired periodic motion is asymptotically achieved [7, 3]. We conjecture the same will be true in the present case. It now falls to the robot control block to ensure that the joint angles, $q(t)$, track the reference signal, $q_d(t)$.

We have implemented a large collection of feedback controllers on the Bühgler Arm, as reported in [17]. We find that as the juggling task becomes more complicated – e.g. simultaniously juggling two balls – that it becomes necessary to move to a more capable controller. We have had good succes with an inverse dynamics control law [17] of the form developed in [11],

$$\tau = C(q, \dot{q})\dot{q} + M(q)[\ddot{q}_d] + K_d(\dot{q} - \dot{q}_d) + K_p(q - q_d). \qquad (22)$$

At the present time, all experiments have all been run with a robot control block that includes the three nodes (10, 11, and 12 in Figure 4). The model based portion of the control algorithms are implemented on processor 11 with update rate of 400 Hz, while the feedback portion (along with uninteresting housekeeping and logging functions) is implemented on 10 with update rate of 1 KHz, and a variety of message passing and buffering processes run on 12 which is really included in the network only for purposes of increasing the number of Transputer links converging on this most busy intersection of the entire network. There are two motivations for this seemingly lavish expenditure of hardware. First, in the interests of keeping the cross latency of the intrinsic data as low as possible [16], the increased number of links permits direct connectivity of the controller block with each node of the actuator block. Second, in the interests of maintaining the "orthogonality" of the real-time and logging data flows, we gain a sufficient number of links at the controller to permit the dedicated assignment of logging channels back toward the user interface.

Nonblocking communication between this collection of processors is implemented through the use small input buffer processes. These buffer processes, which are situated at the inputs to each computational process, allow the various elements of the controller to receive data from each other and other elements of the system asynchronously. Thus far we have found that the reduction of effort required for software development and maintenance resulting from the use of this architecture has outweighed the performance costs imposed by the necessarily increased network latencies.

4.5 Actuator Management

The primary task associated with operating a particular actuator is to present a standard simple interface to the remainder of the system, thereby hiding the often unpleasant details of operating a particular motor from the remainder of the system. Since our interest is in developing systems which respect the dynamics of the robot, it follows that the actuator interface must receive torque commands and report state (position and velocity). In addition to this basic task we have found it desirable to include a strong system of safety interlocks at the lowest possible level, so as to ensure safe operation at all times.

5 Conclusion

This paper provides a comprehensive description of the generalization to three space of our earlier juggling work [7, 6]. To those familiar with that previous body of work, it may be apparent that the generalization begins to shed greater light on what went before. In particular, we are now able explicitly to show how the spatial juggling law and its planar predecessor are based on applying the mirroring notion of the "gedanken" line juggler [5] to the inverse kinematic image of the ball's workspace trajectory. In the case of the line juggler we had shown (using the "Singer-Guckenheimer" theory of unimodal return maps [5]) that the global stability mechanism this mirror law induces is a change of coordinates away from that which we had argued [10] stabilizes Raibert's [12] vertical hopping motion. In the case of the planar juggler, we were able to demonstrate that the mirror law results in local asymptotic stability, but, since an analytical expression for the effective closed loop dynamics (14) remained elusive, no global stability analysis has yet been possible.

The absence of an explicit closed form expression is a consequence of the transcendental system of equations that emerge when "PD" terms are added to the simple gedanken robot's mirror law. The same terms appear in the present algorithm (21). Thus, we do not believe that a global stability analysis will be any easier for our current spatial version of this idea. We believe, however, that an empirically transparent modification of these terms that re-expresses them with respect to the joint space coordinate system (as does the application of inverse kinematics to the ball's flight) may result in tractable closed loop dynamics and, thereby, the possibility of a global stability proof for the more interesting kinematics.

A Vision System Calibration

In the course of getting started with spatial juggling, we have been led to re-formulate a very attractive coordinated camera-arm calibration scheme originally proposed by Hollerbach [2]. At calibration time, one supposes that some point on the robot's gripper (that we willl take to be the origin of the "tool" frame) is marked with a light reflecting material in such a fashion as to produce an unmistakable camera observation — a four vector, $c \in \mathbb{R}^4$ comprised of the two image plane measurements. The problem is to determine the kinematic parameters, $k \in \mathbb{R}^{8+3(m+1)}$, that characterize the chain as well as the relative camera frame relationship and camera focal lengths by comparing measured camera values with the joint space locations that produced them.

The Setting Denote by g_k, the forward kinematic transformation of the kinematic chain that expresses the robot's tip marking with respect to the base frame (that we take to be the frame of the "right" hand camers with no loss of generality). According to the Denavit-Hartenburg convention, the parameter vector, $(k_1, ..., k_{m+1}) \in \mathbb{R}^{3(m+1)}$, that characterizes this function appears in the form

$$g_k(q) = \left(\prod_{i=1}^{m+1} H_i(\theta_i) \right) \begin{bmatrix} 0 \\ 0 \\ 0 \\ 1 \end{bmatrix}; \qquad H_i(\theta_i) \triangleq \exp \left\{ \theta_i \sum_{j=1}^{3} J_{ij} k_{ij} \right\},$$

where θ_i is a joint variable and J_{ij} is a constant 4×4 array whose exponent yields the homogeneous matrix representation of the unit screw scaled by parameter k_{ij}. If these $3(m+1)$ parameters were known then g_k would yield for every jointspace location, $q = (\theta_1, ..., \theta_n)^T \in \mathcal{Q}$, the homogeneous representation of the tool frame origin in base frame coordinates.

Now denote by H_0 the homogeneous matrix representation of the screw relating the "left" hand camera frame to the base frame,

$$H_0 = \exp \left\{ \sum_{j=1}^{6} k_{0j} J_{0j} \right\},$$

where J_{0j} constitutes an arbitrary basis for the Lie Algebra corresponding to the group of rigid transformations and $\bar{k}_0 \in \mathbb{R}^6$ parametrizes the relative camera frame transformation matrix accordingly. The camera transformation is now characterized by the parameters $k_0 = (k_{00}, k'_{00}, \bar{k}_0) \in \mathbb{R}^8$ that appear in the the stereo projective transformation, $\mathbf{p}_{k_0} : \mathbb{R}^3 \to \mathbb{R}^4$, that for a given camera pair associates with each spatial point a pair of ("left" and "right" camera-) planar points. Specifically, let Π, π denote the projections from $\mathbb{R}^4$ that pick out, respectively, the first two, and the third coordinate, of a homogeneous representation of a point. The camera transformation may be written as

$$\mathbf{p}_{k_0}(w) = \begin{bmatrix} \Pi(w)/k'_{00}\pi(w) \\ \Pi(H_0 w)/k_{00}\pi(H_0 w) \end{bmatrix}.$$

This function admits a family of pseudo-inverses $\mathbf{p}_{k_0}^{\dagger} : \mathbb{R}^4 \to \mathbb{R}^3$, whose effect on the camera image plane, $\mathbf{p}_{k_0}(\mathbb{R}^3) \subset \mathbb{R}^4$, returns the original spatial point — that is $\mathbf{p}_{k_0}^{\dagger} \circ \mathbf{p}_{k_0}$ is the identity transformation of $\mathbb{R}^3$ — and whose effect off the camera image plane is to return the "closest" spatial point to that four-vector with respect to a suitable metric.

A Modified Procedure Hollerbach's proposed procedure tested in simulation of a planar arm, [2], calls for recording some number of joint-space/camera-image pairs, $\mathcal{D} = \{(q_l, c_l)\}_{l=1}^n$, and then performing a Newton-like numerical descent algorithm on the cost function

$$\sum_{l=1}^{n} \|\mathbf{p}_{k_0}^{\dagger}(c_l) - g_k(q_l)\|^2.$$

When we attempted to implement this procedure for the three degree of freedom Bühgler arm, we found that the procedure was extremely sensitive numerically.

Instead, we have had great success with a variant on this idea that substitutes a cost function in the stereo camera image space,

$$\sum_{l=1}^{n} \|c_l - \mathbf{p}_{k_0} \circ g_k(q_l)\|^2,$$

for the previously defined workspace objective. We have been using this procedure on average several times a month (the experimental apparatus is frequently torn down and put back together again to incorporate new hardware, necessitating continual re-calibration) for the last six months with very good results. Starting from eyeball guesses of $k = (k_0, k_1, k_2, k_3, k_4)$, we have been able to achieve parameter estimates that give millimeter accuracy in workspace after two or three hours of gradient descent farmed out on a network of eight 1.5 Mflop microcomputers (Inmos T800 TRAMS). We have experienced similar reliable convergence properties with a variety of algorithms — standard gradient descent; Newton Raphson; Simplex descents — none of which seemed to avail (either singly or in more clever combination) using the original objective function.

References

[1] R. L. Andersson. *A Robot Ping-Pong Player: Experiment in Real-Time Intelligent Control.* MIT, Cambridge,MA, 1988.

[2] D. J. Bennett, J. M. Hollerbach, and D. Geiger. Autonomous robot calibration for hand-eye coordination. In *International Symposium of Robotics Research*, 1989.

[3] M. Bühler , D. E. Koditschek, and P.J. Kindlmann. A Simple Juggling Robot: Theory and Experimentation. In V. Hayward and O. Khatib, editors, *Experimental Robotics I*, pages 35–73. Springer-Verlag, 1990.

[4] M. Bühler , D. E. Koditschek, and P.J. Kindlmann. Planning and Control of Robotic Juggling Tasks. In H. Miura and S. Arimoto, editors, *Fifth International Symposium on Robotics Research*, pages 321–332. MIT Press, 1990.

[5] M. Bühler and D. E. Koditschek. From stable to chaotic juggling. In *Proc. IEEE International Conference on Robotics and Automation*, pages 1976–1981, Cincinnati, OH, May 1990.

[6] M. Bühler, D. E. Koditschek, and P. J. Kindlmann. Planning and control of a juggling robot. *Int. J. Rob. Research*, (submitted), 1991 .

[7] M. Bühler, D. E. Koditschek, and P.J. Kindlmann. A family of robot control strategies for intermittent dynamical environments. *IEEE Control Systems Magazine*, 10:16–22, Feb 1990.

[8] M. Bühler, N. Vlamis, C. J. Taylor, and A. Ganz. The cyclops vision system. In *Proc. North American Transputer Users Group Meeting*, Salt Lake City, UT, APR 1989.

[9] M. Bühler, L. Whitcomb, F. Levin, and D. E. Koditschek. A new distributed real-time controller for robotics applications. In *Proc. 34th IEEE Computer Society International Conference — COMPCON*, pages 63–68, San Francisco, CA, Feb 1989. IEEE Computer Society Press.

[10] D. E. Koditschek and M. Bühler. Analysis of a simplified hopping robot. *Int. J. Rob. Research*, 10(6), Dec 1991 .

[11] Daniel E. Koditschek. Natural motion for robot arms. In *IEEE Proceedings 23rd Conference on Decision and Control*, pages 733–735, Las Vegas, Dec 1984.

[12] Marc H. Raibert. *Legged Robots That Balance*. MIT Press, Cambridge, MA, 1986.

[13] A. A. Rizzi and D. E. Koditschek. Progress in spatial robot juggling. Technical Report 9113, Yale University, Center for Systems Science, New Haven, Connecticut, USA, October 1991.

[14] J. L. Synge and B. A. Griffith. *Principles of Mechanics*. McGraw Hill, London, 1959.

[15] John A. Thorpe. *Elementary Topics in Differential Geometry*. Springer-Verlag, New York, 1979.

[16] L. L. Whitcomb and D. E. Koditschek. Robot control in a message passing environment. In *Proc. IEEE International Conference on Robotics and Automation*, pages 1198–1203, Cincinnati, OH, May 1990.

[17] Louis L. Whitcomb, Alfred Rizzi, and Daniel E. Koditschek. Comparative experiments with a new adaptive controller for robot arms. In *Proc. IEEE Int. Conf. Rob. and Aut.*, pages 2–7, Sacramento, CA, April 1991. IEEE Computer Society.

Section 6: Mobile Robots

Mobile robots play an increasingly important role in robotics research. They are indeed a very good paradigm of the integration of perception, decision, and action, and of operation in non structured environments. The following papers cope with different aspects pertaining to mobile robotics.

Real-time obstacle avoidance is a key feature of mobile robots. Each research laboratory has implemented its own algorithm, the most popular being based on the potential field approach. The idea of R. Liscano, R. Manz, and D. Green was to actually implement three of these methods on their own robot and compare their behaviors. Their paper reports on the performance measure of these methods in different situations.

The three other papers in this section deal with the problem of control systems design and implementation. This is a central question that determines the ability of the robot to correctly and efficiently manages its functions to accomplish tasks and to be reactive to events.

H. Chochon presents an original and very promising object-oriented approach for the design of a control structure for a mobile robot. The object-oriented viewpoint enables to define the robot functions as a hierarchy of classes, and methods to access them. A client/server mechanism is used for implementing the exchanges between the functional modules. The architecture also includes dedicated control modules for monitoring, resource management, conflict resolution, etc. This approach is implemented on a mobile robot and experimental results are presented.

The paper by J. L. Crowley, P. Reignier and O. Causse describes the architecture of a robot control system for indoor navigation. This navigation was extensively experimented with a mobile robot. The procedures for environment perception using sonars, trajectory generation and execution, etc. are described and provide a very good example of mobile robot programming.

The last paper in this section by S. Yuta, S. Suzuki and S. Iida describes the control architecture of Yamabico, the mobile robot family developed at the University of Tsukuba. Its implementation is detailed as well as the language that was developed to specify robot tasks.

A Comparison of Realtime Obstacle Avoidance Methods for Mobile Robots[1]

Allan Manz[2] Ramiro Liscano David Green
Autonomous Systems Laboratory , Institute for Information Technology
National Research Council of Canada, Ottawa, Ont, Canada K1A 0R6

Abstract Recently several new dynamic approaches have been developed to enable a vehicle to autonomously avoid obstacles in its environment in real time. These algorithms are generally considered as reflexive collision avoidance algorithms since they are continuously using the latest update in sensory data and computing from this data error signals to drive and/or steer the vehicle away from a collision with the environment. This paper experimentally tested three of these methods, potential fields, generalized potential fields and vector field histograms, using a uniform set of hardware and software modules. The types of tests chosen were typical of navigation in an indoor environment and consisted of avoiding a single obstacle at high speed, travelling through a narrow hallway and passing through an open doorway. The main issues observed during the testing were, the maximum speed at which the vehicle could accomplish the test, the nature of the path taken by the vehicle during the test and any difficulties that arose in the process of implementing any of the algorithms.

1. Introduction

Recently several approaches have been advocated to enable mobile robots to avoid obstacles in real time. These methods are different than conventional "path planning" algorithms in that they take into account the dynamics of the vehicle as new avoidance paths are computed continuously using sensory data. In general most traditional path planning algorithms are known as "global" path planners. These map the environment using a computer representation and then compute an optimal route so that the robot can achieve its desired goal. Several researchers have implemented these path planning methods on vehicles using real sensory data to construct models of the environment which were then used by the the path planning algorithms to compute new paths to the destination point. Some of this original work was reported by Moravec [13] that used a cart equipped with a sliding camera to sense the environment. Range values to a set of candidate points were computed using nine stereo images taken by the camera. These were arranged on a 2-D grid so that a path could be computed around the objects. The cart would follow a repetitive cycle in which it would take a picture while stationary, plan its next move, advance 3 feet, stop, and then repeat the cycle. Several researchers have implemented path planning methods that essentially rely on modelling the environment and which then compute new paths when a collision seemed imminent [14,15,16]. The main criticism of these systems is that they have been developed from a top-down approach using a classical *sense-think-act* [17] cycle which first senses the environment, then reasons about the situation, and finally moves to avoid a collision with the obstacle. Because of the sequential nature of this cycle, these systems introduce a long latency in the computation of an alternative path and therefore do not operate in real time. In contrast, several methods [1,3,5,18,25] have been reported that are based on a bottom-up

[1] NRC No. 31826
[2] Graduate student, University of Saskatchewan, Saskatoon Sk. S7N 0W0

approach using reflexive navigation concepts in which real sensory data is used immediately to modify the behaviour of the vehicle.

This paper is concerned with the implementation and comparison of realtime obstacle avoidance methods. Three approaches to realtime obstacle avoidance that have been discussed in the literature are: (1) the classical potential field approach [1,2,10], (2) the generalized potential field approach [4,5,6], and (3) the vector field histogram approach [3]. Testing of these methods under operational conditions, utilising the same software architecture and hardware, provides an effective and comparative performance evaluation. The criteria for comparison are: (1) The difficulty associated with implementing and testing the algorithms, (2) The success of each algorithm in passing the tests and the limiting conditions in each case, and (3) The nature of the path that the vehicle takes during each test.

In this paper, a modular system for controlling a mobile robot is described which facilitates the testing of the three methods for collision avoidance described above. The system was designed using a multitasking, multiprocessing realtime operating system called Harmony[3] which was developed at the National Research Council, and which has been used to control robots in a variety of experiments [12]. The principal advantage of Harmony is that it allows the user to design a number of independent tasks that reside in a multiprocessor environment and which communicate with each other by message passing. This leads to a system structure that is modular and which provides a uniform methodology for intertask communications which simplifies the design of complex systems such as those used for controlling a mobile robot [19].

Three types of tests were chosen that were considered representative of the hazards often encountered by mobile robots in an indoor environment. They were: (1) the ability of the vehicle to avoid a single object while operating at maximum speed, (2) the ability of the vehicle to pass through a narrow corridor, and (3) the ability to pass through a doorway. It was felt that the most challenging task is the doorway traversal since it appears to be a combination of the other two problems. Obstacle avoidance capability is required to avoid a collision with the walls adjoining the doorway while at the same time, the algorithms have to search for the free space of the doorway. Little experimental testing has been reported on these algorithms. Recently Koren and Borenstein [11,24] performed mathematical and experimental tests on the potential field methods and have described the limitations of these methods in traversing narrow passages. We have been able to verify some of their claims and we have performed the same tests using the generalized potential field and vector field histogram methods.

This paper is divided into the following sections: (1) the experimental setup, which describes both the hardware and system structure used, (2) the collision avoidance algorithms, which gives a brief overview of the potential field, generalized potential field and vector field histogram approaches, and (3) the results which discusses the observations and conclusions produced during the course of the experiments.

2. Experimental System

It was important while conducting these tests that consistency in hardware and system structure be maintained to facilitate a comparison of the results. In the following section we describe the experimental hardware used and the modular system structure

---

3 Harmony is a mark reserved for the exclusive use of Her Majesty the Queen in right of Canada by the National Research Council Canada

that allowed us to simply replace the collision avoidance module by the particular algorithm being tested at that time.

2.1 Hardware

The vehicle used in the experiments was equipped with a 0.6 m diameter ring of 24 Polaroid ultrasound transducers [20] controlled by three ultrasound controllers from Denning Robotics [21] which were each connected to eight transducers. These three controllers were in turn connected to a custom built command interpreter that communicated via a serial line to the main processor [22]. The command interpreter then allowed a user the flexibility to configure the ultrasound sensors and firing pattern in several different ways. For our experiments the controllers were configured so that each transducer had an effective beamwidth of approximately 16° and a range of operation of from 0.1 m to 1.0 m. To improve performance, the system fired three sensors, each offset by 120°, simultaneously.

The platform is commercially available from Cybermotion Inc. [23] and utilises a unique syncro-drive mechanism for driving and steering the vehicle. This configuration provides good vehicle performance and allows us to use the platform's odometer for reliable measurement of the position of the vehicle. For safety reasons, the platform has been equipped with a physical bumper that increases the maximum width from 0.81 to 1.2 m. The platform motors are controlled by algorithms implemented by the manufacturer using a Z-80 microprocessor. Communications with this controller uses a serial line and is synchronised to the 10 Hz vehicle servo rate.

Both the vehicle controller and the ultrasound command interpreter are interfaced using separate serial lines to an on-board 4-processor 68020 VME based multiprocessor system that runs under the Harmony operating system. This multiprocessor is the computing platform that was used for testing the collision avoidance algorithms.

2.2 System

The system structure is designed as a number of self-contained tasks that communicate with each other through message passing. The key tasks in the system are the *coordinator*, the *sonar controller*, the *certainty grid*, and the *algorithms*. Figure 1 is a layout of the main tasks in the system and their associated connections to each other.

The sonar controller task receives 24 range values from the ring of ultrasound sensors approximately every 400 ms. In theory, taking into account the speed of sound and the time to transmit the data serially (52 ms), it should only take 100 ms to receive 24 values from the sensors if the sensors are fired three at a time. This large difference in time is due to a 43 ms delay that has been introduced into the firing sequence. This was necessary in order to reduce the number of false readings that were otherwise obtained. It is believed that these false readings were mostly caused by cross-talk between the sensors, i.e. a sensor receiving an echo from a previously fired sensor. No effort was taken to eliminate this cross-talk and therefore slow scanning speeds were used which limited the maximum velocity of the vehicle during the tests. It was felt that as long as the test conditions were identical between algorithms then they would be equally affected by the low scan rate.

Because of the large number of false readings acquired from the sonar [7,8], the raw sonar data was mapped onto a certainty grid. The certainty grid task calculates and updates the certainty grid, using the Histogramic In-Motion Mapping (HIMM) approach developed by Borenstein [3,9]. Two different types of grids are used in conjunction with each other to compute the certainty values of the cells. The first grid consists of a square

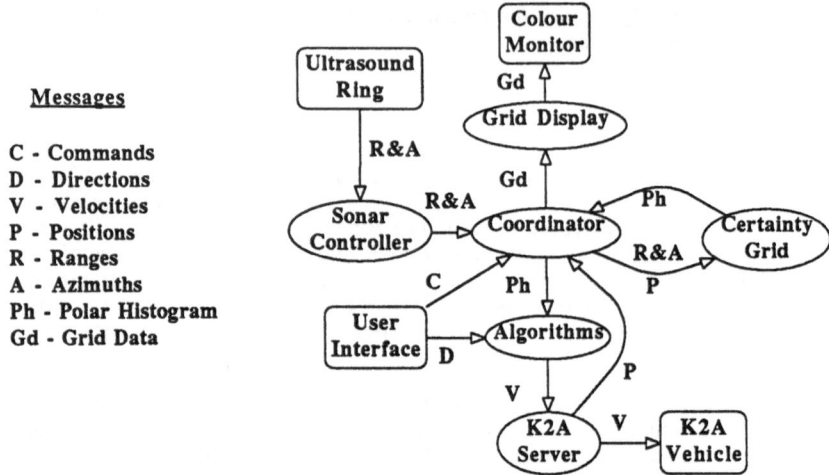

Figure 1. System software layout showing the main tasks and their connections.

2.6 m on a side subdivided into 676 squares each 0.1 m long. The square grid pattern is easier to update for local vehicle motion. The second grid is polar in shape and consists of 48 pie shaped sectors, each spanning 7.5°, and 10 concentric rings of 0.1 m depth starting from a 0.3 m radius out to 1.3 m centered about the vehicle. It was chosen because it conforms directly to the sonar data which is obtained in polar coordinates. Figure 2 shows an 180° portion of the polar grid overlaid on top of the corresponding half of the square grid so that the centers of the grids are aligned.

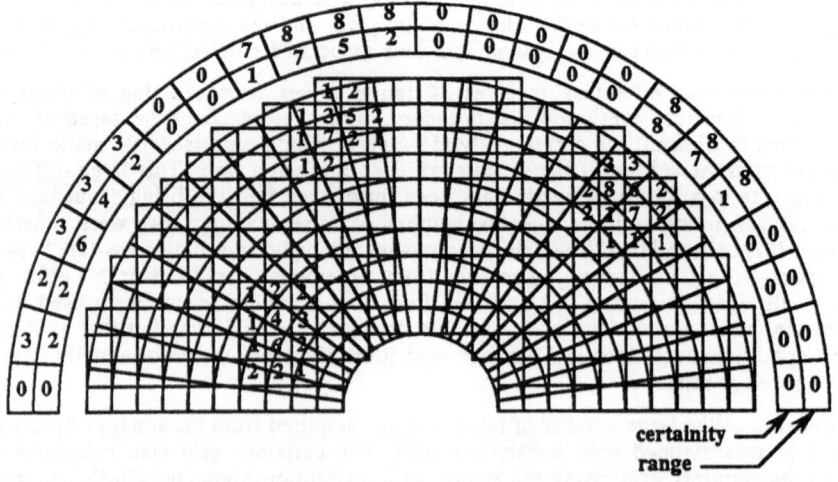

Figure 2. Certainty values computed from detected objects shown on a square certainty grid overlayed with a polar grid. Polar histogram also shown.

For every scan of the ultrasound the range readings are converted into discrete values from 0 to 9 which correspond to a polar cell in the polar grid where the object was detected. The certainty values in the square cells whose centers lie within this polar cell are then increased by a value of 2 up to a maximum value of 10. Similarly the square cells which contain certainty values that do not have to be re-enforced are decremented by 1 down to a value of 0. The increment and decrement values have been determined experimentally by observing the response of the certainty grid in a real environment.

A polar histogram, that is derived directly from the certainty grid, and which displays the range and certainty value for every polar sector is also shown in Fig. 2 as a ring around the certainty grids. The square cells in each azimuth sector are searched to find which square cell has the highest certainty value. The range and certainty of this square cell are placed in this histogram and represent the entire sector. It is the information in this histogram that is used by the algorithms task to compute an alternate path for the vehicle.

3. Collision Avoidance Algorithms

The collision avoidance task uses a user specified direction of motion for the vehicle and combines this with the histogram data from the certainty grid task to compute an obstruction-free direction of travel for the platform. It is this task which was replaced during testing in order to demonstrate the different collision avoidance methods. In this section we give an overview of these methods with an emphasis on the problems that were encountered and on changes that were made to the implementation of the algorithms as compared to the earlier work of other researchers.

For each of the methods, a binary decision is applied to the certainty value to determine if range information in the histogram represents a real obstacle. The threshold value that is used to make this decision is chosen by taking several considerations into account. The first is the limited space in which the vehicle is to be operated. This affects two operating criteria: the sensing range, and the maximum speed of operation. In order to avoid having the walls influence the operation and yet still allow the vehicle sufficient room to move while conducting tests, the range of the sonars was limited. Limiting the sonar transducer range to 1 m and taking into account the width of the vehicle creates a sensing area that is 2.6 m in diameter and allows a lateral motion of about 1.5 m before the walls become a factor.

Another consideration in determining the certainty value that indicates valid range data is the performance of the sonars. It was mentioned above that a full sonar scan took about 400 ms. With this constraint, a certainty value of 3 or more is used to determine if the range data represents a true object. Selecting a value greater than 2 requires at least two sets of sonar data to map an object into the same location in order to confirm the presence of that object. The threshold value of 3 limits the maximum speed of the vehicle to approximately 0.35 m/s. This limit is calculated based on the following considerations. Three sonar scans, which takes approximately 1.2 s, are allocated for the initial confirmation of the presence of an obstacle. allocating enough time for three sonar scans when only two are required improves the robustness of the system in the presence of map discretization and sonar sensing errors. Approximately 100 ms are required to compute new vehicle velocities and the vehicle has an additional internal time lag of about 400 ms before it begins to react to a new velocity command. Consequently, the maximum time elapsed between the presentation of an object until the vehicle can react to that object is about 1.8 s. Allowing a distance of approximately 10 cm for the required course correction and using the fact that the vehicle extends 30 cm beyond the sonar transducers reduces the effective range to 70 cm. With this information a maximum operating speed of 0.35 m/s has been calculated.

For these tests, a target destination point is not used, rather direction and speed are specified for the platform. The magnitude of the attraction vector is set to a constant value and is independent of the velocity of the vehicle. Furthermore, the vehicle drive velocity is kept constant throughout the test.

3.1 Vector Field Histogram

This method for collision avoidance was first introduced by Borenstein [3] as a realtime obstacle avoidance algorithm for a mobile platform in cluttered environments. Promising results were reported for a vehicle traversing a cluttered environment with speeds up to 0.78 m/s [24] and being able to pass through openings as narrow as 1.2 m in width [24].

A brief description of the method is warranted to appreciate the complexities associated with the algorithm. The basic idea of the method is to extract from the certainty grid a direction in which there is open space to navigate and which is close to the desired direction of travel. This is accomplished by integrating the range and certainty data from all the square cells within a sector into a representative value for that sector. the collection of these values for all the sectors is then modified using a low pass filter. The filter is used to compensate for the width of the vehicle by distributing information to the neighbouring sectors. examining the numeric value that represents each sector and comparing it to a threshold value determines which directions allow collision free motion. Defining functions, similar to that used by Borenstein [3], that would provide a consistent number of contiguous free sectors as the vehicle approached a gap proved difficult. This inconsistency causes undesirable vehicle performance when the gap is only slightly larger than the vehicle. functions designed to eliminate gaps that are too small for distant ranges incorrectly eliminate gaps of sufficient size at close range. In other cases gaps that are too small are incorrectly selected as passable until the vehicle draws close to the objects. A lookup table is used with an entry associated for each discrete range value to compensate for the vehicle width. the values in this table are used to eliminate sectors on both sides of an occupied sector as possible paths and thus ensure collision free motion. The table values were originally computed from the angle subtended by the width of the vehicle at the appropriate range, but were modified, using experimental information, to allow the platform to traverse an opening of 1.4 m. to determine which directions are safe for travel an array is used in which each element representes one sector. If the certainty value in the histogram is greater than the threshold value a one is placed in the appropriate array element and then the range information is then used to determine the number of adjoining elements that must also be filled with ones. Figure 3 shows the results of taking the histogram from Fig. 2 and operating on it with the filtering process to compensate for the width of the vehicle. In this type of transformation, the original certainty values are discarded since they are no longer required.

Navigation is accomplished by searching the array for open sectors. if the sector containing the desired direction is occupied and there are two alternative paths on opposite sides of the desired direction, then the path chosen is based on the following equation,

$$\text{Dir} = W_f (\Delta\Omega_{dn} - \Delta\Omega_{dp}) + (\Delta\Omega_{pn} - \Delta\Omega_{pp}) \tag{1}$$

where

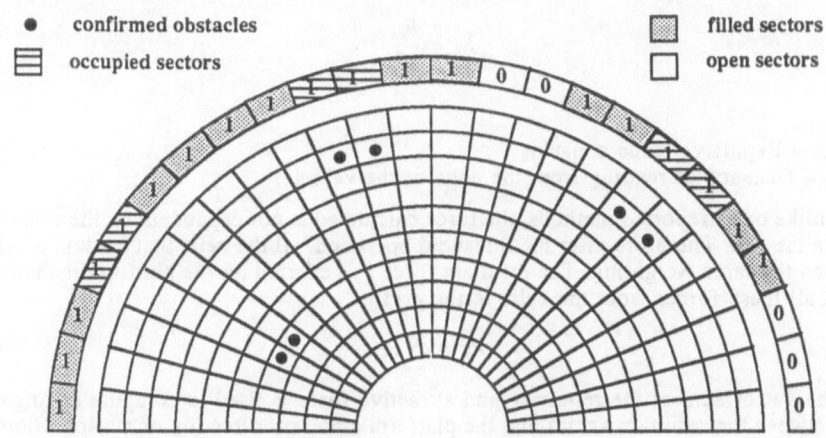

Figure 3. Polar grid showing detected objects after thresholding. Also polar histogram after the filtering process to compensate for the width of the vehicle.

$$\Delta\Omega_{dn} = ABS(A_d - V_n)$$
$$\Delta\Omega_{dp} = ABS(A_d - V_p)$$
$$\Delta\Omega_{pn} = ABS(A_p - V_n)$$
$$\Delta\Omega_{pp} = ABS(A_p - V_p)$$

W_f = Weight factor.
A_d = Specified direction of travel (polar sector).
A_p = Present direction of travel (polar sector).
V_p = Closest polar sector valley from A_d in the positive direction.
V_n = Closest polar sector valley from A_d in the negative direction.

In this equation the decision as to which sector should be used for navigation is based not only on the desired direction of travel but also on the present direction of travel. This is not the case in many other path planning methods. This can prevent situations where the platform may try to steer entirely across an object to find an opening on the other side when one may exist much closer to the present direction of motion. The decision to take an opening closer to the desired or present direction of travel can be influenced by the weighting factor W_f, which was set experimentally.

The sign of Dir determines which open sector is to be used for navigation. Once a sector has been chosen, the steering velocities are calculated using the difference between the current direction and the new chosen direction.

3.2 Potential Fields   The potential field method for collision avoidance has been described elsewhere [1,2,10,25]. To be consistent with the other algorithms the input to this method is also 48 pairs of range and certainty values corresponding to each of the 48 sectors of the polar grid. First a threshold operation is applied to the certainty values. The data that remain mark the presence of objects and are then used to generate potential fields. For each object i,

$$F_i = \frac{F_{cr}}{d_i^2} \qquad (2)$$

where

F_{cr} = Repulsive force constant
d_i = Distance of reading from the edge of the vehicle.

Unlike other reported methods, the force calculated is not dependent on the certainty value in the cell. Therefore after the threshold operation, all the cells that contain a value are given the same weighting. The resultant force F_r, exerted on the platform is then the sum of all these forces from the cells in the grid.

$$F_r = \sum F_i \qquad (3)$$

The vector sum of the repulsive and attractive force is used to compute an angular error between the resultant vector and the platform's present direction of motion. Similar to the vector field histogram approach this error in direction is used directly as the steer velocity error to the vehicle. No attempt was made to add any dynamic enhancements to the steering due to drive velocities that have been computed for the platform. This keeps the algorithms as simple as possible and consistent with each other for testing reasons.

In the potential field approach it is necessary to set the potential gain factor known as F_{cr} in equation 2. This force constant influences the size of the repulsive vectors which in turn influence the steering error fed to the motors. If set to a high value the platform will react more rapidly to objects in its path but will be prone to oscillations. The value for F_{cr} was determined experimentally by picking the smallest value that would just barely steer the vehicle away from an obstacle at the maximum driving speed (0.35 m/s) of the vehicle as shown in Fig. 4. It is then possible to drive the vehicle at lower speeds in narrow areas that are prone to unstable motions and at the same time guarantee that in open areas the maximum speed of the vehicle can be attained with the assurance that the vehicle will react appropriately to the presence of an unknown object.

The post used was 15 cm in diameter and was offset by 20 cm from the center line of the platform's direction. This offset was introduced when it was discovered that if the object was in the direct path of the platform the repulsive force vector was 180° opposite to the attraction vector. In this situation the platform would continue its forward direction until the resultant vector suddenly pointed away from the object. At times this would not occur until 600 ms after the object was detected which would be too late to change course. another problem occured just as the vehicle began to turn away from the object. Here, small changes in the direction of the resultant vector would cause the sign of the steering velocity to oscillate even though the magnitude was large. both potential field methods are subject to these problems. Possible solutions have been considered but have not been implemented so as to prevent undesirable side effects in these tests. However, if the object is not symmetrical about the direction of motion then the resultant force will contain a force component normal to the vehicle's path. This will generate a steering error and the platform will begin to turn as soon as the object is detected.

3.3 Generalized Potential Fields

This method, developed by Krogh [5], is an adaptation of the potential fields approach described above. Instead of computing the potential fields based solely on the geometry of the environment, the velocity of the vehicle towards the objects is also considered. Again for each object i detected in the certainty grid the avoidance potential

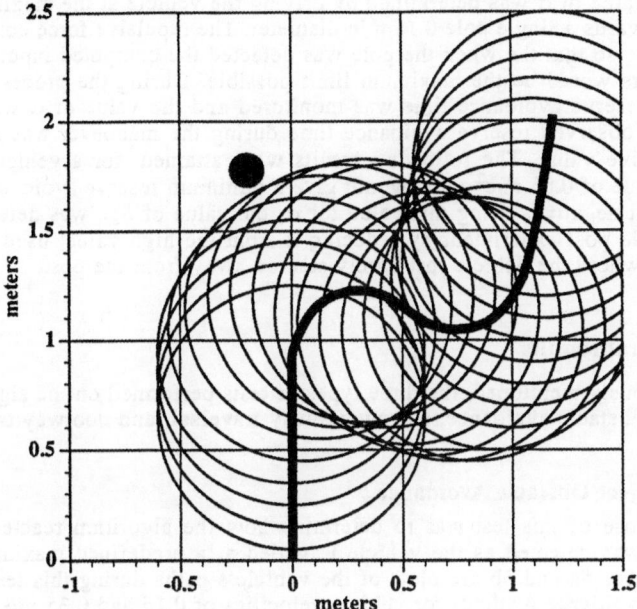

Figure 4. Determination of repulsive force constant for potential field method.

generated by that object onto the vehicle is based on a concept known as the reserved avoidance time $(T_i - t_i)$. The time t_i is known as the minimum avoidance time and is based upon the time it would take to slow the vehicle from an approach velocity v_i to 0 using the vehicle's maximum acceleration α. The time T_i is known as the maximum avoidance time and it is the time it would take to slow the vehicle from an approach velocity v_i to 0 using an acceleration lower than α. The details of this method are given in Krogh's paper [5] and it suffices to list the equation for computing the potential field as,

$$P_i(x,v) = \begin{cases} 0 & \text{if } v_{i,} = 0 \\ \dfrac{\alpha v_i}{(2d_i\,\alpha - v_i^2)} & \text{if } v_i > 0 \end{cases} \tag{4}$$

The overall repulsive force vector acting on the vehicle is then the sum of all the avoidance vectors generated from the certainty grid cells that have certainty values above the threshold value,

$$F_r = F_{cr}\sum P_i \tag{5}$$

where

F_{cr} = Repulsive force constant.

As is the case for the potential field approach, several gain factors have to be determined before the algorithm can be implemented. In this case two gain factors are to be determined, F_{cr} the repulsive force constant and α the acceleration rate of the vehicle. The value of α was determined by driving the vehicle at the maximum velocity of 0.35m/s towards a single pole 0.15m in diameter. The repulsive force constant was set to a high value so that the when the pole was detected the computed input signal to the steering motors would be the maximum limit possible. During the process of avoiding the pole the reserve avoidance time was monitored and the value of α was set so that the minimum observed reserve avoidance time during the manoever was near zero but never a negative value. The following results were attained, for a vehicle velocity of 0.35 m/s a value of 0.85 m/s^2 for α would give a minimum reserve avoidance time close to zero but not negative. Using this value for α, the value of F_{cr} was determined using the same single post experiment by reducing it from the high value, used to determine α, to a value where the vehicle just barely steered away from the post.

4. Experimental Results

As previously mentioned, the three types of tests performed on the algorithms were single object obstacle avoidance, narrow hallway traversal, and doorway traversal.

4.1 Single Object Obstacle Avoidance

The purpose of this test was to determine how the algorithm reacted to a single obstacle that was detected as the vehicle travelled at a predefined maximum speed of 0.35 m/s. Figures 5a and 5b are plots of the vehicle's paths during this test for each of the collision avoidance methods for vehicle velocities of 0.15 and 0.35 m/s respectively.

The post used as an obstacle was offset slightly from the center line of vehicle travel to compensate for the previously mentioned problem associated with approaching a target directly. This problem does not arise when using the vector field histogram approach since it searches for openings. However a similar indecision can occur in which for a period of time the chosen direction of motion may oscillate from an opening on one side of the object to one on the other side. Therefore all the algorithms used an offset post during the tests.

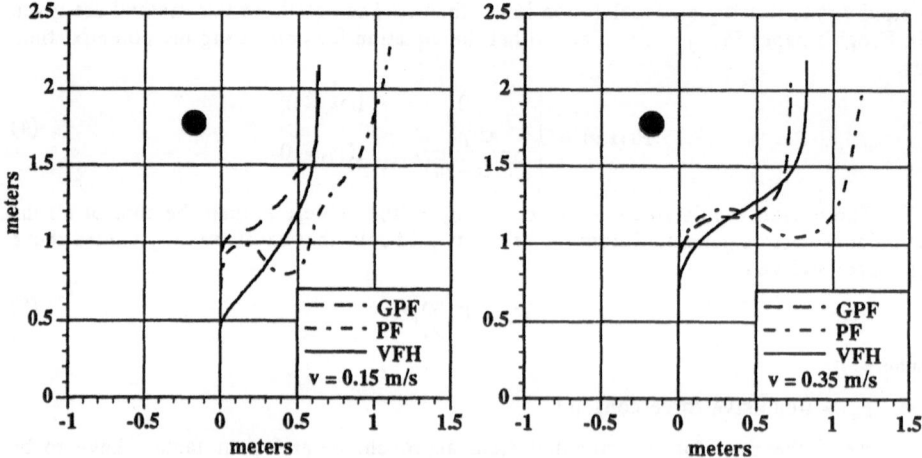

Figure 5. Single obstacle avoidance at 0.15 m/s and 0.35 m/s

Of particular interest in the figures is the shape of the path taken by the vehicle during each test. The potential field method shows very large fluctuations away from the object followed by motion back to the commanded direction of motion. The generalized potential field method is slightly more stable while the vector histogram method shows a well behaved smooth transition. What is perhaps more interesting is to compare the tracks at a high speed to those at a low speed. The potential field method still shows a very erratic behaviour which is to be expected since the potentials generated are not dependent on the speed of the vehicle. In contrast, the generalized potential field example shows a motion that fluctuates due mainly to the fact that when the vehicle veers away from the object, the potential drops off. This in turn causes the vehicle to turn towards the object which in turn increases the potential. Finally, the vector field histogram approach shows only a slight deviation in this situation.

4.2 Narrow Corridor Navigation

With the previously determined gain values the algorithms were tested for stability in traversing a narrow corridor 1.4 m wide and 2.5 m long. The idea here was to determine the maximum speed that could be attained by the platform while it was still able to navigate the hallway. The results of this test are shown in Fig. 6 for both the maximum velocity attained by the vehicle and the minimum velocity of 0.15 m/s. The results show that it was possible to traverse the corridor at 0.35 m/s using the vector field histogram approach and at 0.3 m/s using the generalized potential field method. It was not possible to traverse this hallway using the potential field method at any of the test velocities because the vehicle oscillated enough to collide with the sides of the hallway. In these figures the two darker vertical lines represent the walls of the hallway if the vehicle were to be shrunk to a point. Therefore the trajectory of the vehicle must stay between the two darker lines to signify a good traversal through the hallway. As expected, using the generalized potential field approach the oscillations diminish at lower speeds. In contrast, the vector field histogram approach shows very little change in the path as the speed changes.

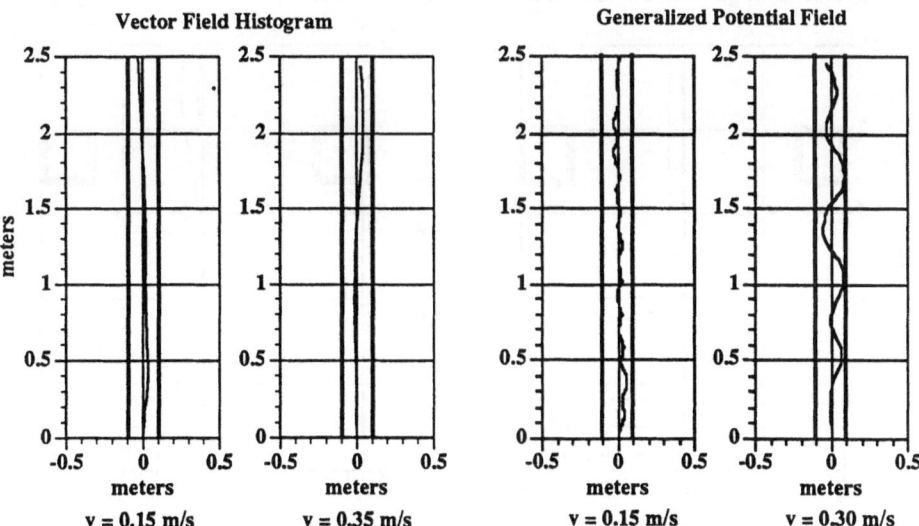

Figure 6. Narrow corridor navigation

312

4.3 Doorway Traversal

In this test we wanted to determine the maximum speed that could be attained before the vehicle failed to pass through a 1.4 m door opening. In the first trial, the motion of the platform was normal to the door opening and in a direction that would drive it through the center of the opening. In concept, the vector field histogram approach is based solely on the geometry of the environment and thus the detection of the opening is independent of speed and failure to pass through an opening is mainly due to the dynamics limitations of the vehicle. For both the potential field and generalized potential field methods failure to pass through an opening can either be due to high repulsive forces or dynamic limitation of the vehicle The results of the experiment are shown in Fig. 7 and successful attempts in traversing through the doorway were possible at 0.35 m/s for both the vector field histogram and generalized potential field approaches. Using the potential field approach the vehicle was unable to detect the existence of an opening. In fact the vehicle usually made contact with one or both of the objects defining the gap. Again results are shown for a vehicle velocity of 0.15 m/s for comparison with the paths at the higher velocity.

In some of the doorway traversal tests in which the vehicle was originally aligned with the centerline of the doorway, a vehicle movement biased to the right was observed. This may be caused by the delay between sonar transducer firings and the order in which the transducers are fired. It is possible that the obstacle on the left was detected one sonar scan sooner than the obstacle on the right and thus the vehicle is deflected away from the left object sooner.

This test was further expanded by changing the vehicle's approach to the doorway by an offset from the center axis of the doorway of 1/4 the platform's width (0.3 m). The vehicle now has to avoid the oncoming wall and try to enter through the doorway while maintaining constant speed. Figure 8 is similar to Fig. 7 except that the vehicle's starting point was offset to the left of the doorway's center axis by 0.3 m. Both the generalized and vector field histogram methods managed to veer towards the center of the doorway and traverse through it at the maximum speed of 0.35 m/s.

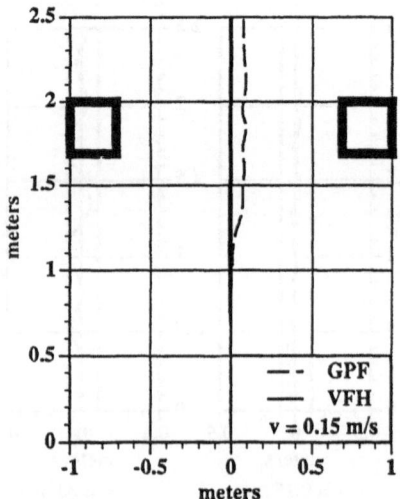 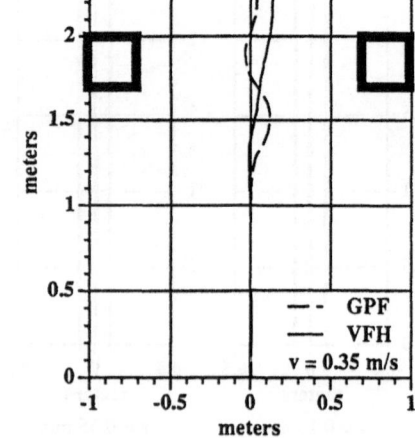

Figure 7. Doorway traversal results

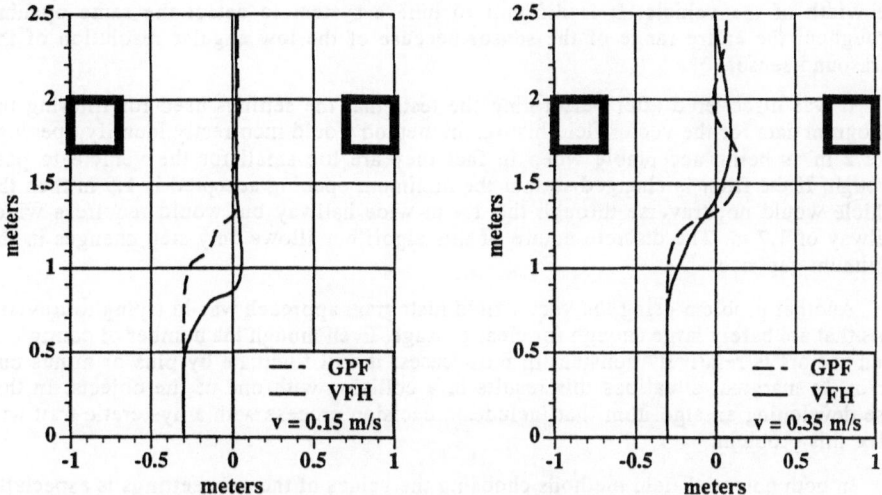

Figure 8. Doorway traversal results with a starting offset of 0.3 m.

5. Discussion

The purpose of these experiments was to improve our understanding of the difficulties encountered in the implementation of these collision avoidance algorithms and to see how the algorithms performed under a series of realistic tests. Many comments have appeared throughout the paper. We would like to summarise these problems here.

5.1 Sensing

One major problem with the ultrasound sensors is the delays that had to be introduced into the system to compensate for the large number of false readings that were originally observed. While this has been reported by a number of researchers the real significance of this limitation only becomes clear when a realtime implementation is attempted. An attempt to reduce the effect of spurious reflections poses a dilemma. Increasing the data sampling rate by decreasing the delay time between sonar transducer firings reduces the response time of the system and therefore increases the maximum velocity at which the vehicle can be operated. However more spurious reflections are produced. This can be offset by using the certainty grid idea, i.e. a low pass filter, but this requires sequential scans and hence more time is taken to achieve the same degree of confidence as to the presence of objects. In our system a reliable result is not generated by the sensors until about 0.8 s after the object has been originally detected. Without this problem, the scan time could be reduced to 100 ms, this is 8 times faster what we were able to achieve in practice.

5.2 Implementation Difficulties

Perhaps the biggest difficulty in these algorithms is the proper setting of the gain factors in the algorithms.

In the vector field histogram approach, much time and effort was spent in determining a method for filtering the histogram values from the certainty grid to compensate for the width of the vehicle. It is difficult to tune a system to detect the same opening throughout the entire range of the sensor because of the low angular resolution of the ultrasound sensor.

It was discovered after performing the tests that the settings used for filtering the histogram data for the vector field histogram method would incorrectly identify openings of 1.2 m as being acceptable when in fact they are too small for the vehicle to pass through. If the filter is changed so that the minimum opening accepted is 1.5 m then the vehicle would not traverse through the 1.4 m wide hallway but would require a wider hallway of 1.7 m. The discrete nature of this algorithm allows only step changes in the minimum gap size.

Another problem using the vector field histogram approach was in trying to traverse gaps that are barely large enough for clear passage. Even though the number of contigious open sectors is relatively constant in most cases, it will fluctuate by plus or minus one sector. In marginal situations this results in a collision with one of the objects. In this case developing an algorithm that includes a decision process with a hysteretic trait will solve this problem.

In both potential field methods choosing the values of the gain settings is especially problematic. It is possible to make the algorithms pass any test by proper tuning. In these tests it was decided to set the gains based on the ability to detect an object during the traversal of free space at the vehicle's maximum speed. Perhaps it is better to set the gains to pass the difficult tasks like passing through a doorway and then reduce the maximum speed to that which allows the vehicle to miss hitting a single object in front of it.

The potential field method fails due to the fact that the magnitude of the avoidance vector normal to the objects defining the gap is greater than the magnitude of the attraction vector. Contact with the objects is a result of the vehicle placing itself in a situation that is even more demading than that encountered during the single obstacle high speed test. the shape of the field produced from the narrow opening is in effect a blind alley. Only a 180° turn will allow the vehicle to escape. Since this is only possible at very low speeds, the vehicle veers and collides with the doorframe. Decreasing the gain will allow passage through the opening but would cause the vehicle to fail the single obstacle avoidance test.

The generalized potential field method, with the chosen gains, was unable to determine that an opening was too narrow. Doubling the gain and reducing the opening to 1 m failed to prevent it from attempting to pass through the opening. Obstacles that lie at a large angle from the direction of motion of the vehicle have small field strengths associated with them and these field strengths are not strong enough to repel the vehicle. The basic problem is one of not taking into account the width of the platform during the calculations. Krogh's method [5] requires the edges of the obstacle to be located. We have not taken this into account.

A problem that occurs with both potential field methods is that different strength fields are generated in the case of a solid wall as compared to the case of a discontinuous wall composed of a number of smaller obstacles with gaps between them.

Another serious problem that has to be addressed when using potential fields is the indecision caused by the direct opposition of the attractive and repulsive forces that occurs when approaching an object along one of the axes of symmetry. This could lead to high gains being used if the indecision is not dealt with as a special condition.

5.3 Experiments

The results of these tests raise questions about the suitability of using the potential field method without further enhancements to the algorithm. This method is very sensitive to disturbances and requires the addition of filters to stabilise it. The vector field histogram algorithm was not influenced as much by disturbances in the environment. This is mainly because it is based on the detection of free space rather than physical objects. This process is not as sensitive to vehicle velocity, position and sensor disturbances as is the computation of vectors from potential fields.

6. Concluding Remarks

From the tests performed on the vector field histogram, potential field and generalized potential field methods the potential field method appeared to be much more unstable when compared to the other two methods. The vector field histogram method showed the best stability under the ranges and conditions in which it was tested. Unfortunately it is more difficult to implement than the potential field method. It also appears that it is better to set the parameters of the algorithm using the doorway traversal test rather than the single object avoidance test that was used here. The speed of the vehicle should then be limited to the maximum used for traversing through the door. Further investigation is required into increasing the rate of data acquisition so that the speed of the vehicle and the reliability of the algorithms can be improved.

References

[1]   M. D. Adams, H. Huosheng, and P. J. Probert, "Towards A Real-time Architecture for Obstacle Avoidance and Path Planning in Mobile Robots," *Proc. IEEE Robotics and Automat. Conf.*, (Cincinnati, OH), May 1990, pp. 584–589.

[2]   J. Borenstein and Y. Koren, "Real_time Obstacle Avoidance for Fast Mobile Robots," *IEEE Trans. Syst. Man Cybern.*, vol. SMC-19, no. 5, pp. 1179–1187, 1989.

[3]   J. Borenstein and Y. Koren, "Real-time Obstacle Avoidance for Fast Mobile Robots in Cluttered Environments," *Proc. IEEE Robotics and Automat. Conf.*, (Cincinnati, OH),May 1990, pp. 572–577.

[4]   P. N. Kacandes, A. Langen, and H. Warnecke, "A Combined Generalized Potential Fields/Dynamic Path Planning Approach to Collision Avoidance for a Mobile Autonomous Robot Operating in a Constraint Environment," *Proc. Intell. Autonomous Syst. 2*, (Amsterdam, Netherlands), December 1989, pp. 145–154.

[5]   B. H. Krogh, "A Generalized Potential Field Approach to Obstacle Avoidance Control," *Proc. SME Conf. Robotics Research: The Next Five Years and Beyond*, (Bethlehem, PA), August 1984 , pp MS84-484.

[6]   B. H. Krogh, and C. E. Thrope, "Integrated Path Planning and Dynamic Steering Control for Autonomous Vehicles," *Proc. IEEE Robotics and Automat. Conf.,* (San Francisco, CA), April 1986, pp. 1664–1669.

[7]   L. Matthies and A. Elfes, "Integration of Sonar and Stereo Range Data Using a Grid-Based Representation," *Proc. IEEE Robotics and Automat. Conf.* , (Philadelphia, PA), April 1988, pp. 727–733.

[8]   H. P. Moravec and A. Elfes, "High Resolution Maps from Wide Angle Sonar," *Proc. IEEE Robotics and Automat. Conf.* , (St. Louis, Missouri), March 1985, pp.19–24.

[9]   U. Rascke and J. Borenstein, "A Comparison of Grid-type Map-building Techniques by Index of Performance," *Proc. IEEE Robotics and Automat. Conf.*, (Cincinnati, OH), May 1990, pp. 1828–1832.

[10]   R. B. Tilove, "Local Obstacle Avoidance for Mobile Robots Based on the Method of Artificial Potentials," *Proc. IEEE Robotics and Automat. Conf.*, (Cincinnati, OH), May 1990, pp. 566–571.

[11]   Y. Koren and J. Borenstein, "Potential Field Methods and Their Inherent Limitations for Mobile Robot Navigation," *Proc. IEEE Robotics and Automat. Conf.*, (Sacramento, CA), April 1991, pp. 1398–1404.

[12]   W. M. Gentleman, C. Archibald, S. Elgazzar, D. Green, D., and R. Liscano, "Case studies of realtime multiprocessors in robotics," *Proc. Second Int. Specialist Seminar Design and Applications of Parallel Digital Processors*, (Lisbon, Portugal), April 1991, In press. NRC 31813.

[13]   H. Moravec, "Obstacle Avoidance and Navigation in the Real World by a Seeing Robot Rover," Technical Report CMU-RI-TR-3, Carnegie-Mellon Robotics Institute, Pittsburgh, PA, 1980.

[14]   J. Crowley, "Navigation for an Intelligent Mobile Robot," *IEEE J. Robotics Automat.*, vol. RA-1, no. 1, pp. 31–41, 1985.

[15]   G. Gilbreath and H. R. Everett, "Path Planning and Collision Avoidance for an Indoor Security Robot," *Proc. SPIE, Advances Intell. Robotics Syst.: Mobile Robots III*, (Cambridge, MA),SPIE vol. 1007, November 1988, pp. 19–27.

[16]   C. Balaguer and A. Marti, "Collision-Free Path Planning Algorithm for Mobile Robot which Moves Among Unknown Environment," *Proc. 2nd IFAC Symp. on Robot Control*, October 1988, pp. 261–266.

[17]   C. Malcolm, T. Smithers, and J. Hallam, "An Emerging Paradigm in Robot Architecture," *Proc. Intell. Autonomous Syst. 2*, (Amsterdam, Netherlands), December 1989, pp. 545-564.

[18]   R. A. Brooks, "A Robust Layered Control System for a Mobile Robot," *IEEE J. of Robotics and Automat.*, vol. RA-2, no. 1, pp. 14–23, 1986.

[19]   D. Green and R. Liscano, "Real-Time Control of an Autonomous Mobile Platform using the Harmony Operating System," *Proc. IEEE Int. Symp. Intell. Cont.*, (Albany, NY), September 1989, pp. 374–378, NRC 30554.

[20]   Ultrasonic Ranging System, POLAROID Corporation, Ultrasonic Components Group, Cambridge, MA.

[21]   Range Transducer Control Module Product Specification, Denning Mobile Robotics Inc., Boston, MA.

[22]   S. T. Lang, L. W. Korba, and A. K. C. Wong, "Characterizing and Modelling a Sonar Ring," *Proc. SPIE, Advances Intell. Robotics Syst.: Mobile Robots IV* , (Philadelphia, PA), November 1989, pp. 291–304, NRC 30928.

[23]   K2A Mobile Platform, Cybermotion, Roanoke, VA, 1987.

[24]   J. Borenstein and Y. Koren, Y, "The Vector Field Histogram—Fast Obstacle Avoidance for Mobile Robots,"*IEEE Trans. Robotics and Automat.*, vol 7, no. 3, pp. 278–288, 1991.

[25]   R. C. Arkin, "Motor Schema-Based Mobile Robot Navigation," *Int. J. of Robotics Res.*, pp. 92–112 , 1989.

Object-oriented design of mobile robot control systems *

Hélène Chochon
Alcatel Alsthom Recherche
Route de Nozay, 91460 Marcoussis

1 Introduction

In this paper, we address a problem that remains critical when one is concerned with the development of an autonomous mobile robot : how to design and implement control in such a complex system ? Mobile robot control systems are characterized by decisional capabilities, by context-dependant reactivity to events, by real-time interactions with a variable and uncertain environment, by concurrency between perception, decision-making and action, and by a distributed implementation.

Control design is addressed by many authors in Robotics : they intend to develop general control methods, but their approaches give stress to some specific control features. Among these works, we can cite the following significant approaches :

- Brooks at M.I.T. emphasizes *reactivity* and builds robot control systems by incremental integration of elementary behaviors [2];

- NASREM [1] is a standard *hierarchical* architecture with several predefined control levels, that are decomposed into three modules : "planning and execution", "world modelling", "sensory processing";

- CODGER, a control system developed to drive the Navlab [5], is based on a *blackboard* architecture with a set of dedicated modules that access asynchronously data in the blackboard;

- TCA, a general-purpose architecture is developed at Carnegie Mellon University [3]: TCA relies on a *centralized* control that enables parallelism between planning and execution, achievement of multiple goals with temporal reasoning and context-dependent exception handling.

We present in this paper an *object-oriented method* that is suited for an easier design and implementation of complex robot control systems. This method enables a *transparent*

---

*This work is developed within RAMINA, a research project performed by Alcatel Alsthom Recherche, the corporate research center of Alcatel Alsthom, in collaboration with FRAMATOME.

distribution of control among well-identified functional subsystems that are described and implemented as objects. The functional subsystems are made distributed by means of a general server / client mechanism. Several *control modules* can be locally linked to the functional subsystems and provide the following basic control functions : data monitoring, command spying, state and resource management, temporal constraints and exception handling.

This work is performed within the RAMINA project. This project aims at developing an autonomous mobile robot that achieves navigation tasks specified at a symbolic level, in partially unknown indoors environments. The object-oriented method has been experimented for implementing the robot motion controller, a subsystem that performs concurrently motion servoing and obstacle avoidance using both ultrasonic sensors and a laser stripe range-finder. In this paper, we examine how the object-oriented method is applied to this subsystem and we describe the obstacle avoidance algorithm that has been developed.

2 Object-oriented design of a control system

In this section, we present the object-oriented method that has been defined to implement a mobile robot control system. It is important to distinguish between functional and control viewpoints while designing a mobile robot system. Hence, we define hereafter some abstract notions that are introduced to clarify the interface between functional and control viewpoints.

2.1 Functional viewpoint

In a functional architecture, data, functions and tasks are grouped into well-identified subsystems. Figure 1 represents the functional architecture for RAMINA with the main subsystems and knowledge representations. It is organized into several abstract layers :

- the *sensor and effector layer* includes a set of subsystems that provide logical interfaces to sensors and wheel motors;

- the *motion layer* performs concurrently trajectory generation, feedback control and obstacle avoidance using both ultrasonic sensors and a laser stripe range-finder;

- the *geometry layer* updates the robot position and a geometric map of the environment and plans feasible trajectories and motions in order to follow a topological path;

- the *topology layer* is in charge of maintaining the free space with a topological graph representation and of planning optimal paths related to mission objectives;

- the *semantics layer* interprets symbolic navigation tasks into objectives and controls their execution with respect to specified constraints.

We do not aim at specifying control within the functional architecture : there is no control or data flow represented in figure 1. The layers are defined to group functional subsystems that are reasoning at the same level of abstraction. But, in the approach that is presented in this paper, a functional subsystem can communicate directly with any other subsystem, and control can be implemented in a distributed manner.

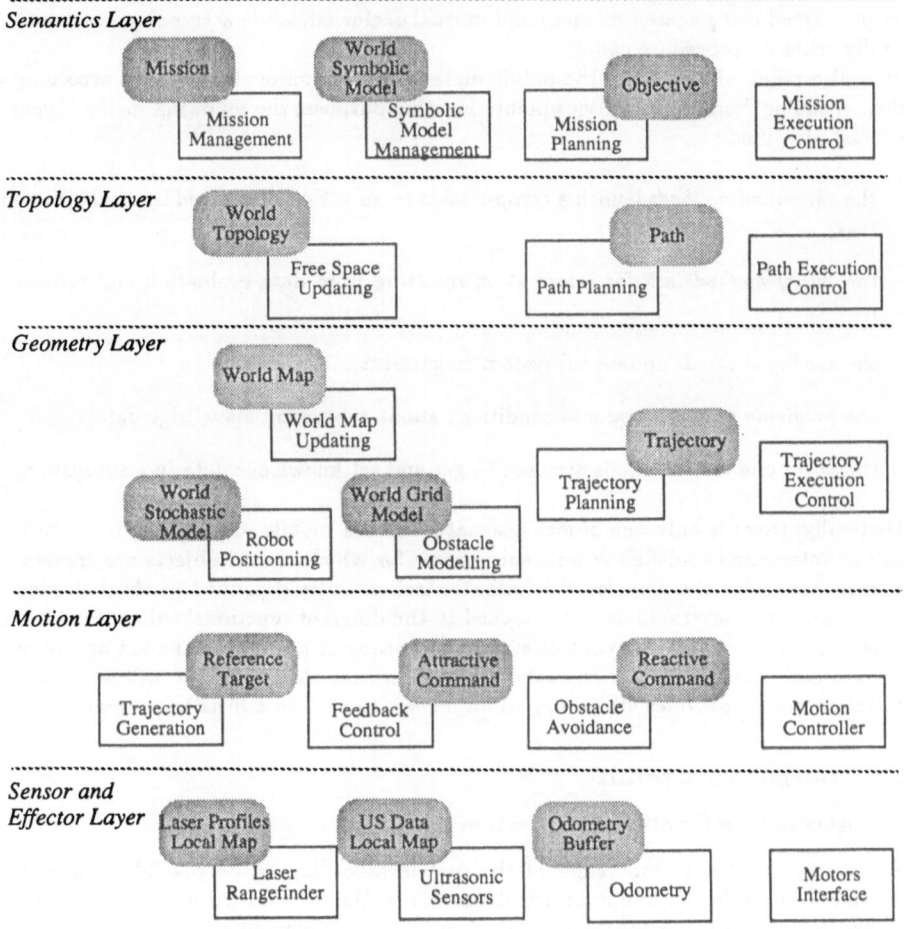

Figure 1: Main functional subsystems of RAMINA

In our system that is based on an object-oriented approach, a functional subsystem is represented by a class with a set of *public methods* that clearly defines the subsystem interface. In addition, the object-oriented approach enables to implement a subsystem as a hierarchy of classes : this mechanism enables to define several interface levels, and to use polymorphism facilities.

Data encapsulated in a functional subsystem correspond either to *knowledge* managed by the subsystem (such as the odometry buffer or the world map), or to *constraints* used by subsystem methods (for instance, the gains used by feedback control). This distinction is important for two reasons. From a control point of view, knowledge updating operations have to be monitored to detect events, whereas control is not concerned by constraint modifications. From an implementation point of view, access to subsystem knowledge is likely performed using shared memory and mutual exclusion, while access to a constraint generally relies on procedure call.

It is also required to classify the public methods of a functional subsystem according to their meaning from a control viewpoint. For this purpose, the following *method types* have been identified :

- the *command methods* launch a computation or an action performed by a subsystem task;

- the *query methods* ask for information resulting from data evaluation and returns it;

- the *modify methods* update subsystem constraints;

- the *predicate methods* evaluate conditions about subsystem knowledge data;

- the *write and read* methods are used to get and set knowledge data in a subsystem.

Generally, there is only one object associated with a mobile robot subsystem. But, it may be interesting to define generic subsystems for which several objects are created. This is in particular the case for the basic control modules described in the following section : they have several instances attached to the different functional subsystems.

The functional design of a control system by means of an object-oriented approach provides a clear specification of the subsystems interfaces. Moreover, we will see in the next section how it provides also a way to implement control in a distributed manner.

2.2   Control viewpoint

2.2.1   Access to a functional subsystem

Access to any functional subsystem of the mobile robot has to be possible from any geographic location in the computer architecture, and the distribution must be managed transparently in order to make system reconfiguration possible and easy.

We propose to use a server / client mechanism that can be attached to any functional subsystem and that enables to dynamically create *client objects* of this subsystem anywhere in the computer architecture. A client object has the same public interface than its functional subsystem, but it runs the methods remotely, sending messages to the server object or having access to global shared data.

The server / client mechanism is built upon a communication layer that enables the exchange of typed messages on an heterogeneous hardware architecture. Communication is performed by bidirectional links between two *ports*. A port is a basic communication object that provides a common interface to send, wait for, and receive typed messages. Several communication protocols are implemented and the choice of a protocol can be optimized according to the port locations. Communication links are established dynamically by means of a *port server*. It is also convenient to broadcast a message to several subsystems or to receive messages that may be issued by different subsystems. In order to meet this requirement, we use a communication object called a *port group*.

A *server object* includes a functional subsystem, a port server and a port group, and an evaluation method called by the *server interpretation task*. This task waits for messages on the port group and evaluates them either by accepting a connection, or by calling the associated subsystem method. The evaluation method requires a description of the subsystem public interface. For each method, the programmer must specify the following information : its identifier, its type (command, query, modify, predicate, etc...), the list of parameters types and the return type. The object-oriented design provides a way to make a subsystem distributed by creating a server object associated with this subsystem. Now, we will show how basic control modules can be locally attached to a subsystem thanks to the server object.

2.2.2 Control modules attached to a subsystem

We define the following basic *control modules* that can be associated to any functional subsystem separately.

- A *monitoring module* watches for events that occur when a condition on data is satisfied. For instance, when the robot speed becomes superior to 0.5, or when the robot enters a polygonal region, actions on any subsystem can be performed.

- A *spying module* detects the calls to specific command methods, for which a set of actions is defined. It enables for instance to inform the motion controller subsystem that an emergency stop has been sent directly to the motors interface and to interrupt the current motion.

- A *state machine module* ensures the coherence between commands issued by different subsystems. It maintains a subsystem state and filters the received commands according to the current state. A set of actions can be specified that are achieved when a command is rejected.

- A *resource management module* checks if the resources required for the execution of a command are available. Commands that cannot be achieved immediately are buffered. For instance, if the current motion is not finished and if a new motion command arrives, it is buffered (in this example, the resource corresponds to robot motors).

- A *temporal constraints handling module* is in charge of checking some temporal constraints (starting and ending date limits, duration bound) associated to a command. If a constraint is not fulfilled, a list of actions is performed. For instance, if part of the path is not achieved before a date, mission replanning is required.

- The *exception handling module* specifies what to do when a command fails. For instance, if no path is found to achieve a robot task, the operator is informed and can decide what to do.

The control modules are implemented using classes as other functional subsystems. They are made distributed by creating server objects. These control modules provide appropriate methods that enable a clear and dynamic specification of control.

- The monitoring module is implemented as a 0+ rule-based system. The rules can be dynamically defined, removed, activated or inhibited. The left hand side of a rule is a set of conditions related to a variable value ("isEqual x 2"), or to subsystem knowledge data through predicate methods such as ("isRobotMoving ?") or ("isRobotOrientation equal to 3.14 ?"). The right hand side of a rule is a set of actions that enable to modify a variable value ("set x 2"), to update a rule state ("activate rule number 10"), or to call a subsystem method ("motion controller, stop").

- The spying module manages a table that describes the set of actions (i.e. subsystem method calls) associated to each command method : any subsystem can add or remove an action associated to a command.

- The state machine module has a description of the allowed commands that achieve a state transition, and of the forbidden commands associated to each state. For each forbidden command, a list of actions is provided that is performed if the command is rejected. Allowed commands, forbidden commands and their associated actions can be added or removed.

- The resource management module is based on the specification of resources required for each subsystem command method. Methods are provided to manage the buffers attached to each resource in order to remove one or all waiting commands.

- The temporal constraints handling module uses a set of temporal constraints attached to each command. For instance, one can specify the maximal duration of a command : if the command is not finished on time, a set of actions is performed.

- In the exception handling module, a list of actions is associated to the status returned by a command. A command status is a subsystem variable that is set by a modify method. A query method enables the exception handling module to get the status at the end of a command execution and to perform the actions attached to the status value.

The dynamic specification of control is well suited for implementing a mobile robot control system that includes decisional capabilities. For example, while executing a path, the robot must check that its position remains inside a computed region around the path and that no unexpected obstacles appear that intersect the path. Rules are created and activated appropriately to monitor these conditions and if one condition becomes not satisfied, the robot must plan another path.

For each control module, an interpretation method is defined. This method is called by the server interpretation task of the functional subsystem when required :

- after calling a writing method, for monitoring modules;

- before calling a command method, for spying, state machine, resource management and temporal constraints handling modules;

- after the end of a command method, for resource management, temporal constraints handling and exception handling modules.

A control module can be attached to several subsystems. This is convenient for some monitoring modules, when they are used to check conditions about different subsystems. For instance, in Figure 2 two subsystems are represented : the ultrasonic sensors and the odometry subsystems. The monitoring module M2 is attached to both subsystems and enables to create rules such as : " if the robot is inside a region and if the distance returned by an ultrasonic sensor is inferior to a threshold, then stop".

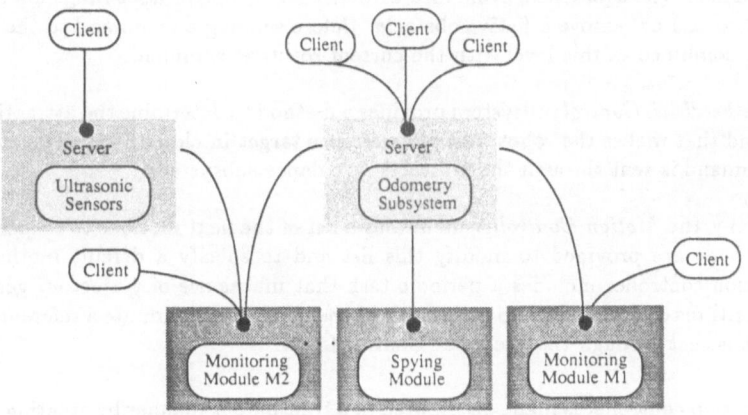

Figure 2: Examples of control modules attached to different functional subsystems

The generic control modules are currently developed in C++. The object oriented method has been experimented for implementing the sensors and effectors layer and the motion layer. We present in the following section this implementation and we describe the algorithm that has been defined to get a reactive obstacle avoidance, concurrently with motion servoing.

3   Implementation of the robot motion controller

3.1   Object-oriented design of the motion controller

For a mobile robot, the motion controller is an important functional subsystem : it is in charge of achieving various types of motion such as free motion (i.e. no trajectory is specified), linear or circular motion, door crossing, object following, etc... While performing a motion, the robot must have a reactive behavior in order to avoid static or moving unexpected obstacles without stopping. The motion controller is implemented as a hierarchy of classes that provide several interface levels for sending a command (wheel speeds) to the motors.

- The *Command Filtering* subsystem provides a method to send a command after checking that the robot dynamic constraints (maximal wheel speed and acceleration) are satisfied. If not, the closest admissible command corresponding to a similar radius of curvature is computed. The dynamic constraint values can be modified at any time.

- The *Obstacle Avoidance* subsystem manages a periodic task which computes a reactive command taking into account the robot state (configuration and speed), the last sensor data and some fictive obstacles (i.e. borders that the robot must not cross during a motion). This subsystem level provides a way to specify the avoidance mode (no avoidance, avoidance with ultrasonic and/or laser range finder data), and to add or remove a fictive obstacle. Before sending a command to the motors, it is combined at this level with the current reactive command.

- The *Feedback Control* subsystem provides a method to determine the attractive command that makes the robot follow a reference target in closed loop. The attractive command is sent through the Obstacle Avoidance subsystem.

- Finally, the *Motion Controller* maintains a list of the next motions to be performed. Methods are provided to modify this list and to specify a default motion. The motion controller includes a periodic task that makes use of trajectory generation algorithms corresponding to each motion type, in order to compute a reference target that is sent through the Feedback Control Layer.

The motion controller is made distributed in a transparent manner by creating a server object. This mechanism improves the openness of the motion controller subsystem : it can be accessed from anytwhere and at any time for modifying some parameters, adding motions, or interrupting the current motion. For instance, a robot navigation task may include some constraints such as stopping one second when the robot is in front of a window, or such as slowing down inside a corridor.

The motion controller is a client of the Ultrasonic Sensors, the Laser Range Finder, the Odometry and the Motors Interface subsystems. Reading methods are called to get the last odometry and sensor data.

The following basic control modules can be attached to the motion controller :

- a state machine module that is required to guarantee the coherence of commands that can be issued by different subsystems (for instance, if a motion is interrupted, motion recovery requires a specific command);

- a spying module that can be convenient to indicate to the operator when given commands are performed;

- a monitoring module that can be used to know when the robot does not follow the target anymore, or when obstacle avoidance is effective;

- a temporal constraints handling module in order to bound the duration of a motion.

Our experience with the implementation of the motion controller shows clearly that the object- oriented approach brings openness and enables an easy and dynamic control implementation. This is well suited for integrating other subsystems such as trajectory planning and execution control, path planning and execution control, etc. To get a reactive behavior of the robot, we have developed an obstacle avoidance algorithm that is briefly presented in the next section.

3.2 Obstacle avoidance algorithm

Obstacle avoidance can be performed in a reactive manner with the potential method [4] and with a straightforward interpretation of the more recent sensor data. The ultrasonic sensor data are interpreted as obstacle points, and obstacle segments are extracted from the laser range finder profiles.

One of the potential method shortcomings is the impossibility for the robot to come near obstacles. To overcome this problem, the influence distance of an obstacle must take into account the robot current state (configuration and speed). Indeed, an obstacle must have an effect upon the robot motion, if and only if the obstacle is inside the area that would be swept by the robot before stopping. If the robot moves slowly, the influence area becomes small and the robot can come near obstacles. More precisely, we define the influence condition of an obstacle by the following constraint where d is the minimal distance to the obstacle, and ds the security distance : $\dot{d} < -\sqrt{2A_{max}(d - ds)}$.

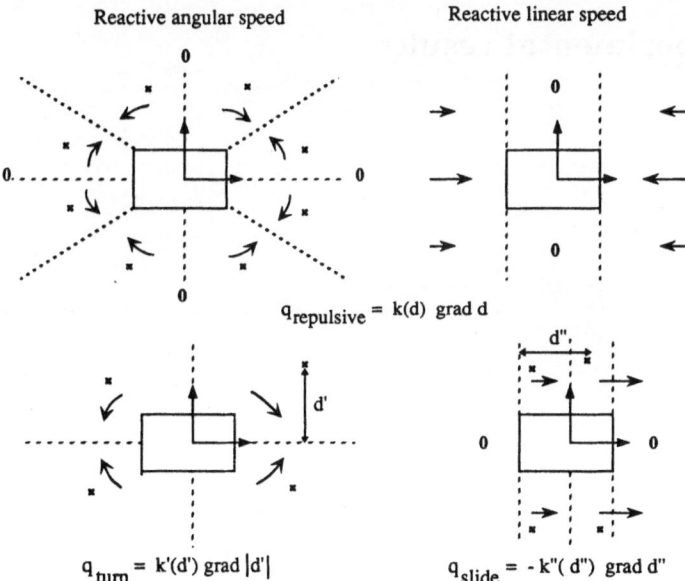

Figure 3: Computation of the reactive command

The computation of repulsive forces with the potential method must also to be adapted to take into account the kinematics constraints of a two motorized wheels vehicle. Indeed, this vehicle is not omnidirectional : when avoiding an obstacle, the robot must leave it besides instead of trying to maximize the distance to the obstacle. If the robot follows the gradient of the minimal distance d to the obstacle, it will turn the wrong way and reach a potentiel minimum position where the obstacle is just in front of it (see Figure 3).

We propose to compute a *reactive command* that is a combination of a repulsive command $q_{repulsive}$ corresponding to the classical potential method, and of two active avoidance commands q_{turn} and q_{slide}.

- q_{turn} makes the robot turn correctly to leave the obstacles besides : when the robot is just in front of an obstacle, a default avoidance side (left or right) is used to determine qturn.

- q_{slide} makes the robot slide along the obstacles when they are beside.

With this reactive command, the robot is able to avoid small obstacles and to pass between two obstacles without stopping. But, the reactive command resulting from perceived obstacles induce a perturbation of the attractive command. Therefore, a filter is required to comply with dynamic constraints such as the wheel speed and acceleration bounds. Moreover, the distance between the robot and target configurations may become large during an obstacle avoidance. Feedback control cannot guarantee that the robot will catch the target again. Hence, trajectory generation algorithms must compute a target configuration that takes into account the current robot configuration.

4 Experimental results

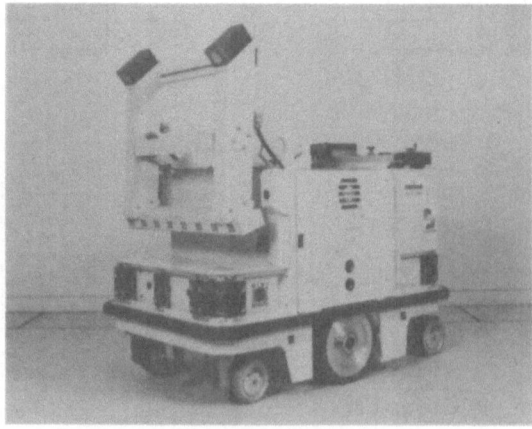

Figure 4: RAMINA mobile robot testbed

The RAMINA control system is developed on a mobile robot testbed (see Figure 4) : it has two motorized wheels and is equipped with a ring of 16 ultrasonic sensors and a laser stripe range- finder. Robot control software runs on a VME multi-processor hardware supporting a real-time operating system.

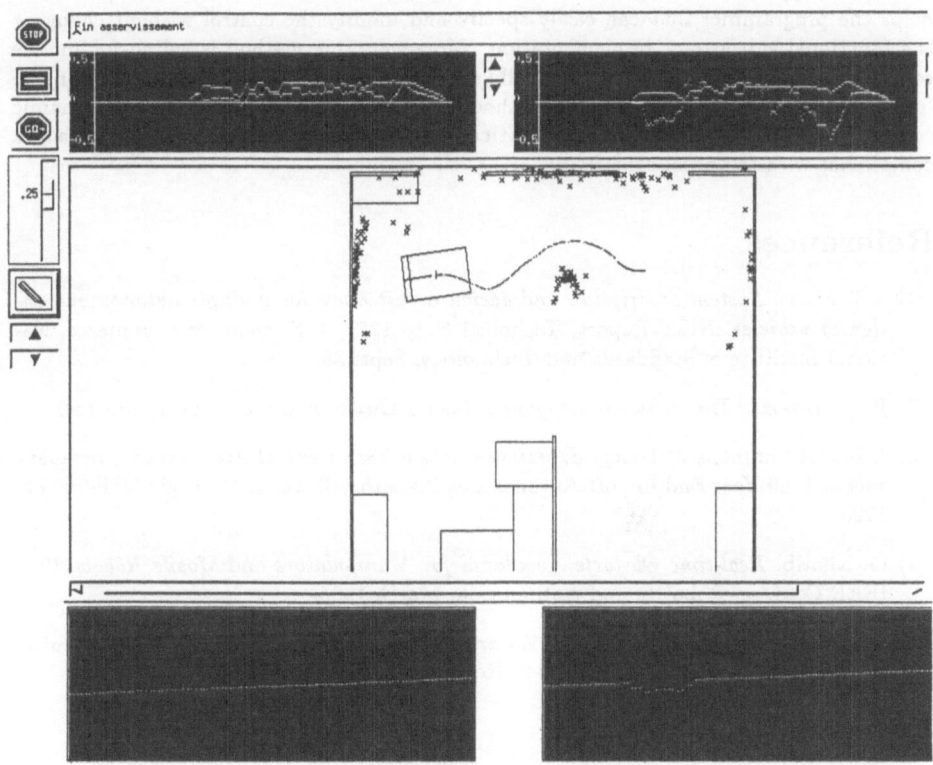

Figure 5: Avoidance of a chair during a free motion

The motion controller has been tested on the mobile robot : Figure 5 represents the operator interface and the result of a robot motion. In the upper windows, the curves indicate the evolution of robot speeds during motions (speeds of the wheels in the left window, linear and angular robot speeds in the right window). The central view depicts the robot in its environment, with all the ultrasonic data collected (black crosses) and the trajectory followed during the last motion. In this particular motion, the robot avoids an unexpected chair and passes between the chair and a wall without stopping. In the lower windows, we can see the last profiles obtained by the two cameras of the laser stripe range finder. The interface enables the operator to modify the robot linear speed while moving.

5 Conclusion

The object-oriented approach with the basic control modules has been defined and experimented for the design and implementation of a robot motion controller : it appears that this method brings openness to the system, allowing a concurrent access for any subsystems. This approach clarifies the interface between functional and control viewpoints : it helps the programmer that can easily specify and modify the control without changing the functional subsystems implementation. Moreover, this method is suited for mobile robots because control can be dynamically specified. But, it would be interesting in the future to relate this object- oriented method to a formal represention of such a dynamic control that could be used to validate critical features such as absence of dead-locks and compliance with temporal constraints.

References

[1] J.S. Albus, *System description and design architecture for multiple autonomous undersea vehicles*, NIST Report, Technical Note 1251, U.S. Dept. of Commerce, National Institute of Standards and Technology, Sept. 88.

[2] R. A. Brooks, *The Behavior Language; User's Guide*, A.I. Memo 1227, April 90

[3] L Lin, R Simmons, C Fedor, *Experience with a Task Control Architecture for mobile robots*, 1989 Year End Report Autonomous Planetory Rover at Carnegie Mellon, Feb. 1990.

[4] O. Khatib, *Real-time Obstacle Avoidance for Manipulators and Mobile Robots*, 1985 IEEE Conf. on Robotics and Automation, March 1985.

[5] S. Shafer, A Stentz, C.E. Thorpe, *An architecture for sensor fusion in a mobile robot*, Technical Report CMU-RI-TR-86-9, Robotics Institute, 1986.

Action Level Behaviours for Locomotion and Perception[1]

James L. Crowley, Patrick Reignier, Olivier Causse, Frank Wallner
LIFIA (IMAG), 46 Ave Félix Viallet, 38031 GRENOBLE

1. Introduction

Autonomous navigation in an indoor environment is not difficult provided that:

1) the building is described by a network of named places and routes,
2) a map of the walls, furniture and other limits to free space is available, and
3) the size of free space is large compared to the size of the robot.

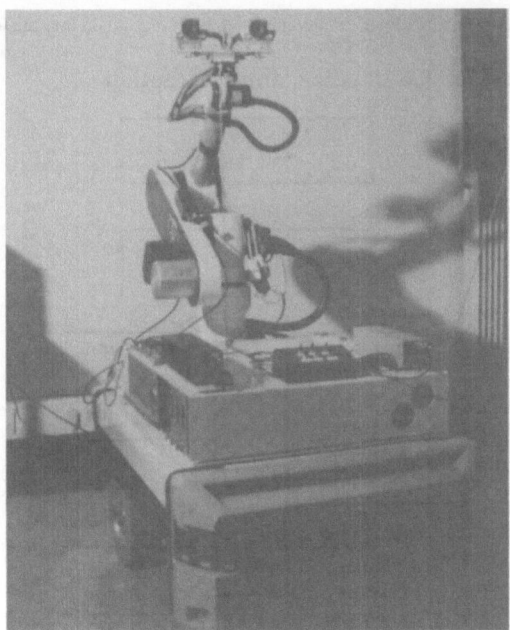

Figure 1.1 The experimental robot vehicle on which the techniques have been implemented.

[1]This work was sponsored by Project EUREKA EU 110: MITHRA and ESPRIT BR 3038 "Vision as Process".

This paper describes a system which is capable of autonomous navigation when the above three conditions are true. It then describes the problems encountered and the modifications that have been necessary to make the system function in a cluttered laboratory environment. The system and experiments described in this paper have been implemented on a Robuter mobile platform equipped with 24 ultra-sonic range sensors, shown in figure 1.1.

Section 2 establishes the basis for the paper by describing the layered architecture, and reviewing the supervisor, vehicle controller and world modeling system. The elements described in section 2 are necessary for the development of the techniques developed in the rest of the paper.

In section 3 we develop a set of locomotion "actions" based on the principle of servoeing toward a target point. We first present a basic algorithm for locomotion which we call "Follow()". We then show how the algorithm can be specialized to for navigating over a network of places by developing the procedures for "GoTo()" and "GoThrough()". Follow requires the verification of a free path, provided by interrogating the local model with procedures "FreePath()" and TurnPath()". We conclude this section by describing a local path planning procedure called "FindPath()".

In chapter 4 we describe how reflex level commands have been integrated into the follow algorithm.

2. Layered Control of Locomotion and Perception

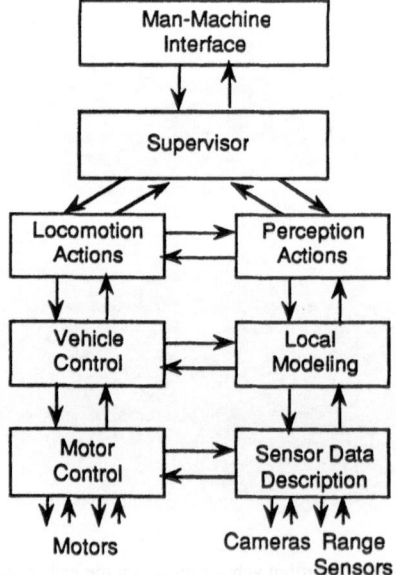

Figure 2.1. The System Architecture: Parallel Architecture for Navigation and Perception Controlled by an Knowledge Based Supervisor.

2.1 Layered Control of Navigation

The organization of a hierarchical control system for a mobile surveillance robot is illustrated in figure 2.1

[Crowley 87]. Twin hierarchies for navigation and for perception are organized as a set of levels according to the abstraction of the information which is processed and the speed with which responses are required. These levels are known as the signal level, the effector level and the action level.

The Signal Level: At the lowest level, each hierarchy asynchronously processes raw signals. On the perception side, processing involves acquiring sensor signals (in camera or sensor coordinates) and converting these to an initial representation. In the locomotion hierarchy this process involves closed loop control of the motors to maintain a specified velocity, as well as capture of pro-prioceptive sensor signals for estimating position and velocity. The cycle time of our motor controllers is 4 milli-seconds. New sonar data is acquired at a rate of about 60 milliseconds per range measure.

The Effector Level: At an intermediate level, both hierarchies represent their information in terms of the vehicle and its environment. At the center of the locomotion hierarchy is a vehicle level controller [Crowley 89b]. This controller accepts asynchronous commands to move and turn the vehicle. The vehicle level controller also maintains an estimate of the position and velocity of the robot and shares this with the perception hierarchy.

The perception hierarchy projects the description of new sensor signals into a common coordinate system, and uses the projected information to update a composite model of the environment [Crowley 89a]. As a side effect of the update process, errors in the estimated position are detected and relayed to the vehicle controller. The position of a sensor in the common external coordinate system is provided by the composition of the position of the sensor with regard to the vehicle and the configuration of the vehicle.

The cycle time for the vehicle controller is currently 80 milliseconds. The local model is updated as each segment describing the limits to free space is acquired. On the average this is about every 200 milliseconds.

Action Level: An action is a coordinated sequence of effector commands. The concept of an action is central to both the planning and execution of tasks in the supervisor, and to the coordination and control of locomotion and perception. In human terms, the action level procedures correspond to learned "skills".

Supervisor Level: Controlling the twin hierarchies for locomotion and perception involves selecting the appropriate actions in order to accomplish symbolically expressed goals. The reasoning activity within a supervisor is naturally expressed as rules, organized into "contexts". Within each context, rules are triggered by internal facts which represent things such as goals, external events, or descriptions of the environment. The supervisor is a form of symbolic servoe loop, composed of three phases:
   1) Select an action to bring the robot or world to the desired state,
   2) Execute the action, and
   3) Evaluate the consequences of the action.
In this paper we are particularly interested in navigation tasks. The supervisor plans navigation tasks using a form of heuristic search through a network of places.

2.2 Planning Paths with a Network of Places

A mission for our system is "programmed" by specifying a sequence of tasks which are to be accomplished. The basic set of surveillance tasks may be paraphrased as:

"Be at place <P> during a time interval <T>."

"Survey region <R> (set of places) during the interval of time <T>."
"Signal the detection of event <X> within a region <R> during the interval <T>."

Before execution, each task is decomposed into a sequence of subgoals which comprise a plan. Advance planning permits the system to estimate the resources required for the mission. Because the environment is not perfectly known in advance, the decomposition of the task is not sure to succeed. To successfully execute a mission, the supervisor must monitor the execution of each task and dynamically generate the actions required to accomplish its goals.

Both planning and plan execution are knowledge intensive processes which are adapted to a forward chaining production system. The supervisor on our surveillance robot has been implemented using a version of the OPS-5 language [Brownston et. al. 85] to which we have added an asynchronous message passing facility [Crowley 87]. The supervisor plans navigation actions using a network of named places connected by named routes. Each place contains its location in an local (building based) reference frame, as well as a list of places which are directly accessible by a straight line routes. Each route is described by a data structure which contains information about appropriate navigation procedures, speeds, path length, and surveillance behaviours which are appropriate for the route.

The supervisor commands navigation "actions" in order to accomplish the tasks in its mission. The set of navigation actions are based on the use of a vehicle level controller which independently servos translation and rotation.

2.3 Control of Displacement and Rotation: The Standard Vehicle Controller

Robot arms require an "arm controller" to command joint motors to achieve a coordinated motion in an external Cartesian coordinate space. In the same sense, robot vehicles require a "vehicle controller" to command the motors to achieve a coordinated motion specified in terms of an external Cartesian coordinate space. The locomotion procedures describes below is based on the LIFIA "standard vehicle controller" which provides asynchronous independent control of forward displacement and orientation [Crowley 89b]. A production version of this controller is in everyday use in our laboratory. We will limit the presentation here to the minimum required for the rest of the paper.

The vehicle controller operates in terms of a standardized "virtual vehicle" and thus may be easily adapted to a large variety of vehicle geometries. The control protocol permits independent commands for rotation and translation. When a parameter is not specified the previous value is taken by default. The seven commands which make up the vehicle control protocol are described as follows.

Move ΔD, V_d, A_d Move a maximum distance of "ΔD" meters, with speed V_d meters/sec and an acceleration of A_d meters/sec^2.

Turn $\Delta\alpha$, V_α, A_α Turn a maximum rotation of $\Delta\alpha$ degrees with an angular speed of V_α deg/sec and an angular acceleration of a deg/sec^2.

Stop   Stop all motion and hold the current position and orientation.

Commands which refer to the estimated position and velocity of the vehicle are:

GetEstPos	Return the estimated position (x, y, α) and its 3 x 3 covariance C_p.
GetEstSpeed	Return the estimated speed of the vehicle, V_d, v_α.
CorrectPos	Correct the position and orientation and their uncertainties by Δx, Δy, $\Delta \alpha$ using the Kalman filter gain matrix K_p.
ResetPos	$(x, \ y, \ \alpha, \ C_p)$

Position estimation includes a model of the errors of the position estimation process. Thus the estimated position is accompanied by an estimate of the uncertainty, in the form of a covariance matrix. This covariance matrix makes it possible to correct the estimated position using a Kalman Filter. This correction is provided as a side effect in the process of dynamic modeling of the local environment.

2.4 Dynamic Modeling with Ultrasonic Range Sensors

Perception serves two fundamentally important roles for navigation:
1) Detection of the limits to free space, and
2) Position estimation.
Free space determines the set of positions and orientations that the robot may assume without "colliding" with other objects. Position estimation permits the robot to locate itself with respect to goals and knowledge about the environment.

Our navigation system employs a dynamically maintained model of the environment of the robot using ultrasonic sensors and a pre-stored world model [Crowley 89b]. We call such a model a "composite local model". Such a model is "composite" because it is composed of several points of view and (potentially) several sources. The model is local, because it contains only information from the immediate environment.

The modeling process is illustrated in figure 2.2. Raw sensor data is processed to produce an immediately perceived description of the environment in terms of a set of geometric primitives expressed in a common external coordinate system. Data are currently provided from the ultrasonic range sensors. We have also developed a real time stereo system, and are currently waiting for delivery of the hardware implementation [Crowley 91]. Sonar interpretation uses coherence in the data to determine a subset of the measurements which are reliable.

The composite model describes the limits to free space which are currently visible to the robot. The contents of the composite model are never directly available to the other parts of the system. Instead the contents are accessible through a set of interface procedures which interrogate the composite model and return both symbolic information and geometric parameters.

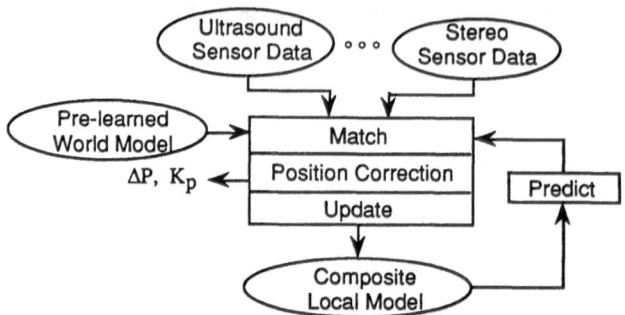

Figure 2.2 An abstraction description of sensor data is used to dynamically update a composite model of the limits to free space.

3 Locomotion and Perception Actions

The basis for locomotion actions in our system is an algorithm called "Follow". This algorithm dynamically servoes the orientation and translation of the vehicle towards a goal point while testing for and avoiding obstacles. Specialized forms of "Follow()" exist for different "kinds" of goals. In particular, for a static goal point, we employ procedures "GoTo()" and "GoThrough()".

The basic perceptual action of our system is "FreePath" which detects obstacles within a rectangular region. A variation of procedure FreePath, called "TurnPath" permits the detection of obstacles within the region traced by the robot when it turns. Procedures FreePath is uased as a building block for the construction of procedure FindPath, described in section 3.3. FindPath, FreePath and TurnPath are described in detail in [Reignier-Crowley 91].

3.1 Procedure Follow

Many researchers have worked on the problem of "path planning" as one of generating an "optimal" path within a static, known environment [Lozano Pèrez 81]. Our experience is that in the real world:

1) Knowledge of the local structure of the environment is partial and imprecise,

2) The structure is often not static,

3) The execution of a planned trajectory is imprecise because of perturbations and because of delays in the motor and vehicle servoeing.

As a result of these factors, we believe that a much greater degree of robustness is provided by letting the robot chooses its own local path by servoeing towards a goal. In the case where a desired trajectory exists, as in road following, the goal can be a point on the trajectory a certain distance in front of the vehicle. We assert that the procedural action of "chasing a point" is basic to locomotion.

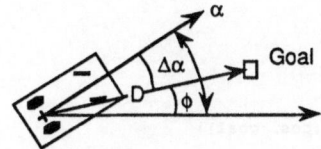

Figure 3.1 The distance and heading to a goal are defined by D and φ.

Our work on this problem was first inspired by Wallace et. al. [Wallace et. al. 85] who were concerned with the problem of road following. We have formulated the problem in a somewhat different manner. Let us refer to the vehicle heading as α and the direction towards the current goal point as φ, as shown in figure 3.1. In order to approach the goal, the vehicle must turn by an amount $\Delta\alpha = \alpha - \phi$, and then travel a distance D. Thus, the simplest locomotion algorithm would be to turn the robot by an amount $\Delta\alpha$ and then travel an amount D. We call this the Stop-and-Turn algorithm. Anyone who implements such an algorithm immediately observes a number of problems.

The first problem is that the estimated position and orientation are imprecise. If we operate a perception system while the vehicle moves, the vehicle may suddenly discover an error α_e and P_e in estimated position requiring it to stop and turn. A second, more serious problem, can occur if the acceleration of the power wheels is not perfectly synchronized, or if vehicle loading leads to the wheels having an unequal wheel diameter. These problems can result in a movement that is not straight line. Wheel slip during accelerations can also contribute to this problem. Such a problem was very evident in Moravec's "Pluto" robot. The problem was even more severe in the "Sofito" painting robot. These and other early locomotion systems demonstrated the need to servoe orientation during movement.

The "standard" vehicle controller is designed to provide combined servoeing of orientation and translation. The action level procedure simply calls the commands "move" and "turn" at each cycle, setting new target values for orientation and displacement. At each cycle, new values for $\Delta\alpha$ and ΔD are specified based on the difference of the current heading and position and the goal heading. rotation and translation velocities can be specified as a proportion of the error, but this is not necessary as the vehicle controller includes a PID servoeing of these velocities.

The locomotion action "Follow" is composed of three states, which we have named "follow_turn", "follow_move" and "follow_halt". Follow() verifies that there is a freepath to the goal and then determines the current state, based on the error in distance and orientation.

```
int follow()
{
    POSITION * goal, * pos;    /* data structure for x, y, alpha */
    double  delta_alpha, delta_D, error;
    double va, vd;             /* current speed */
    int obstacles;             /* Nearest obstacles */

/* get latest goal and position and speed */

    goal = get_goal();
    pos = get_est_pos();
```

```
    get_speed(&vd, &va);

/* verify FreePath */

    if (obstacle = free_path(pos, goal))
    {
        push_goal(goal);
        goal = find_path(pos, goal, obstacle);
        if (goal == NULL) return(follow_halt(FAILED));
    }

/* compute distance and angle to goal and maximum error from path */

    delta_D = distance(pos, goal);
    delta_alpha = angle(pos, goal);
    error = delta_alpha * (vd/va) * sin(delta_alpha/2);

/* determine behaviour */

    min_dist = get_goal_tolerance();
    if (delta_D < min_dist) return(follow_halt(SUCCESS));

    path_tolerance = get_path_tolerance();
    if (error > path_tolerance)
        return(follow_turn(pos, delta_alpha));

    else return( follow_move(delta_D, delta_alpha));
}
```

The behaviour "follow_move()" issues commands of move and turn for the vehicle controller to servoe the vehicle toward the goal.

```
int follow_move(delta_D, delta_alpha);
double delta_alpha, delta_D;
{
    move(delta_D);
    turn(delta_alpha);
    return(ACTIVE);
}
```

A large change in angle can lead to a large deviation from a straight line path. In a cluttered environment, this can be dangerous. Large path deviation can be avoided by introducing a state in the locomotion action such that whenever the error in orientation is too large, the robot will stop forward motion, while continuing to turn. The perpendicular error for the path is computed as a function of change in angle. This formula is developed in the next section.

The behaviour follow_turn() is used whenever the change in angle will lead to a deviation from the path which is too large. Procedure follow_turn() makes a call to "move()" with a distance of 0 in order to halt forward motion of the vehicle. The procedure then uses procedure turn_path() to verify that it can turn toward the goal without a collision. Finally it calls the vehicle control function "turn()" to turn by the angle $\Delta\alpha$.

```
int follow_turn(pos, delta_alpha);
POSITION * pos;
double   delta_D;
{
    move(0);
    if (turn_path(pos, delta_alpha))
         return(follow_halt(TURN_BLOCKED));
    turn(delta_alpha);
    return(ACTIVE);
}
```

By changing the procedure which determines the goal, procedure follow() can be easily adapted to going to a point, going through a point, following a trajectory, chasing a moving target or traveling toward a landmark or beacon.

3.2 Procedures GoTo and GoThrough

Two widely used action level procedures are "GoTo()" and "GoThrough()". Procedure GoTo() will servoe the vehicle to stop at a specified point. Procedure GoThrough() will servoe towards a point until it gets within a certain distance. GoThrough will then switch to servoeing to the next goal point. The distance at which GoThrough() must switch target points can be based on the error from a straight-line trajectory which can be tolerated. This error can be approximated by a relatively simple relation, which can be derived as follows.

Let ΔT represent the cycle time for the servoe loop for Follow(), and let V_α represent the rotational velocity. The vehicle will deviate from a straight line path by a distance which is proportional to $\Delta\alpha$, V_α and ΔT as shown in figure 3.2.

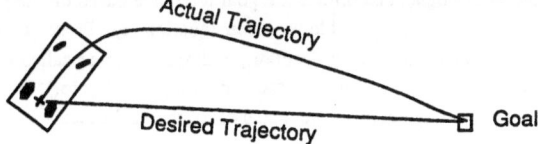

Figure 3.2 changing paths can lead to a deviation from a straight line trajectory.

If we assume constant velocities for rotation and translation, it is fairly easy to determine a formula for the maximum perpendicular error, E_p, from a straight line trajectory. Consider a straight line path along the x axis to a goal at $\phi = 0$. The point of maximum deviation will occur when the vehicle orientation reaches the direction to the goal, $\alpha = 0$, as shown in figure 3.3. After this point, the vehicle reduces it distance from the straight line path. The time to turn by an angle $\Delta\alpha$ is determined by the rotational velocity, V_α.

$$T_c = \frac{\Delta\alpha}{V_\alpha}$$

The distance traveled during this time, D_c, depends on the translation velocity, V_D.

$$D_c = V_D\, T_c = V_D \frac{\Delta\alpha}{V_\alpha}$$

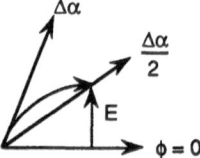

Figure 3.3 Maximum perpendicular error, E, is determined by $\Delta\alpha$ and V_D / V_α.

For constant rotational velocity, the trajectory is an arc of constant radius spanning an angle of $\Delta\alpha$. The direction spanned by the arch is given by average direction, $\Delta\alpha/2$. Thus the perpendicular distance from the path is approximated by the distance traveled times $\mathrm{Sin}\,(\Delta\alpha/2)$. From this we deduce the formula for the perpendicular error given in equation (1).

$$E_p \approx D_c\, \mathrm{Sin}(\frac{\Delta\alpha}{2})$$

or

$$E_p \approx \Delta\alpha\, \frac{V_D}{V_\alpha}\, \mathrm{Sin}(\frac{\Delta\alpha}{2}) \tag{1}$$

Note that equation (1) assumes a constant ratio of translation and rotational velocities. In the real world these velocities are never constant, and so the relation becomes an approximation.

Procedures GoThrough() and GoTo() can be developed based on procedure Follow(). GoTo is initialised by init_goto, which sets the goal and specifies the tolerances for stopping at the goal and for path deviations. Procedure init_goto() then schedules the procedure "goto()" to be executed every T milliseconds. Procedure goto calls procedure follow(). When follow() returns a status which is not active, goto() removes itself from the scheduler and signals the supervisor.

Procedure Procedure init_gothrough() determines the goal tolerance based on the path tolerance and the transition angle between paths, $\Delta\alpha$, using a table based on equation (1). Procedure init_gothrough() then sets up a stack of goal points and schedules procedure go_through() to be called every T milli-seconds. Procedure gothrough() calls procedure follow() in the same manner as goto(). However, when follow() returns a non-active status, go_through pops the next goal off of the goal stack.

3.3 Procedures FreePath and TurnPath

Procedures FreePath() and TurnPath() verify that a command move or turn can be safely executed, by searching for the overlap of an obstacle in the local model with an approximation of the path which is swept by the vehicle. In the case of FreePath, the path is approximated by a rectangle, as shown in figure 3.4.

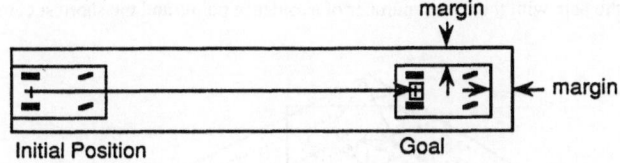

Figure 3.4. Procedure FreePath() verifies that a rectangular path is free of obstacles.

Procedure FreePath is implemented by applying a "clipping" algorithm to the list of segments maintained in the local model. A counter-clockwise scan is made of the of the boundary segments of the path. For each boundary segment, all model segments entirely to the right (outside the path) are eliminated from the list. Any segments remaining in the list after this scan are considered as obstacles. The identity of the nearest segment in the list of obstacles is returned.

Procedure TurnPath() detects intersections in a path swept by rotation of the vehicle. This path is somewhat more difficult to model. It is modelled as a intersection of two discs and a wedge. A segment must simultaneously intersect with all three of these forms to be considered as an obstacle to turning.

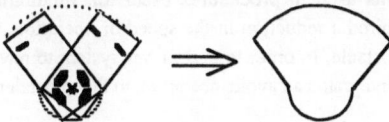

Figure 3.5 The region for procedure TurnPath()

3.4 Procedures FindPath

Procedure FindPath is a recursive path planning procedure. FindPath operates by calculating avoidance points on either side of an obstacle. The avoidance points are determined by a perpendicular projection of each end-point of the obstacle segment away from the original path, as shown in figure 3.6.

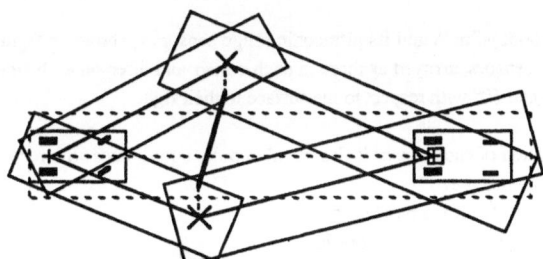

Figure 3.6 Procedure FindPath generates avoidance paths based on projecting the end-points of the obstacle perpendicularly from the path.

Each avoidance path is tested using procedure FreePath. If an obstacle is found by procedure FreePath, the FindPath will call itself recursively. Findpath is limited to a maximum of three levels of recursion. The result can be a new pair of avoidance points as shown in figure 3.7. Findpath generates a list of possible paths and chooses the path with the fewest number of avoidance points and the shortest cumulative length.

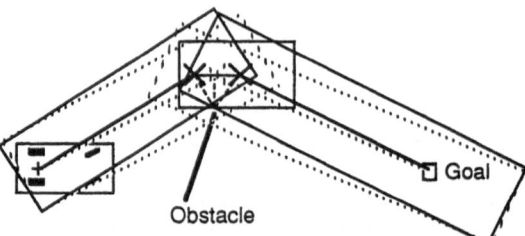

Figure 3.7 A recursive call to procedure Findpath permits the robot to refine its avoidance points.

4 Reflexes for Navigating in Encumbered Spaces

Our navigation system was originally debugged using a realistic simulation of the robot operating in a rather sparse environment. When we moved the procedures on to the real robot, we initially operated in relatively open spaces in the laboratory. As we attempted navigation in increasingly encumbered spaces we were required to reduces the margins used in procedures FreePath, FindPath and GoTo. However, each reduction of these margins required a reduction in the speed of operation, and an increase in the risk of collision with an unobserved obstacle. In order to permit our system to navigate robustly in encumbered spaces, we have added recently integrated an avoidance reflex to the procedure follow().

The composite local model contains line segments extracted from three or more sensor readings. If an individual sensor detects an object which is not seen by the neighboring sensors, the object will not be modeled in the local model. The move_reflex protects against a collision by temporarily halting the robot whenever any of the forward sensors detects a range reading which is inferior to a stopping distance. The turn_reflex causes the vehicle to be repelled to the side of nearby objects seen by individual side looking sensors. To explain these reflexes, we must examine the configuration of range sensors on our vehicle.

4.1 Configuration of Range Sensors

The geometry of our robot vehicle and its ultrasonic range sensors is shown in figure 4.1. The vehicle is equipped with 24 range sensors, arrayed as three in each corner and three on each side. The corner sensors are oriented at 30°, 45° and 60° with respect to the surface to their right.

The position and orientation of each sensor is described in polar coordinates with respect to the origin of the robot. The position parameters are:
  m   Distance of the sensor from the origin.
  θ Direction to the sensor from the origin.
  ϕ Orientation of the sensor.

This configuration is based on the hypotheses that
  1) most surfaces give lambertian reflection of ultra-sound for incident angles of less than 15° from perpendicular, and that

  2) the effective beam width of an ultrasonic range sensor is approximately 15°.

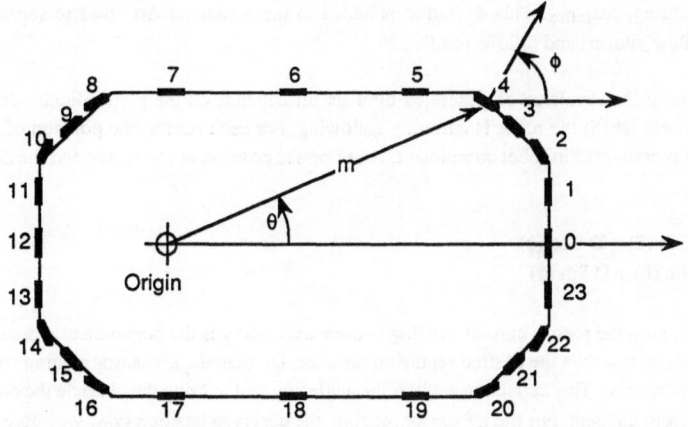

Figure 4.1 Configuration of Ultrasonic Range Sensors on our robot vehicle.

Our experience is that the first hypothesis is essentially correct for concrete, brick and cloth surfaces. This incident angle for specular reflection for painted fiberboard is closer to 8 degrees from perpendicular. For metal, the specular angle approaches 5 degrees. Thus, the robot is able to model walls made of concrete or cloth from just about any angle. To model fiberboard or metal, the vehicle must be nearly aligned with the wall.

The second hypothesis is just wrong. While the emitted energy for a Polaroid sensor has a 3 dB width of nearly 15°, the effective beam-width is closer to 7 degrees, depending on the range and the material. Thus there are blind regions between the corner sensors and at near distances between the side and forward looking sensors. These blind spots are a particular danger in an environment in which the principle obstacle is a table leg.

4.2 Integration move_reflex() and turn_reflex() to follow().

A "reflex" is an unplanned command based on raw sensor data. We have implemented reflexes to maintain a safety distance in front of the robot, and as a potential-field torque which turns the vehicle away from nearby obstacles.

The safety distance is composed from three terms. The first term is the distance required to halt the vehicle at the current acceleration. The second term is the distance traveled by the vehicle at its current velocity, during one cycle of follow, T_{follow}. The third term is a safety distance, SAFE_DIST, usually set to 10cm. Thus the stop distance, D_s, computed by a function "get_stop_distance() is given by a polynomial in V_D

$$D_s = \frac{V_D^2}{2 A_D} + V_D T_{follow} + 10.$$

The repulsive torque is computed as sum of the torques for each of the 8 forward looking ultrasonic range sensors, $C = \{1, 2, 3, 4, 20, 21, 22, 23\}$. The sum of the individual torques determines a deviation from the current heading, $\Delta\alpha_{reflex}$. This deviation is added to the parameter $\Delta\alpha$, used to servoe heading in procedures follow_move() and follow_turn().

The calculation of the repulsion of generated by a sensor depends on the perpendicular distance of the obstacle from path which the robot is currently following. For each sensor, the position of the observed object (x_0, y_0) is computed in robot coordinates, based on the position of the sensor and the depth returned by the sensor.

$$x_0 = m \cos(\theta) + D \cos(\phi)$$
$$y_0 = m \sin(\theta) + D \sin(\phi)$$

In robot coordinates, the robot's current heading is the x axis, and y is the perpendicular distance from the current path. If y is less than the desired repulsion distance, D_r, then the kth sensor reading will deflect the robot by an amount $\Delta\alpha_k$. This deflection angle is the angle required to bring the obstacle the obstacle further than R_DIST from the path. For the kth sensor reading, the observed position (x_{ok}, y_{ok}) gives a deflection angle of :

$$\Delta\alpha_k = \operatorname{Tan}^{-1}\left(\frac{y_0 - D_r}{x_0}\right)$$

The total deflection angle is the sum of the individual deflection angles for the set C = {1, 2, 3, 4, 21, 22, 23}

$$\Delta\alpha_{reflex} = \sum_{k \in C} \Delta\alpha_k.$$

The deflection angle is integrated in to the procedure follow_move() in the following manner:

```
void follow_move(delta_D, delta_alpha);
double  delta_alpha, delta_D;
{
    if (move_reflex() < get_stop_distance()) move(0);
    else  move(delta_D);
    turn(delta_alpha + turn_reflex());
    return(ACTIVE);
}
```

5. Conclusions

In this paper we have described a system which permits an experimental robot vehicle to navigate in a cluttered laboratory. This system demonstrates the robustness and flexibility which are provided by viewing the navigation problem as a problem of feedback control of layers corresponding to signals, effectors, actions and tasks.

The basic navigation action in our system is servoeing toward a goal point, provided by procedure "follow()". The procedure Follow() has been specialized to provide procedures GoTo() to arrive at a goal point, and GoThrough() to pass through a sequence of points. All of these functions rely of perception actions FreePath, TurnPath an FindPath.

The system we have described relies on 24 ultrasonic range sensors. This numbers of sensors is inadequate to properly model an environment containing unknown small objects, such as table legs. In order to improve the robustness of the system we have added reflexes to stop in front of an obstacle, and to be repelled from an obstacle.

The system continues to experience occasional problems with un-modeled table legs. An additional source of problems are important visitors who wear suit-pants made of highly specular materials such as silk. We expect to solve this problem in the near future by installing a hardware implementation of our real-time vertical line stereo system.

Bibliography

[Brownston 85] Brownston, L., Farrell R., Kant, E., et Martin, N., Programming Expert Systems in OPS5, Addison Wesley, Reading Mass., 1985.

[Crowley 85]   Crowley, J. L., "Navigation for an Intelligent Mobile Robot", IEEE Journal on Robotics and Automation, 1 (1), March 1985.

[Crowley 87] Crowley, J. L., "Coordination of Action and Perception in a Surveillance Robot", IEEE Expert, Vol 2(4), pp 32-43 Winter 1987. (Also in the 10th I.J.C.A.I., Milan 1987).

[Crowley 89a] Crowley, J. L., "World Modeling and Position Estimation for a Mobile Robot Using Ultrasonic Ranging", 1989 IEEE Conference on Robotics and Automation, Scottsdale AZ, May 1989.

[Crowley 89b] Crowley, J. L., "Asynchronous Control of Orientation and Displacement in a Robot Vehicle", 1989 IEEE Conference on Robotics and Automation, Scottsdale AZ, May 1989.

[Crowley 91] Crowley, J. L, . P. Bobet et K. Sarachik , "Dynamic World Modeling Using Vertical Line Stereo", Journal of Robotics and Autonomous Systems, Elsevier Press, Juin, 1991

[Nilsson 80] Nilsson, N. J. Principles of Artificial Intelligence, Tioga Press, 1980.

[Lozano Pèrez 79] Lozano-Perèz, T. and M. A. Wesley, "An algorithm for Planning Collision-free Paths among Polyhedral Obstacles", Communications of the ACM, 22(10):560, Oct. 1979,

[Reignier-Crowley 91] Reignier, P. and J. L. Crowley, "Raissonnement Géométrique et Contrôle d'exécution pou la mobilité", Internal Research Paper, LIFIA, 1991.

[Wallace et. al 85] Wallace, R. W. et. al., "First Results in Road Following", IJCAI 85, Los Angeles, August 1985.

Implementation of a small size experimental self-contained autonomous robot — sensors, vehicle control, and description of sensor based behavior

Shin'ichi Yuta, Shooji Suzuki, and Shigeki Iida

Intelligent Robot laboratory, Institute of Information Science and Electronics,
University of Tsukuba, Tsukuba, 305 JAPAN

Abstract

A small size self-contained autonomous mobile robot has been implemented as a platform for experimental research of sensor based behavior. This robot can move in two dimensional environment using two driving wheels and have optical and ultrasonic range sensors. Each hardware function of the robot is modularized as a single board computer and master module located in center to control the whole system.

In the real environment, the autonomous robot must get environment information from sensors of itself, while it is moving, because the environment is not completely known before moving. In the behavior program, it must monitor sensor information and select the suitable motion. Action mode representation is proposed to describe such behavior.

Keywords: autonomous mobile robot, function distribution, centerized decision-making, sensor based behavior, action mode representation

1   Introduction

The implementation of a practical experimental robot is one of key issue in the research on autonomous mobile robots. Many research and development have been made in the experimental mobile robots after 'Shakey' in 1984 [1], and many researchers are still interested in realizing experiments of the mobile robots. Recently, several autonomous robots have been reported with a good experimental results including indoor or outdoor environment [2] [3]. The importance of the experiments in robotics is not only as a verification of the theoretical analysis or application of the theory, but also as a base for the problem formulation in theoretical consideration. For these purpose, the capability for the easy experiments is important, and so, light and small robots may be convenient than the big and heavy ones.

The authors have been developed small size self-contained autonomous robots for indoor environment as experimental platforms for the mobile robot research [4] [5]. These robots are named 'Yamabico' and they are always growing both in hardware and software, to have more capability in behaving in more complexed environments.

In this paper, the current status of the authors' robots and recent basic concept for the designing of them are reported. The authors' main interests of the research is to find the basic idea for;

(1) To realize a real time robot behavior in the real environment which includes the moving and sensing error, and which may be different from the given model.

(2) To design the robot system with enough performance to cope with real environment which has the capability of incremental development of such multi function robot system. (i.e. to design a good architecture of the autonomous robot.)

The authors give the distinguished definition on the words 'behavior' and 'motion' for discussion. 'Behavior' is a task or a job to achieve a certain purpose. So, the behavior includes the decision making of motion in it. An example of the autonomous mobile robot's 'behavior' is map based navigation. 'Motion' is an element of 'behavior'. Examples of 'motion' are; running with given velocity and trajectory, tracking the line drown on the floor, following the wall, and stopping at the given position. Motion is executed by the sense of "control".

The autonomous robot moves in the real environment after behavior planning using the environment model. However, even in the simple behavior, execution of the behavior is not easy to cope with the real environment. Because the modeled environment is not completely same to the real one. Therefore, the autonomous robot must move with getting environment information by its sensors to detect differences between the model and the real environment. And if the differences are detected, the robot is expected to deal with the detected environment by itself and continue its action autonomously.

For realizing such autonomous behavior, the control manner or the program to describe the robot's behavior is very complicated. So, the description method of the robot's behavior in the real environment must be considered. At the same time the controller architecture of the robot control system must be considered much to discuss the description method.

The other side, for realizing incremental development of many functions of the robot. The modularity or distribution of these functions both in hardware and software with the capability for possible expansion of new functions are important. So, the central decision making and functional modularization are the basic concept of the architecture in these robots.

2　A design concept of the autonomous mobile robot

2.1　Controller architecture of the robot system

The authors have proposed to divide the robot control system into two stages; **the function level** (system level) and **the behavior level** (application level) [6]. The function level is a part to realize the general functions for wide purpose, while the behavior level is a part to realize the certain behavior for the given concrete objectives. In this division, the behavior level controls the total robot behavior by executing the behavior program, and the function level realizes robot functions by controlling actuators and sensor devices. Figure 1 shows the conceptual architecture of the robot control system.

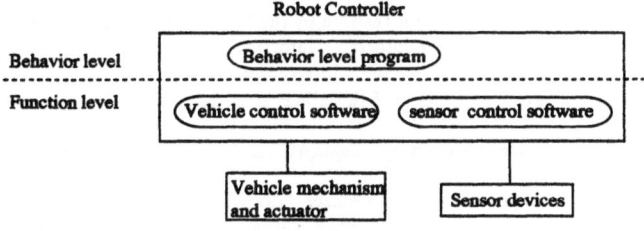

Figure 1　Behavior level and Function level

2.1.1 function level — functional distribution —

The autonomous mobile robot must have many kind of sensors and actuators. So in the proposed architecture, the function level of the robot is modularized to be 'function modules' in both hardware and software for each function of the robot. Each function module has its own processor and the peripherals in hardware. The merits of the modularization by functional distribution are;

1. increasing the efficiency of development,

2. good maintainability,

3. possibility of gradual grade-up or incremental development of the system [4]

The function modules are classified into two types; sensor controller and actuator controller. The sensor controller module continuously and autonomously drives sensor devices, processes sensory data and provide the sensor information for use by behavior level. And the actuator controller module controls actuators according to the given control modes and parameters by behavior level.

2.1.2 behavior level — centralized decision-making —

Only one functions module has the role of the behavior level and decide the behavior of the whole system. It is called 'master module', The behavior program of the robot executed on master module must monitor the status of the robot and the environment information around the robot by sensors. According to the result of the monitoring, the behavior program decides the action of the robot and gives the control information to actuator controller modules.

2.1.3 interface between behavior level and function level

In the proposed architecture, the modules form the star-connection, in which the master module locates in the center and is connected with all function modules. No inter-function module communication is needed because of centralized decision making. To execute the behavior program in master module, the method of communication with the function level is important. Two communication methods are defined for interface to the actuator controller and interface from sensor controller.

The interface to the actuator controller is realized by **control command**. The behavior program gives the instruction to the actuator controller to move itself or perform a concrete task. The important point on this interface is that, the execution of giving the command is completed instantaneously without delay, but the issued command gives the effect on the robot moving after that. This command gives the parameters for the continuously working actuator controller, so the execution of the command is not over by itself, and it continues until the issuing of the next command which is overwritten on the previous command.

The interface from sensor controller is realized by the **State Information Monitoring Panel(SIMP)**. SIMP is the window, on which all the detectable information are displayed and updated in real time. The behavior program can know the status and situation of the robot and its environment by only looking this panel. The typical property of the SIMP is as follows,

1. The latest result of sensing is always displayed on the panel, whether it is used or not.

2. When the behavior program use the sensor information, it only looks the panel without any side effect on the environment or the robot status and motion.

Figure 2 shows the interfaces between function level and behavior level.

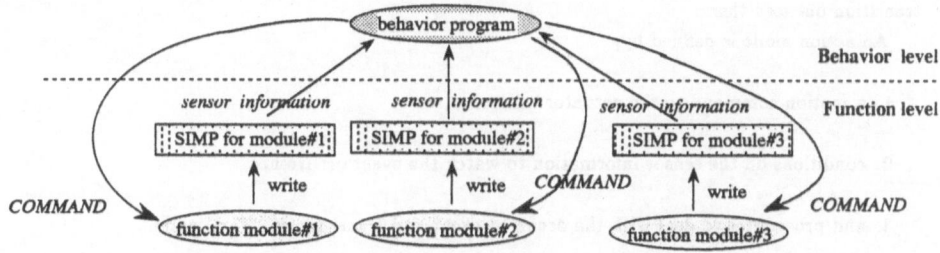

Figure 2. Interface between behavior level and function level

2.2 Analysis and description of the sensor based behavior

2.2.1 features of the sensor based behavior

The first step to describe the behavior program of the robot is the analysis of the robot's behavior. Generally, it's analized or planned as a simple sequence of motions under the environment model. If the real environment is completely same to the model, the mobile robot can achieve the given purpose by executing the planned simple sequence of motions.

However, the mobile robot will detect the difference between the assumed environment(i.e. the model) and the real one. For example, obstacles which didn't exist at planning stage may appear at the execution stage. In such case, the robot can't continue executing the planned motion sequence, and, the robot is expected to modify the motion sequence to deal with actual situation autonomously. To achieve such behavior, detection and treatment for such events are important. The autonomous mobile robot must select and change the suitable motion, when it detects an event occurrence by its sensors.

The behavior of the autonomous mobile robot is planned based on the model, but the execution of the behavior is done based on the sensor information. Such sensor based behavior can not be represented by the simple motion sequence. And the behavior representation which includes the dynamic motion selection according to the real situations is required.

The feature of the sensor based behavior is that the motion and the timing of motion change can not be determined before the robot moving. It is determined by event occurrences. When an event is detected by sensors, the suitable motion must be selected based on the sensor information, assumed environment, and current robot motion.

One method to express such behavior is 'if-then' type rule expression. However, another feature of the robot behavior is that the result of the behavior execution is analized as a motion sequence in retrospect. So, the required representation method of the sensor based behavior has both two feature; (1)ability to select a suitable motion based on the robot status and sensor information, (2)and ability to follow easily the sequential flow of motion select by the robot.

2.2.2 a method for representation of the sensor based behavior

We proposed **action mode representation** to represent the sensor based behavior [7][8]. An **action mode** is an unit of behavior, and the robot's behavior is represented by many action modes and transition between them.

An action mode is defined by;

1. a motion command to the actuator controller,

2. conditions on the sensor information to watch the event occurrence,

3. and procedures to deal with the occurred event and transition of action mode.

Each action mode gives one motion command to the actuator controller to control the robot motion. Only one action mode is processed in each time. The motion change is controlled by transition of action mode in the behavior program.

An action mode includes many conditions to watch the event occurrence. A condition includes sensor information to know the status of robot or the real environment. All conditions in the action mode are evaluated continuously and concurrently until certain condition is met. When a condition is filled, the suitable motion will be selected to deal with the occurred event. The motion command for the next motion is sent after action mode transition, because the conditions to watch events and treatment for the occurred events are considered for each robot's motion.

A state transition diagram is useful to represent the robot's behavior, where action modes are considered as states. Figure 3 shows a sample of action mode representation with a state transition diagram.

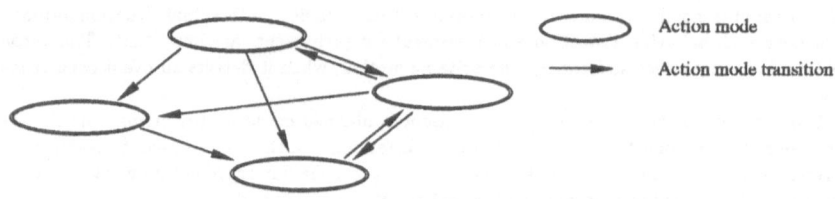

Figure 3. Action mode representation

3 Implementation of a small size self-contained mobile robot "Yamabico"

The authors have been implementing small size self-contained mobile robots for their experiments. The implemented robots are named "Yamabico"(Figure 4). The Yamabico has two wheels driven by DC motors and two casters for locomotion, dead reckoning system for position estimation, and four directional ultrasonic sensors and an laser range sensor for getting the environment information. In this section a recent version of authors' self-contained robots are described in following topics; (1)the architecture of the robot controller, (2)description and execution of the robot's behavior program, (3)implementation of controllers for sensors and locomotion function of the robot.

Figure 4 The self-contained robot "Yamabico"

3.1   The robot controller

The architecture of the new Yamabico is constructed based on the authors' conceptual architecture, **function distribution** and **centralized decision-making**[6]. The constructed robot has four function modules; (1) master, (2) ultrasonic sensors, (3) a laser range sensor, (4) and vehicle. So it has a distributed multi-processor architecture.

The controller is constructed by the star connected multi processor system, where, master module locates in the center. Each module has 68000 CPU , ROMs and RAMs as its own processor and realized as a single board computer system including necessary peripheral circuits. The CMOS devices are used to save the power for battery operation and each module consumes two or three hundreds mili-amperes. The robot has Ni-Cd battery of 6,000 mili-amperes, so, it lives more than two hours after fully charging.

Since the interface from the master module to the function modules is realized by command sending, so the communication traffic is small and any type of communication is applicable. On the other hand, State Information Monitoring Panel(SIMP) must be realized as an interface from the sensor part in the function level to the behavior program. On the SIMP, the relatively much data must be always updated and prepared to be monitored. Considering these points, the shared memory realized by dual-port-memory devices is adopted as the communication channel between the modules. The necessary size of shared memory depends on the number of the defined parameters in the SIMP, and it may be less than 1K byte for the simple sensory system. And this shared memory is also used to send command from master module to function modules controlling actuators. The constitution of the controller is shown in Figure 5.

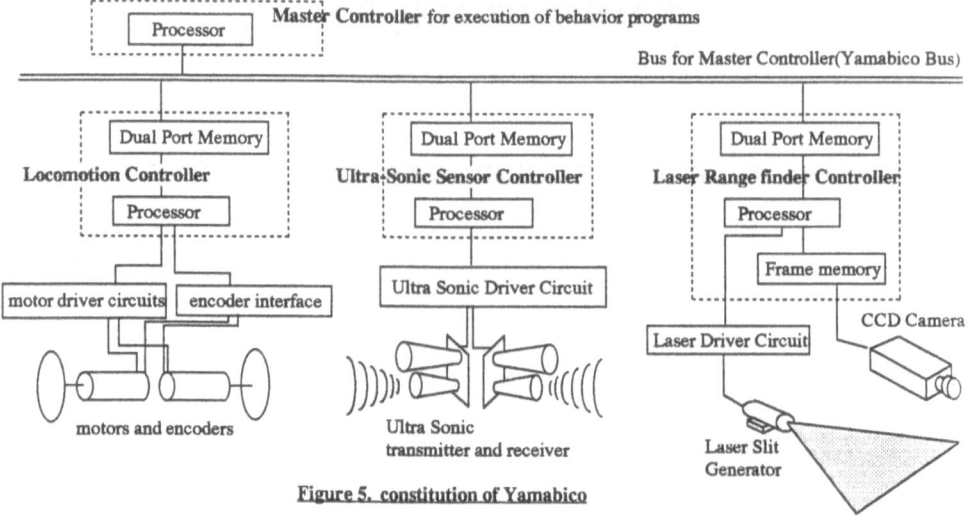

Figure 5. constitution of Yamabico

3.2 Execution of the behavior program

3.2.1 method to use functions of the robot

The behavior program of the robot is executed in the master module. So, the method to communicate with other function modules must be defined to use robot's sensors and actuators. Such communication methods in the behavior program are called **robot function accessers**. In authors' robot control system robot function accessers are implemented as library functions of C language. Robot function accessers are classified two types; (1) sensor accessers (2)and actuator accessers. Sensor accessers get sensor information from sensor subsystems by reading data from shared memory. Actuator accessers gives control command to actuator control subsystems through shared memory. The main feature of robot function accessers is that the execution time is shorter than the controlling time of actuators or calculation time of sensor data.

3.2.2 description of the behavior program

The authors designed a programming language named **ROBOL/0** for describing sensor based robot behavior program, with action mode representation [7][8]. The action mode representation of robot behavior can be translated easily to ROBOL/0 program with available motion commands. On the authors' robot, ROBOL/0 compiler is implemented as a pre-processor for C language. So, the method to use robot functions from the behavior program is provides by robot functions accessers mentioned above.

In the ROBOL/0 program, an action mode is written as shown in Figure 6. The action mode is defined with unique name <mode_name> in the program, which is used to be referred by other action modes when transition of action mode happens. *Motion_command* includes actuator accessers which send commands to vehicle subsystem. Execution of *motion_command* ends instantaneously and doesn't wait for end of the command control. After *motion_command*, several conditions showed as *condition-1, condition-2, ..., conditions-n* are evaluated continuously and concurrently. *Condition-i* is the logical expression including sensor accessers. When a condition come true, the following statements *procedure-i* (a procedure call) is executed , and the motion of the robot is changed by transition of action mode <mode_name_i>.

```
MODE  <mode name-1>
   procedure-0;
   WAIT
         WHEN  condition-1  EXEC procedure-1 NEXT <mode name-i >
         WHEN  condition-2  NEXT <mode name-k >
         WHEN  condition-3  EXEC procedure-3 NEXT <mode name-j>
                 .
                 .
                 .
         WHEN  condition- n  EXEC procedure-n  NEXT <mode name-i>
MEND
```

<u>Figure 6. Description of an Action mode in ROBOL/0</u>

3.3   Sensor subsystems and sensor accessers

The Yamabico has two kind of external sensors; ultrasonic sensors and laser range sensor. Both sensors are installed as a function module and work independently from other subsystems. And they get and process sensor information continuously and export it to the master module.

3.3.1   ultrasonic sensor subsystem

Yamabico has four directional(front, rear, left and right) ultrasonic sensors[9]. The hardware of this ultrasonic sensor subsystem is shown in Figure 7. The works of this module are follows;

(1) In every 30 milliseconds, the sensor system transmits ultrasonic waves into 4 directions, detect time interval until to receives the echos, calculates the distance for each direction.

(2) export the calculated distance information through dual-port memory.

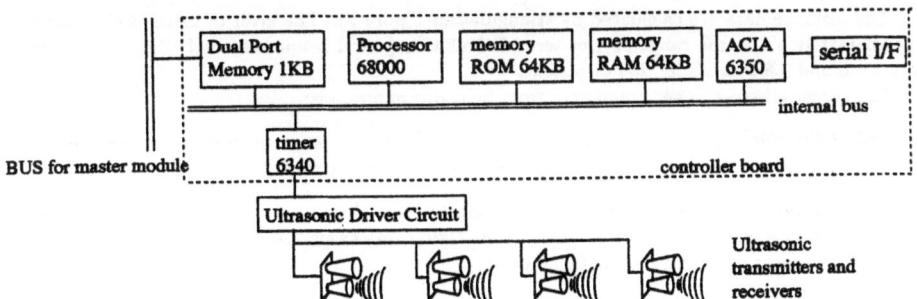

<u>Figure 7. Hardware Structure of Ultrasonic Sensor Module</u>

Following two sensor accessers are provided for the program in behavior level.

us_dist(*direction*): this sensor accesser returns distance value in *direction*(left, right, front or back).

us_mask(*direction*): this accesser sends the command to inhibit transmitting ultrasonic wave in *direction*(left, right, front or back).

3.3.2 laser range sensor subsystem

Ultrasonic sensor can measure only one directional distance and can detect stably only when the object is an all flat and perpendicular wall. Thus, the data from ultrasonic sensors is not enough for autonomous exploring for map generation or obstacle avoidance task, since such tasks require more detail information especially the shape information of objects in the environment. So an active optical range sensor system using laser slit and CCD camera has been developed and installed on Yamabico to get a directional distance information with higher resolution.

This sensor system is a function module which consists of a laser infra-red diode and optical slit collimetor, a laser driver circuit, one small-size CCD video camera, frame memories, and a micro processor for image processing and distance calculation(Figure 8). This sensor system has view angle of 50 degrees and it can get distance information for 255 directions(at most) from one image without mechanical scanning.

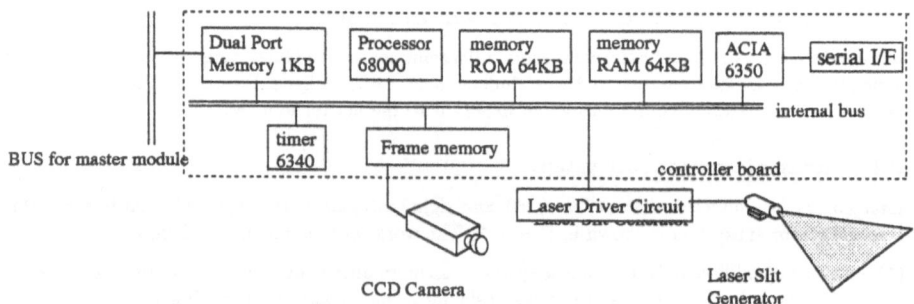

Figure 8 Hardware Structure of Laser range finder Module

The distance data are calculated by triangulation(Figure 9). The error of calculated distance is approximately 4% in the range of 4 meters. Calculation period is approximately 0.4 seconds for 50 directions and 1.8 seconds for 255 directions.

The sensor accesser to get range data from behavior program is;

IAS_dist(*direction*): this accesser returns most recent distance value in *direction*. *Direction* is the angle(degree) in the sensor view; from -25 degree to 25 degree.

IAS_thdist(*data_array*): this accesser returns most recent distance values in all directions in the sensor view angle by storing to *data_array*. *Data_array* keeps 256 number(i.e. every 0.2 degree in the sensor view angle) of distance value.

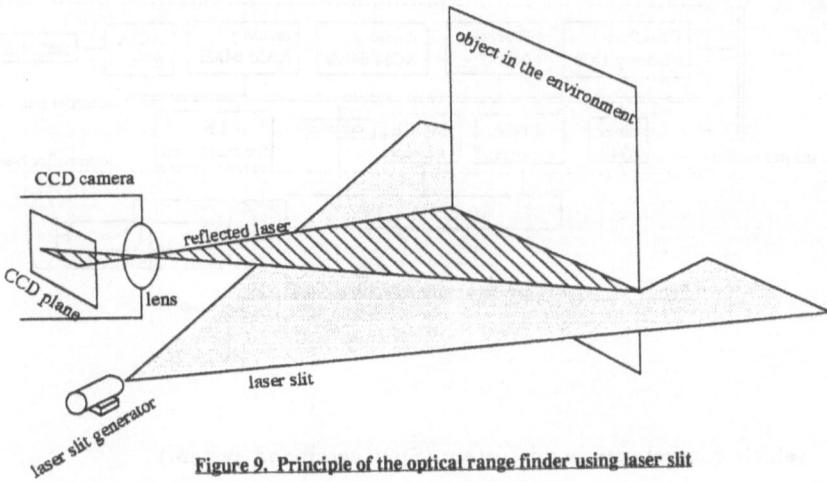

Figure 9. Principle of the optical range finder using laser slit

3.4 Vehicle control subsystem

3.4.1 vehicle controller hardware structure

Figure 10 shows the vehicle control mechanism of "Yamabico". It has two driving wheels connected to the DC servo motors through the gears in both left and right sides, and two casters in front and back to support the vehicle. Its steering is controlled by the differential velocities of the both side of the wheels. In order to control the DC servo motors, the Pulse Width Modulation (PWM) switching method is adopted to save the electric power.

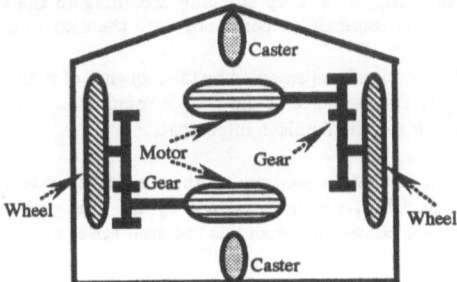

Figure 10. Locomotion mechanism of Yamabico

Figure 11 shows the controller structure of the vehicle control subsystem. This vehicle control subsystem has two roles; one is control of locomotion of the robot, another is the position estimation by dead-reckoning.

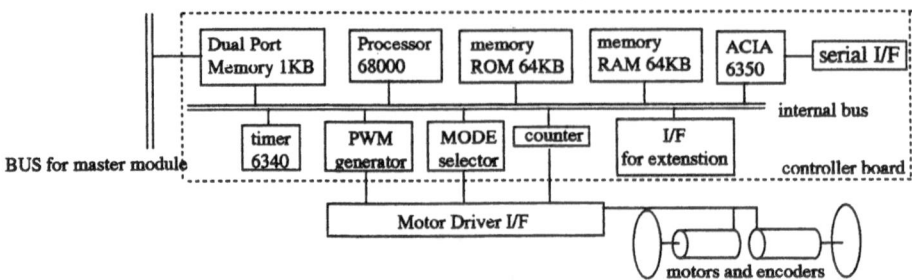

Figure 11. Hardware structure of vehicle controller

3.4.2 vehicle control command system (Spur command system)

The following commands system and its execution program is designed and implemented. This command system is named "Spur".
(1) Function of vehicle control commands
The main feature of the vehicle control command system and control system are:

(a) The mobile robot moves on the two dimensional $x - y$ coordinate plane, and the vehicle commands give the designed trajectory or desired stop position using the $x - y$ coordinate to the vehicle control subsystem.

(b) The vehicle control subsystem performs the feedback control in real time. It changes the trajectory to track along or the position to keep stopping according to the given commands. The feedback control based on the given command is continued until the next command is given.

(c) The sequence of the robot motion is represented by the sequence of vehicle commands. However, it is not enough just only to give the sequence of the vehicle commands, but to define the timing of the vehicle command execution in real time axis is important.

(d) A behavior program or a navigation program on the master subsystem can get the estimated current robot position from the SIMP, which is realized by the dual-port-memory on the vehicle controller, and can give the vehicle control command as the need arises at the suitable timing.
(2) Coordinates
The mobile robot moves on the two dimensional plane, and the designed trajectory or desired stop position is represented using the $x - y$ coordinate. The robot may have three coordinates systems —global(GL), local(LC) and robot-front-side(FS) coordinates, and the trajectory can be represented on any coordinate. Figure 12 shows Spur coordinate system.

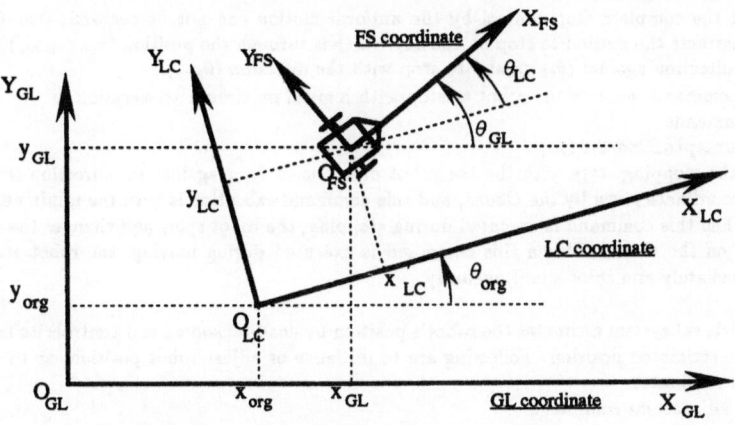

Figure 12. Spur coordinate system

3.4.3   actuator accessers to send vehicle motion commands

In the vehicle control subsystem, motion commands are controlled such as following a line, moving on a circular arc, stopping, and so on. Actuator accessers are defined for each motion command controlled in the vehicle subsystem. Actuator accessers gives the motion and parameters like a straight line to track along is represented by its start point and orientation, a circular arc is represented by its center point and the radius, and the position to stop at is represented by the point and the direction on the $x - y$ coordinate.

Following actuator accessers are defined.

(a) Straight line tracking commands

<u>Format:</u> **Spur_line_Coord**, $x_{line}, y_{line}, \theta_{line}$.

This accesser gives a new straight line for the robot-trajectory through a start point (x_{line}, y_{line}) with the direction (θ_{line}) against the x axis on the coordinate given by the *Coord*, and the robot tracks along the given line. *Coord* is GL, LC or FS.

(b) Circular arc tracking commands

<u>Format:</u> **Spur_arc_c_Coord**, x_{cent}, y_{cent}, r.

<u>Format:</u> **Spur_arc_t_Coord**, $x_{tan}, y_{tan}, \theta_{tan}, r$.

It give a new circular arc denoted by the center point (x_{cent}, y_{cent}) and the radius $(|r|)$, or the circular arc which touchs to the tangent line through the point (x_{tan}, y_{tan}) with the direction (θ_{tan}) and which has the radius $(|r|)$, on the coordinate given by the *Coord*, and the robot tracks along the arc. The rotational direction is the counter-clock-wise in the case of the positive value of r, and the clock-wise in the case of the negative value of r.

(c) Stop commands

<u>Format:</u> **Spur_stop_Coord**, $x_{stop}, y_{stop}, \theta_{stop}$.

<u>Format:</u> **Spur_stop_q**.

First command gives a position (x_{stop}, y_{stop}) to be stopped at, and the direction (θ_{stop}) at the stop position on the coordinate given by the *Coord*. Since it is difficult to control the detail stopping position and direction under the non-holonomic constraint condition such as the wheeled vehicle control, and the complete stop control by the uniform motion can not be realized, therefore, this command instructs the motion to stop on the line which is through the position (x_{stop}, y_{stop}) and with the vertical direction against (θ_{stop}), and to stop with the direction (θ_{stop}).

Second command requests the robot to stop with a given maximum acceleration.

(d) Spin commands

Format: **Spur_spin_Coord** , θ_{spin}.

It requests the stopping state with the the robot direction of θ_{spin} against the direction from the x axis of the coordinate given by the *Coord*, and this command execution is with the minimum moving distance. When this command is executed during stopping, the robot spins and changes the direction to the θ_{spin} on the *Coord*. When this command is executed during moving, the robot starts stop motion immediately and spins simultaneously.

The vehicle subsystem estimates the robot's position by dead-reckoning and controls its movement based on the estimated position. Following are to initialize or adjust robot position, ar to define or modify the coordinates.

(e) Position adjustment commands

Format: **Spur_adjust_pos_Coord**, $x_{adj}, y_{adj}, \theta_{adj}$

This command gives the robot current position ($x_{adj}, y_{adj}, \theta_{adj}$) on the indicated coordinate *Coord*. This command is principally used to adjust the error in the position estimated by the vehicle control subsystem, when the external sensor gives more accurate position information. This command does not indicate the robot trajectory directly. However, since the robot control is done based on the estimated robot position, the motion of the robot is modified after getting these commands.

(f) Coordinate transformation commands

Format: **Spur_set_Coord1_on_Coord2**, $x_{org}, y_{org}, \theta_{org}$

Format: **Spur_set_pos_Coord**, x'_0, y'_0, θ'_0

First command gives the new coordinate *Coord1* on the *Coord2*. The origin of the new coordinate *Coord1* is x_{org}, y_{org}, and the direction is θ_{org} on the coordinate *Coord2*.

Second command gives the new coordinate *Coord* using the value of the current position on the new coordinate *Coord*. So, the robot current position is changed to x'_0, y'_0, and direction θ'_0.

3.4.4 sensor accessers to get the estimated position

Sensor accessers are defined to get the estimated position on each coordination defined and used in Spur command system. Followings are sensor accesser to know the estimated position in Spur command system.

Format: **Spur_get_Coord**, $x_{line}, y_{line}, \theta_{line}$.

This accesser returns current estimated position in the indicated coordinate. *Coord* is GL, LC or FS mentioned above.

4   A sample of the robot behavior - real time navigation with obstacle avoidance -

As an example of the robot behavior, the authors first realized real time navigation using map, estimated position by dead reckoning system, and environment information from sensors on Yamabico. This navigation program is called 'route runner' and moves the robot along the planned routes described in the map. Map also includes the environment information around the routes, so that it will be used to check the current robot position by comparing with the actual sensor information.

In first step, the behavior is considered without using laser range sensor and without the function of obstacle avoidance. Route runner has following functions;

(1) moves robot to the goal position along the route information in the map
(2) change the robot's course to follow the routes
(3) check the robot position in the environment
(4) find another point to change the course if some object exist at the planned changing point
(5) try to find the current position if the position is lost
(6) moves the robot to reduce the found position error
(7) find the object on the robot's course by ultrasonic sensor and stop until it is removed.

This behavior is designed and represented by 13 Action modes with 36 conditions.

As next step, the obstacle avoidance function using the laser range sensor is added. More complicated situation must be considered as follows;

(8) find the avoidance course by laser range finder if objects exist on the robot course
(9) moves the robot following the avoidance course
(10) return to planned route after obstacle avoidance
(11) position checking is not done during obstacle avoidance
(12) change the course to follow the planned route if the robot reaches at the changing position during obstacle avoidance

The combined behavior is represented by 21 Action modes with 69 conditions(Figure 13) and realized robust robot navigation behavior.

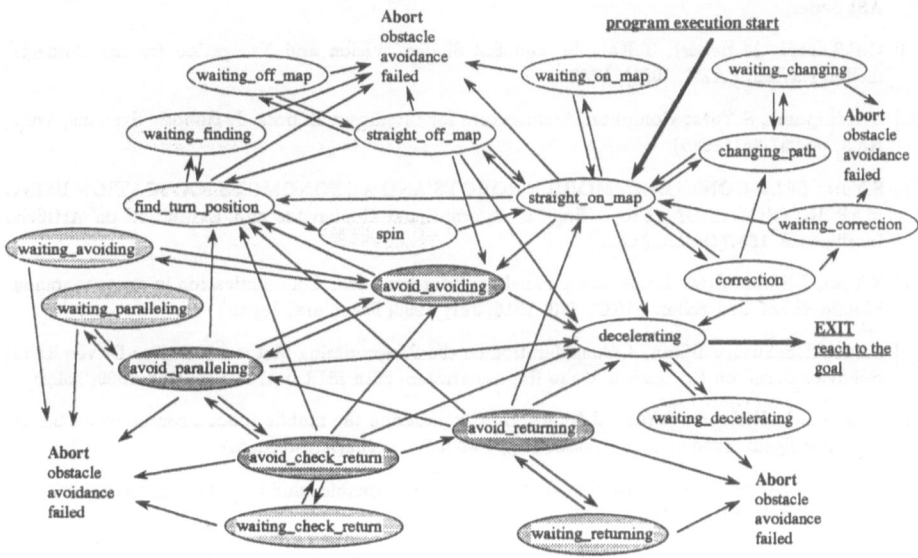

Figure. 13 transition between Action modes for autonomous navigation

In experiment, Yamabico moved 30 meters of distance in the corridor world in authors' institute using map information while avoiding many unexpected obstacles.

5 Conclusion

In the filed of robotics, implementation and experiment are important as same as theorys and concepts. As a basic concept the authors developed an architecture for the autonomous robot controller and action mode representation for describing the sensor based behavior. The authors have been implementing a small size self-contained mobile robots "Yamabicos" based on the proposed concepts.

The experiment of execution of sample behavior shows an effectivity of the proposed architecture. This is only first step, and more robustness and flexibility to cope with much situation is required. It is important to design and implement new sensor and locomotion functions of the robot, and to consider about more powerful description method of the robot behavior. The authors believe that "Yamabico" is a good platform for such robotics researches.

References

[1] N.J.Nilson: Shakey the Robot, SRI International, Technical Note 323.

[2] R.Chatila, G.Giralt: TASK AND PATH PLANNING FOR MOBILE ROBOTS, Machine Intelligence and Knowledge Engineering for Robotic Applications, pp299-330, Springer-Verlag NATO ASI Series,

[3] C.E.Thorpe, M.Hebert, T.Kanade, and S.A.Shafer: Vision and Navigation for the Carnegie-mellon Navlab, PAMI, 10(3), 1988

[4] Y.Kanayama, S.Yuta: Computer Architecture for Intelligent Robots, J. Robotic Systems, Vol.2, No.3, pp.237-251(1985)

[5] S.Yuta: SELF-CONTAINED MOBILE ROBOTS AND AUTONOMOUS NAVIGATION USING MAP INFORMATION, Proceedings of International Conference and Exhibition on Artificial Intelligence, 1987(Osaka, Japan)

[6] S.Yuta, J.Iijima: State Information Panel for Inter-Processor Communication in an Autonomous Mobile Robot Controller. , IROS '90, 1516, July 1990(Tsuchiura, Japan)

[7] S.Suzuki, S.Yuta, J.Iijima, A Consideration on the Programming Method of Sensor-Driven Robot Behavior Based on the Action Mode Representation, 20th ISIR, pp127-134, Oct. 1989(Tokyo)

[8] S.Suzuki, M.K.Habib, S.Yuta, J.Iijima, How to describe the mobile robot's sensor based behavior?, Intelligent Autonomous System 2, pp78-88, Dec. 1989(Amsterdam)

[9] M. SONG and S. YUTA: Autonomous Mobile Robot Yamabico and its Ultrasonic Range Finding Module, '89 KACC, pp711-714, Oct. 1989, Seoul(Korea)

[10] S.Iida, S.Yuta: "Control of a Vehicle Subsystem for an Autonomous Mobile Robot with Power Wheeled Steerings.", in Proc. of IEEE Workshop on Intelligent Motion Control, pp.859-866, 1990.

[11] S.Iida, S.Yuta: "Vehicle Command System and Trajectory Control for Autonomous Mobile Robots", IEEE Workshop on Intelligent Robots and Systems '91(to appear). 1991.

Section 7: Sensing and Perception

The level of decisional autonomy, of adaptation and reactivity to the environment is directly related to the sensing capacities of the robot and its ability to build the correct representations. The papers in this section deal with several aspects of sensing.

The paper by E. Krotkov addresses sensing for the Ambler, a walking robot. In order to plan its motion, such a robot needs models of rugged terrain. The basic representation is an elevation map, i.e., a numerical description, built from laser range-finder data. It is hence important to have a good experimental model of the sensor.

A different aspect of perception is identification and localization of an object in a scene. A structured description is necessary for the object (planes, edges in this case). The problem is furthermore complicated if the object has a variable shape. M. Devy and J. Colly present an approach and results for the localization of an articulated object with a mobile multisensory system, using a laser range-finder and vision, with observations from several viewpoints.

M. Accordino, F. Gandolfo, G. Sandini and M. Tistarelli present a system integrating visual perception with manipulation. Indeed, object manipulation requires perception to close the loop. Hand-eye calibration is then the first necessary step, wich is performed by observing the end effector through the stereovision system. The approach adopted by the authors relies on the analysis of the optical flow for extracting qualitative information (e.g., equilibrium) or object motion estimation.

The paper by N. Andersen, O. Ravn and A. Sorensen reports on the use of vision for real-time motion control. Image processing being usually time consuming, they reduce it by dynamical selection of windows in the image. Furthermore, in order to cope with lighting constraints, an adaptative thresholding is used. Three different experimentations validate the work.

Mapping Rugged Terrain for a Walking Robot

Eric Krotkov

Robotics Institute, Carnegie Mellon University

Pittsburgh, PA 15213 USA

Abstract

This paper briefly reports on progress in the design and performance of a mapping system that uses laser rangefinder input to construct elevation maps of rugged, natural terrain. The paper emphasizes performance; the rationale for the design is presented elsewhere. The performance of two rangefinders is considered, and found to be the most significant limitation on the constructed maps. In addition, the accuracy and precision of the maps are quantified.

1 Introduction

The primary function of the currently implemented perception system for the Ambler walking robot (Figure 1) is to build maps of rugged terrain from sequences of range images. Planning modules use these maps to select footholds and to plan collision-free leg and body trajectories over the terrain. This paper outlines the current state of the perception system, focusing on *performance* rather than rationale, which is presented elsewhere [4, 6].

We have divided the perception system into modules that communicate via a central node. The implemented perception modules, slightly simplified for clarity of presentation, are the following.

Local Terrain Mapper— Builds elevation maps in a specified polygonal region of a specified reference frame. The module uses as many range images as necessary to build the requested map, processing the most recent images first. The maps are not stored between body moves.

Global Terrain Mapper— Builds elevation maps in a specified region of the global coordinate system. For convenience during indoor experiments, this reference frame is attached to the building.

Sensor Interface Manager— Each time the body moves an appreciable amount, this module acquires images from laser scanners and cameras and stamps them with the pose of the Ambler at the time of image acquisition. It then sends the images to the Image Queue Manager for storage.

Image Queue Manager— Maintains a doubly-linked list whose nodes are images tagged with properties. Handles insertion and deletion of records, plus traversal.

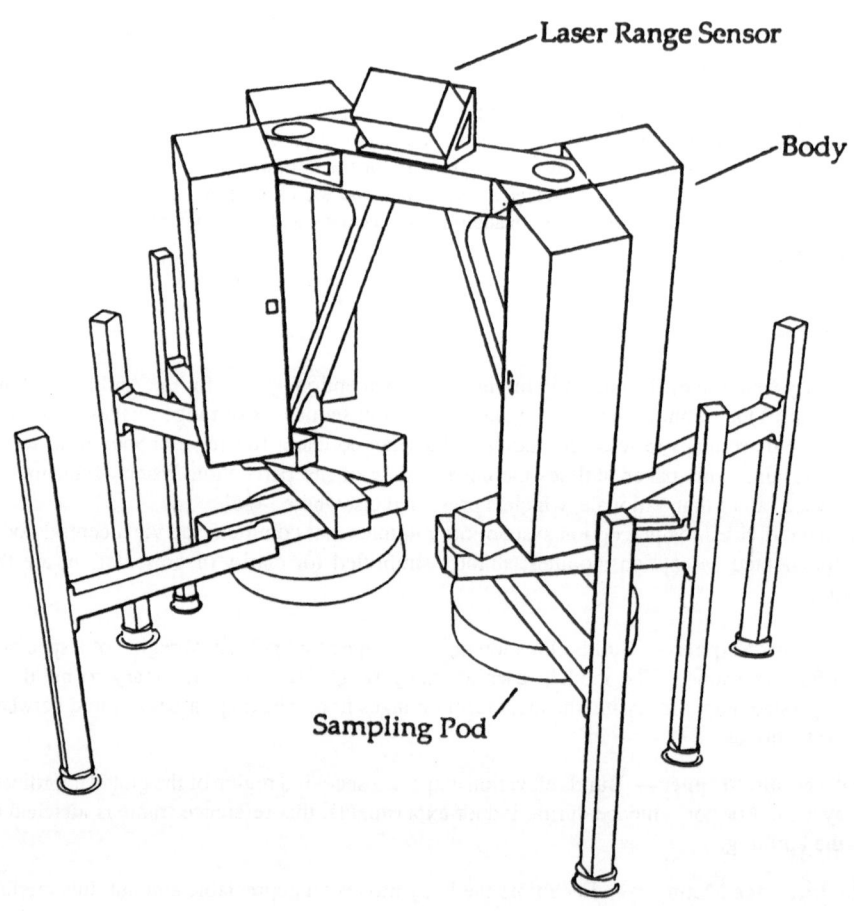

Laser Range Sensor

Body

Sampling Pod

Figure 1: Ambler walking robot

The figure shows the six-legged Ambler, and the Perceptron laser scanner mounted on the bridge between the two leg stacks.

Visual Position Estimation— Determines the pose of the Ambler from a single image and a pre-defined map marking locations of landmarks. This module employs the interpretation tree approach described in [2].

User Interface— Allows the user to communicate with external modules, view debugging information, select graphic displays, and control the flow of processing.

These modules have been implemented in C, and have been tested extensively during walking experiments. They are imperfect, but their performance—accuracy and precision—in constructing maps has proven to be sufficient for the Ambler to negotiate rugged, boulder-strewn, sloping terrain.

2   Scanning Laser Rangefinders

During walking experiments, we have employed two scanning laser rangefinder sensors, one manufactured by Erim, the other manufactured by Perceptron. In both cases, we have found the most significant limitations on the performance of the mapping system to be imposed by the sensors. Both exhibit some very desirable properties (high bandwidth and accuracy compared

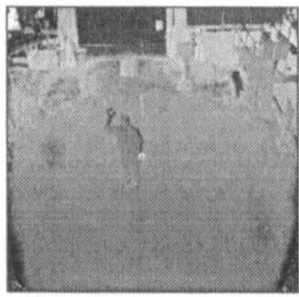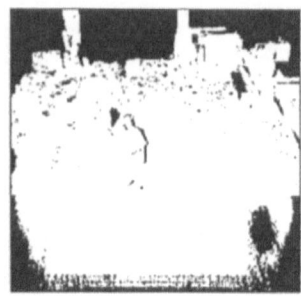

Figure 2: Perceptron images: reflectance (left), range (center), filtered

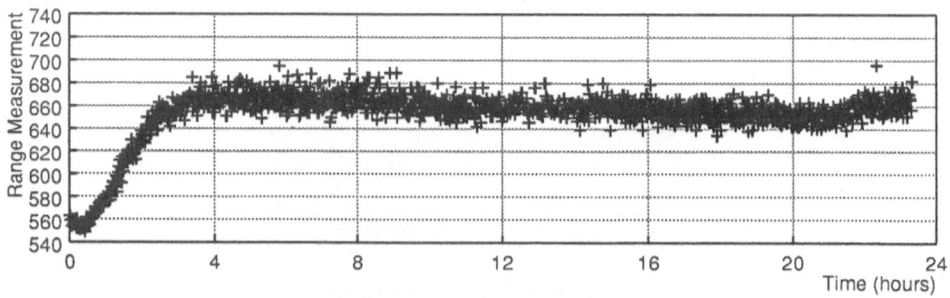

Figure 3: Range drift over 24 hours

The units of the ordinate are grey-levels, which correspond approximately to 1 cm. One measurement was acquired per minute with the Perceptron scanner. The target was a planar cardboard construction lying 6 m from the scanner.

to standard computer vision techniques, ability to function outdoors and in rugged, unknown terrain), and both suffer from some significant practical deficiencies requiring compensation.

We have systematically explored those drawbacks, and have experimentally characterized them [5]. There are more than a dozen known problems revealed in Figure 2. To illustrate one of them, consider the top central portion of the images, which correspond to a wooden garage door painted a dark green color. In the range image, grey-level is proportional to range; therefore, more distant objects should appear to be brighter. But the garage door appears to be darker than closer objects, contrary to our expectations. We have determined that the cause of this anomaly is the low infrared reflectivity of the dark wooden surface coupled with sensor processing artifacts. We have analyzed a number of similar problems with the range images, and have developed reasonably robust methods for handling them. Figure 2 illustrates the results of applying these methods.

Another problem is drift over time. Figure 3 illustrates the variations of the range measurements to a stationary target over a full day of observations. We compensate for this by changing the coefficients of the mapping from grey-levels to metric units.

3 Mapping Rugged Terrain

The mapping modules take as input the processed range images, and construct elevation maps from them using the *locus method* [1]. Figure 4 illustrates examples.

We determined the performance of the mapping modules by investigating the results of executing a calibration procedure that identifies the rigid transformation relating the sensor and body (or world) reference frames [3]. For both Erim and Perceptron sensors, the relative accuracy (accuracy in computing the dimensions of an object) is 5–10 cm, and the absolute accuracy (accuracy in computing the position of an object in an external reference frame) is 10–20 cm. The precision (repeatability) is 2–5 cm for both sensors. These figures do not take into account the inaccuracy of the Ambler navigation module, which consisted only of dead reckoning at the time of testing.

The achieved mapping performance has proven to be sufficient for the walking experiments conducted to date [6]. One reason for the success is that the errors are inferior to the size of the Ambler feet (30 cm diameter). For a more petite vehicle, the achieved performance may not have been adequate.

Acknowledgements

This research was sponsored by NASA under Grant NAGW-1175. We would like to acknowledge contributions by the entire CMU Planetary Rover research group. We express special thanks to K. Arakawa, P. Balakumar, M. Blackwell, M. Hebert, R. Hoffman, and T. Kanade.

References

[1] M. Hebert, E. Krotkov, and T. Kanade. A Perception System for a Planetary Explorer. In *Proc. IEEE Conf. on Decision and Control*, pages 1151–1156, Tampa, Florida, December 1989.

[2] E. Krotkov. Mobile Robot Localization Using a Single Image. In *Proc. IEEE Intl. Conf. Robotics and Automation*, pages 978–983, Scottsdale, Arizona, May 1989.

[3] E. Krotkov. Laser Rangefinder Calibration for a Walking Robot. In *Proc. IEEE Intl. Conf. Robotics and Automation*, pages 2568–2573, Sacramento, California, April 1991.

[4] E. Krotkov, J. Bares, M. Hebert, T. Kanade, T. Mitchell, R. Simmons, and W. Whittaker. Ambler: A Legged Planetary Rover. *1990 Annual Research Review, The Robotics Institute, Carnegie Mellon University*, pages 11–23, 1991.

[5] I. Kweon, R. Hoffman, and E. Krotkov. Experimental Characterization of the Perceptron Laser Rangefinder. Technical Report CMU-RI-TR-91-1, Robotics Institute, Carnegie Mellon University, Pittsburgh, Pennsylvania, January 1991.

[6] R. Simmons and E. Krotkov. An Integrated Walking System for the Ambler Planetary Rover. In *Proc. IEEE Intl. Conf. Robotics and Automation*, pages 2086–2091, Sacramento, California, April 1991.

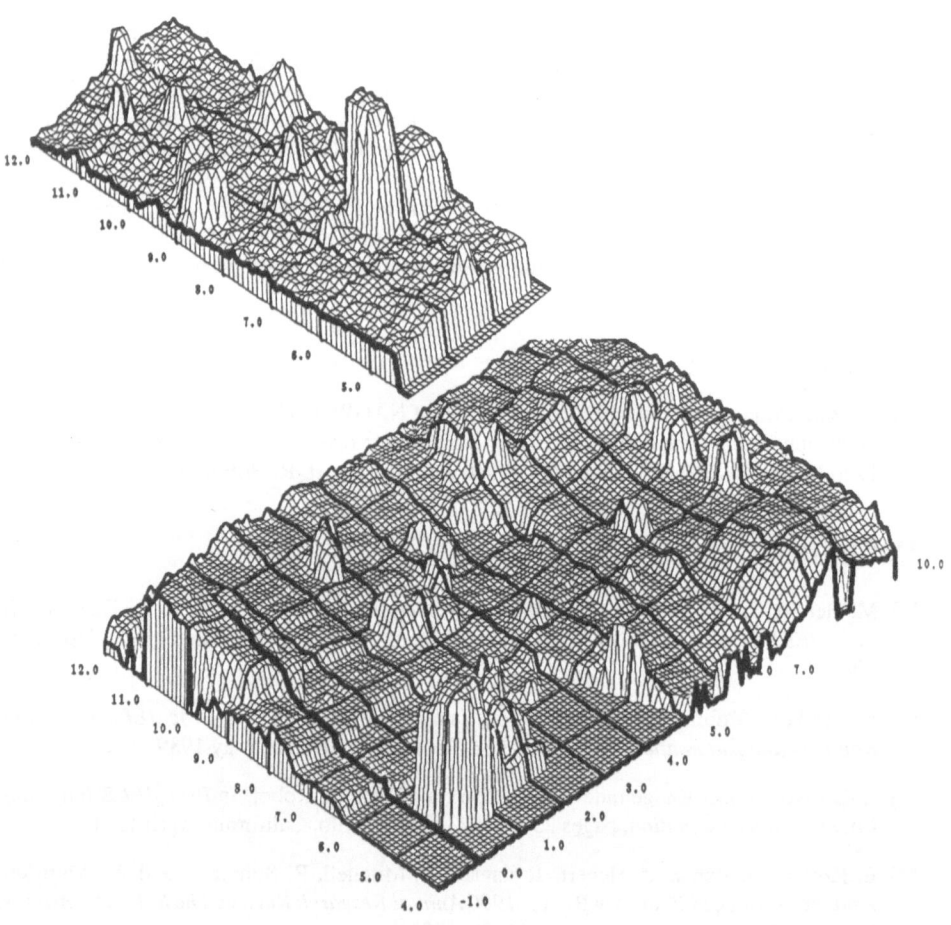

Figure 4: Maps constructed from Erim (left) and Perceptron imagery
The units are meters. The map resolution is 10 cm.

Localization of a multi-articulated 3D Object from a mobile multisensor system[*]

Michel Devy   Joël Colly

L.A.A.S. / C.N.R.S.

7, Avenue du Colonel Roche 31077 Toulouse Cedex FRANCE

e-mail: michel@laas.fr, colly@laas.fr

Abstract

This paper details a model-based method for identifying and locating 3D polyhedral multi-articulated objects from a multisensory perceptual system; this system is composed of a 3D laser range finder and a CCD camera, and is mounted on a mobile robot. The method implies first to identify and locate at least one body of the object from the initial robot position , then to search the configuration of the other bodies from other positions if needed.

The initial body recognition algorithm is based on the region-edge-vertex models and the aspect graphs of the bodies, which compose the object; it relies first on a generation of hypotheses performed only with 3D data extracted from the depth map, then on a verification of each pertinent hypothesis, using only 2D data extracted from the reflectance image. Localization of the other bodies can imply to move the perceptual system around the object in order to perceive the joints and to infer relative positions between connected bodies.

In the paper, we describe algorithms developped in order to extract pertinent 3D features from such a perceptual system, to identify and localize a body, and to estimate joints configuration; experimental results performed on our mobile robot Hilare, allow to validate this approach.

1 Introduction

The issue of 3D object recognition has already been extensively addressed in the literature [2]. In this paper, we deal with a difficult problem which is less addressed: recognition of 3D multi-articulated 3D object, by the use of a mobile multisensory perceptual system, without any assumption on a restricted viewpoint and on the object attitude according to the sensor reference frame. The perceptual system is composed by a laser range finder (or LRF) and a monochromatic CCD camera. The final application we must cope with, concerns the identification, localization and inspection of satellites from a shuttle [5].

We assume that the object we must localize, can be represented as a polyhedral 3D object, made up with several bodies connected through joints; the object model requires the following off-line processings:

---

[*]This work was carried out under CNES fundings

1. CAD model acquisition: for standardisation reasons, a CAD modeler must be used to build geometrical representation of the object.

2. Face-edge-vertex model extraction: we must retrieve this model from the CAD database. This model is structured and contains both geometrical and relational information. An object is composed of bodies and articulations:

   - a body corresponds to a rigid object subpart, and is represented by three lists, for faces, edges and vertices;

   - an articulation corresponds to a prismatic or rotoid joint between two object bodies, and is known by references on the concerned bodies, and on faces which are into contact, by the translational or rotational axis and by maximum and minimum movement parameters.

3. Aspect-graph computing: the above model is completed by the object aspect-graph, which is computed separately for each body. The aspect graph of a polyhedral object is a graph where each node represents a set of simultaneously visible faces of the object; it will be used

   - by the recognition algorithm in order to reject some hypotheses which correspond to impossible combinations of faces (for example, two opposite faces of a cube);

   - by the control system of such an application, in order to search optimal viewpoints from which a given joint could be perceived.

In order to compute the aspect graph of each body, we use an exact method for convex polyhedral bodies, using the algorithm described in [7]; some heuristic considerations allow to take in account non-convex bodies; we do not cope with the combination of these aspect graphs in order to know the possible object viewpoints.

In order to validate our algorithms, we have used our mobile robot Hilare on which a multisensory system LRF/camera has been set up. We can acquire raw perceptual data from the scene: a depth map and a grey-level image. In the paper, we will describe an experiment, the aim of which is to localize a 3D object, to search the configuration of its mobile subparts. For the different steps of this process, quick description of the perception algorithms will be presented and emphasis will be put on the experimental results; we do not deal here with other algorithms needed when the robot moves (viewpoint determination, path generation, path execution).

2 Extraction of a high level scene description

From the raw perceptual data, we must provide a high level representation on which interpretation will be easier:

- *from the depth map,* 3D planar regions are extracted using a classical region growing algorithm [12] [8] [10]; region seeds are coplanar patches given by 4 connected points of the depth map; the criterion used to merge a point in a region, relies on the distance of this point to the corresponding plane; the plane parameters are fit by a Kalman filter, using the variances on each point of the depth map. It is well-known that the supporting plane of such a region can be approximated with a good accuracy, but on the other hand, its boundary is very uncertain, because of the poor resolution of the LRF (and particularly, farther is the physical face from the sensor, or more important is the specularity, worse is the sensor resolution).

- *from the reflectance image,* the discontinuities are extracted and approximated by straight-line segments, from which polygonal chains can be extracted (sequence of connected segments).

A symbolic fusion of these features is then performed, resulting in the higher level scene description : a set of meta-faces. This routine requires that an off-line calibration has been done, in order to evaluate the relative position of both sensors, and their position on the robot. A meta-face is built from

- a 3D planar region extracted from the depth map;

- the segments which have been extracted from the image and which can match with the projection of the boundary of this 3D region or of non geometrical informations beared by the corresponding face (for example, logos, colored stripes,...)(see figure 1);

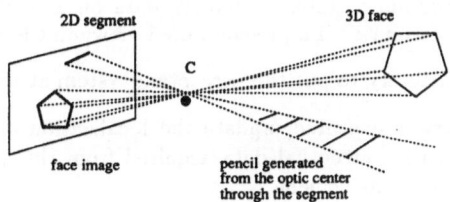

Figure 1: 2D and 3D features

- an attribute which tells if the face is fully seen, or if it is partially occluded.

A limited numerical fusion step, allows us to enhance the 3D position of several features : for example, on the figure 2, a 3D segment can be provided by :

- the intersection between the supporting planes of 2 faces, when they correspond to 2 adjacent physical faces of the object (orientation discontinuity in the depth map),

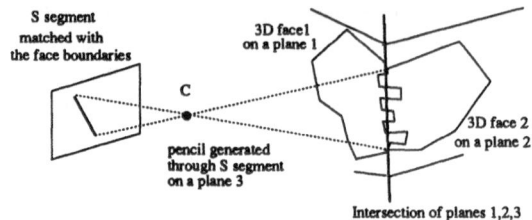

Figure 2: Possible numerical fusion to obtain a 3D segment

- or by·the intersection between such a plane and a planar sector generated from the optic center through a 2D segment, when this segment matches the boundary of this region.

We do not detail here this method [6], because meta-faces are not used for the recognition algorithm. For this task, where we have an a priori knowledge on the perceived scene, *fusion comes from interpretation*: 3D features and 2D features which match the same object entity, will be fused after the recognition.

3 Identification, localization of a rigid body

From an initial position of the robot on which the perceptual system is mounted, an accurate localization of a body according to the robot, must be computed. During the following steps, we could use this position in order to:

- select a given viewpoint suitable to acquire data on a joint, the configuration of which we want to estimate , if a pre-computed viewpoint is not accessible;

- generate a safe trajectory to put the perceptual system at this point;

- once this trajectory is executed, update the localization of the object according to the robot, from the perceptual data acquired from this new position, using an estimation of the robot movement.

In this experiment, we do not tackle with the two first functions: how to search the best 3D viewpoint of an object, taking the sensor models into account (perception field, specularity,...)? How to generate a trajectory in a known but uncertain 3D environment, taking cinematic constraints of the robot into account?

From the high level scene description, we apply a 3D object recognition algorithm, in order to localize the main body of the object. The same decisionnal process can be applied whatever sensor is available to acquire the perceptual data; we can performed a 3D object identification based

- only on matchings between features extracted from a single image (straight-line segments, or polygonal chains of segments) and faces of the object model [11],

• or only on matchings between the 3D faces built from the depth map and faces of the object model [1].

So our algorithm can stand a failure or bad acquisition conditions for one sensor (darkness for the camera, or too high specularity for the LRF). If we can acquire good perceptual data by the two sensors at the same time, we have developped a simple approach where we can use 2D features from the image, and 3D features from the depth map for different steps of the identification process. At this moment, we don't use explicitly the meta-faces built from the fusion process. Figure 3 illustrates the flow chart for this object recognition algorithm. Left side corresponds to the scene processing, which must generate features (3D faces, 2D segments, meta-faces); right side shows model processing.

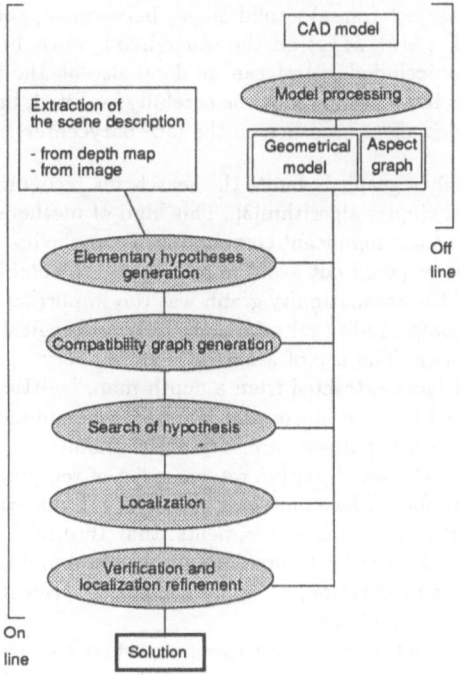

Figure 3: Our approach for the object identification

Our system relies on the search of hypotheses in a compatibility graph, in which each node corresponds to a matching between a scene feature and a model face, and each arc stands for the compatibility between two matchings.

The first problem is to generate these elementary hypotheses, that are (scene face SF, model face MF) associations. This initial matching relies on 2 rules:

• if SF is not occluded, then, SF and MF must have equivalent surfaces;

• if SF is occluded, then SF must have a surface lower than MF.

Note that we don't use criteria based on the boundaries of the scene face, like for example perimeter, convexity, edge number...Surface criterion is sufficient to generate a few number of elementary hypotheses : at this time, face boundaries are too uncertain in order to refine hypotheses, or to reduce their number.

The compatibility between two elementary hypotheses (SF_i, MF_i) and (SF_j, MF_j), requires two kinds of constraint :

- **topological constraints** : connexity (if SF_i and SF_j are adjacent in the scene, then MF_i and MF_j must be connected in the model), and visibility (we must refer to the body aspect graph in order to verify that MF_i and MF_j can be seen together).

- **numerical constraints** , on measures which are invariant from any reference frame (sensor frame or object frame) : solid angles between support planes of (SF_i, SF_j) on the one hand, (MF_i, MF_j) on the other hand, must be equivalent; if neither of these faces are occluded, a test can be done also on the distance between their barycenters. This last criterion must be carefully evaluted, because uncertainties on the face boundaries affect accuracy on the face barycenters.

Once the compatibility graph is built, the search for recognition hypotheses is performed by the maximal cliques algorithm[3]. This kind of methods are very expensive in computing time, due to their significant combinatorial complexity; with only a monocular perceptual system [11], we could not avoid in many cases, a combinatorial explosion, because the dimension of the compatibility graph was too important (too many elementary matchings, too weak compatibility criteria, since only topological constraints are sound without 3D informations). The use of a 3D LRF allows very significant improvements, because the number of faces extracted from a depth map, is little related to the number of 2D segments extracted from an image, and also because numerical constraints to stand for compatibilities between hypotheses are very discriminant.

Finally, the maximal cliques algorithm returns a list of recognition hypotheses, sorted using the decreasing number of face matchings (or nodes in the graph). If we only search one object, we can certify from our experiments, that the right solution is allways the first hypothesis in the list; in order to break up an eventual ambiguity, or more often, in order to refine the object localization, we apply a classical verification routine, using only the segments extracted on the image.

For each pertinent hypothesis we want to refine, a first localization based on the face matchings, is computed. Using the relative position between LRF and the CCD camera, we can predict the object position in the image and infer (scene segments, model edges) matchings. If such matchings are not found, the confidence rate on this hypothesis must be reduced; otherwise, it can be increased, and a more accurate localization can be computed using Kalman filtering algorithm on the 2D segments.

4 Localization of the articulated bodies

Once a rigid body is identified and localized, we must tackle the search of the joints between this body and other bodies of the object; this procedure will be repeated as far as all bodies are located. For our final application, it is simpler to identify first of all, a given body (the main satellite body), because other ones (solar panels, antennas,...) present important symetries, and their identification would provide strong ambiguities. Other solutions on the general case would be:

- to modify the identification method in order to take in account articulations, and to estimate together object position and the configuration parameters, which give positions of its articulated subparts, like in [9],

- or, if the object has several unambiguous bodies, apply the first identification step simultaneously for each of them. It is possible to identify in the same process, many rigid bodies, to locate separetely each of them and to deduce the configuration parameters.

Whatever the strategy we apply for the first step, it will be necessary to go along several viewpoints in order to perceive all the articulations (for example, we cannot see on the same image, two joints which are on opposite faces of a cube). It's the reason why we must move perceptual system around the object; a planification could be computed off-line, from the object and the sensor models, with some strategies in order to minimize robot movements and to optimize the quality of the perceptual data (no specularity...).

In the process shown hereafter, we try to identify and localize all the bodies, as soon as they appear, with an opportunistic strategy:

<u>begin</u>
    acquire perceptual data;
    identify/locate a given body B_0;
    initialize L with the list of body linked to B_0;
    <u>while</u> L not empty <u>do</u>
        <u>for</u> <u>each</u> B in L <u>do</u>
            <u>if</u> inspection possible with current data <u>then</u>
                inspect position B;
                add bodies linked to B in L;
                remove B from L;
        <u>if</u> L not empty <u>then</u>
            search a viewpoint from which inspection could be made
            generate a safe trajectory to go to this viewpoint;
            acquire new perceptual data;
            update B_0 localization;
<u>fin</u>

In this process, we use a list L of bodies not yet inspected, or not yet located according to the already known bodies. It allows to go through the frame graph which gives relative position of bodies related to their neighbours.

Note that once a given body has been located in the first step, we can predict the possible position of the other bodies, and apply a classical verification process in order to identify them. If a movement has been done, the prediction phase must use estimation of this movement. If this estimation is too bad, we must update the localization in an iterative process; while the good viewpoint is not reached, we must use perceptual data to update relative positions between robot and already located bodies, and if necessary, we must generate a new trajectory in order to approach this good viewpoint.

In the worse case, if prediction on the object position is not good, we must return to the first identification process in order to locate a body without any previous knowledge. The figure 4 shows a flow chart where we detail this operation.

5 Experimental results for inspection

In this section, we describe experimental results. We try to localize an object, made with a main polyhedral body and two panels linked on two opposite faces. Then the object model is made up with 3 bodies and 2 rotoid joints. At this time we have not integrated in the experiment, automatic viewpoint search and path generation [4]; viewpoints and trajectories are given throught a simple man machine interface.

In the knowledge base, we first initialize the object model, with the panels perpendicular to the main body (see fig. 5); in the real world we put the panels in any position. We put Hilare in an initial position, where the object can be perceived; figures 6 and 7 show the perceptual data acquired from this position by LRF and by CCD camera.

The scene description extracted from these data is shown on figure 8 for the 2D segments found in the image, and on figure 9 for the 3D faces found in the depth map. In fact, we show here a 3D display of the face boundaries, projected on their corresponding support plane, with a display viewpoint at the same place than the LRF.

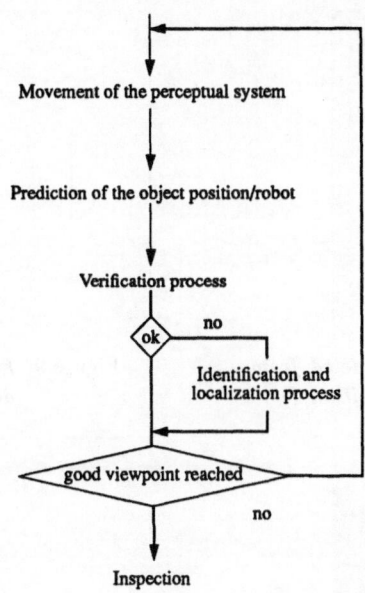

Figure 4: Synoptic of our approach to update object localization

The first identification and localization process gives a transform between the reference frame of the object main body and the laser frame; for example, in this case, because object reference frame is on the floor, it is almost 1m under the laser frame ($Z = -0.93m$); perceptual system is almost 1.5m in front of the object ($X = 1.32m$).

$$T_{cm} = \begin{pmatrix} 0.307370 & 0.951321 & 0.022653 & 1.323444 \\ -0.951503 & 0.307576 & -0.006203 & -0.175673 \\ -0.012868 & -0.019647 & 0.999724 & -0.937072 \\ 0.0 & 0.0 & 0.0 & 1.0 \end{pmatrix}$$

We can execute the inspection process from these same perceptual data. The list L is initialized with bodies associated at the panels; joints between the main body and the panels are searched; only one is visible from this initial position. The joint is found on the 3D data, by the a matching between a predicted position and an orientation discontinuity (Joints is supported by the intersection of the planes which support the region matched with the panel -the one we search-, and a region matched with a given face of the main body); once the joint is found, the panel is identified . A configuration parameter is then computed, by the angle between these two support planes (here, we find -0.856703 radians).

The knowledge base can be updated and we show on figure 12 the 3D display of the extracted face boundaries, superposed with the updated object model, with a display viewpoint above the object; we can verify that the superposition of the panel model is satisfactory.

Figure 8: *Segments extracted from the image (position 1)*

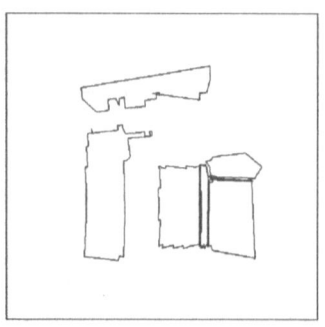

Figure 9: *Faces extracted from the depth map (position 1)*

Figure 10: *Image (position 2)*

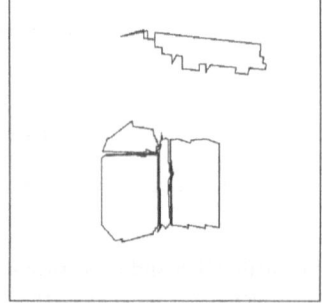

Figure 11: *Faces extracted from the depth map (position 2)*

$$T_{cm} = \begin{pmatrix} -0.414468 & 0.909960 & 0.013711 & 1.768545 \\ -0.909890 & -0.414637 & 0.013297 & 0.716526 \\ 0.017785 & -0.006964 & 0.999818 & -0.941465 \\ 0.0 & 0.0 & 0.0 & 1.0 \end{pmatrix}$$

transform after only the verification process

From this viewpoint, the system can inspect the configuration of the second panel, by the same algorithm than for the first one. We find an angle equal to 0.858616 radians; we can again update the knowledge base and display the object model superposed with the 3D scene description (see figure 13).

Note that our method to inspect relative position between the main body and the panel implies that the joint is visible. In some cases, such an inspection would be possible, without seeing the joint, but only with an occluded view of the panel. We consider that the right way to inspect such an articulation is to search the edge matched with the joint, otherwise the two bodies could be unfastened but near; such a situation must be detected.

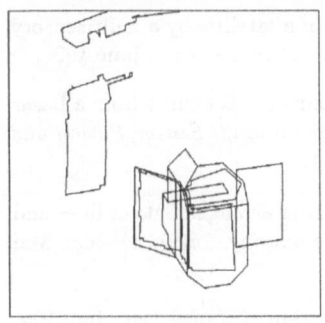

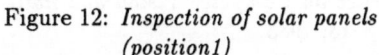

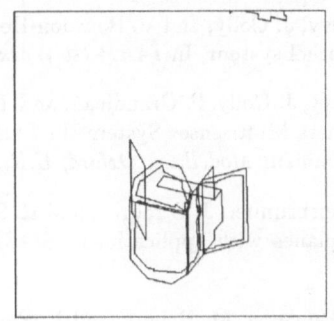

Figure 12: *Inspection of solar panels (position1)*   Figure 13: *Inspection of solar panels (position2)*

We must point out that performance of these algorithms is satisfactory; a sequence with identification, localization and inspection is executed in 3s on a Sparcstation. Segmentation of the depth map uses more execution time, but this kind of algorithm can be adapted to be executed on a parallel architecture.

6 Conclusion

We have described in this paper, an perception application related to identification, localization and inspection of a multi-articulated 3D polyhedral object; perceptual data are acquired by a multisensory system, made with a LRF and a CCD camera. Results have been commented; they show the validity of our work. Other research works will be done in order to improve the perceptual algorithms, especially for a better use of the multisensory system, and to take in account more complex objects (multi-resolution model, not only polyhedral objects, ...).

References

[1] F. Arman and J.K. Aggarwal. Object Recognition in Dense Range Images Using a CAD System as a Model Base. In *IEEE International Conference on Robotics and Automation, Cincinnati, (USA)*, pages 1858–1863, Mai 1990.

[2] P.J. Besl and R.C. Jain. Three dimensional object recognition. *ACM computing surveys*, 17(1), Mars 1985.

[3] R.C. Bolles. Robust Feature Matching Through Maximal Cliques. In *Proceedings of the SPIE's technical symposium on Imaging applications for automated industrial inspection and assembly*, volume 182, pages 140–149, 1979.

[4] J. Colly, M. Devy, M. Ghallab, and L.H. Pampagnin. Sensory Control for 3D Environment Modeling. In *IARP 1st Workshop on Multi-Sensor Fusion and Environment Modeling, Toulouse (France)*, Octobre 1989.

[5] M. Devy, J. Colly, and V. Bourdon-Henry. Inspection of a satellite by a multisensory perceptuel system. In *IARP 1st Workshop on Robotics in Space, Pisa*, june 1991.

[6] M. Devy, J. Colly, P. Grandjean, and T. Baron. Environment Modelling from a Laser / Camera Multisensor System. In *IARP 2nd Workshop on Multi-Sensor Fusion and Environment Modelling, Oxford, U.K.*, Septembre 1991.

[7] H. Edelsbrunner, J. O'rourke, and R. Seidel. Constructing arrangements of lines and hyperplanes with applications. *SIAM Journal on Computing*, 15(2):341–363, Mai 1986.

[8] O.D. Faugeras, M. Hebert, and E. Pauchon. Segmentation of Range Data into Planar and Quadratic Surfaces. In *IEEE Conference on Computer Vision and Pattern Recognition, Washington, D.C. (USA)*, pages 8–13, 1983.

[9] D.G. Lowe. Three-dimensional object recognition from single two-dimensional images. *Artificial Intelligence*, 31:355–395, 1987.

[10] M. Oshima and Y. Shirai. Object Recognition Using 3D Information. *IEEE Transactions on Pattern Analysis and Machine Intelligence*, 5(4):353–362, Juillet 1983.

[11] L.H. Pampagnin and M. Devy. 3D Objet Identification based on Matching Betwen a Single Image and a Model. In *IEEE International Conference on Robotics and Automation, Sacramento, (USA)*, 1991.

[12] B. Parvin and G. Medioni. Segmentation of Range Images into Planar Surfaces by Split and Merge. In *IEEE Conference on Computer Vision and Pattern Recognition, Miami Beach, Florida (USA)*, pages 415–417, 1986.

Object Understanding through Visuo-motor Cooperation

M. Accordino, F. Gandolfo, A. Portunato, G. Sandini and M. Tistarelli

University of Genoa
Department of Communication, Computer and Systems Science (DIST)
Integrated Laboratory for Advanced Robotics (LIRA - Lab)
Via Opera Pia 11a - 16145 Genova, Italy

Abstract

The topic of this paper is to illustrate some experimental results on the acquisition of physical characteristics of real objects. These features have been extracted by observing the reactions to planned motor actions applied to the objects by a robot arm.

A system architecture for visuo-motor cooperation is presented [1], based on the integration of sensory skills both as an aid to the manipulation process and as a mean to extract information from the scene [2, 3]. Its main characteristics are the ability to cope with an unmodelized environment and the closure of the loop between sensing and acting providing though a mean of validating the effects of the robot actions. The architecture embeds a reflex mechanism to deal with potentially hazardous situations.

Two aspects are emphasized in this paper: the implementation of a self-calibration strategy and the extraction of object features from images acquired during exploratory manipulative actions.

1 Introduction

The use of vision in robotics has been mainly devoted to localization, reconstruction, and matching problems. Tasks like 3D reconstructions of shape (for manipulation) or free-space (in case of navigation) have attracted considerable research efforts along with the problem of recognition either model-based or driven by a-priori knowledge. This paper presents a different perspective in relation to the use of vision for monitoring robot actions and robot control. This research activity reflects an exploratory approach to object understanding and manipulation, driven by the observation that, apart from trivial situations, the structure of the environment is never perfectly known. Therefore, the planning of motor actions cannot be based entirely on pre-stored models. To stress these ideas, the scenario envisaged in our experiments is that of an environment composed of unknown objects laying on a flat surface (e.g. a table).

The adopted sensory-motor system has much in common with the "active vision" approach proposed by [4, 5, 6, 7, 8], the "qualitative/purposive vision" paradigm proposed by Aloimonos [9], and also with the "animate vision" paradigm proposed by Ballard [10]. In these approaches the ability to "move" is exploited to simplify visual computations, but also make visual information richer. Moreover, purposively planned movements of the observer, driven by visual information, facilitate the motor control of robot actions. For example, in a very simple case represented by a camera(s) motion with stabilized direction of gaze, not only it is possible to simplify the computation of environmental measurements like depth or time-to-impact from visual data, but also the fixation point can be used as a reference point in the environment for driving the robot motion. This behaviour is also coupled by saccadic motions which are used to change the focus of attention.

The major difference with the approach presented in this paper, is related to the "use" of visual information which, in the former approaches, is mainly devoted to the "exploration" of the environment, whereas, in our approach is devoted to monitoring the execution of motor actions. The commonality arises not only from the visuo-motor cooperation but also by the fact that monitoring also provides information about the environment (and, in this sense, is similar to active exploration): if some unexpected event occurs, then the assumptions made on the environment were wrong or incomplete (for example if a moving object enters the camera field of view) and the system knowledge can be updated (learning from mistakes). This assert can be formalized in terms of a "theory of expectations", and, in our case, as "anomalies" in input visual patterns [3].

In order to explain our approach some experiments have been performed in the field of object manipulation. The outcome of the experiments is that, not only it seems possible to detect unexpected events, but it is possible to control the arm in such a way to achieve a complex task like pushing an unknown object along a predetermined trajectory.

2 Hand-Eye Self-Calibration

In order to extract the features of interest from an object being manipulated, it is necessary to rely on an architecture able to coordinate the sensory and the motor system. The architecture which has been developed is shown in Fig. 1.

The sensory modality which has been taken into account is vision. Since the vision system refers to a reference frame and the robot system works in another one, it is necessary to provide a mean by which to convert measurements between the two frames. This can be achieved through a calibration phase. The first step towards the implementation of the architecture is then the self-calibration procedure.

The hardware set up used for the experiments is composed of:

- a six degrees of freedom robotic arm;

- a stereo pair of CCD cameras;

- a VDS 7001 Eidobrain workstation for image processing.

The first step of the procedure is the calibration of the cameras, performed using the method described in [11]. Only the focal length, the vergence angle, the conversion factor between mm and pixels, and the distance between the cameras are needed to be able to recover a point's depth from the stereo images. Then a correspondence is sought between the cameras reference frame and the robot reference frame; this latter phase is the self-calibration process. It basically consists of the estimation of the transfer matrix between the two systems. It is computed by moving the robot's end effector in three different positions in turn and by letting the stereo system observe them. In order to simplify the visual processing, a LED is mounted onto the robot's end effector. In this way, it can be located on the images via thresholding, and the disparity computation is limited to a single point on each image. Using the geometric relationships established in the calibration phase, the 3D position of the end effector is recovered.

Such information can also be obtained from the robot's joint positions (derived using proprioceptive sensing). Using the direct kinematic, it is possible to compute the end effector's position in the robot's reference frame.

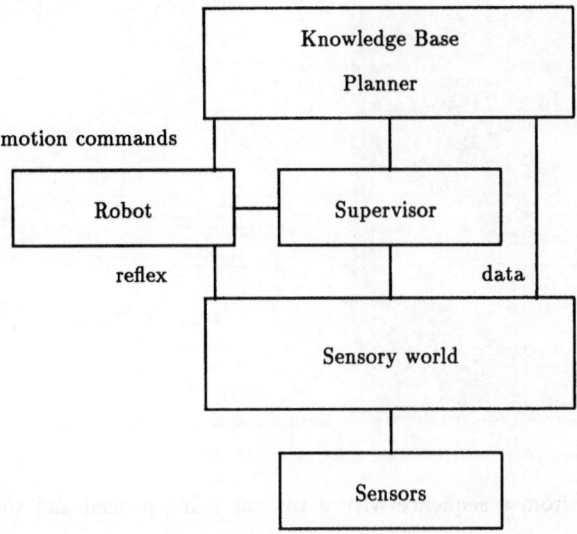

Figure 1: Schema of the architecture

By repeated measurements of the end effector's position in both reference frames, it is possible to compute a matrix allowing to switch between the systems. Such matrix is most accurate close to the calibration points, but its efficiency reduces away from them due to the non-linearities of both the robotic and the vision system. To cope with this problem, the matrix can be updated periodically during the robot motion; in this way, the matrix estimate is improved through repeated measurements. Moreover, if the robot changes its work area, the matrix is updated during the arm motion providing a suitable matrix for the new work space.

3 Sensory-Motor Strategies for Object Understanding

A multi-sensorial robotic system can execute various manipulation tasks on different kind of objects; the actions we take into account in this paper are basically pushing and tapping or impulsive pushing. Since we want to deal with unknown objects, the manipulation task can be performed only if it is preceded by the acquisition of a sufficient amount of information about the object. In this work we define some strategies to extract useful object's features and to characterize the object's motion as a reaction to external actions by using cooperatively a TV camera observing the scene and a robotic arm to push various kind of objects. Particular emphasis has been given to pushing actions [12, 13, 14] with the final goal of pushing an unknown object along a predefined trajectory [3]. The developed strategies are devoted to the extraction of several object features useful for manipulation. The methodology which has been chosen to extract these features is the computation of the optical flow, giving particular attention to the detection of anomalies with respect to the predicted velocity field [8]. An example of a moving object and the optical flow extracted from it, is shown in figure 2. Some qualitative and quantitative features which have been extracted by these means are the object's equilibrium properties, its opposition to motion, its 2D silouhette and its motion parameters when pushed.

Figure 2: One image from a sequence with a toy cat being pushed and the computed optical flow.

3.1 Evaluation of object's equilibrium

The first object property which can be easily determined by the observation of its reaction to manipulation is its stability with respect to external forces applied on its surface. The object's equilibrium can be determined by performing a tapping action on the object's surface and then observing its motion: if the object is in stable equilibrium it will stop very shortly after the touching. On the image plane, the number and amplitude of optical flow vectors will gradually decrease after tapping until the object stops. In this case, it is possible to characterize this behaviour by computing the histogram of the mean amplitude of optical flows versus time. The standard case, which corresponds to a stable equilibrium, is determined by an approximately bell-shaped histogram, where the maximum is very close to the time instant of the contact.

Two possible anomalies can be detected (corresponding to two different events):

- if the object falls (from unstable equilibrium) after the touching, the histogram will have a steep slope (corresponding to the progressive acceleration of the object) with the maximum far from the touching instant and then will suddenly fall, (corresponding to the quick stop of the object on the ground);

- if the object oscillates (from unstable equilibrium, but with the impulsive force applied on a particular point of the object's body), then the histogram will be double-bell shaped, the two peaks corresponding to the two opposite motions in two different time instants.

An experiment has been carried out, in which a tapping action is performed on two different points of an object lying, in unstable equilibrium, on a flat table. The same experiment was then performed with the same object in stable equilibrium. The histograms relative to these experiments are shown in figure 3, 4, 5.

3.2 Estimation of motion opposition

The second feature of interest is what we call "motion opposition". This feature is qualitatively similar to friction in that it gives an idea of how an object should behave when pushed with a known force while it lies onto a particular surface. Two methods can be

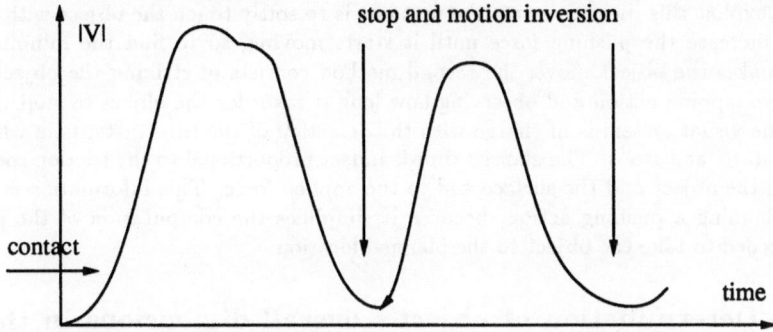

Figure 3: Histogram of mean |V| with an oscillating object.

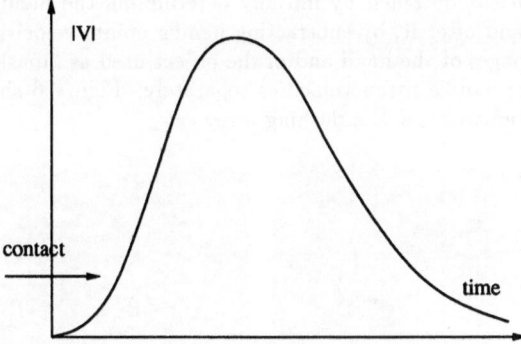

Figure 4: Histogram of mean |V| with an object in stable equilibrium.

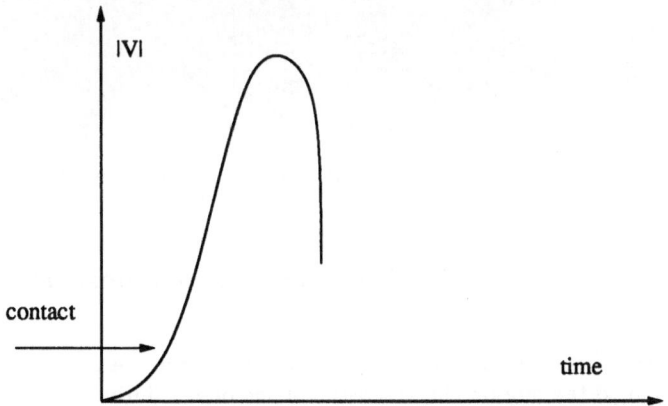

Figure 5: Histogram of mean |V| for a falling object.

used to exploit this measurement: the first one is to softly touch the object with a hand and to increase the pushing force until it starts moving, so to find the minimal force which makes the object move; the second method consists of striking the object by an impulsive tapping action and observing how long it takes for the object to stop. In both cases, the visual system is in charge with the detection of the time instants in which the motion starts and stops. The elapsed time is in fact proportional to the friction coefficient between the object and the surface and to the applied force. This information is helpful when planning a pushing action, because it simplifies the computation of the pushing force needed to take the object to the planned location.

3.3   Determination of object's overall dimensions in the 2D image

Having found a good pushing point and the force to push the object, we can begin to study its behaviour during the pushing. To obtain this it is necessary to segment the object, that is, to separate its points from the hand's ones in the 2D binary image obtained by optical flow. This is obtained by initially determining the hand's shape and speed before the contact and after it, by subtracting hand's points velocities from the current image. From the images of the hand and of the object used as "masks" we can compute the object's and the hand's mean velocities separately. Figure 6 shows the silouhettes computed while a human hand was pushing a toy cat.

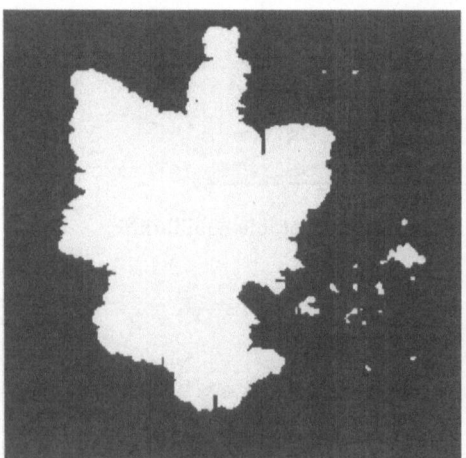 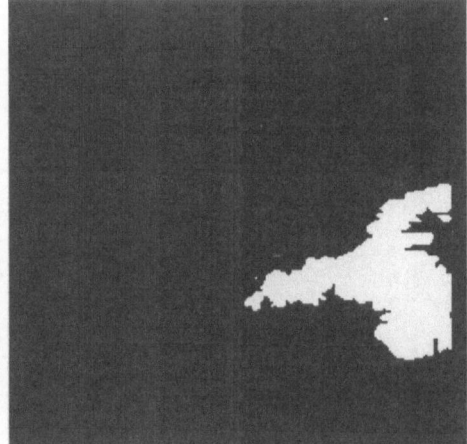

Figure 6: Silouhettes of a cat and of the hand pushing it.

3.4   Analysis of object's trajectory in response to a pushing action

The other features which we now consider are not intrinsic properties of the object since they are related to the motion characteristics of the object during a particular pushing experiment. They are not absolute properties, but they are a measure of the behaviour of an object at a time instant during a manipulation action. A feature easy to detect is the relative motion between the pusher and the pushed object. It can be computed by comparing the directions of the mean velocity of the hand and the object. If the two vectors are almost aligned, then the object and the hand are translating concordingly.

If this is not the case, it is necessary to discriminate between an object's translation along a direction different from that of the hand (this may happen if it is sliding along an eventual obstacle or, more in generally, if the pusher is sliding along the object not causing it rotate) and an object's roto-translation (this happens when the direction of pushing does not pass through the object's centre of friction [12, 15, 13]. This information can be used to correct the pushing point so to make the object follow a predefined trajectory. If the object is rotating, this means that its center of mass is not lying along the pushing direction and the counter/clockwise rotation provides an hint on the semi-space on which the center of mass lies, guiding though the choice of a new pushing point. By iterating this process, a suitable pushing point is found.

3.5 Quantitative measurement of rotation

Since it is possible to detect a rotational component in the object's motion, it may be useful to compute some of its parameters (like the angular velocity and the centre of instantaneous rotation) to help the pusher to correct the pushing point.

The centre of instantaneous rotation can be located by exploiting two characteristics of a rotating object: the object's points lie at a distance from the centre of instantaneous rotation which is directly proportional to their mean velocities, and their displacement vectors are perpendicular to the line joining them with the centre of rotation. From these properties, it is possible to compute the centre of rotation relative to the supporting plane (that is, the intersection between the axis of instantaneous rotation and the plane). Figure 7 shows the position of the centre of rotation for a planar object. The first image shows the center of rotation and the second shows the optical flow relative to the rotating object.

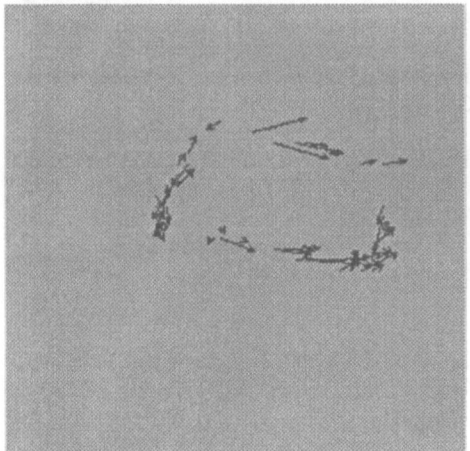

Figure 7: Optical flow for a rotating object and the computed centre of rotation superimposed to one image of the sequence.

4 Conclusions

To manipulate objects in the real world, it is necessary to have a sensory apparatus to perceive what is going on. The sensory apparatus must be able to interact with the motor system in order to make its information useful; a calibration phase is then necessary to

386

tune the two systems so to make them cooperate. The sensory system is in charge of the exploration of the world. The visual processing technique we employ for the dynamic evaluation of object motion is the optical flow analysis, [16, 17, 18, 19, 20, 21] performed on data extracted from images acquired during a manipulation action. The processing does not involve the recovery of exact data from optical flows, but rather the exploitation of some qualitative properties of the motion field. The acquisition of image sequences is done in real time while some of the processing is performed off-line. The experimental set up we have been using is composed of a pair of CCD cameras, a 6 d.o.f. robotic arm and a VDS Eidobrain 7001 workstation. Experiments have been performed to extract the described features. An experiment has been carried out to perform a pushing action and a correction of the pushing trajectory, based on the visual information processed during the pushing action.

Acknowledgements

This work has been partially supported by an ESPRIT Basic Research Actions project (P3274 - FIRST) of the European Community and by the Special Project on Robotics of the Italian National Research Council.

References

[1] F. Gandolfo, G. Sandini, and M. Tistarelli. Towards vision guided manipulation. In *ICAR Fifth Int. Conference on Advanced Robotics*, Pisa, Italy, June 1991.

[2] R. Bajcsy and C. Tsikos. Perception via manipulation. *Proc. of the Int. Symp. & Exposition on Robots*, pages 237–244, 1988. November 6-10.

[3] G. Sandini and M. Tistarelli. Understanding optical flow anomalies. Technical report, DIST University of Genoa - LIRA lab., Genoa, Italy, May 1990.

[4] R. Bajcsy and C. Tsikos. Assembly via Disassembly: A case in Machine Perceptual Development. In *Preprints of the 5th. Int. Symp. of Robotic Research*, pages 303–309, 1988.

[5] J. Aloimonos, I. Weiss, and A. Bandyopadhyay. Active vision. *Intl. Journal of Computer Vision*, 1(4):333–356, 1988.

[6] P. Morasso, G. Sandini, and M. Tistarelli. Active vision: Integration of fixed and mobile cameras. In *NATO ARW on Sensors and Sensory Systems for Advanced Robots*, Berlin Heidelberg, 1986. Springer-Verlag.

[7] R. Bajcsy and C. Tsikos. Perception via manipulation. In *Proc. of the Int. Symp. & Exposition on Robots*, pages 237–244, Sydney, Australia, November 6-10 1988.

[8] G. Sandini and M. Tistarelli. Active tracking strategy for monocular depth inference over multiple frames. *IEEE Trans. on PAMI*, PAMI-12, No. 1, January 1990.

[9] J. Aloimonos. Purposive and qualitative active vision. In *Intl. Workshop on Active Control in Visual Perception*, Antibes, France, April 27 1990.

[10] D.H. Ballard, R.C. Nelson, and B. Yamauchi. Animate vision. *Optics News*, 15(5):17–25, 1989.

[11] E. Grosso, G. Sandini, and C. Frigato. Extraction of 3d information and volumetric uncertainty from multiple images. *Proc. ECAI-88*, pages 683–688, August 1988. Munchen.

[12] Z. Balorda. Reducing uncertainty of objects by robot pushing. In *1990 IEEE Int. Conf. on Robotics and Automation*, pages 1051–1056, May 1990. Cincinnati, Ohio.

[13] D.T. Pham, K.C. Cheung, and S.H. Yeo. Initial motion of a rectangular object being pushed or pulled. In *IEEE Int. Conf. on robotics and automation*, pages 1046–1050, May 1990. Cincinnati, Ohio.

[14] M. A. Peshkin and A. C. Sanderson. The motion of a pushed, sliding workpiece. In *IEEE Journal of Robotics and Automation*, volume RA-4,6, Dec 1988.

[15] M. Mason. Mechanics and planning of manipulator pushing operations. In *International journal of robotic research*, pages 53–70, 1986.

[16] B. K. P. Horn and B. G. Schunck. Determining optical flow. *Artificial Intelligence*, 17 No.1-3:185–204, 1981.

[17] H. Nagel. Displacement vectors derived from second-order intensity variations in image sequence. *CVGIP*, 21:85–117, 1983.

[18] E. C. Hildreth. *The Measurement of Visual Motion*. MIT Press, Cambridge, USA, 1983.

[19] G. Sandini M. Tistarelli E. De Micheli and V. Torre. Estimation of visual motion and 3d motion parameters from singular points. In *Proc. IEEE Workshop on intelligent Robots and Systems*, pages 543–548, 1988.

[20] A. Verri F. Girosi and V. Torre. Constraints for the computation of optical flow. In *IEEE Int. Conf. on Robotics and Automation*, pages 116–124, 1989.

[21] D. Shulman and J. Herve'. Regularization of discontinuous flow fields. In *IEEE Int. Conf. on Robotics and Automation*, pages 81–86, 1989.

Real-time Vision based Control of Servomechanical Systems

Nils A. Andersen, Associate Professor
Ole Ravn, Associate Professor
Allan Theill Sørensen, M. Sc. E. E.

Institute of Automatic Control Systems, Servolaboratoriet,
Technical University of Denmark, Build. 326,
DK-2800 Lyngby, Denmark

Abstract

This paper presents properties of vision used in real-time control systems. Two laboratory experiments have been carried out demonstrating the various aspects of vision used on servomechanical systems.

The paper describes the basis for using vision in real-time servomechanical control systems including achievable accuracy, sample rates and methods for reducing the computational load. In real-time control systems the features of the sensor relating to maximum achievable sample rate and the quantization of the measurement are of great importance in relation to the performance of the system.

The paper describes the theory and implementation of the following methods:
- Dynamic sub-picture selection as a means of minimizing computing time.
- Adaptive thresholding to cope with varying illumination of the scene.

These aspects have been investigated by implementing 2 systems with vision feedback in the laboratory: An inverted pendulum and control of a labyrinth game.

The experiments have shown that by using vision, it is possible to solve new problems and the versatility of vision provides a greater functionality with less instrumentation effort than more conventional sensors.

1 Introduction

Through a number of years image processing has been used in off-line applications such as analysis of satellite and air photographs etc. The rapid decrease of prices of vision hardware combined with the raising performance have in recent years made it possible to use vision in real-time applications. It seems that vision has a great potential for use as a multipurpose sensor in real-time control systems.

The paper describes three important features of vision in real-time control systems. Firstly the sampling process is described, the achievable sampling frequency using different techniques is derived as is the special design precautions needed when using vision instead of more traditional sensors in real-time applications. Secondly the achievable accuracy is determined and some simple measures that makes it possible to enhance this are discussed. Further some uses of a priori knowledge that enhances performance are discussed. A number of practical experiments have been implemented in the laboratory. One of them, an automated labyrinth game, is described in some detail with emphasis on the vision sensor and the algorithms used here. All the results presented are calculated based on the european video standard, however they should be easy to adapt to the US standard.

2 Sampling

The choice of sampling time when using vision in real-time control systems depends on the following conditions:

- The dynamics of the controlled object as in conventional control systems.
- The frame rate of the camera.
- The delay extracting the needed information of the image.
- The delay by calculating the controller output signal.

The picture from the camera is updated every 40 ms. Most cameras use interleaving meaning that by turn the part of the image composed by the even lines and the part composed by the odd lines is updated. Thus a new half-picture is present every 20 ms. Depending on the possibilities of access to the video memory of the specific vision system during a frame grab operation this means that a minimum sample time of 20 ms is achievable. The vision system used during this work does not allow access to the video memory during a frame grab operation. This means that only every second half-picture can be used leaving 20 ms between the half-pictures for extracting the needed information from the picture. Thus the minimum sample time is 40 ms which gives an upper Nyquist frequency of 12.5 Hz. It is not possible to use an anti-aliasing filter so it is important to ensure that the objects in the picture does not move in an undesirable way. The effect is similar to that known from western movies where the wheels of a stage coach seem to rotate backwards.

The minimum sample time is only achievable if it is possible to perform the necessary image processing and calculate the controller output during the 20 ms of free time between the half-pictures. In the vision system used it takes approximately 125 ms just to access all the pixels of one half-picture, thus to obtain the minimum sample time measures must be taken to minimize the time used for image processing. Possible means are:

- Using effective image processing algorithms.
- Using a simplified scene.
- Using a limited area of interest.
- Using binary image features.
- Using active signalling and a priori knowledge.

Using a priori information of the dynamics of the object and the control signal, the position of the controlled object may be predicted and used for limiting the area of interest used by the image processing algorithm. A good contrast between the object and the background enables the use of a binary image facilitating fast image processing. It may be possible to introduce information into the scene which makes e.g. position detection very simple. One example could be marking the object with a lamp with greater intensity than the immediate surroundings thus ensuring good local contrast making it simple to identify the position of the object. Active signalling e.g. blinking with the lamp could also be used. This makes it possible to distinguish multiple objects.

When using vidicon cameras it is important to realize that all the lines in the picture are not sampled at the same time. As the picture is scanned from the top this means that the lines at the bottom of the picture are the most recent when access is allowed to the video memory after a frame grab operation. Thus a variable delay is introduced into the system. The size of the delay is between 0 ms and 20 ms depending on the location in the picture of the object, it is easily calculated as it is proportional to the distance from the object to the bottom of the picture. In tightly coupled real-time control systems with small sample times the stability and quality of the controller may be affected if the delay is not taken into account when designing the controller. One way of designing the controller considering the delay is to use the modified z-transform [Tou 1959]. Having obtained the modified z-transform of the object and the current delay the appropriate controller may be designed. By dividing the picture into areas with almost the same delay the corresponding controller can be designed and at run-time the controller parameters are then determined by gain scheduling based on the position of the object.

The transfer function of the delay is e^{-sT} which has a constant amplitude of 1 and a phase lag of $2\pi fT$, T being the delay and f the frequency. As T is never more than 20 ms the maximum phase of the delay at the phase margin frequency may be calculated and if possible the controller may be designed with sufficient phase margin to preserve the necessary performance and stability even when the object is in the top of the picture.

Other considerations should be made when using CCD cameras as they perform the pixel sampling simultaneously in the entire picture. However a transmission delay may be introduced [Corke & Paul 1989].

3 Quantization

The quantization of a 512 x 512 pixel image corresponds to that of a 9 bit A/D-converter. To avoid loss of precision precautions must be taken to ensure that the whole range of this virtual A/D-converter is utilized. By using zoom optics the picture may be scaled to ensure that the interesting part of the scene takes up most the picture. In the case of position measurement the achievable accuracy may be increased by using circular markings and determining the position as the centre of mass of the markings. The calculation of centre of mass makes it possible to achieve subpixel accuracy determining the position. The obtained effect is similar to that known from the vernier scale on a slide gauge. If square markings are used the achievable accuracy depends on the orientation of the marking with respect to the axes of the digitized picture while the use of circular markings eliminates that effect. Figure 1 shows the mean absolute error versus the radius of the marking measured in pixels. The result is found by computer simulation.

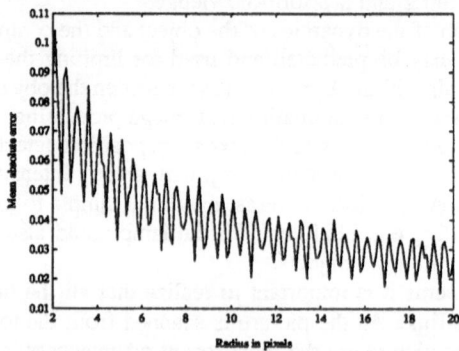

Fig. 1. Error in pixels by digitizing circular markings.

An experiment has been carried out to determine the accuracy of position measurement with the equipment used in this work. The experiment was based on determining the position of a white spot with a diameter of 20 pixels in the digitized picture. The position was measured 20 times at each position in 29 x 26 point grid covering the entire area viewed by the camera. The position is measured by calculating the centre of mass of the white spot after converting the picture to a binary image. The 20 measurements at one point were made with a delay allowing the noise terms to be considered independent. The experiment was carried out with the position measurement based on both half-pictures and full pictures. The experiment showed virtually no deviation related to the position of the white spot in the picture. Figure 2 shows the deviation from the true position for all the measurements made using a full picture. Assuming that the measurements are normally distributed the standard deviations are:

	Full Picture	Half Picture
x-direction	0.21 pixel	0.29 pixel
y-direction	0.14 pixel	0.18 pixel

Comparing figure 1 and figure 2 shows that the error introduced by the digitization (figure 1) is small compared to the error occurring because of noise. Figure 2 show that the accuracy in the y-direction is better than in the x-direction. This is explained by the fact that the horizontal scan in the vidicon happens at a greater speed than the vertical scan. One would expect a symmetric deviation when a CCD camera is used.

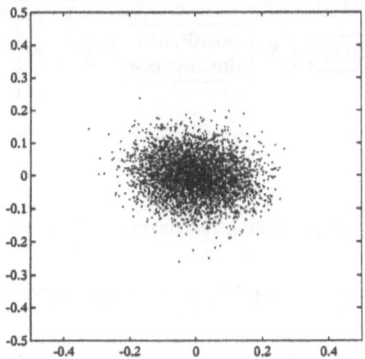

Fig. 2. Deviation in pixels from true position (experiment).

4 Adaptive Image Processing

As pointed out in the previous sections the use of vision in closed loop control introduces a time delay in the loop. This delay consists of two parts one from the sampling and digitization of the image and one from the image processing of the digital image. This time delay will in many systems be the factor limiting the obtainable dynamic behaviour of the system. It is therefore of crucial importance to minimize the delay introduced by the image processing. One of the most effective ways of minimizing the image processing time is reducing the size of the processed image as the calculation time of most algorithms is proportional to the number of pixels in the image. The idea of dynamic subpicture selection is to use knowledge of the system to select only the interesting parts of the picture for image processing thus reducing the necessary processing time. (The soundness of the idea is supported by its use in one of the most advanced vision systems namely the human vision). Another important way to reduce the introduced delay is to split the image processing in two parts, a fast part extracting just the features necessary for the control at the sample times and a more elaborate part which is used to update a model of the scene and calculate parameters for the fast algorithm. A structure supporting these ideas is shown on figure 3. The basic structure is similar to the one proposed by Weiss et. al. [1987] for visual servoing.

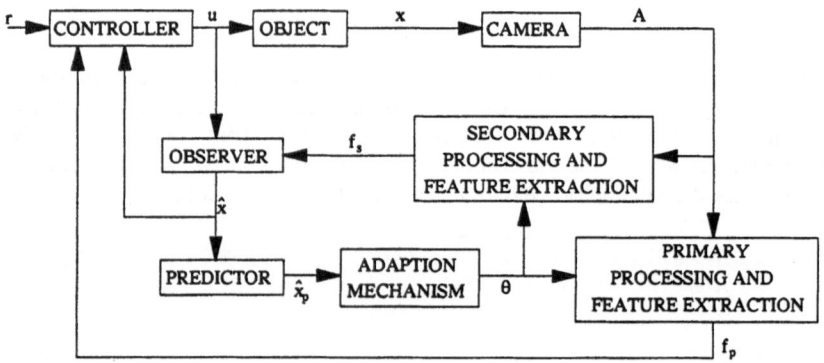

Fig. 3. Structure of adaptive image processing for vision based control.

The object to be controlled is described by a nonlinear time variant continuous time state space equation:

$$\dot{x}(t) = G(x(t), u(t), t, w(t)), \quad x \in \mathbf{R}^n \tag{1}$$

- x(t) is the state vector of the object, it may contain information about visual features for the object such as illumination.
- u(t) is the control vector produced by the controller.
- w(t) is the process noise term.

Equation 2 describes the mapping from the state vector to the image plane done by the lens and the sampling process performed by the camera.

$$A(n) = C(x(t), n, v(t)) \quad A \in Q^{N \times M}, \ Q \in [0, \text{maxgreylevel}] \tag{2}$$

- A(n) is the picture array at the discrete time n.
- v(t) is the measurement noise term.

The primary and secondary image processing and feature extraction blocks of figure 3 are shown in equation 3 and 4. The difference between the primary and secondary block is that the features extracted in the primary block is used for generating the control signal and the execution time of this block should therefore be kept at an absolute minimum as it is in the control loop. The secondary block is executed after the control signal is generated and the features extracted here are used in the prediction.

$$f_p(n) = F_p(A(n), \theta(n), n) \tag{3}$$

$$f_s(n) = F_s(A(n), \theta(n), n) \tag{4}$$

The functions F_p and F_s are selected depending on which features $f_p(n)$ and $f_s(n)$ are wanted. The functions are tuned by the adaption mechanism based on predictions of the state vector $\hat{x}_p$. Threshold level and subpicture size and position are examples of parameters which could be included in $\theta(n)$.

The observer generates an estimate of the state vector based on the features $f_s(n)$ the old values of $\hat{x}(n)$ and the control signal $u(n)$. The estimate could be used in the control law and in the prediction (6).

$$\hat{x}(n) = H_{obs}(f_s(n), \hat{x}(n-1), u(n), n) \tag{5}$$

-   $\hat{x}(n)$ is an estimate of the state vector of the object but it does not necessarily contain estimates of all the states of the object.

The predictor is used to predict the state of the object based on the observer state vector. The predicted state vector $\hat{x}_p(n+1)$ is used in the adaption mechanism to tune the parameters of the image processing and feature extraction blocks. The predictor is shown in equation 6.

$$\hat{x}(n+1) = H_{pred}(\hat{x}(n), u(n), n) \tag{6}$$

The control signal is generated based on the reference signal $r(n)$, the estimated state $\hat{x}(n)$ and the features extracted in the primary image processing.

$$u(n) = K(r(n), f_p(n), \hat{x}(n), n) \tag{7}$$

To illustrate the use of the structure the labyrinth game example will be presented. The state vector contains the position, the velocity and the brightness of the centre of the ball. The primary feature extraction calculates the position of the ball by calculating the centre of gravity of a binarized subpicture. The θ-vector contains the threshold value and the position of the subpicture used in the primary feature extraction. The secondary feature extraction calculates updates for the observer. The observer and the controller designs are described in 5.2.1.

5 Examples

Two practical experiments have been implemented in the laboratory using vision as a sensor. An inverted pendulum and a labyrinth game. The inverted pendulum is an example of a tightly coupled real-time system. The labyrinth game is an example of an inherently visual application where control using traditional sensors is very difficult. It will be described in greater detail below.

5.1 Equipment

The central unit used in this work is a standard off-the-shelf vision system. It consists of a real-time frame grabber digitizing the incoming video signal into 512 x 512 8 bit pixels. It has an on-board MC68020 microprocessor with a MC68881 floating-point coprocessor. It has 1 MB video memory for storing two digitized pictures and two images of overlay graphics and 128 kB local RAM for programs and data. The vision system is mounted in a VME-bus rack with a MC68030/MC68882 based host computer. The video memory and RAM of the vision system is accessible from the host computer via the VME-bus. This facilitates synchronization and data transfer between the two units. The host computer is running OS-9 with the C programming language. On the VME-bus is also mounted a DA-converter for output of control signals for the labyrinth. The video signal for the vision system is obtained from a Philips black/white vidicon surveillance camera.

5.2 Labyrinth Game

The labyrinth game consists of a wooden board mounted on gimbals making it possible to tilt the board in all directions by turning two handles. The board is equipped with walls and holes defining a maze. A route is drawn through the maze and the purpose of the game is to direct a steel-ball along the route by turning the two handles avoiding that the ball falls into one of the holes.

To demonstrate the versatility of vision systems as few changes as possible have been made on the original labyrinth. The two handles of the labyrinth have been replaced by DC-servomotors and the angle of the maze is measured by servo-potentiometers. The only visual change is that the shiny steel-ball has been painted white to obtain good contrast to the wooden board of the maze.

The labyrinth game has been chosen because it is well suited for demonstrating the power of vision in automation and control systems. A non-interfering measurement of the position of the ball is necessary to obtain satisfactory control of the ball, and it is very difficult to conceive such a measurement without using vision. In addition to the position measurement the vision system is used during initialization of the automation system and to perform error detection. The problem of controlling the ball in the maze is interesting, because it is necessary to control an unstable object in two dimensions.

5.2.1 The Control Problem

To simplify the control of the ball two analogue controllers control the angle of the maze. This divides the control problem into two parts: a simple problem of controlling the angle of the maze and a more complicated problem of controlling the movement of the ball. Thus the output of the controller of the ball-movement becomes the reference for the two controllers of the angle of the maze.

The axles of the gimbals of the maze are perpendicular and each of the two analogue controllers are controlling one of the axles. This means that each of the analogue controllers only affects the movements of the ball in one of the two directions. Thus the movements of the ball in the two directions are decoupled and this means that the problem of controlling the ball simplifies into a problem of designing two (identical) single-input/single-output controllers, one for each of the two dimensions. The input for each controller should be the position of the ball in that direction and the output the angle of the maze in that direction.

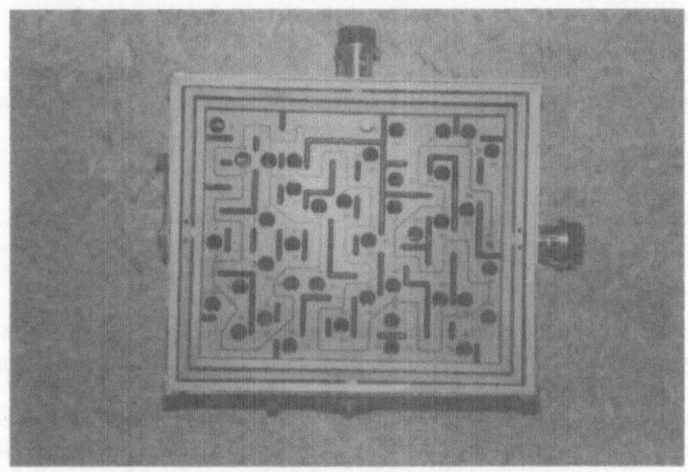

Fig. 4. Labyrinth game seen from the top.

Fig. 5. Experimental setup.

The dynamical equation of the movement of the ball in one direction is:

$$\ddot{x} = 5/7 g \sin(\theta) + \dot{\theta}^2 x \tag{8}$$

x being the position of the ball and θ the angle of the maze. This is a non-linear equation. To obtain a linear model for the controller design the equation is linearized:

$$\ddot{x} = 5/7 g \theta \tag{9}$$

The transfer function from the angle of the maze to the position of the ball is:

$$H(s) = \frac{x(s)}{\theta(s)} = 5/7 g \frac{1}{s^2} \tag{10}$$

Thus the acceleration of the ball is directly controllable by the angle of the maze.
Theoretically the system (10) may be stabilized to any bandwidth but practically several conditions affect the controller design:
- The model (8) of the ball-movements assumes that the ball is always in contact with the board of the maze. This means that no point of the board can accelerate faster than the acceleration of gravity without violating the model.
- The noise on the position measurement of the ball. This noise originates from: the quality of the camera, the transmission of the video signal from the camera to the frame grabber and from the quantization in the frame grabber.
- The chosen sampling frequency according to Shannon's theorem.
- The speed of the ball in the maze. The achievable speed depends on the bandwidth of the controller.
- The condition of the board of the maze. The board is not completely even and that causes disturbances in the resulting angle of the board as it is seen from the ball.
- The walls of the maze. The model (8) of the ball-movement ignores the walls and the model fails if the ball runs into one of the walls.

Considering these conditions a state feedback controller based on a kalman filter has been designed. The kalman filter is designed on the basis of the model shown in figure 6.
The model has three states: position and speed of the ball and the disturbance of the angle occurring from the board not being level. The disturbance of the angle is assumed to be lowpass filtered white noise (w(t)). $w_2(t)$ is the noise affecting the measurement of the position of the ball which is assumed to be white noise. An experiment has been carried out in the laboratory to determine the intensity of $w_2(t)$ while the parameters of the model of the disturbance have been tuned during experiments with the controller on the maze. The parameters have been chosen to make the suppression of the disturbances as good as possible. The feedback gains of the kalman filter states are chosen on the basis of weights on the states and on the controller output signal. Only the position of the ball (state x_3) and the controller output is weighted and thus the coefficient between the weights on these two signals determine the feedback gains. The weights has been chosen on the basis of experiments to give the controller as high a bandwidth as the disturbances permit.

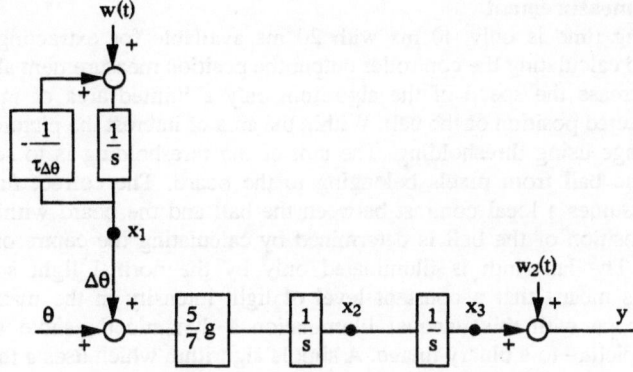

Fig. 6. Model used designing Kalman filter.

The kalman filter and the state feedback controller are combined into a single-input/single-output controller and the final system of labyrinth and controller has a bandwidth of 1.3 Hz. The system has a phase margin of 44° at the bottom of the picture being reduced to 37° at the top due to the delay described above. The Bode plot of the controller is seen on figure 7. The controller consists of elements known from traditional PID-controller design: a PD-element stabilizing the object, a lowpass filter to remove the noise on the position measurement and an integrating network which handles the disturbances on the angle of the board.

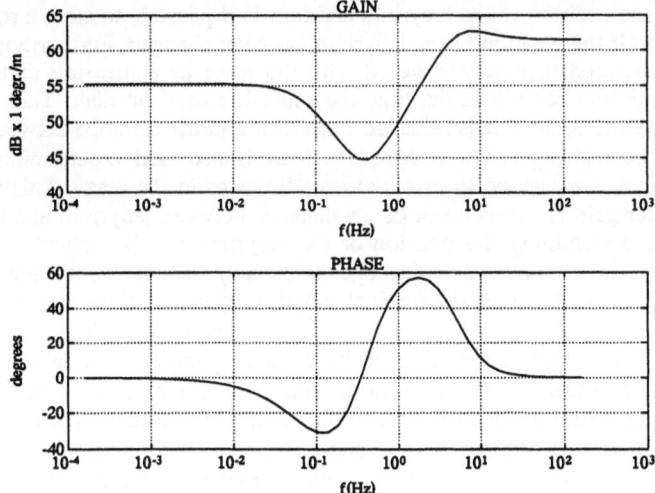

Fig. 7. Bode plot of the controller.

5.2.2 Position measurement

As the sampling time is only 40 ms with 20 ms available for extracting the position information and calculating the controller output the position measurement algorithm must be fast. To increase the speed of the algorithm only a limited area of interest is used around the expected position of the ball. Within the area of interest the picture is converted to a binary image using thresholding. The aim of the thresholding is to separate pixels belonging to the ball from pixels belonging to the board. The correct function of the thresholding assumes a local contrast between the ball and the board within the area of interest. The position of the ball is determined by calculating the centre of mass of the binary image. The labyrinth is illuminated only by the normal light sources of the laboratory. This means that a constant level of light intensity in the maze can not be expected. To cope with this unequal illumination a dynamic threshold value is used converting the picture to a binary image. A simple algorithm which uses a threshold value equal to the intensity of the centre pixel of the ball from the previous sample minus a constant has been used. Although it is simple this algorithm has proven very effective. Using the algorithm it is possible to run the system even if the intensity of the board in one side of the maze is greater than the intensity of the ball in the other side. The algorithm also copes with light fluctuations caused by shadows of people moving around the labyrinth.

5.2.3 Start up routines

To make the system independent of the position of the labyrinth and a constant distance between camera and labyrinth an algorithm has been designed to determine the position of the labyrinth in the digitized picture. The algorithm determines the coordinates of the rectangle limiting the board of the maze. The upper left corner of the rectangle is used as offset of the coordinates of the reference route for the ball. This makes the coordinates independent of the position of the labyrinth and thus is it possible to store a route on disk for use later even if the position of the labyrinth has been changed. Furthermore the width of the board measured in pixels is stored with the route as a measure of the distance between labyrinth and camera at the time the route is stored on disk. This enables the system to rescale the route if it is reloaded from disk and the distance between labyrinth and camera has been changed. The width of the board is also used to determine the scaling factor from distances measured in pixels to true distances in the maze. This is to ensure that the controller gain is independent of the distance between labyrinth and camera.

The problem of determining the position of the labyrinth can be solved very easily if characteristic markings are placed on the edges of the labyrinth. But as the aim of the work was to solve the task changing the original labyrinth game as little as possible is was demanded that the designed algorithm worked on a labyrinth without changes. The algorithm performs two transformations of the digitized picture of the labyrinth. First an edge emphasizing Sobel transform is done and then a simplified Hough transform [Rosenfeld & Kak 1982] to identify lines in the picture. The Sobel transform recognizes edges parallel to the x- or y-direction best and as the labyrinth is expected to be oriented parallel to the edges of the picture the Sobel transform gives a good detection of the frame of the labyrinth game. Since the walls of the maze are parallel to the x- or y-direction they also give rise to detected edges in the Sobel transformed picture. As the aim of the algorithm is only to determine the position of the rectangle limiting the board of the maze, these edges can be considered as background noise. The edges arising from the outer frames of the board are the most distinct edges in the picture, which makes it possible to identify these edges. The left and right edge of the frame of the board is found simply by calculating a histogram of cumulated pixel values of each column in the Sobel transformed picture. The columns containing the most distinct edges parallel to the y-direction give rise to the largest values of the histogram. A similar histogram is calculated to determine the upper and lower edge of the board. To cope with the background noise occurring from the walls of the maze some additional filtering of the histogram is necessary to ensure that the edges identified by the algorithm are indeed the correct edges of the board.

As the algorithm is based on detecting significant edges parallel to the x- and y-direction it becomes sensitive to light falling on to the labyrinth from one of the sides. This light can cause shadows and the edge of a distinct shadow could erroneously be identified as one of the outer edges of the labyrinth. Identifying a wrong edge would be fatal for the control of the labyrinth as the coordinates of the reference route and the controller gain are determined on the basis of the coordinates of the frame of the board. To avoid control of the labyrinth based on wrong coordinates of the frame the ratio of height and width of the identified frame is calculated. This ratio is constant no matter the distance between camera and labyrinth. The correct ratio can be calculated off-line during design of the system and if the calculated ratio is wrong the operator can be advised of the error and could then either try to change the illumination of the labyrinth or enter the position of the frame manually. The possibility of entering the position of the frame manually ensures that it is possible to the run the system even under bad lighting conditions.

Another start up routine is necessary for initializing the position measurement of the ball. The position measurement described above uses a limited area of interest around the expected position of the ball. The thresholding of the picture within the area of interest is then based on a local contrast between the ball and the board of the maze. Because of the possibility of uneven illumination of the labyrinth a global contrast between ball and maze can not always be assured. This means that the ball can not automatically be identified in the maze by searching for the largest pixel value of the digitized picture. To cope with this an algorithm has been designed to identify the ball in the picture even if it is in a shadow thus not giving rise to the largest pixel value of the picture. The first step of the algorithm is an adjustment of the level of the digitized picture. The aim of the adjustment is to level out global differences of illumination while local contrasts are saved. The adjustment is done by adjusting the value of each pixel proportional to the mean pixel value of an area of interest around the pixel. Thus the pixel value is divided by the mean pixel value of the local area of interest and multiplied by a constant reference level. This operation causes a certain smoothing of the picture. The operation also emphasizes areas of the board surrounded by dark walls so after this operation the largest pixel of the picture is not guaranteed to belong to the ball. Consequently the picture is then filtered through a match filter emphasizing structures of the picture matching the structure of the white ball on the background of the board. After applying the match filter the ball is identified by searching for the largest pixel value within the maze.

5.2.4 Performance
Running the ball through the maze shows a good performance of the kalman filter based state feedback controller. About 75% of the attempts to run the ball succeeds while the rest of the attempts fail because of one of two reasons: either the ball falls into a hole or it gets stuck in the maze because of dust on the board. The first situation is fatal and it is

typically caused by the board of the maze is not 100% even. The unevenness of the board causes sudden changes in the resulting angle of the board as it is seen from the ball. Despite the fact that the kalman filter estimates the disturbance caused by the unevenness the controller is quite sensitive to changes in the angle of the board. Simulation shows that a sudden 1° change of the angle of the board causes a 0.7 cm deviation of the position of the ball from the reference route. Certain places the distance from the reference route to a hole is not more than 0.5 cm, so around these places only small disturbances can be suppressed by the controller without the ball falling into a hole. A higher bandwidth of the controller would be necessary to obtain a better rejection of the disturbances but as mentioned earlier this is not possible because of the noise affecting the system.

When the ball getting stuck because of dust in the maze the situation is handled by leaving the system in a stable state. If the normal control algorithm is continued the position error will increase because of the reference still moving along the reference route. The increasing error causes the output of the controller (the angle of the board) to increase. Eventually the ball will loosen but if that happens at a large position error the ball is very likely to fall into a hole before the controller regains control. To avoid that situation the position error is monitored and if the error exceeds a predefined limit the ball is thought of as being stuck. The normal advancing of the position reference along the route is stopped and the position reference is moved back to the position of the stuck ball and kept there until the controller is stopped by the operator.

As the model (figure 6) ignores the walls of the maze the state estimates of the kalman filter are incorrect if the ball runs into a wall. These incorrect estimates lead an output from the state feedback controller that is inappropriate for the actual state of the system. When this happens a resonance is observed, the controller keeps bouncing the ball against the wall. Although the phenomenon looks fatal the resonance is normally damped quickly and very seldom causes the ball to fall into a hole. Thus the phenomenon does not limit the ability to control the ball through the maze.

As mentioned above the start up routine determining the position of the labyrinth is sensitive to the illumination. Combined with the check of the ratio of height and width of the identified frame the algorithm is useful anyway because it has the ability of identifying the frame under the right conditions of lighting while it gives the operator the possibility of entering the route manually if the lighting conditions are insufficient for the algorithm to work correctly.

The routine for initializing the position measurement is very stable and it fails only if the ball is hidden in a very dark shadow which usually does not happen under normal lighting conditions.

5.3 Inverted Pendulum

The inverted pendulum has been chosen because it is an unstable system which makes it well suited for demonstration of vision used in a tightly coupled real-time system.

The system consists of a 1 m long aluminium pole placed on a vehicle which is driven by DC-motor through a wire drive. The pendulum is free to fall in the moving direction of the vehicle. See figure 8.

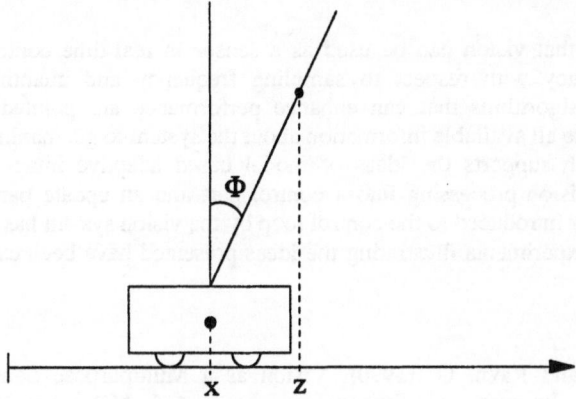

Fig. 8. Inverted pendulum.

One way to control the pendulum is to measure the position of a point on the pendulum and feed this into a controller which controls the position of the vehicle. If the point on the pendulum is placed 2/3 of the pendulums length from the supporting point linearizing the equations of motion gives the following transfer function:

$$H(s) = \frac{z(s)}{x(s)} = \frac{-a^2}{s^2-a^2} \tag{11}$$

z being the position of the point on the pendulum,
x the position of the vehicle,

$$a = \sqrt{\frac{3g}{2l}} \, ,$$

g the acceleration of gravity and
l the length of the pendulum.

The object is stabilized by a conventional PD-controller with the zero of the controller cancelling the stable pole of the object and a differentiating bandwidth of a factor 3.
The position of the point on the pendulum is measured directly with the vision system. The sample time of the system is 40 ms and to obtain this nearly all the measures mentioned above have been used. The scene has been simplified as much as possible by using a white circular spot on the pendulum and making the background black, the area of interest has been reduced to 20 lines of 80 pixels length centred around the last position of the white spot. The position of the white spot is found by calculating the centre of mass of the area of interest. To speed up this calculation the picture has been converted to a binary image using a proper threshold thus saving a multiplication per pixel in the area of interest.

6 Conclusion

We have shown that vision can be used as a sensor in real-time control systems. The achievable accuracy with respect to sampling frequency and quantization has been determined and algorithms that can enhance performance are pointed out. It is very important to utilize all available information about the system to get maximal performance. A structure which supports the ideas of model based adaptive image processing and dividing of the vision processing into a control part and an update part as a means of reducing the delay introduced to the control loop by the vision system has been presented. Two laboratory experiments illustrating the ideas presented have been carried out.

References

Andersen, N.A. and Ravn, O. (1990). Vision as a Multipurpose Sensor in Realtime Systems. ISMM International Symposium. Mini and Microcomputers and Their Applications.

Corke, Peter I. and Paul, Richard P. (1989). Video-Rate Visual Servoing for Robots. Proc. of The First International Symposium on Experimental Robotics. Montreal, Canada, June 1989.

Rosenfeld, A. and Kak, A. C. (1982). Digital Picture Processing. Second edition Academic Press.

Tou, J. T. (1959). Digital and Sampled-data Control Systems. McGraw-Hill Book Company, Inc.

Weiss, L. E. and Sanderson, A. C. (1987). Dynamic Sensor-Based Control of Robots with Visual Feedback. IEEE Journal of Robotics and Automation. Vol. RA-3. No. 5.

Section 8: Compliance and Force Control

Compliance and force control has developed as one of the central fields in robotic research for many years now with a lot of theoretic insight gained meanwhile, but still with fairly little help for potential users in industry that would need kind of standard algorithms.

The first paper of Lindsay, Funda and Paul on contact operations using a compliant wrist is strongly motivated by the teleprogramming goal, i.e. simulation and building up of force-conrolled tasks in a "virtual" environment, and sending them to a remote robot for later execution. This means that for good coincidence of presimulation and real data, a very robust set of rules e.g. for collision and error detection, contact operations, stopping conditions etc. must be developed. The paper shows how to approach these goals.

Although there are many papers in literature treating force control, only few deal with the transient from non-contact to contact motion. This problem is in the focus of a paper by Allotta, Buttazzo, Dario and Guglielmelli. They allow an acceptable value for the impact force overshoot and use ultrasound and infrared proximity sensing techniques for estimating the "time to impact". In addition they compare performance with only force sensing to the case of proximity and force sensing.

Arocena, Daniel and Elosegui report on the design of a three-degree of freedom compliant arm, using direct drives and compliant transmission shafts. They compare controllers based on rigid body assumptions with those based on straing-gauge feedback loops for compliance compensation. Motor torque ripples representing a non-negligible problem in direct drive systems are addressed as well as time delays in the electronics.

C-surface theory as originally proposed by Mason needs efficient techniques for - preferably automatically - defining the position and sensor controlled subspaces. The paper of Merlet addresses these kind of problems, e.g. finding C-surface normals from sensory data, and propose solutions for realistic problems e.g. in the assembly of different parts of a washing machine. An interesting feature of his approach is the allowance of a fairly high initial uncertainty of the location of parts.

Contact Operations Using an Instrumented Compliant Wrist[†]

Thomas Lindsay, Janez Funda, and Richard Paul

University of Pennsylvania, Philadelphia, PA, USA

Abstract

Teleprogramming was developed as a solution to problems of teleoperation systems with significant time delays [5]. In teleprogramming, the human operator interacts in real time with a graphical model of the remote site, which provides for real time visual and force feedback. The master system automatically generates symbolic commands based on the motions of the master arm and the manipulator/model interactions, given predefined criteria of what types of motions are to be expected. These commands are then sent via a communication link, which may delay the signals, to the remote site. Based upon a remote world model, predefined and possibly refined as more information is obtained, the slave carries out commanded operations in the remote world and decides whether each step has been executed correctly.

Contact operations involve the remote site manipulator interacting with the environment, including planned and unplanned collisions, and motion with contact with the environment. A hybrid position/force control scheme using a instrumented compliant wrist has been demonstrated to be very effective for these types of operations. In particular, switching between position and force modes (when contacting a surface, for example) does not present problems for the system. A brief introduction of teleprogramming and contact operations is presented, including a model of sliding motions and early experimental results. Problems with these early experiments are presented, and solutions discussed. The criteria for an object to slide rather than tip over are presented, relating to the geometry of the object and the applied forces. Finally, methods are presented to match the experimental results to a simple model, to help the remote manipulator to quickly and robustly sense collisions.

1 Introduction

Teleoperation systems are important for the execution of tasks in hazardous and unstructured environments. Hazardous environments range from those extremely dangerous to humans, such as contaminated nuclear power plants and hazardous waste sites, to those such as space and deep sea that can be made safe to humans only for short periods

---

[†]This material is based upon work supported by the National Science Foundation under Grant No. BCS-89-01352, "Model-Based Teleoperation in the Presence of Delay." Any opinions, findings, conclusions or recommendations expressed in this publication are those of the authors and do not necessarily reflect the views of the National Science Foundation.

at great expense. Completely autonomous activity and manipulation is impractical in unstructured environments with state of the art artificial intelligence.

When delays in excess of one second occur, direct force reflecting teleoperation becomes difficult to impossible [6, 13]. Delays can occur on the order of 2-8 seconds for communication with a remote site orbiting the earth (shallow space),and up to 20 seconds for subsea operations (communicating via acoustic link). In order to solve problems associated with communication delays, we have developed a teleoperation structure called teleprogramming [10]. At the master site, a human operator works with a 6-DOF master manipulator to guide a simulated slave manipulator in a geometric model of the remote site. The model provides for monitoring of contacts, and feeds back information to the master arm to give the operator kinesthetic feedback, lacking in most of the current work involving time delays [2, 8, 12, 13].

The master system generates commands based upon the motions and manipulator/model interactions. This information is sent to the remote site, which interprets and executes these command steps. Each step is executed autonomously, and the resulting motion of the slave manipulator is analyzed as to whether it succeeded or failed. If it succeeds, an acknowledgment is sent to the master and the slave continues with the next command. Commands from the master are sent continuously, so there is no delay between commands at the remote site if they are executed without errors. If a command fails, information about the error state is sent to the master, and then the slave waits for the human operator to send a new set of commands that will correct the error.

At the remote site, the slave interprets small execution model steps that make up individual motions. Each execution model step contains information about how long and how far to move, information about contacts and contact forces, and information about what conditions the slave should expect to terminate the motion. For example, a typical move could command the slave manipulator to slide along a surface, pushing against it with a given force, and stop when a wall is encountered. Errors in this example could include falling off the surface, failing to find the specified wall, and encountering an unexpected obstacle.

We are using an instrumented compliant wrist for sensory feedback [17, 9]. The compliance is extremely beneficial for the interactions (expected and unexpected) between the manipulator and the environment [11]. However, the compliance makes sliding motions more complex. Depending on the surface friction and the applied forces, the object on the surface may tend to tip over instead of sliding. Control and other problems lead to non-constant steady state forces in the normal and tangential directions. Peaks in these forces, which are used to determine expected and unexpected collisions, can cause a false identification of an error state. The level of system 'noise' is partially a function of manipulator configuration and direction of movement, so that constant limits that would work successfully in one direction will not work in another. A more robust method of detecting collisions while performing contact operations is necessary.

The rest of the paper is organized as follows. First, the experimental teleprogramming testbed is presented. Next, contact operations in the task environment are examined, from both a model based and experimental based perspective. Criteria for an object to slide rather than tip over are presented. Finally, methods to relate the model and experimental results are examined, with an emphasis on a more general and robust method for interpreting sensor readings.

2 Experimental Setup

The GRASP Lab [1] teleprogramming testbed is shown in schematic form in figure 1. The operator's station and the remote workcell are physically separated. The system can be divided into the master site (operator's station), the remote workcell, the communication link between these sites, and the task environment.

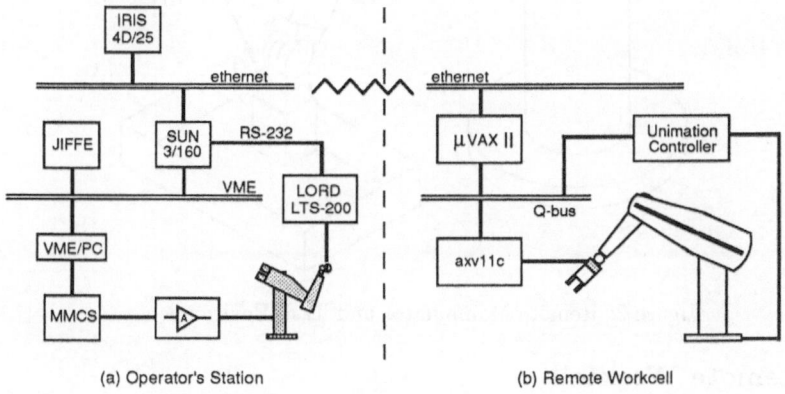

(a) Operator's Station   (b) Remote Workcell

Figure 1: The experimental teleprogramming testbed

2.1 Master Site

The master site is composed of a Unimation Puma 250 robot, acting as a 6-DOF backdrivable input device, and several computers. The Puma hardware is controlled by PC-bus based Modular Motor Control System (MMCS) [4]. There is a 6-d.o.f. force/torque sensor (LORD Corp., LTS-200) mounted at the tip of the 250, which measures the directional input from the human operator. Joint and cartesian level control for the master is performed by JIFFE - a 20 Mflop VME-based floating point co-processor [1]. JIFFE communicates with its host (Sun 3/160) via shared memory and with the graphical workstation (Iris 4D/25) via the Sun and ethernet socket connection. The Iris runs a modeling environment for 3-D manipulation of articulated figures, provided by the Computer Graphics Laboratory at the University of Pennsylvania [3]. This software provides the operator with a graphical model of the remote manipulator and its environment. Manipulator/environment interaction is monitored, and is fed back to the master manipulator. This provides the operator with kinesthetic feedback, which is an important part of the teleprogramming system. The master system contains no information about the dynamics or friction at the remote site.

---

[1] General Robotics and Active Sensory Perception Laboratory, University of Pennsylvania Dept. of Computer and Information Science, Philadelphia, PA. Ruzena Bajcsy, Director.

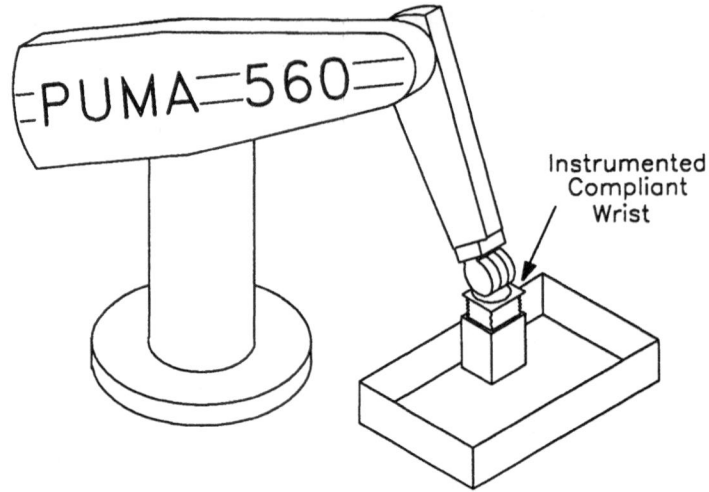

Figure 2: Remote Manipulator and Task Environment

2.2 Remote Workcell

The remote manipulator is a Puma 560 robot, linked to a MicroVax II. The robot uses a cartesian-based hybrid position/force controller, built upon the low-level RCI robot interface [7]. The hybrid controller has been shown experimentally to be stable in the operating region we are using, as long as the task frame origin is located relatively close (within 20 cm) to the robot wrist point. We use a 6-DOF instrumented compliant wrist for force/torque measurements.

The compliance of the wrist simplifies interactions between the robot and the environment. This is especially beneficial in dealing with the impact forces generated when the robot makes the transition from free space motion to motion in contact with the environment. Both natural and active damping help absorb the energy of impact [15]. Also, the compliance of the sensor helps to make the force control more responsive [11].

2.3 Communication

Communication from master to slave is composed of execution models (EMs), which are automatically generated at the master site. Each EM step can contain information about the working task frame, the hybrid modes, contact forces, and movement information. Information not supplied in a given EM step is assumed to carry over from the previous step, thus communication time is reduced by elimination of repetition of known information. The EM step does not contain information about the dynamics or friction of the environment.

The communication between the robots in the lab, using an ethernet connection, is virtually instantaneous. Therefore, communication delays are emulated by software. Currently, we are using a delay of 3 seconds for our experiments.

2.4 Task Environment

We are currently experimenting with very simple contact operations. A small box attached to the manipulator is maneuvered into and around a larger box, as shown in figure 2. Elements of tasks include free-space motion, transitions between free-space (position mode) and constrained space (force mode), and constrained motion. In this environment we can test overall system performance, error detection, and error recovery. Within this task environment, our command language and teleprogramming concept have been shown to be effective. Problems between theory and experimental work have also been examined, and in many cases we have modified how commands are interpreted at the remote site. Complex procedures can be built using the commands we can now generate. Current work includes creating a new task environment which requires more complex motions.

3 Contact Operations

Although many tasks include free-space motion, most tasks require interaction with the environment. Free-space motion is a relatively simple operation; there is no need for feedback at the operator's station, and therefore a telerobotic scheme that has only visual feedback (in real time) would be adequate (as with JPL's predictive display). Most of our work concentrates on contact operations, where the manipulator interacts with the environment.

For two reasons, contact operations are executed semi-autonomously. First, the communication delays make force feedback to the operator impossible. Therefore, the remote site must close the feedback loop locally. Second, there may be inaccuracy in the graphical model at the master site. If the geometry of the environment is known only to a tolerance ϵ, the remote site must locally deal with this inaccuracy. A fully autonomous system would have to understand all possible problems and deal with them appropriately; this is beyond the scope of modern artificial intelligence. When the remote system runs into a problem that it cannot correct, it simply sends back information to the human operator, who can reason through the problem and create a suitable correction.

Due to slow and often unreliable (esp. with acoustic links) communications, the commands sent to and from the remote site need to be minimal. The remote site receives only information about the kinematics of the system. Dynamics and friction are dealt with locally at the remote site. Further, the remote site must keep pace with the master site, albeit delayed by the communications. The slave therefore has no opportunity to explore the environment beyond the scope of the commanded actions. Thus the system can only gain information about the remote environment, such as friction, while it is also trying to discern expected and unexpected changes from the sensor data. Within these constraints, the remote system must interact with the environment and robustly sense forces, contacts, and collisions.

3.1 Contact Model

Motion of the robot/sensor/environment interaction can be simulated for one degree of freedom using a second order model. The second order model is a mass-spring-damper

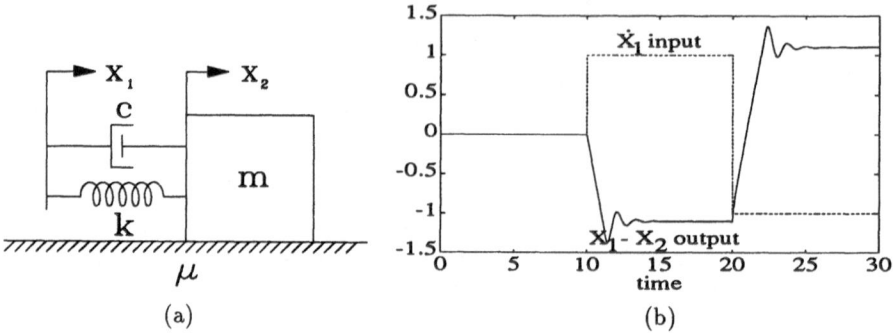

Figure 3: Second Order Model: (a) System; (b) Response.

system with a velocity input and coulomb friction. The equation for this model is shown below:

$$m\ddot{x}_2 + c\dot{x}_2 + k(x_2 - x_1) = f \tag{1}$$

where f represents the coulomb friction.

$$f = \begin{cases} \mu_{sl} N(\frac{\dot{x}_2}{|\dot{x}_2|}) & \text{if } |\dot{x}_2| > v_s \text{ and } F > \mu_{sl}N(\frac{\dot{x}_2}{|\dot{x}_2|}) \\ \mu_{st} N(\frac{\dot{x}_2}{|\dot{x}_2|}) & \text{if } |\dot{x}_2| < v_s \text{ and } F > \mu_{st}N(\frac{\dot{x}_2}{|\dot{x}_2|}) \\ F & \text{otherwise} \end{cases} \tag{2}$$

where $F = c\dot{x}_2 + k(x_2 - x_1)$, and N is the surface normal force. The value of v_s is the cutoff velocity that defines where, for simulation purposes, μ_{st} (static friction) no longer applies, and the value μ_{sl} (sliding friction) is used. Values for m, c, and k are selected to model the wrist behavior, but do not represent the exact physical parameters of the wrist.

The spring-damper subsystem models the wrist sensor. The output from the sensor will be the change in spring length, Δx, and can be interpreted explicitly as a position deflection, or implicitly, using Hooke's law $F = k\Delta x$, as a force. Figure 3(a) illustrates the second order model. Figure 3(b) displays data from a simulation of the second order model, with the velocity input shown. For a given mass and input velocity, the rise time and the output level are a function of the spring constant and the coulomb friction. Overshoot is a function of the coulomb friction value and the viscous damping term.

3.2 Experimental Data

Data from the system has been collected and compared to the simulation model. In this section, some of the data will be presented, along with an introduction to some of the problems that were encountered while using the system. One problem was that the box being moved had a tendency to tip over while being pushed. Also, there were many problems associated with sliding along a surface until a wall was encountered. False interpretation of sensor readings, due to uneven frictional force and noise caused by sensor

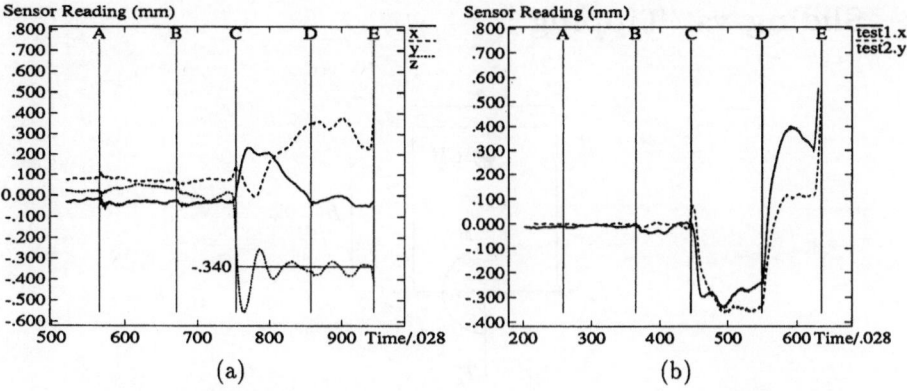

Figure 4: Sensor Readings: (a) Typical Move; (b) Direction Dependent.

electronics and by arm control, cause the system to stop before hitting a surface or to press on the surface with a large force before deciding to stop. Methods to overcome these problems are presented later in this paper.

Figure 4(a) shows the sensor readings, for translational directions, of a typical move. The robot moves at approximately 2 cm/sec. Section A-B is a free-space motion. There is a small amount of noise at the beginning of move A-B, which is caused by the transition from the previous free-space motion. Section B-C is a guarded move. At the end of move B-C, the robot comes into contact with the environment. Here, there is a large change in the z-direction sensor reading (note that the contact is smooth and stable, and the robot never breaks contact with the environment). Move C-D is a standard sliding motion, with the robot in contact with the environment and moving in the negative y-direction. The robot tries to maintain a normal force (z-direction) of approximately 4.2N (.34 mm) while sliding. There are large, unexpected changes in the x and y sensor readings during this section of the motion. Theoretically, there should be no forces in the x-direction, and the y-direction should have a constant frictional force of μN. The sensors, however, show that the tangential force (y-direction) has a minimum below zero, and a maximum of approximately 2.5N. Section D-E is another guarded move, and the robot comes into contact with a wall of the box. The slope of the y-direction sensor reading is high, but the actual value of the reading when the robot touches the wall is not significantly higher than other readings in the D-E section.

Figure 4(b) illustrates one of the inaccuracies of the sensor readings. Data "test1" and "test2" are from similar moves. Section B-C shows the robot coming into contact with the environment. In section C-D, the robot moves slightly away from a wall, and in section D-E the robot moves into contact with the wall. Motion is in the x-direction for the "test1" data, while test2 motion is in the y-direction. Although the normal force for both tests are nearly identical, section D-E in figure 4(b) shows very different tangential forces. The cause of this direction dependent phenomena is unclear, but a method to overcome the problem must be found in order to correctly monitor collisions and contacts.

The data presented above was collected after the tipping problem, presented below, was overcome.

4 Sliding vs. Tipping

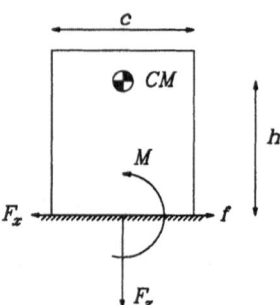

Figure 5: Forces on box in tool tip coordinates

When sliding in contact with the environment, the robot sometimes has the tendency to tip the box over. There are many factors that contribute to this tipping phenomenon. Three factors discussed here are the height to width ratio of the box, the normal to tangential force ratio, and the effect of rotational compliance upon sliding stability.

Expressing the applied forces in tool tip coordinates [16], which for this case will be the bottom of the box, the conditions for the box to tip over in the positive and negative Y-directions are found by summing the moments about the center of mass. The normal force N will act at the left side of the box if it is tipping about the negative Y-direction (into the page in figure 5), and the criteria for a 2-dimensional box not to tip is:

$$(F_x - f)h + M + N(\frac{c}{2}) \geq 0 \tag{3}$$

The normal force will act at the right side if it is tipping in the positive Y-direction. The criteria for the box not to tip in this direction is:

$$(F_x - f)h + M - N(\frac{c}{2}) \leq 0 \tag{4}$$

where $N = F_z + mg$, and $f = \mu N$ (See figure 5). Reorganizing,

$$hF_x - (\mu h - \frac{c}{2})F_z - mg(\mu h - \frac{c}{2}) \geq -M \tag{5}$$

$$hF_x - (\mu h + \frac{c}{2})F_z - mg(\mu h + \frac{c}{2}) \leq -M \tag{6}$$

If we assume that mg is negligible compared with applied forces and moments,

$$hF_x - (\mu h - \frac{c}{2})F_z \geq -M \tag{7}$$

$$hF_x - (\mu h + \frac{c}{2})F_z \leq -M \tag{8}$$

In terms of h/c, these equations become

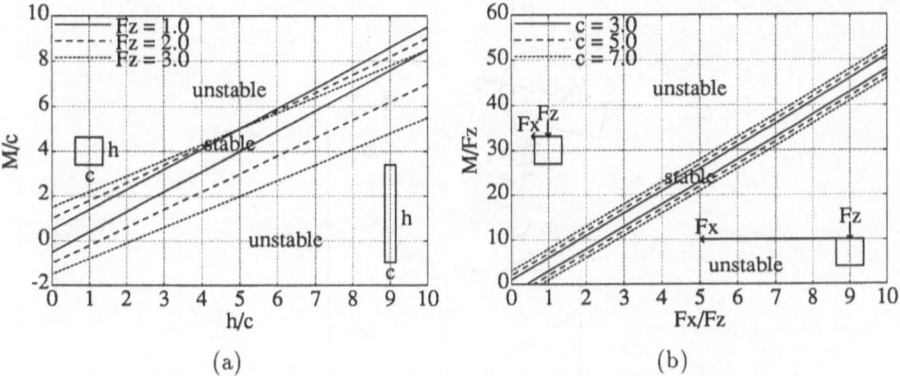

Figure 6: Tipping Criteria: (a) (-M/c) vs. h/c; (b) (-M/Fz) vs. Fx/Fz.

$$\frac{h}{c}(F_x - \mu F_z) - \frac{c}{2}F_z \geq -\frac{M}{c} \qquad (9)$$

$$\frac{h}{c}(F_x - \mu F_z) + \frac{c}{2}F_z \leq -\frac{M}{c} \qquad (10)$$

These equations are plotted in figure 6(a), with parameters: $F_x = 1.0, \mu = .1$, and F_z as shown. Note that for a given F_z, if h is large compared to c, then a moment must be applied for the box to remain stable. Also note that as F_z increases, the box will not tip for a greater range of applied moments. It is therefore more stable.

In terms of F_x/F_z, equations 9 and 10 become

$$h\frac{F_x}{F_z} - (\mu h - \frac{c}{2}) \geq -\frac{M}{F_z} \qquad (11)$$

$$h\frac{F_x}{F_z} - (\mu h + \frac{c}{2}) \leq -\frac{M}{F_z} \qquad (12)$$

$$(13)$$

These equations are plotted in figure 6(b) with parameters: $h = 5.0, \mu = .1$, and c as shown. For a given value of c, if F_x is large compared with F_z, a moment must be applied for the box to remain stable. As c increases, the range for the applied moment becomes greater, and the box becomes more stable.

The conditions above are intuitive and easy to compensate for. However, in our experimentation the box still tends to tip. The reason pertains to the rotational compliance, and with transforming the applied forces and moment into the tool tip frame.

To transform the forces and moment, the compliance values of the wrist are needed. There are two parts to the compliance that are important here: the physical compliance and the control compliance. The physical compliance is inherent in the structure of the wrist and its compliant elements. The control compliance is a result of the gains used in the control of the system. A stiff wrist can be made more compliant with higher gains, if

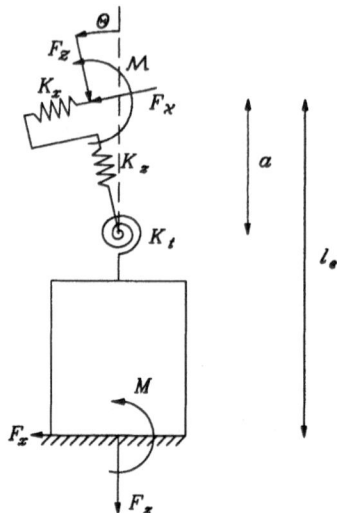

Figure 7: Transformation of forces from application to tool tip coordinates

it remains stable. The important point is that we can change the control compliance to suit our needs.

To transform the applied forces and moment for the two-dimensional wrist, the following equations are needed (see figure 7)

$$F_x = F_\chi \cos(\Delta\theta) - F_Z \sin(\Delta\theta) \tag{14}$$

$$F_z = F_\chi \sin(\Delta\theta) + F_Z \cos(\Delta\theta) \tag{15}$$

$$M = F_x(l_e - \Delta z) + F_z(\frac{c}{2} - \Delta x) + \mathcal{M} \tag{16}$$

$$\Delta x = F_\chi / K_x \tag{17}$$

$$\Delta z = F_Z / K_z \tag{18}$$

$$\Delta x = (\mathcal{M} + F_\chi a) / K_t \tag{19}$$

Substitution yields:

$$F_x = F_\chi \cos(\frac{\mathcal{M} + F_\chi a}{K_t}) - F_Z \sin(\frac{\mathcal{M} + F_\chi a}{K_t}) \tag{20}$$

$$F_z = F_\chi \sin(\frac{\mathcal{M} + F_\chi a}{K_t}) + F_Z \cos(\frac{\mathcal{M} + F_\chi a}{K_t}) \tag{21}$$

$$M = F_x(l_e - \frac{F_\chi}{K_x}) + F_z(\frac{c}{2} - \frac{F_\chi}{K_x}) + \mathcal{M} \tag{22}$$

$$\tag{23}$$

Equation 20 is plotted in figure 8. The constants in the equation are chosen to approximate the behavior of the wrist: $K_x = 7.29$ N/mm, $K_z = 12.36$ N/mm, $l_e = 25$ cm, $c = 10$ cm,

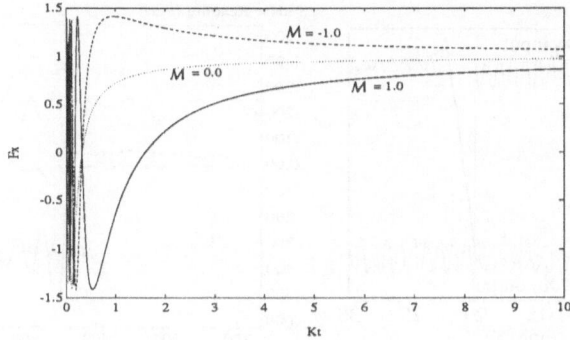

Figure 8: Fx vs. Kt for different values of M

$F_X = 1$ N, $F_Z = 1$ N, and $\mathcal{M}$, in N-m, as shown. Variable a was chosen to be 25 cm, which would correspond to the case where the center of compliance is at the tool tip (bottom of box, here), although the actual value is smaller. The physical value for K_t is 6.93 N-m.

The plot shows how the transformed forces and torque vary from those applied to the wrist. It is obvious that small values of K_t did not yield satisfactory performance. By decreasing the control compliance (increased K_t), the box became much more stable. Note here that figure 8 also shows that the tool tip forces are never the same as the applied forces. It is important to the stability of the box that the tool tip forces are controlled accurately, and that the compliance of the wrist must be compensated for in the control. By examining equations 20 and 21, it is seen that the smaller the distance from the applied forces to the center of compliance (a), the less effect that the force F_x has upon changing the values of the peg tip forces. Better results would be obtained for the operation of sliding if the center of compliance coincided with the applied forces. This is much different than the conclusions for peg insertion operations with RCC devices by Whitney [16], for which the center of compliance should be located at the tool tip.

5   Robust Stopping Conditions

The data presented in section 3.2 deviates from the predicted second order model behavior of the system. The deviations have many causes, and as a whole will be termed "noise".

Noise from the sensors is inherent in any system. Experiments suggest noise that may be dependent on complex phenomena that may be difficult or impossible to model. Such phenomena include non-homogeneous friction, static friction, sensor coupling (coupling of compliant directions in the sensor), orientation instabilities (tipping, as presented above), and sensor-based hysteresis. These phenomena are all responsible for sensor "noise".

As the manipulator slides around the environment, it attempts to maintain a constant normal force. With a constant normal force and velocity, the sliding friction should also be constant, assuming homogeneous surface friction. Contact with a side wall of the box thus could be determined by even a small increase in the tangential force. However, experiments have shown that a small threshold value causes the system to stop on noisy

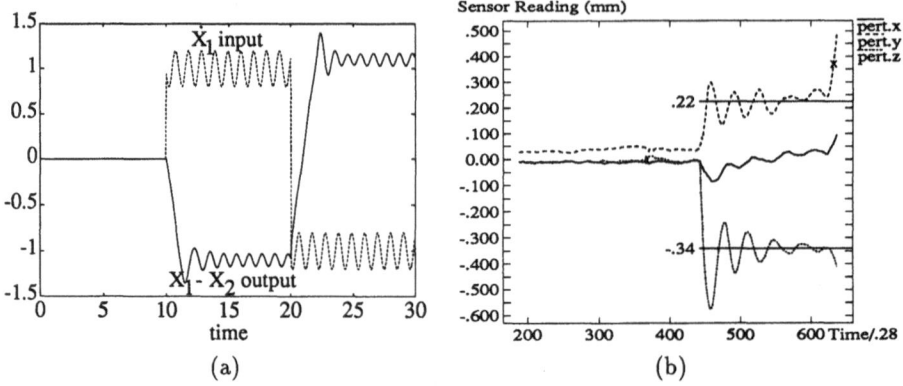

Figure 9: Motion Perturbation: (a) Model Behavior; (b) Typical System Behavior.

data. Using a constant threshold based stopping condition, a high threshold is needed to keep the noisy data from interfering with normal stopping criteria. Too high of a threshold may cause the system to interpret an actual contact with the wall as mere noise. Also, a high threshold causes the box to impact the environment with much more force than is wanted.

The following sections present attempts at developing more robust methods for determining stopping conditions, including ways to reduce the effects of the sensor noise, and to determine stopping criteria under noisy conditions.

5.1   Torque Preloads

Some of the control noise could be a result of the box being on the verge of tipping over. In order to reduce this noise, a torque preload could be used to make the box more stable. The preload direction is computed as $F \times v$. Using this preload unfortunately does not reduce the control noise, and does not significantly improve the performance of the system. However, it will make the box stable under more adverse conditions, at little computational cost.

5.2   Motion Perturbation

By perturbing the motion of the manipulator with small amplitude sine waves, some of the effects of noise phenomena can be actively reduced. Specifically, static friction problems can be overcome.

Figure 9(a) shows the simulated output of the second order model with a velocity input as in figure 3(b), with a superimposed sine wave with an amplitude of 1/5 the constant velocity input. The output is quite similar to that of figure 3(b), superimposed with a very small amplitude sine wave. The sine wave perturbation causes no adverse effects to the output as long as the frequency is not near the natural frequency of the system.

Experimental results from motion perturbation are shown in Figures 9(b) and 10(a). Figure 9(b) can be compared with figure 4(a) to show the improvement of the sensor

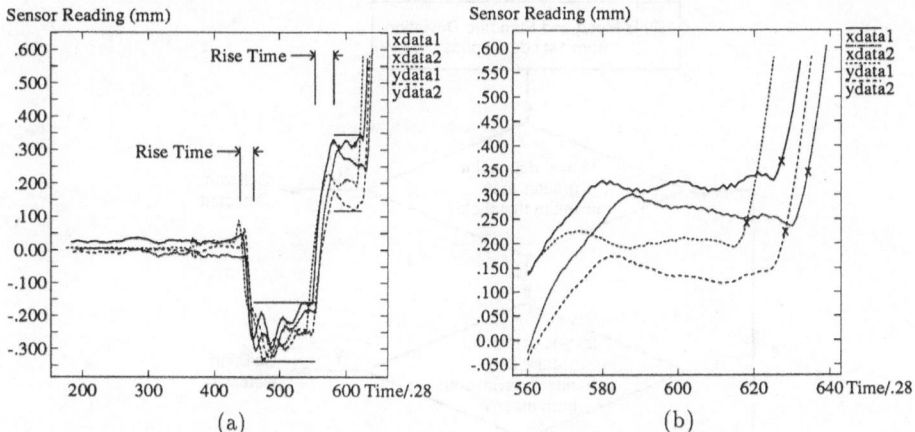

Figure 10: Motion Perturbation: (a) Experimental Data; (b) Magnified View of Collision Data.

output with motion perturbation. The z-direction output is similar for both moves, but in the perturbed motion move, the x-direction (normal to motion) output remains close to zero. Further, the y-direction (motion direction) output has the characteristics of a 2nd order system. As motion in the y-direction begins, the sensor output rises to a peak value, and then oscillates about a steady state value. Contact with the wall is indicated with a distinct rise in the sensor output. Figure 10(a) compares multiple moves. The data spread for the steady state value of the output is much smaller than similar moves without perturbation.

With data as shown in figure 10(a), the use of a constant threshold value for determination of contact can be revised. If a move that is known to terminate in contact is long enough to create a model, contact with the wall can be determined by a data point that falls outside n standard deviations computed from data collected after the rise time of the move. A simple collision detection algorithm is shown in figure 11. The "X" in figure 9(b) indicates where a collision would have been detected using this algorithm, with $n = 3.5$. Figure 10(b) shows a closeup of the data from figure 10(a) where the wall is encountered. The "X" marks indicate where the wall would have been detected.

Two refinements to this algorithm can be made. The first is to retain an absolute maximum constant threshold, so that if the data readings are very noisy, or if an obstacle is encountered before an adequate model can be built, the robot can still stop on a given force. This would eliminate the possibility of damage to the robot and the environment. Second, the algorithm as it stands uses all of the data points after the rise time to compute the mean and standard deviation. Computing the mean and standard deviation from only the previous N data points would provide a local standard deviation that would be more useful for non-homogeneous friction.

Motion perturbation, while improving the performance of the system by reducing the effects of static friction, still does not overcome the "noise" associated with non-homogeneous friction. It does, however, produce sensor output that conforms well to a

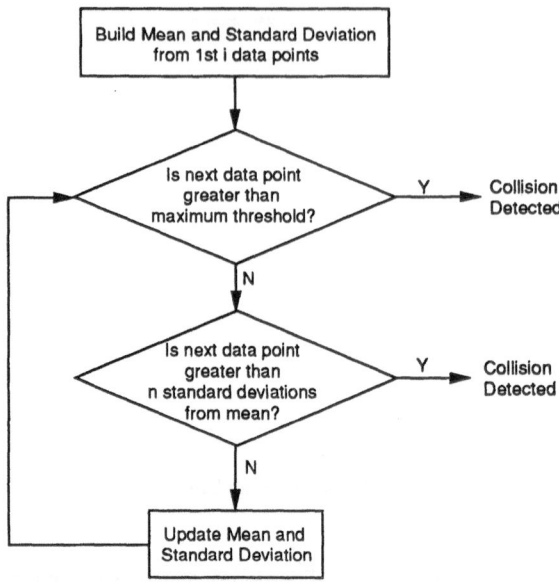

Figure 11: Collision Detection Algorithm

system model. The collision detection algorithm is also beneficial when the environment contains other surfaces with different coefficients of friction.

5.3   Exploratory Procedures

In some instances, surface conditions may impede command execution to the point that contact operations are impossible under the current model of the environment. An example of this would be trying to detect collision with a foam rubber wall while sliding across a very rough surface. In cases like this, it may become necessary for the remote site to autonomously explore surface conditions while the human operator waits.

A more refined model of the environment obviously leads to more accurate analysis of sensor data. The operator works in a model world dealing with kinematics only. As the slave manipulator operates in the real world, data about the environment can be gathered, analyzed, and used to refine new incoming data. Many surface attributes can be recovered through normal operation of the manipulator, including penetrability, hardness, compliance, compressibility, deformability, and surface roughness [14]. These criteria may be enough to refine the environment model to the point where contact operations can again be accomplished using the same types of commands from the master site as before. However, there may be surfaces where the current paradigm of contact operations cannot be used. At this point, the human operator must adapt the motion strategies to reflect the surface attributes. Instead of sliding along a surface to find a wall, for example, the operator may have to move above the surface, poking the surface occasionally to make sure that "contact" has not been lost, until the wall is encountered.

6 Conclusions

Although the criterion for contact operations, including collision and error detection, appear to be simple, it is shown that using real world sensors and control, a much more robust set of rules must be used. By utilizing robust criterion for error detection, limited execution model commands can be successfully carried out, and actual error states can be differentiated from spurious data.

References

[1] R. L. Andersson. Computer architectures for robot control: a comparison and a new processor delivering 20 real mflops. In *Proc. of the IEEE Int. Conf. on Robotics and Automation*, pages 1162–1167, 1989.

[2] Forrest T. Buzan. *Control of Telemanipulators with Time Delay: A Predictive Operator Aid with Force Feedback*. PhD thesis, Massachusetts Institute of Technology, 1989.

[3] C.B.Phillips and N.I.Badler. Jack: a toolkit for manipulating articulated figures. In *Proc. of ACM/SIGGRAPH Symposium on User Interface Software*, Banff, Alberta, Canada, 1988.

[4] Peter I. Corke. *A New Approach to Laboratory Motor Control: The Modular Motor Control System*. Tech. Report, Univ. of Pennsylvania, Philadelphia, PA, 1989.

[5] Janez Funda. *Teleprogramming: Towards Delay-Invariant Remote Manipulation*. PhD thesis, Univ. of Pennsylvania, 1991.

[6] B. Hannaford and W.S. Kim. Force reflection, shared control, and time delay in telemanipulation. In *Proc. of the IEEE Int. Conf. on Robotics and Automation*, pages 133–137, 1989.

[7] Vincent Hayward. *RCCL User's Manual*. Tech. Report TR-EE-83-46, Purdue Univ., 1983.

[8] B. Hirzinger, J. Heindl, and K. Landzettel. Predictive and knowledge-based telerobotic control concepts. In *Proc. of the IEEE Int. Conf. on Robotics and Automation*, pages 1768–1777, 1989.

[9] T. Lindsay and R.P. Paul. *Design of a Tool-Surrounding Compliand Instrumented Wrist*. Tech. Report MS-CIS-91-30, GRASP LAB 258, Univ. of Pennsylvania, 1991.

[10] R.P. Paul, J. Funda, T. Simeon, and T. Lindsay. Teleprogramming for autonomous underwater manipulation systems. In *Intervention '90*, pages 91–95, The Marine Technology Society, June 1990.

[11] R.K. Roberts, R.P. Paul, and B.M. Hillberry. The effect of wrist force sensor stiffness on the control of robot manipulators. In *Proc. of the IEEE Int. Conf. on Robotics and Automation*, pages 269–274, April 1985.

[12] P. S. Schenker and A. K. Bejczy. Workspace visualization and time delay in telerobotic operations. In *13th Annual AAS Guidance and Control Conf., Aerospace Human Factors Session*, February 1990. Keystone, CO.

[13] Thomas B. Sheridan. Telerobotics and human supervisory control. To be published as a book.

[14] P.R. Sinha, Y. Xu, R. Bajcsy, and R.P. Paul. *Robotic Exploration of Surfaces With a Compliant Wrist Sensor*. Tech. Report MS-CIS-90-92, GRASP LAB 244, Univ. of Pennsylvania, Philadelphia, PA, 1990.

[15] R. Volpe and P. Khosla. Experimental verification of a strategy for impact control. In *Proc. of the IEEE Int. Conf. on Robotics and Automation*, pages 1854–1860, 1991.

[16] Daniel E. Whitney. Quasi-static assembly for compliantly supported rigid parts. In M. Brady, J.M.Hollerback, T.L.Johnson, T.Lozano-Perez, and M.T.Mason, editors, *Robot Motion: Planning and Control*, pages 429–462, MIT Press, 1982.

[17] Yangsheng Xu. *Compliant wrist design and hybrid position/force control of robot manipulators*. PhD thesis, Univ. of Pennsylvania, 1989.

Controlling Contact by Integrating Proximity and Force Sensing

B. Allotta, G. Buttazzo, P. Dario, E. Guglielmelli

ARTS Lab*

Scuola Superiore S. Anna, Pisa, Italy

June, 1991

Abstract

This paper describes a sensor–based robot control method for handling *non contact to contact* transitions in a partially unknown environment. The proposed approach is based on the integration of proximity and force/torque sensing capabilities. Proximity information is used not only to avoid undesired contacts during robot operation but also to improve efficiency in terms of speed and safety, when the contact is desired. By detecting in advance the target surface to be reached, it is possible to plan the trajectory in order to optimize the approach velocity profile, as well as to maintain the impact forces at the desired level by means of force control.

1   Introduction

In many situations of practical interest both in industrial and advanced robotics, it is important for a robot to control the transition between free and constrained motion. Although recognized as critical, this problem has not been addressed in detail so far: in fact, whereas the problems of obstacle avoidance and of contact control have been widely investigated separately [1] [3] [4], criteria for optimal transition between free and constrained motion are available in the literature, but only from the point of view of force controllers.

In this paper we propose a simple approach to this problem, based on the use of proximity information for free motion trajectory planning and control, in order to limit the impact velocity without affecting the performance in terms of execution time.

It is worth to notice that, in order to accomplish safety, stability and execution time requirements, by using traditional pure force control methods [5], the maximum approaching velocity of the robot to the object, immediately before the contact, must be bounded so that the rise time of the contact force is compatible with the response time of the robot control system. This fact limits severely the performance of robot systems working in a partially unknown environment, where object locations are known with some

---

*Advanced Robotics Technology and Systems Laboratory

degree of uncertainty. Limiting the velocity of the robot when an impact is expected is extremely important when the system bandwidth is not wide enough, as in the case of typical *inner–outer loop architectures*, where no joint torque information is available and the exteroceptive force control loop provides reference inputs to a standard PD position controller at a low sampling rate [2].

Although the same approach for handling contacts could be pursued by means of stereo vision [6] or very complex multisensory systems, we have elected to investigate simple ultrasound (US) and infrared (IR) active sensing techniques, which are much faster and cheaper than vision. For istance, US time of flight sensors directly provide range information without any further processing. Our aim is to show how an integrated proximity–force control strategy performs better than traditional force control methods, in tasks involving desired impacts with the object to be reached.

Integration of US and IR proximity information allows to compensate some of the drawbacks of these two sensing techniques. In fact, because of specular reflection, airborne US sensors perform poorly when the target is tilted relative to the axis of the acoustic source; distance evaluation is influenced by air conditions, such as temperature/pressure, relative humidity and local turbulences. On the other hand, IR range finding methods are more complex than simply evaluating the time of flight, as they are based on techniques like triangulation, interferometry or reflectance; moreover, the material of the target, its colour and the background lighting widely affect IR absorption coefficient. For the mentioned reasons, we have deviced to develop a compound US/IR sensor module.

The redundant proximity information obtained from the compound sensor mainly guarantees great safety in detecting various kinds of surfaces indipendently of their colour and orientation.

The first two paragraphs of this paper describe the theory of trajectory planning and control for non–contact/contact transitions. Then the experimental set–up used for the implementation of the proposed theoretical approach is described, along with the control architecture. Different control strategies, namely pure force control and the integrated proximity/force control method, are compared in various situations. Finally, the good agreement between experimental results and indications from theory is discussed, and indications on future work are given.

2 Trajectory planning

We have focused our attention on a one dimensional problem. The robot end–effector, equipped with force and proximity sensors is placed above a table. The task can be described as follows:

"from a rest initial condition, detect the table and reach it in the shortest time compatible with the conditions $\|v(t)\| < VMAX$, $\|a(t)\| < AMAX$, and establishing a contact with a desired force of X Newton, maintaining the force overshoot at 10% of X".

In our first experiments we assumed $VMAX = 100\ [mm/s]$ and $AMAX = 150\ [mm/s^2]$, then we performed more experiments with $VMAX = 600\ [mm/s]$ and $AMAX = 800\ [mm/s^2]$, which are close to the real limits of the used hardware.

We have choosen quintic polynomials in the cartesian space as reference trajectories for the robot end–effector during free motion (i.e. before the contact). This is because quintic polynomials are the minimum–degree ones which allow the fulfilment of initial and final conditions on position, velocity and acceleration [13]. In this way, it is possible to obtain continuity of the acceleration profile also at the boundaries of the trajectory, and this results in a limited jerk [6]. A quintic polynomial can be written as:

$$\begin{aligned}
p(t) &= a_5 t^5 + a_4 t^4 + a_3 t^3 + a_2 t^2 + a_1 t + a_0 \\
v(t) &= 5a_5 t^4 + 4a_4 t^3 + 3a_3 t^2 + 2a_2 t + a_1 \\
a(t) &= 20a_5 t^3 + 12a_4 t^2 + 6a_3 t + 2a_2 \\
t_i &< t < t_f
\end{aligned} \tag{1}$$

where t_i and t_f are the initial and final time. In the following, without loosing generality, we will assume $t_i = 0$. By solving the linear system:

$$\begin{aligned}
p(0) &= a_0 \\
v(0) &= a_1 \\
a(0) &= 2a_2 \\
p(t_f) &= a_5 t_f^5 + a_4 t_f^4 + a_3 t_f^3 + a_2 t_f^2 + a_1 t_f + a_0 \\
v(t_f) &= 5a_5 t_f^4 + 4a_4 t_f^3 + 3a_3 t_f^2 + 2a_2 t_f + a_1 \\
a(t_f) &= 20a_5 t_f^3 + 12a_4 t_f^2 + 6a_3 t_f + 2a_2
\end{aligned} \tag{2}$$

it is possible to obtain the symbolic expressions of the trajectory as a function of t, and t_f and the boundary conditions on $p(t)$, $v(t)$, and $a(t)$. It is also possible to calculate the expressions of the maximum and minimum values of $v(t)$ and $a(t)$ as a function of t_f. For the moment, we have concentrated on the study of trajectories with zero initial and final velocity and acceleration. For the case of initial zero velocity and acceleration we obtain the following expressions:

$$\begin{aligned}
\|v_{max}(t_f)\| &= 15\frac{\|p(t_f)-p(0)\|}{t_f} \\
\|a_{max}(t_f)\| &= 5.77\frac{\|p(t_f)-p(0)\|}{t_f^2}
\end{aligned} \tag{3}$$

One method to determine the coefficients of the quintic is to solve separately the two following equations:

$$\begin{aligned}
\|v_{max}(t_f)\| &= VMAX \\
\|a_{max}(t_f)\| &= AMAX
\end{aligned} \tag{4}$$

thus obtaining two different values for t_f:

$$\begin{aligned}
t_{f_v} &= t_f(VMAX) \\
t_{f_a} &= t_f(AMAX)
\end{aligned} \tag{5}$$

and choosing the biggest between t_{f_v} and t_{f_a}:

$$t_f = max\{t_{f_v}, t_{f_a}\} \tag{6}$$

By using equation 6 in order to calculate t_f, for "short" paths we obtain a quintic trajectory which maximizes the exploitation of the available acceleration resources, without reaching the maximum velocity. For "long" paths we maximize the exploitation of the velocity resources, although the maximum velocity is reached only at the center of the trajectory.

Instead of using equation 6 in any case, we calculate the "critical" length d_{cr} of the path that allows the exploitation of both acceleration and velocity resources. d_{cr} is a function of $VMAX$ and $AMAX$.

- If the initial distance d is smaller than d_{cr}, we choose $t_f = t_{f_a}$;

- if $d = d_{cr}$, $t_f = t_{f_a} = t_{f_v}$ and we obtain a trajectory $p_{cr}(t)$ that maximizes the exploitation of both acceleration and velocity resources;

- if $d > d_{cr}$ we use $p_{cr}(t)$ until $p(t) = d - \frac{d_{cr}}{2}$, then we follow a trajectory at constant velocity $(v(t) = VMAX)$ until $p(t) = \frac{d_{cr}}{2}$, and finally, for $\frac{d_{cr}}{2} < p(t) < 0$ we follow $p_{cr}(t_*)$ again, in order to decelerate with the maximum deceleration allowed. t_* is equal to t delayed of the time necessary to reach $\frac{d_{cr}}{2}$.

The resulting velocity and acceleration profiles in the three cases are shown in fig. 1:

3 Control strategy

3.1 Free motion control

As described in section 4.4, a simple model of the robot dynamic behaviour has been derived in the neighbourhood of the configuration used for the experiments. In order to make the robot to follow the desired trajectory during free motion, we have divided the control signal u into two parts, namely u_1 and u_2. u_1 depends on the desired trajectory and on the robot model, whereas u_2 is error–driven:

$$u(k) = u_1(k) + u_2(k) \tag{7}$$

The model–based part (u_1) of the control signal is calculated as follows. By manipulating the discrete–time open loop transfer function $H(z)$, we obtain:

$$H(z) = \frac{X(z)}{U(z)} = \frac{1-p}{z(z-1)(z-p)} = \frac{(1-p)z^{-3}}{1-(1+p)z^{-1}+pz^{-2}} \tag{8}$$

where $X(z)$ is the distance from the target, and $U_1(z)$ is the input velocity reference command to the position controller.

Therefore:

$$U(z) = X(z)\frac{z^3 - (1+p)z^2 + pz}{1-p} \tag{9}$$

By backtransforming equation 9, we finally obtain:

$$u(k) = \frac{x(k+3) - (1+p)\,x(k+2) + p\,x(k+1)}{1-p} \tag{10}$$

Since the robot trajectory is already planned, as described in paragraph 2, depending on the current distance from the table, we use equation 10 to feed–forward the trajectory to be executed.

$$u_1(k) = \frac{x_d(k+3) - (1+p)\,x_d(k+2) + p\,x_d(k+1)}{1-p} \tag{11}$$

By using equation 11, we partially eliminate the effect of the finite delay present in the external loop.

424

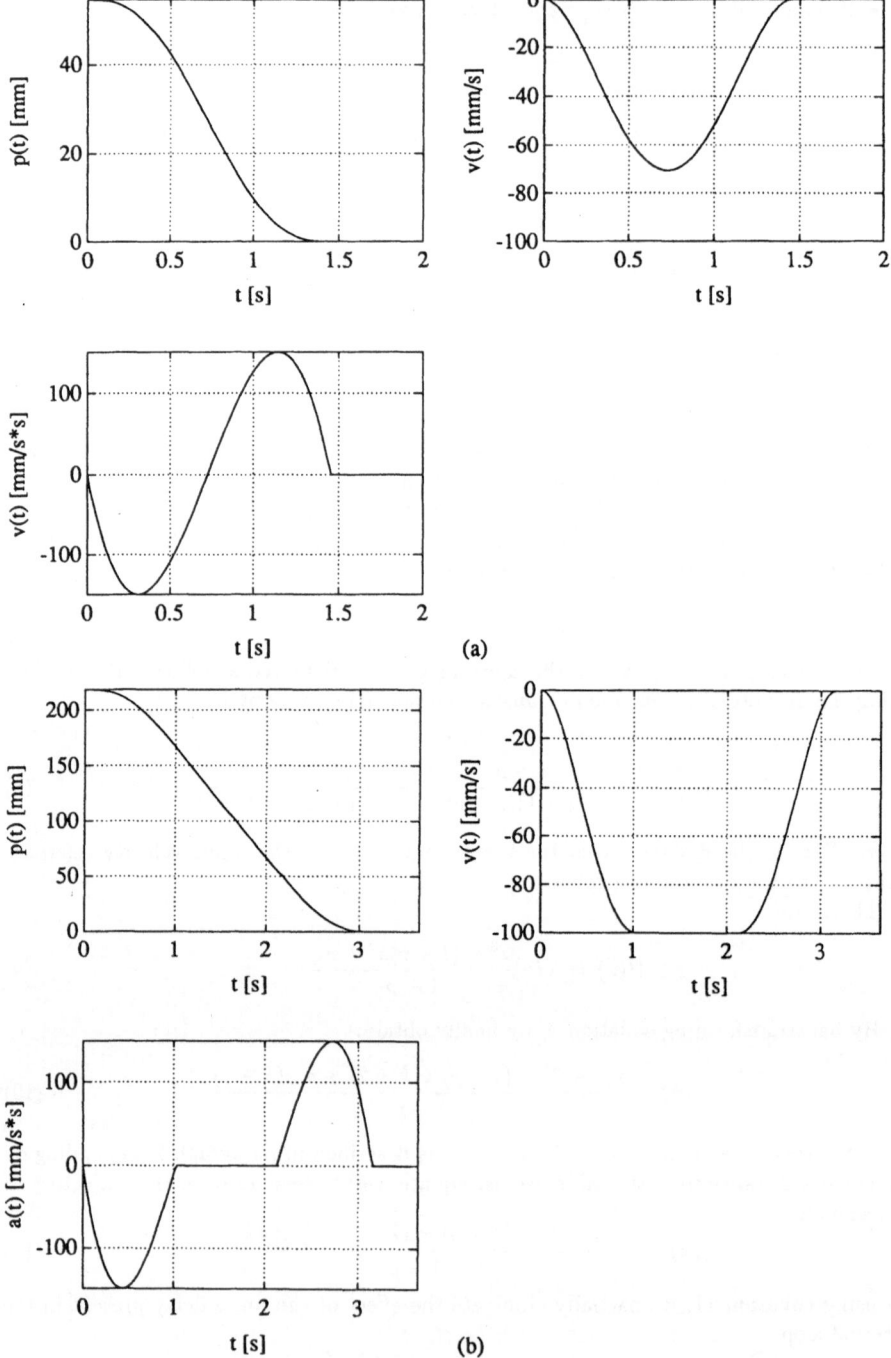

Figure 1: Trajectory calculated for (a) $d < d_{cr}$ and (b) $d > d_{cr}$

The error–driven part (u_2) of the control signal is calculated by using the following equation:

$$u_2(k) = k_p(x_d(k) - \hat{x}(k)) + k_v(v_d(k) - \hat{v}(k)) \qquad (12)$$

where $\hat{x}(k)$ and $\hat{v}(k)$ are the current extimates of $x(k)$ and $v(k)$, based on the proximity sensory information. The tuning of k_p and k_v has been first made by means of simulations, and then the stability of the real system has been verified experimentally.

3.2 Force control

To manage the contact interactions between robot and environment, an exteroceptive force controller has been designed, based on a typical inner–outer loop architecture. A passive compliance is placed between the wrist F/T sensor and the robot end–point (namely the point that "touches" the table during the contact). The importance of having such a compliance is discussed in [1] [2]. The resulting stiffness of the other elements of the closed kinematic chain (composed by the robot and its base, the sensor, the table, and the inner joint position loops) can be assumed to be at least one order of magnitude bigger than the stiffness of the passive compliance.

3.3 Non–contact/contact transition

In order to obtain a smooth transition from the free motion trajectory tracking to the contact, dominated by the force controller, we devised to use a function $\lambda(x)$ of the measured distance as a weight for mixing the two control laws.

The control signal used for the integrated proximity/force control strategy is then given by:

$$u(k) = (1 - \lambda(x))u_p(k) + \lambda(x)u_f(k) \qquad (13)$$

where u_p is the component of the control signal relative to the free motion trajectory tracking, and u_f is the component generated by the force controller.

$\lambda(x)$ must accomplish the following specifications:

- $0 < \lambda(x) < 1$;

- $\lambda(0) = 1$;

- $\lambda(\infty) = 0$.

- $\frac{d\lambda(x)}{dx} < 0$ for $x > 0$.

Many different options are left to the control designer in order to choose $\lambda(x)$. Basically, the parameter that has to be taken into account when choosing $\lambda(x)$ is the standard deviation of the error of the proximity sensor in the neighbouring of the contact, in order to ensure that the robot does not touch the target before the velocity is low enough. The function $\lambda(x)$ we chose is the following:

$$\lambda(x) = \frac{1}{1 + 1.5x^4} \qquad (14)$$

where x is expressed in $[mm]$. The graph of $\lambda(x)$ is shown in figure 2.

lambda

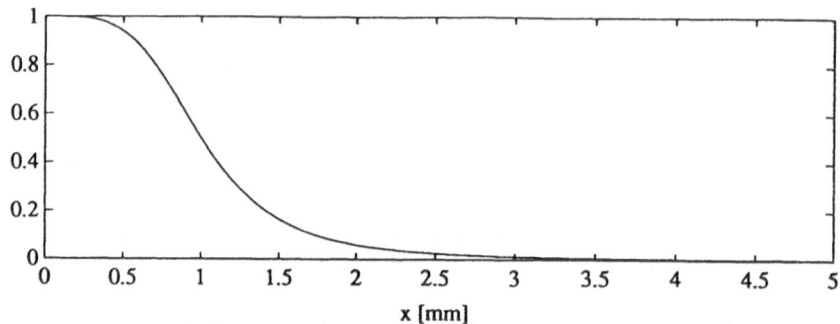

Figure 2: The function $\lambda(x)$

4 System description

4.1 The proximity sensor

In spite of recent developments of vision–based sensory systems, ultrasound and infrared proximity sensors are still widely used in robotics, mainly because they are cheap, fast and reliable. However, this kind of sensors are affected by many typical drawbacks which severely limit their sensing capabilities in presence of unfavourable conditions. By comparing the performances of these two sensing techniques in various working conditions the idea of extracting a unique proximity information from a compound IR/US module was considered.

The proximity sensor we have designed, fabricated and used for our experiments has been devised to be compact and cheap, and its reliability is mainly based on the integration of US and IR sensing. One commercial PZT US transmitter, surrounded by three PVDF semicylindrical receivers, located at 120° one from each other, is coupled with a pair of IR emitter/receiver diodes. By using US PVDF receivers, together with low noise input stage amplifiers, a large sensitivity is achieved. The US transducers have a resonance frequency of 80 Khz while bursts of radiation modulated at 40 Khz drive the IR emitters; as a result a very low cross–talking between the two sections (US/IR) is obtained.

4.2 The sensory data acquisition system

In order to interface the sensor module with the host computer (a PC 386), which is dedicated to the control of the manipulator, a hardware board has been devised and developed. This board must support the electronics to:

- generate the sequence of bursts of known frequencies which are needed to drive the transmitters;

- convert the received echoes in an acceptable form for the input parallel port of the computer;

- guarantee high rejection of any possible noise signal.

US section – The US transmitted signal is directly supported by a pulse generator (PANAMETRICS).

This section has been devised by using the SING AROUND technique to guarantee a better performance while approaching an obstacle. In fact a new pulse is emitted as soon as an echo is received; as a result the frequency of acquisition increases as the distance of the target becomes smaller and smaller.

The three PVDF US receivers drive low noise input amplifiers. The reflected echoes are compared with an exponentially decreasing signal which is synchronous with the TX emissions, and by using 12-bit binary ripple counters, an evaluation of the time of flight, i. e. of the distance of the target, is obtained. The digital output signals from the counters are sent to the parallel input port (8255) of the PC.

IR section – The IR transmitted signal is a sequence of bursts at 40 Khz with a constant duty cicle. The LED is driven by a power final stage which supports the needed pulsed current (peak value = 2A). The received echoes are primarily filtered and then coherently detected. In order to reject external noise, the clock which drives the sampling and hold stage (synchronous detector) is directly obtained from the transmitters stages. The resulting low frequency output signal is acquired through an A/D converter from the host computer.

4.3 Control architecture

The robot arm used in these experiments is a 6 d.o.f. PUMA 562, equipped with its MARK III controller and VAL II operating system. The F/T sensor is a strain gauge–based wrist sensor, provided by a microprocessor unit, whose function is to transform the row data acquired from the sensor in 6 cartesian force/torque components. Data transmission is performed through a 16-bit parallel port at the maximum rate of 100 Hz.

The host computer is an IBM compatible personal computer (PC), with Intel 80386 microprocessor, 80387 math coprocessor, and MS–DOS operating system. As system supervisor, the PC reads the force signals from the F/T sensor and distance signals from the composite proximity sensor, realizes the exteroceptive control loop, handles the real-time communication with the MARK III robot controller, and manages the user interface. Software flexibility is provided by using the *Harems* programming environment [8].

The host computer communicates to the robot controller through a standard RS232 serial line, at 19200 baud rate, by using the *Alter* protocol, provided by the VAL II system for real–time path control. This feature allows the host to modify the robot trajectory in real–time, by sending position commands every 28 [*ms*]. Position increments can be given to the robot in cartesian space either in absolute or cumulative fashion.

The exteroceptive control loop is realized around the internal position control loop, according to a two level hierarchical architecture, often identified as "inner–outer loop architecture" [1]. The outer control loop handles data acquisition and generates position commands to the inner loop, according to a sensor–based control low, whereas the inner loop guarantees the correct execution of the robot trajectory.

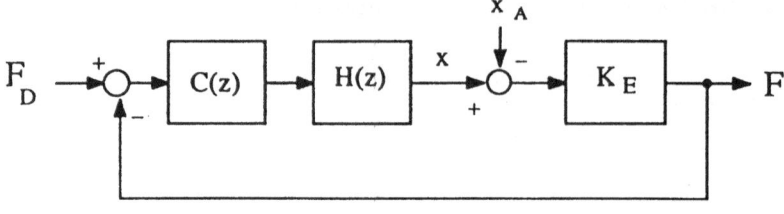

Figure 3: System block diagram

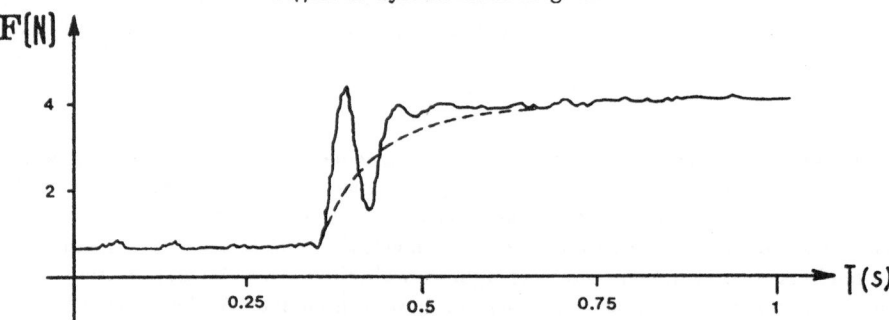

Figure 4: Open loop force step response

4.4 Robot model identification

The model we have developed has been derived by using classical techniques of control theory such as step response and root locus. The good reliability of the model has been extensively prooved by the consistency observed between the theoretical simulations and the experimental results. Thanks to our model, we have experimented different force control strategies, such as explicit force control, hybrid control, and impedance control, that we have tested on several applications of practical interest, such as autonomous assembling, surface tracking, fruit harvesting, and tiling.

In order to perform the identification of a simple model for the robot arm, we used a one–dimensional strain gauge force sensor, consisting of a flexible steel cantilever beam. A second–order model of this sensor has been experimentally derived by measuring its pulse response. Once the sensor's transfer function has been evaluated, we used the sensor as a probe for studying the dynamic behavior of the robot arm. Before performing experiments on the real set–up, a simple "black box" model of the position loop of the robot has been identified by means of digital simulation, in order to estimate the stability of the system under exteroceptive force control.

The block diagram of the system is shown in fig. 3. In order to perform the robot model identification, open loop force steps were imposed to the robot, already in contact with the environment by means of the elastic probe, in various arm configurations. A typical open loop force step response is shown in fig. 4.

The transient response contains a mode due to unmodeled dynamics at a frequency of about 13 $[Hz]$ (the intensity of this oscillation is depending on the configuration of the arm) and a "slow" pole at 2.2 $[Hz]$. Since the low frequency pole is about six times smaller than 13 $[Hz]$, and as the closed loop bandwidth is only few Hz (as it can be seen below), we used a "dominating pole" approximation with a pole at 2.2 $[Hz]$ to describe the behaviour of the arm. The continuous time transfer function, taking into account also the finite delay existing in the loop (about 60 $[ms]$), is:

$$H(s) = X(s)/V_d(s) = \frac{a}{s(s+a)}e^{-2Ts} \tag{15}$$

The corresponding discrete–time transfer function results:

$$H(z) = \frac{1-p}{z(z-1)(z-p)} \tag{16}$$

where:
$a = 2\pi 2.2\ [Hz] = 14\ [rad/s]$ is the dominating pole;
$2T = 0.056\ [s]$ is the finite delay of the direct chain.

5 Experimental results

In our first experiments we assumed $VMAX = 100[mm/s]$ and $AMAX = 150[mm/s^2]$. Then we assumed $VMAX = 600[mm/s]$ and $AMAX = 800[mm/s^2]$, which are close to the ratings of the used hardware.

All experiments were carried out by using an end-effector mounted on the F/T wrist sensor, specifically designed for grasping and placing ceramic tiles on a planar surface. The tool has a stiffness (measured along the vertical axis) of about $10N/mm$. This allows us to consider it as the sole source of compliance in the system.

The algorithm we used for force control is simply:

$$u(k) = K_f(F_d - Fz(k)) \tag{17}$$

where $u(k)$ is the reference velocity input sent to the internal position controller, K_f is a proportional gain, F_d is the desired contact force, and F_z is the force measured along the Z axis of the robot tool frame.

We first tuned the gain of the force control loop in order to obtain a stable behavior when the tool is in contact with the table and to maintain the force overshoot during the transition within the specifications. We performed the same task both on soft and rigid surfaces, experimenting different control strategies.

Soft contacts

The first experiments are relative to a "soft" environment, namely a compliant box which allowed us to use a gain 10 times bigger than in the case of the table. It is worth to notice that by using the algorithm described in 17, if F_d is maintained constant during the task, the approaching velocity is proportional to F_d itself. As a consequence, under pure force control, the approaching speed results too low when the desired force is small, and it is too high when the desired force is big. In our case, we used $F_d = 5N$ and $K_f = 0.1$, therefore the approaching velocity resulted of $18mm/s$, but by considering the

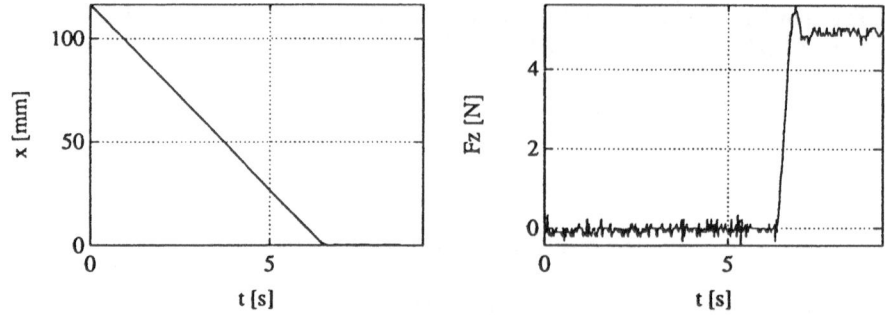

Figure 5: Pure force control for a "soft" contact: $F_d = 5N$, $d_0 = 120mm$.

time necessary to reach the desired force, we have an average speed of $v = 16.5mm/s$, as shown in figure 5.

By using the integrated proximity/force sensor–based approach, in the same "soft" environment, we obtain an average approaching velocity of $v = 80mm/s$ (see fig. 6).

Hard contacts

We repeated the task with a "hard" environment, by establishing the contact with the table of the robot workstation. In this case, a comparison with the case of pure force control has not been made, because the allowable gain is too small to have an acceptable approach velocity. In fact, the gain of the force controller that we used in the "hard" environment was $K_f = 0.01$. This means that the resulting approach velocity is about $1.8mm/s$, i.e. 10 times smaller than that in the "soft" case.

The results of such experiments, carried out by using the integrated proximity/force sensor–based approach, are shown in fig. 7, from which we can see that the average approach velocity, calculated as in the soft case, is $v = 80mm/s$.

One important result is that the more the environment is stiff, the more the proposed technique is faster than a pure force control. This is because the gains of a pure force controller, indipendently from the hardware and the implementation details, must be reduced in order to accomplish the task specifications.

6 Discussion and conclusion

The proposed approach has demonstrated the possibility of improving the performance of a robotic system in tasks involving force control procedures by means of proximity sensing. The experiments that have been performed are relative to a very simple case, but the results seem to be ready for applications where there is uncertainty on the height of the working surface.

It is certainly not so easy to extend the proposed technique to the case of surfaces that are not flat, but the use of IR could help in this sense.

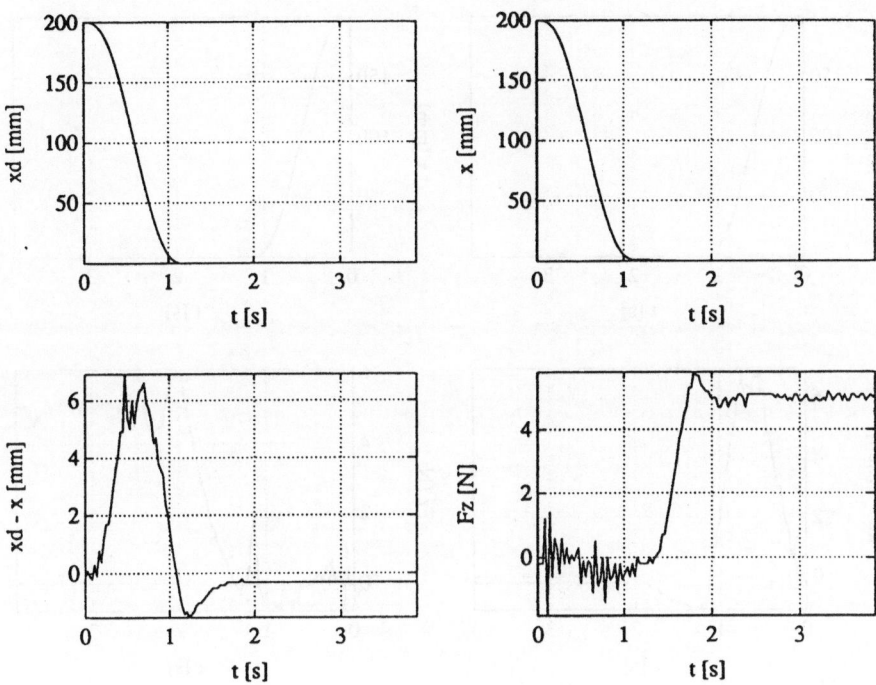

Figure 6: Integrated proximity/force control for a "soft" contact: $F_d = 5N$, $d_0 = 200mm$.

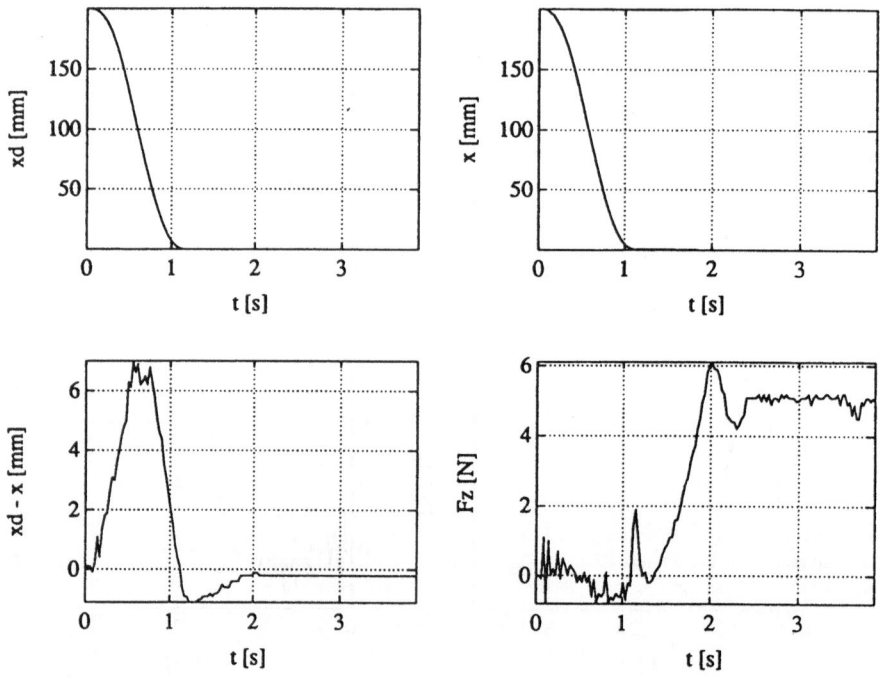

Figure 7: Integrated proximity/force control for a hard contact: $F_d = 5N$, $d_0 = 200mm$.

Further study is currently being addressed to extend the proposed technique to the case of impacts in which uncertainties are also present on the orientation of the surface. In fact our US sensor, having three receivers, is also capable of giving an estimate of the orientation of the reflecting target.

As far as the velocity of the target is concerned, more work is in progress, in order to be able to deal with moving obstacles.

References

[1] De Schutter, J. and Van Brussel, H. – *Compliant Robot Motion* – Int. J. Rob. Res.,Vol. 7, No. 4, pp.3–33, 1988.

[2] De Schutter, J. – *Improved Force Control Laws for Advanced Tracking Applications* – Proc. IEEE Conf. Rob. Automation, pp.1497–1502, 1988.

[3] V. Lumelsky, E. Cheung – *Motion Planning for Robot Arm Manipulators with Proximity Sensing* – IEEE, 1988, pp. 740–745

[4] O. Khatib – *Real Time Obstacle Avoidance fo Manipulators and Mobile Robots* – The International Journal of Robotics, Vol. 5, No. 1, pp. 90–98, Spring 1986.

[5] O. Khatib, J. Burdick – *Motion and Force Control of Robot Manipulators* – Proc. of IEEE Conf. on Robotics and Automation, pp. 1381–1386, San Francisco, CA, April 1986.

[6] R. L. Andersson – *Aggressive Trajectory Generator for a Robot Ping–Pong Player* – IEEE Control System Magazine, pp. 15–20, February 1989.

[7] R. L. Andersson – *Dynamic Sensing in a Ping Pong Playing Robot* – IEEE Transactions on Robotics and Automation, Vol. 5, No. 6, pp. 728–739, December 1989.

[8] G. Buttazzo – *HAREMS: Hierachical Architecture for Robotics Experiments with Multiple Sensors* – Proc. of 5th Int. Conf. on Advanced Robotics ('91 ICAR), Vol. 1, pp. 43-48, Pisa, Italy, June 19–22, 1991.

[9] G. Hirzinger, J. Dietrich – *Multisensory Robots and Sensor–Based Path Generation* – Proc. of IEEE Conf. on Robotics and Automation, San Francisco, CA, April 1986.

[10] B. Espiau – *Sensory–Based Control Robustness Issues and Modelling Techniques Application to Proximity Sensing* – Nato ARW on "Kinematic and Dynamic Issues in Sensor–Based Control", Il Ciocco, Italy, Nov. 1987.

[11] A. Fiorillo, B. Allotta, P. Dario, R. Francesconi – *An Ultrasonic Range Sensor Array for a Robotic Fingertip* – Sensor and Actuators 17, 1/2, 1989, pp. 103–106.

[12] V. Hayward, L. Daneshemend, A. Nilakantan – *Model Based Trajectory Planning Using Preview* – Report TR-CIM-88-9, pp. 1–19, McGill Research Center for Intelligent Machines, Mcgill University, Montreal, March 1988,.

434

[13] V. Hayward, J. Lloyd – *Real–Time Trajectory Generator Using Blend Functions* – Report TR-CIM-88-9, McGill Research Center for Intelligent Machines, Mcgill University, Montreal, IEEE 1991.

End Point Control of Compliant Robots*

J.I. Arocena,[†] R.W. Daniel and P. Elosegui[‡]
Dept of Eng. Science, University of Oxford
19, Parks Road, OX1 3PJ Oxford, England (UK)

Abstract

We describe the design of an experimental three degrees of freedom compliant robot arm. Motivation for the design decisions are given with reference to previous experimental findings on a single link and a two link compliant arm. The arm is direct drive, with three compliant transmission shafts and supported within an external exoskeleton to remove the problem of bending modes. Initial results are reported on comparisons between two controllers, one based on rigid body assumptions, and the other including a strain based feedback loop for compensating the effects of arm compliance. The advantages and disadvantages of adding strain based feedback loops to model based controllers are discussed. The problem of realizing high performance control with the presence of motor torque ripple and time delays within the drive electronics is also discussed.

1 Introduction

Our motivation for our research into flexible manipulators originated with the real industrial problem posed by the need to perform remote maintenance within a nuclear reactor when access is limited to a 20 cm diameter pipe through up to 15 metres of concrete shielding. These arms may be required to perform tasks such as welding or drilling, where sub millimetre relative accuracy may be demanded, with various optical sensors providing information on relative end point position with respect to a (fixed) workpiece. Maintaining stability when the end point is under such close control of a non-colocated sensor can be problematic, and has led us to investigate some of the options for control. Here, the motivation is not the more usual requirement of "lightweight" manipulator arms [3], but the stable *closed loop* control of end point vibrations which may occur as a result of interactions between the robot and its environment, or because of residual vibrations during gross motion.

1.1 Background

Motivated by the above, three arms have been designed at Oxford to investigate the problems of end point sensing in, possibly compliant, robot manipulators. The first was a single link built from a spring steel plate with end point sensing using a mechanical

---

*Project supported by the SERC contract no. GR/G 17516
[†]Supported by the SERC contract no. GR/G 17516
[‡]Funded by the Basque Regional Authorities

follower arm [8]. This enabled us to obtain high frequency data on the end point position without the need for recourse to optical [3] or sonar sensors with their commensurate implementation problems. The use of a mechanical sensor arm for tip position has been maintained throughout our investigations, a two link spring steel arm has been built and controlled using strain gauges with tip position [4]. The principal of operation is a compliant driven arm with a stiff lightweight trailing sensor arm. The design and control of these arms provided us with experience on the design and control of compliant arms leading to the design of a three degrees of freedom robot, compliant in torsion, with direct drive and end point sensing. As for control we have relied on strain feedback to cancel the vibrational modes or at least rigidize the system up to a certain frequency. Then end point control may be applied to a more tractable system for which rigid body assumptions are made [4].

Classic adaptive control, including adaptive predictive control, has also been implemented on the single link and two link arms described above, to control the varying vibrational modes when changing the load inertias [8]. The results suggested that no great improvement was obtained with respect to a frequency response designed strain feedback controller plus an outer loop based on a rigid body assumption.

1.2 Design Problems

The design of the three degrees of freedom arm (the Rotabot) was guided by a number of problems encountered in the simpler spring steel arms: kinematic instability near singularities [4]; dynamic coupling bending mode vibrations and torsional modes in the beam [10], and transmission nonlinearities, gearboxes have low backlash, friction and some unmodelled flexibility [5].

2 Design of Rotabot

Restriction to torsional compliance within an exoskeleton removes the problem of the kinematic coupling experienced in the two link arm and also removes the complication of whirling. Furthermore in going from bending to torsion the spatial Euler-Bernouilli equations for a differential segment dx are reduced from

$$EI(x)\frac{\partial^4 y(x,t)}{\partial \theta x^4} - m_s(x)\frac{\partial^2 y(x,t)}{\partial \theta t^2} = 0 \tag{1}$$

for bending to

$$GJ(x)\frac{\partial^2 \theta(x,t)}{\partial \theta x^2} - I_s(x)\frac{\partial^2 \theta(x,t)}{\partial \theta t^2} = 0 \tag{2}$$

for torsion, here y is the linear displacement; x is the position along the neutral axis of the link; θ is the torsional angle; $I_s(x)$ and $m_s(x)$ are the moment of inertia and mass per unit length respectively; E is the Young modulus of elasticity; G is the shear modulus and t is time. Thus the dynamic equations are transformed from being fourth order in displacement and second order in time to being second order in both.

Our aim was to design a "clean" mechanism, with known degrees of compliance in known positions in the robot structure. The design is based on a basic "Tee" shaped

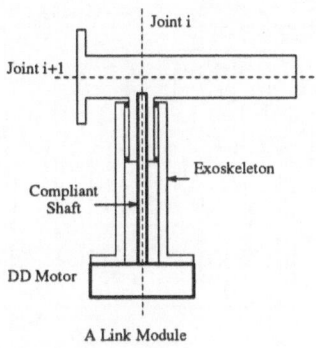

Figure 1: The basic kinematic module.

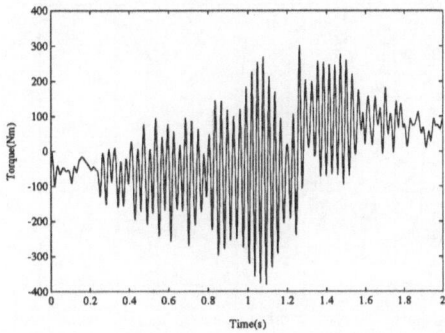

Figure 2: Strain gauge signal for an open loop step demand on the original direct drive motors

module shown in Figure 1 consisting of a direct drive motor (NSK Megatorque series) mounted onto a long thin shaft held inside the rigid housing from the previous joint and driving a rigid link which contains the next compliant element. The resulting arm kinematics, by repeating this module, provides a statically balanced mechanism, able to reach any point in its 3D workspace.

The reason for direct drive was two fold. Firstly we wished to keep the effect of friction to a minimum, and so gearboxes were considered undesirable, and secondly, the extra mass needed by the direct drive motors could be used to statically balance the arm. The whole arm is controlled via a hybrid 68000/Transputer parallel computer with code mostly written in Occam. This allows extra computational modules to be added up to a maximum of 16 1MFLOP T800 processors.

3 Problems in Designing Direct Drive Arms

Direct-drive motors were initially chosen to avoid problems such as backlash, friction, cogging and unknown compliance introduced by gear boxes. However, we soon discovered that direct-drive motors were not a panacea and in fact add other problems which probably make torque control more difficult than with classical geared manipulators. The problems with direct-drive motors are torque ripple, pure time delays, friction, and the torque/velocity characteristic. Initial experiments had such a bad combination of resonance, backlash (from poor in-house motor/shaft coupling design), and torque ripple that the arm was open loop unstable, see Figure 2. A possible explanation is that the higher order dynamics introduced by the compliance destabilised the driver current feedback loop or the inherent motor position and velocity commutation feedback loop within the internal driver look-up table. The original drivers were replaced with a microprocessor based system and the motor/shaft couplings redesigned to at least result in an open loop stable system.

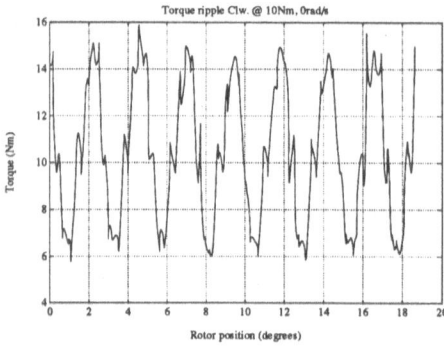

Figure 3: Static torque ripple @ 10Nm.   Figure 4: Static torque ripple @ 80Nm.

3.1 Torque Ripple

The NSK motors are SR (Switched Reluctance) type motors which are inherently nonlinear and require a position sensor and a customised look-up table to linearise the torque characteristics. Unfortunately the linearisation circuits do not produce a perfect torque source and there is considerable residual torque ripple sufficient to excite the resonant modes in the arm, see Figs. 3 and 4. This poses a problem, since torque feedback is now necessary to linearise torque as well as the original aim to rigidize the compliance in the transmission. An alternative to using feedback to reduce the affect of torque ripple is to add harmonic currents into the drive system as described in [9] for DC motors, but this is far too complicated and requires a very special test rig.

3.2 Motor Bearing Coulombic and Viscous Friction

High levels of friction have been measured in the direct drive motors, shown in Figure 5, which corresponds to the NSK RS1010 motor. The curve is obtained by manually moving the motor backwards and forwards for more than a minute and collecting data at 1000 Hz. The data collected consisted of shaft strain, motor acceleration (using an accelerometer), and motor position and velocity. Friction torque was then extracted using the equation $\tau_{\text{friction}} = \tau_{\text{shaft torque}} - \tau_{\text{motor acceleration}}$. The figure suggests that there is some additional hysteresis, but this is due to variations in friction with respect to position. This variation is due partially to residual magnetisation of the motor iron core and partially as a result in variations in bearing characteristics with position.

3.3 Driver Characteristics

The commutation electronics introduces a 1.2ms time delay between the torque demand input and output current. This has been experimentally verified by both using a Hall current probe on the motor windings and during dynamic identification of the motor-driver system using an accelerometer, see Fig. 6. The consequence of this delay is an increased phase lag that severely limits the possible closed loop torque bandwidth.

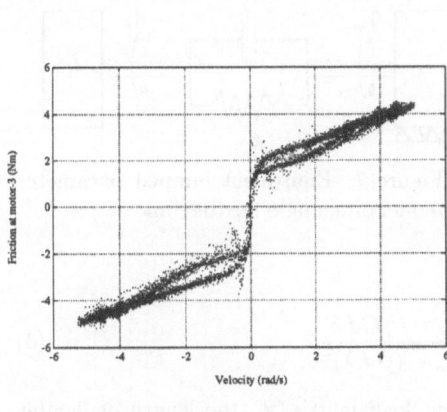

Figure 5: Coulombic and viscous friction characteristics of an NSK-1010 direct-drive motor.

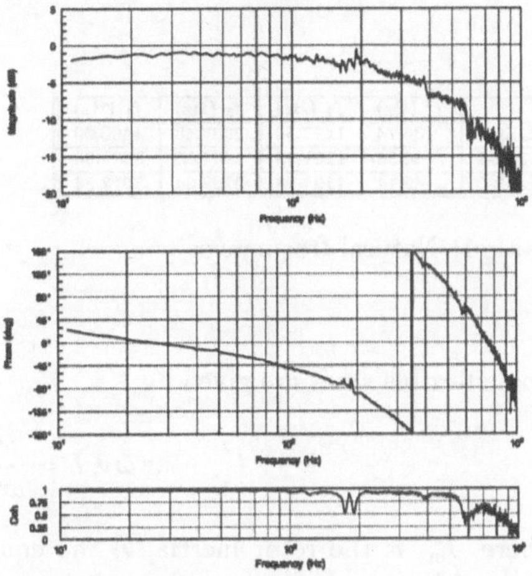

Figure 6: Torque Bode-plot of the NSK-1010 motor/driver system

4   Experimental procedures and results

We will now consider the modelling of the three degrees of freedom arm in two phases: Modelling of a single link (ii) Modelling of the full arm including torsion.

4.1   Modelling of a Single Compliant Link

The mass distribution of the Rotabot is such that the c.o.g. of the load, including the mass of the corresponding shell, is very near the axis of rotation of the transmission shaft, at the motor end of the link. As a consequence, the equivalent load of the link is almost configuration independent and therefore, if the end-tip load is unaltered, the natural frequencies will almost be fixed. This has been derived from the diagonal dominance of the mass matrix and experimentally verified by measuring the first natural frequency of the links for different configurations. This property allows us to consider the compliant model of each link to be treated independently. A distributed parameter model of a single link also shows that the natural frequencies above the fundamental frequency (first resonant frequency) are above 1000Hz. This feature allows further simplification, the system can be approximated as a single resonant system consisting of a mass-spring-mass system.

4.1.1   Distributed Parameter Model of a Single Link

Here we derive an analytical model for the flexible inner links of the Rotabot. Several methods are available, in our case Hamilton's principle for continuous systems was chosen

The robot design is such that bending moments are negligible and each flexible link is modelled as a free-free torsional shaft. It can be shown that the natural frequencies of a

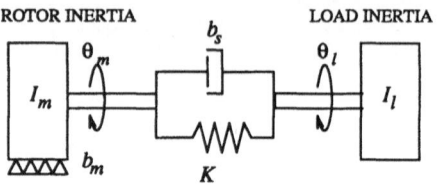

Figure 7: Equivalent lumped parameter model of a single flexible link

	f_1 (Hz)	f_2 (Hz)	f_3 (Hz)	f_4 (Hz)
Link-1	34.74	1493.53	2987.48	4480.60
Link-2	46.36	1201.99	2404.88	3605.98
Link-3	49.14	1104.18	2209.64	3312.54

Table 1: Natural frequencies

single flexible shaft are given by:

$$\tan \omega \lambda L = \frac{\omega^2 \left(I_m + I_l \right) GJ \lambda}{\omega^4 I_l I_m - \left(GJ \lambda \right)^2}. \tag{3}$$

where, I_m, is the rotor inertia, I_l the equivalent load inertia, L, the length of flexible shaft with a constant moment a inertia per unit length, I_s (kg.m). Due to the symmetry of the shaft the torsional stiffness, $GJ(x) = GJ$, is constant, where, J (m⁴), is the polar moment of inertia of the cross sectional area and, G (N.m²), is the material shear modulus. Solving for ω we obtain an infinite set of solutions corresponding to the natural frequencies. Likewise, the mode shapes may be also obtained from,

$$F(x) = -\frac{\omega^2 I_m}{GJ \lambda} \sin \lambda x + \cos \lambda x. \tag{4}$$

where $\lambda = \omega \sqrt{I_s/GJ}$.

A bisection technique, using MACSYMA^{TM}, was used in order to obtain the natural frequencies, roots of Eq. (3). As can be seen from Table 1 the higher order natural frequencies are above 1000 Hz for all the joints and therefore can be neglected.

4.1.2 Simple Resonant Lumped Parameter System

The assumed linear lumped parameter model is shown in Figure 7. To this effect, every part of the Rotabot was weighted, their centres of gravity measured, the moments of inertia were calculated using the two-wire method as in [1], and the stiffness constant of the flexible shafts where also measured and proved to be very linear.

The relationship between between the applied torque and the sensed torque at the load is:

$$\tau_l(s) = \frac{K/I_m}{s^2 + b_S(1/I_l + 1/I_m)s + K(1/I_l + 1/I_m)}(\tau_m(s) - b_m \dot{\theta}_m(s)) \tag{5}$$

where K is the lumped stiffness constant of the shaft; b_s and b_m are the shaft and rotor bearing viscous friction factors respectively.

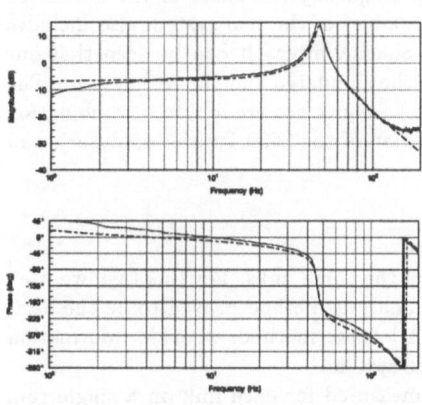

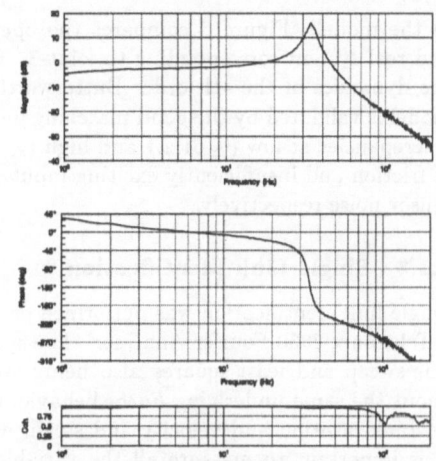

Figure 8: Strain-gauge Bode-plot at link-3, real system (cont.line) versus simulation (discont. curve).

Figure 9: Strain gauge to motor torque Bode plot for joint 1

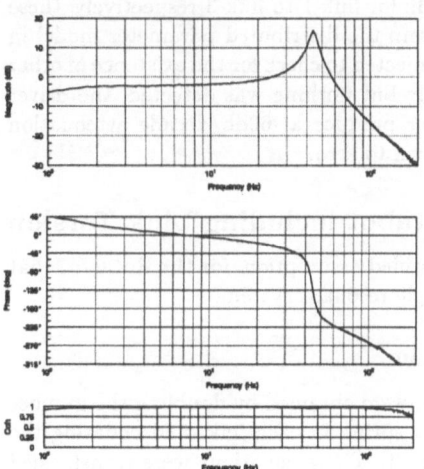

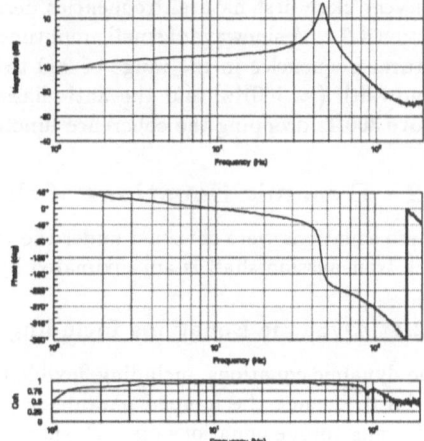

Figure 10: Strain gauge to motor torque Bode plot for joint 2

Figure 11: Strain gauge to motor torque Bode plot for joint 3

The identified motor/driver transfer function (NSK RS-1010 (150Nm), 3rd link), see Figure 6, is also included and is approximately given by

$$\frac{\tau_d(s)}{\tau_m(s)} = \frac{0.9e^{-0.0012s}}{s + 600\pi},\tag{6}$$

where τ_m and τ_d are the demanded and motor torque respectively. The anti-aliasing filter, a 4th order Butterworth filter with a cut-off frequency of 250Hz was also included in the model. Figure 8 compares the open loop frequency responses of the simulated and real system corresponding to joint-3. The dynamics of the real system also includes the dynamics of the 4th order Butterworth anti-aliasing filter. It can be seen that our model is validated by the good matching between the simulated and the real system. The discrepancies at low (≤ 6Hz.) and high (≥ 140Hz.) frequencies are due to a combination of friction and insufficiently exciting input (limitation of the 2630 Fourier analyser), and sensor noise respectively.

4.1.3 Single Link Identification

Single link identification was performed on each of the three links. For this task we used a Tektronix 2630 Fourier Analyser. The spectral analysis method proved to be the best, sine-sweep and least-squares also being tried. All these methods provide information about the same underlying *linear* behaviour of the system.

Link position and velocity and strain where measured for each link on a single run. It is important to measure all the variables on the same run since this provides extra information when one correlates between the plots. The tests were repeated many times to check for repeatability, this proved to be very good.

The Bode plots obtained, shown in Figures 9, 10, 11, verify our hypothesis that each link can be treated as a single resonant lumped parameter system. The first resonant frequencies are approximately at 33, 44 and 46 Hz for link-1 to link-3 respectively, these are very close first natural frequencies derived from the distributed parameter model in section 4.2.1. A sinewave of small amplitude was injected to check for the existence of other natural frequencies in the range of 200 to 1000Hz but nothing was detected, the driver bandwidth (≈ 300Hz) and the anti-aliasing filter produce a -60db/decade attenuation above 300Hz dropping the coherence function down to zero.

4.2 Dynamic Equations of the Rotabot including Link Torsion

These equations were obtained under the well justified assumption, for the Rotabot, that each transmission shaft is approximated by a single resonant system.

4.2.1 Dynamic Equations Including Torsion

The dynamic equations, including flexible torsion, were obtained by doubling the number of dof to include the lumped parameter torsional spring in each link, and equating the balancing torque equations on each side of the shaft. These equations were transformed to the well known Recursive Newton-Euler form and a closed form solution was obtained using the symbolic manipulation language MATHEMATICA^{TM}. This resulted in the two following dynamic equations,

$$0 = M_l(\Theta_l)\ddot{\Theta}_l + M_{lm}(\Theta_l)\ddot{\Theta}_m + V_l(\Theta_l, \dot{\Theta}_l, \dot{\Theta}_m) + K_l(\Theta_l)(\Theta_l - \Theta_m) + G_l(\Theta_l) \quad (7)$$

$$\tau = M_{ml}(\Theta_l)\ddot{\Theta}_l + M_m\ddot{\Theta}_m + V_m(\Theta_l, \dot{\Theta}_l, \dot{\Theta}_m) + K_m(\Theta_l - \Theta_m) \quad (8)$$

where Θ_l and Θ_m are the link and motor position vectors respectively; $M_l(\Theta_l)$ the link upper triangular mass matrix; $V_l(\Theta_l, \dot{\Theta}_l, \dot{\Theta}_m)$, is the link velocity vector; $M_{lm}(\Theta_l)$, is the coupling inertia matrix between link and motor; K_l is the stiffness matrix seen from the

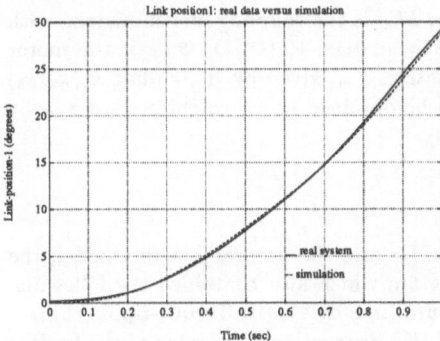

Figure 12: Real versus simulated link position at joint-1.

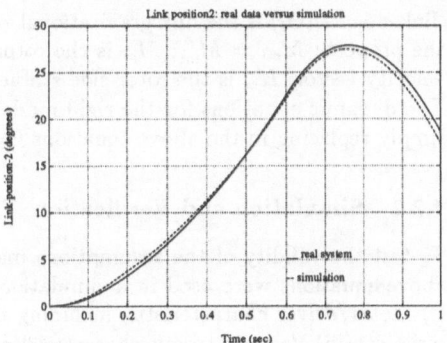

Figure 13: Real versus simulated link position at joint-2.

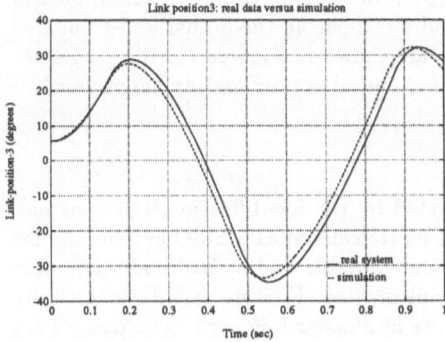

Figure 14: Real versus simulated link position at joint-3.

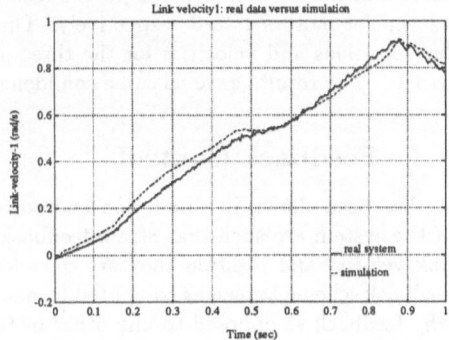

Figure 15: Real versus simulated link velocity at joint-1.

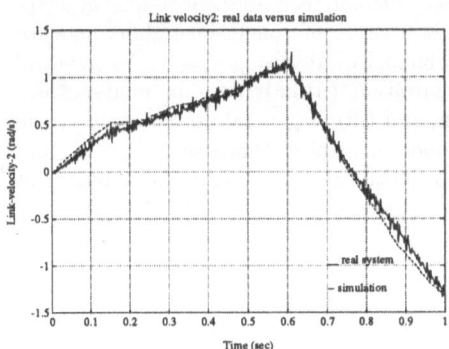

Figure 16: Real versus simulated link velocity at joint-2.

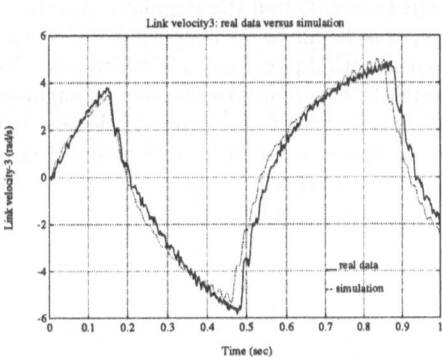

Figure 17: Real versus simulated link velocity at joint-3.

link side; $G_l(\Theta_l)$ is the link gravitational vector; M_{ml} is the coupling inertia matrix, with the property $M_{ml} = M_{lm}^T$; M_m is the rotor inertia matrix; $V_m(\Theta_l, \dot{\Theta}_l, \dot{\Theta}_m)$, is the motor velocity vector; K_m is the rotor-side stiffness matrix, K_m, given by $K_m = \text{diag}(k_1, k_2, k_3)$ The dynamic equations for the rigid model can be obtained in a straightforward way by simply replacing in the above equations $\Theta_l = \Theta_M$.

4.2.2 Simulation and Verification

To test the validity of the assumptions made so far and the accuracy of the model, the above equations were used in a simulation program which also contained the following: (i) Motor/driver characteristic, including the pure time delay, (ii) Torque-ripple characteristics, (iii) Velocity-torque characteristic, and (iv) Proportional damping of the flexible shafts.

The simulation results were compared with real data from the Rotabot, by asking the same demand of both systems, a simultaneous step in demand from rest of 50, 30 and 12 Nm for motors 1 to 3 respectively. The results comparing the actual and simulated link positions and velocities for the three joints are shown in Figures 12, 13, 14, 15, 16, and 17. The results gave us some confidence in our initial approximating hypothesis.

4.3 Feedback Control

As has been shown in simulation and corroborated by the identification, the dynamics of the system are such that strain feedback can be considered as a SISO system. As for link velocity and position they are considered multivariable and a rigid-body nonlinear feedback scheme known as computed torque is implemented. First we would like to clarify why feedback as opposed to any other method, eg feedforward, is used. The reasons are twofold, (i) to achieve a desired input/output characteristic, and (ii) to obtain good disturbance rejection.

After many unsuccessful modern control technique implementations (state space and LQR (Linear Quadratic Regulator)), only frequency domain techniques, ie frequency shaping, was found to be successful. This suggested a series of conclusions about modern control techniques from our experience. The principal disadvantages of modern control methods were found to be: (i) bad engineering intuition, (ii) only provide input/output, no disturbance or sensor noise rejection is given, and (iii) in general do not improve the performance, stability and robustness. Some modern control methods such as SVT (Singular Value Theory) take into account model uncertainty into LQR or LQG but yield very conservative controllers, see [6].

4.3.1 Non-colocated Control Problem

A sensor is said to be non-colocated when the actuator input and sensed variable are connected via a non-rigid connection. For instance motor velocity is colocated and its transfer function, see Figure 7, is given by

$$\frac{\omega_m(s)}{\tau_m(s)} = \frac{1/I_m\,(s^2 + b_s/I_l s + K/I_l)}{s\,(s^2 + b_S(1/I_l + 1/I_m)s + K(1/I_l + 1/I_m))} \tag{9}$$

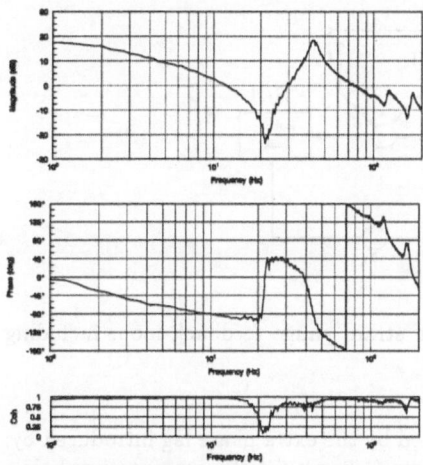 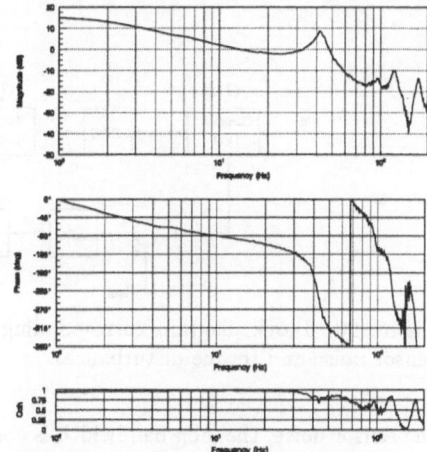

Figure 18: Colocated sensor correspond-
ing to motor velocity at joint-3, notice the
complex zero.

Figure 19: Non-colocated sensor corre-
sponding to link velocity at joint-3

whereas the non-colocated link-velocity is

$$\frac{\omega_l(s)}{\tau_m(s)} = \frac{b_s/I_m I_l(s + K/b_s)}{s\left(s^2 + b_S(1/I_l + 1/I_m)s + K(1/I_l + 1/I_m)\right)} \tag{10}$$

where ω_l and ω_m are the link and motor velocity respectively. The difference between Eqs. (9) and (10) is a that the colocated sensor has a complex zero which limits the phase lag to be above -180° as $\omega \to \infty$ These systems are easily recognisable in a frequency plot because their phase is $\leq$ -180° as $\omega \to \infty$. In our case strain, link velocity and link position are all non-colocated, see Figures 9, 10, 11 for strain. The difference between colocated and non-colocated can be seen in Figure 18 for motor velocity, colocated, versus, link velocity Figure 19, non-colocated. The phase of motor velocity should be going up to -90°, and that of link velocity up to -270°, in our case both figures show a exponential increase of the phase lag, this is due to the pure time delay of the drivers and the anti-aliasing filter. An added problem with non-colocated sensors is the existence of non-minimum phase zeros, associated with the wave propagation delay (transport delay) along the flexible shaft. In our case these are placed at very high frequency and do not affect us. However, for very flexible arms, see [3], they introduce further delay into the system making the control design still more difficult.

4.3.2 Strain Gauge Feedback Design

For the control point of view non-colocated systems offer a challenge when trying to increase the bandwidth beyond the first natural frequency, and this is mainly because of the large phase deficit. The PCT (Pseudo-Continuous Time) method is used to provide an analog model of the sampled data system [7]. In addition to the limitations due to

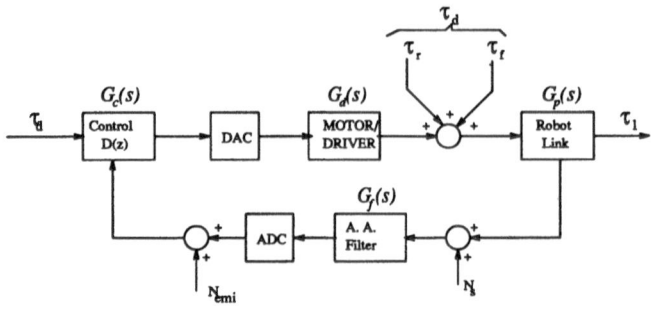

Figure 20: Block diagram corresponding to the strain gauge feedback loop, including sensor noise and torque disturbances.

the sensor noise, the loop bandwidth is constrained by the extra phase lag introduced by: (i)the pure time delay of the motor driver circuitry (1.2ms); (ii) the computational time delay (1.1ms); (iii) the anti-aliasing filters together with the sensor noise.

Our objective is to extend the strain gauge bandwidth to the limit, so as to rigidize the system as much as possible and achieve maximum disturbance rejection, with a minimum affect on the sensor noise amplitude. The closed loop block diagram of the system is shown is Figure 20. Torque ripple has been considered to be uncorrelated to the dynamics of the driver and all the disturbances are bundled under a single general disturbance input. The input-output transfer function is then given by

$$\tau_l(s) = \frac{G_p G_d G_c}{1 + G_c G_d G_p G_f} \tau_{dem}(s) + \frac{G_p}{1 + G_c G_d G_p G_f} \tau_{dis}(s) +$$

$$\frac{G_p G_d G_c G_f}{1 + G_c G_d G_p G_f} N_s(s) + \frac{G_p G_d G_c}{1 + G_c G_d G_p G_f} N_{emi}(s) \tag{11}$$

where τ_l, τ_{dem}, τ_{dis}, N_s and N_{emi} are the link torque (strain gauge), demanded torque, disturbance torque, sensor noise, and E.M.I. noise respectively; G_p, G_d, G_c, G_f are the plant, driver & motor, feedback controller, and anti-aliasing filter to strain transfer functions respectively, where for simplicity their dependency of s has been dropped.

A notch filter is first placed just below the resonant frequency. This boosts the phase advance and cancels the resonant peak. The non-complex poles of the notch are placed at ≈ 100 Hz, where a satisfactory compromise between the noise amplification and phase increase is reached. A small lead of $\approx 35°$ is also placed so that it provides its maximum phase angle at the resonant frequency. This proved to be of crucial importance since it corrected any phase mismatch between the notch zero and the resonant pole. The low frequency regime requires increasing the gain as much as possible with appropriate lag compensators. The limit of the number of lag compensators and their maximum phase lag is dictated by by the phase lag of the open loop transfer function, $G_c G_d G_p G_f$, which must not fall below $-120°$ at the low frequency range, otherwise residual oscillations will appear [2].

4.3.3 Strain Feedback Controller Evaluation

This above design yields a fourth order compensator. Each of the transfer functions contained in Eq. 11 have been separately included in the simulation program and for every design the effects of each of them were evaluated. Figure 21, compares the simulated

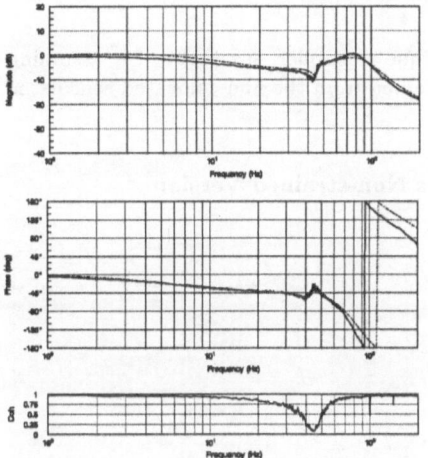

Figure 21: Joint-3 strain gauge feedback: simulation (discont. line) versus real (cont. line) system

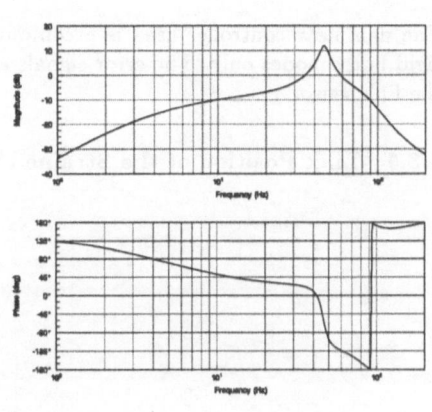

Figure 22: Disturbance rejection at joint-3 with strain gauge feedback

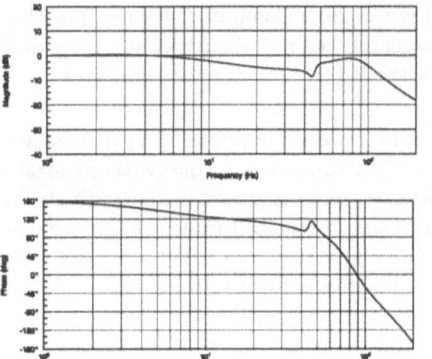

Figure 23: Sensor noise rejection at joint-3 with strain gauge feedback

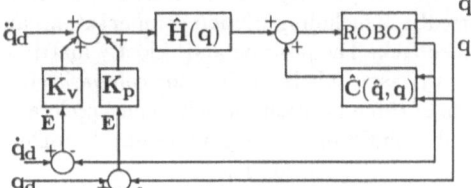

Figure 24: The rigid body computed torque controller

closed loop response versus the real closed loop response, it can be seen the simulated results are very reliable. The resonant peak has completely disappeared and the closed loop bandwidth, if considered as the frequency corresponding to -50°, as suggested by Biernson [2], is around 40 Hz, Fig. 11. For these types of systems it is always difficult to give a bandwidth figure in terms of the magnitude response. Note that the low value of the real system coherence is due to the zero of the notch filter. Figure 22 shows the disturbance rejection, which is quite good up to 30Hz. Likewise, Fig. 23 shows the sensor noise rejection which, as predicted, cannot be improved.

4.3.4 Nonlinear Controller Design

The nonlinear controller used is a computed torque controller, see Figure 24, assuming rigid body modes only. The error signals are obtained from the non-colocated sensors, ie the link sensors.

4.3.5 Link Position of the Strained versus Non-strained Version

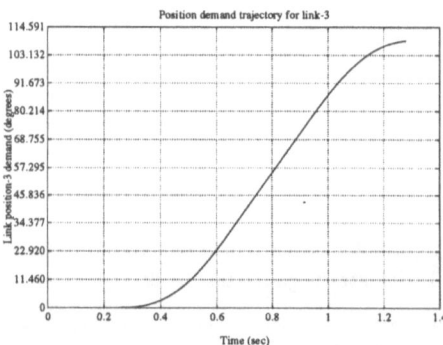

Figure 25: Joint space position demand trajectory corresponding to link-3

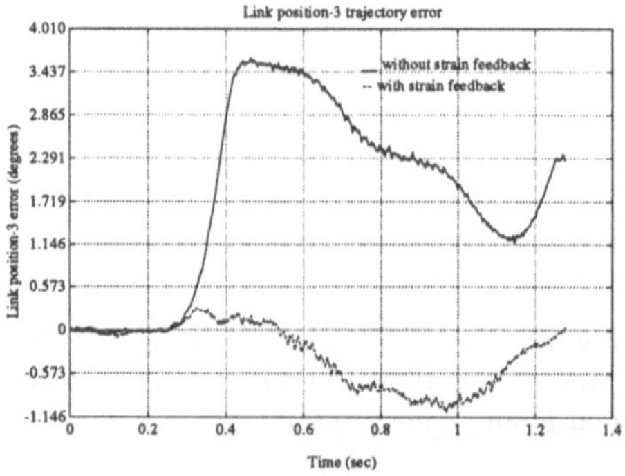

Figure 26: Computed torque position tracking errors with (discont. line) and without (cont. line) strain feedback.

It has already been shown, see Figure 21, that strain feedback improves the performance of the torque loop. Two systems are compared, (i) a computed torque without strain feedback including friction compensation, and (ii) the same computed torque with strain feedback. The position and velocity matrix gains of the computed torque are the same in both cases, this is to allow for comparison. The three joints are run simultaneously with three three position, velocity and acceleration trajectories defined in joint space. So far only joint-3 has the torque feedback implemented. The position trajectory of joint-3 is shown in Figure 25. The cruise speed is 1.9 rad/s and the maximum acceleration demand reaches 8 rad/s^2. The two position trajectory errors are compared in Figure 26 and clearly show an improvement by the strain feedback system. The maximum error for the strain feedback system is 1.03° versus 3.55°, this represents an improvement of 3.44 times with respect to the maximum peak errors. Likewise, the calculated rms value for the strain feedback system is 0.27 versus 1.09 for the non strain feedback system, this represents an improvement of 4.037 times. Both systems have also been compared for a number of other trajectories, including Cartesian Space, and the improvement is more or less of the same value.

5 Conclusions

A novel manipulator with end-point sensing has been presented. Its principal features are an exoskeleton which removes bending modes and provides a mechanical measuring device to obtain relative end-tip position. It has been seen that working in torsion then has its advantages over working in bending. The statically balanced design ensures that the natural frequencies remain almost fixed. One of the consequences of this is that the system behaviour is easier to predict, as demonstrated by the simulation results. It has been shown that by using a carefully designed strain feedback inner loop it is possible to consider the flexible system as a rigid body, still enabling further simplifications. The results so far are very encouraging and in the near future the remaining links will be implemented with strain feedback inner loops as well.

References

[1] B. Armstrong, O. Khatib, and J. Burdick. The explicit dynamic model and inertial parameters of the puma 560 arm. In *Proceeding of the IEEE International Conference on Robotics and Automation*, April 1986.

[2] George Biernson. *Principles of Feedback Control*, volume 2: Advances Control Topics. John Wiley and Sons, 1988.

[3] R.H. Cannon, Jr. and E. Schmitz. Initial experiments on the end-point control of a flexible one-link robot. *The International Journal of Robotics Research*, 3(4), Fall 1984.

[4] R.W. Daniel and M. Irving. Evaluation of control techniques for in-reactor manipulators. Technical Report CEGB Contract R/RER/DSD/TD 464, Robotics Lab. Department of Engineering Science, University of Oxford, June 1987.

[5] M.C. Good, L.M. Sweet, and K.L. Strobel. Dynami models for control system design of integrated robot and drive systems. *Transactions of the ASME, Journal of Dynamic Systems, Measurement and Control*, 107, March 1985.

[6] Isaac Horowitz. Quantitative feedback theory. *Proc. IEE*, 129, Part D(6), November 1982.

[7] C.H. Houpis and G.B. Lamont. *Digital Control Systems: Theory Hardware and Software*. McGraw-Hill, 1985.

[8] M. Lambert. *Adaptive Control of Flexible Systems*. PhD thesis, Univertity of Oxford, 1987.

[9] Benjamin J. Paul. A systems approach to the torque control of a permanent magnet brushless motor. Master's thesis, MIT Artificial Intelligence Laboratory, August 1987. Technical Report 1081.

[10] W.T. Thomson. *Theory of Vibrational Analysis with Applications*. Unwin-Hyman, third edition, 1988.

Use of C-surface based force-feedback algorithm for complex assembly tasks

J.-P. Merlet
INRIA Sophia-Antipolis
2004 Route des Lucioles
06665 Valbonne Cedex
France
E-mail: merlet@cygnusx1.inria.fr

Abstract
We describe the principle of a force-feedback controller based on the C-surface theory. Experimental results on complex assembly tasks are presented and we compare various force-feedback strategies. We show that our force-feedback algorithm can perform these tasks with few a-priori informations and can be used to learn a reference trajectory.

1 Introduction

1.1 Hybrid control

Hybrid control [3] is probably one of the most efficient force-feedback control scheme. In this scheme the configuration space of the task is partitioned in directions where pure positional control is used and directions where there is only force-control (i.e. the input of the control law is the force measurements). In the classical implementation of this scheme [1] the operator has first to choose a configuration space (i.e. a set of parameters necessary to describe the task), a compliance frame in this configuration space and to determine which kind of control is to performed along each direction of this frame, either positional control or force control. Therefore the operator has to determine a *selection matrix* which is diagonal and whose diagonal element is 1 if the direction is force-controlled and 0 otherwise. To obtain good performances with this kind of implementation some assumptions are necessary :

- existence of the compliance frame
- the compliance frame is perfectly known.
- the relation between the contact forces[1] and the positional errors of the manipulator is one-to-one

For simple tasks these assumptions may be verified (for example if the end-point of the robot has to follow a fixed plane). However, even in these simple cases, small errors may yield to an unstable behavior of the robot (for example if the position of the plane is slightly different from its model). Furthermore it has been shown that the compliance frame may be not defined for some tasks [3] or that its position is time-dependent (for example when the robot has to follow an unknown shape). Various tasks for which the assumption about the one-to-one mapping between the force measurements and the positional errors is clearly false has been given in [4].

1.2 Hybrid control with the C-surface theory

The drawbacks of the classical hybrid control scheme can be suppressed by computing the compliance frame according to the informations obtained from the task i.e. from the forces measurements. We have proposed such a method in [5], based on the C-surface theory developed by Mason. Let us consider a configuration space which dimension is equal to the number of degree of freedom necessary to perform the task (we will consider that this dimension is at most

---

[1] In this paper forces will mean forces and torques

6). In this configuration space the manipulator (or the part it moves) is represented by a point P. According to the task the configuration space can be divided into the *free space* where there is no contact between the object grasped by the manipulator and the surrounding and the *contact space* where there is a contact. The frontier between these two volumes is called the *C-surface* of the task. Clearly the only possible motion of point P when there is contact is to slide on the C-surface (we assume that the deformations of the objects can be neglected). Locally this motion occurs along the tangent hyperplane of the C-surface. At the opposite a motion of P along the C-surface normal unit vector N yield to an increase or a decrease of the contact forces. Therefore from the hybrid control view point the C-surface normal is the direction of the configuration space on which a pure force control is necessary and at the opposite a positional control is necessary along the tangent hyperplane. Therefore the determination of the C-surface normal N enables to determine both the compliance frame and the selection matrix. This normal can be estimated geometrically if the task is fully specified but in most real tasks the models errors will be to great to ensure a correct control. Thus a better way to calculate N is to use the force measurements. Indeed we have shown in [4] that the forces measurements vector is exactly the C-surface normal if the velocity of the manipulator is sufficiently low (which will be the case for the tasks considered in this paper). Therefore the force measurements enables to measure locally the C-surface normal.

1.3 Principle of a C-surface based hybrid controller

The configuration space is determined by a set X describing the position of the manipulator (or the object it grasps) and a set Ω of angular positions. Let us assume that the desired velocity of the manipulator (i.e. the magnitude of $\dot{X}$) is V_d, the maximum angular velocity (i.e. the magnitude of $\dot{\Omega}$) is ω_m, the desired force and torque magnitude F_d, M_d. The force measurements vector is $\mathbf{F_m}$ and is divided in its force and torque components $\mathbf{F_m^X}$ and $\mathbf{F_m^\Omega}$. An initial velocity $\dot{X}_i$ is given at the beginning of the task. According to our assumption we have during a contact:

$$N = [N_X \ , \ N_\Omega]^T = \left[\frac{\mathbf{F_m^X}}{\|\mathbf{F_m^X}\|} \ , \ \frac{\mathbf{F_m^\Omega}}{\|\mathbf{F_m^\Omega}\|} \right]^T \tag{1}$$

A control law along N ensures the regulation of the magnitude of the force. For example a proportional control law can be used with:

$$\dot{X}_N \ = \ k_p^X(\|\mathbf{F_m^X}\| - F_d)N_X \quad \text{with} \quad \|\dot{X}_N\| \leq V_d \tag{2}$$

$$\dot{\Omega}_N \ = \ k_p^\Omega(\|\mathbf{F_m^\Omega}\| - M_d)N_\Omega \quad \text{with} \quad \|\dot{\Omega}_N\| \leq \omega_m \tag{3}$$

where k_p^X, k_p^Ω are gains. From the value of N we are able to determine a vector T in the tangent hyperplane of the C-surface such that a motion along this vector will satisfy a goal given by the operator. We have:

$$T = [T_X \ , \ T_\Omega]^T \tag{4}$$

The velocity along T_X and the angular velocity along T_Ω will be :

$$\dot{X}_T \ = \ (\sqrt{V_d^2 - \|\dot{X}_N\|^2})T_X \tag{5}$$

$$\dot{\Omega}_T \ = \ (\sqrt{\omega_m^2 - \|\dot{\Omega}_N\|^2})T_\Omega \tag{6}$$

The desired velocities of the manipulator are given by :

$$\dot{X} = \dot{X}_T + \dot{X}_N \qquad \dot{\Omega} = \dot{\Omega}_T + \dot{\Omega}_N \tag{7}$$

In the case where the magnitude of $\mathbf{F_m^X}$ is too low we will have :

$$\dot{X} = \dot{X}_i \tag{8}$$

A problem may arise when the C-surface normal is not unique, for example when P lie on an edge of the C-surface. In that case the motion resulting from our scheme may yield either to some disturbance of the force signal or to loose the contact with the C-surface. In the first case the disturbance will disappear as soon as point P leaves the edge of the C-surface. In the second case either the velocity $\dot{X}_i$ enables to perform correctly the task or a new contact will appear.

2 Experimental results

2.1 The test-bed

The purpose of our experiments is the assembly of parts of a washing machine. The test-bed can be seen on Figure 1. The robot is an IBM 7576 Scara robot with a parallel manipulator as a wrist (in these experiments this wrist is used only to ensure some passive compliance to the robot). The parts are fixed on a table which lie on an AICO 6 componants force sensor. The robot is controlled through an AICO controller with 3 CPU. The experimental programs are written in C on a SUN workstation and downloaded through a serial link in the controller. The sampling rates of the position loop of the robot and the force measurement are 5ms.

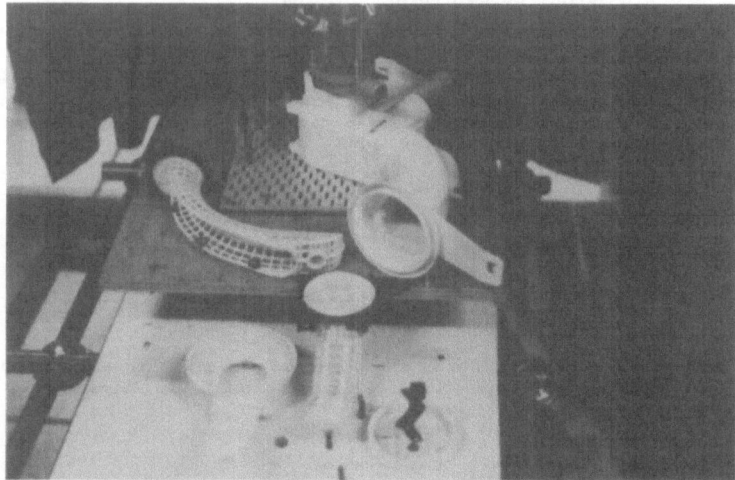

Figure 1: The assembly test-bed. We use an IBM 7576 Scara robot with a parallel manipulator as a wrist. The parts are fixed on a table which lie on an AICO 6-componants force sensor.

2.2 Preliminary experiment

A first experiment was necessary to calibrate the force sensor i.e. to find the position of the force sensor frame with respect to the robot frame in order to get the force measurements in this last frame. At the same time our purpose was to verify our assumption about the relation between the C-surface normal and the force vector.

For this purpose a ruler is fixed on the table and the robot is programmed so that it hits the ruler at various points (the contact was detected by the force sensor). At each point the current position of the robot together with the force measurements are recorded. We assume that we know the position of the center of the force sensor in the robot frame and that both z axis of the force sensor and robot frame are parallel: under these assumptions the calibration unknown is the rotation angle α between the two frames. A least-square algorithm is used to find the equation of the line associated to the ruler and from this equation we deduce the normal vector N to the ruler. Our basic assumption is that the force measurement vector is perpendicular to the ruler. To verify this assumption we use another least-square algorithm to find the rotation angle α which ensure that a criteria J is minimized with:

$$J = \sum_{i=1}^{i=n} ||R(\alpha)\frac{\mathbf{F_i}}{||\mathbf{F_i}||} - \mathbf{N}||^2$$

where n is the number of contact points, $\mathbf{F_i}$ the force vector at contact point i and $R(\alpha)$ the rotation matrix of angle α around the z axis. If our assumption is verified this criteria will be low at the end of the process. The robot hits the ruler at ten contact points and after completion of the least square algorithm the value of J is 0.072555. The angles between N and the force vector at each contact point are given in the following table.

Point	1	2	3	4	5	6	7	8	9	10
Angle	6.4664	-0.4738	2.7	2.6067	-2.8425	2.1147	-1.4375	-4.0476	-2.0729	-3.0375

Their average value is -0.002398 degree and the standard deviation is 3.172530 degree. The very small average value shows that our basic assumption seems to be reasonable.

2.3 Phase 1 : Assembly of the filter and the plug

These two parts are described in Figure 2. This operation is a zero-clearance assembly task because the diameter of the hole in the plug is less than the distance between the two pins on the filter. Therefore this assembly can be performed only by using the elasticity of these pins.

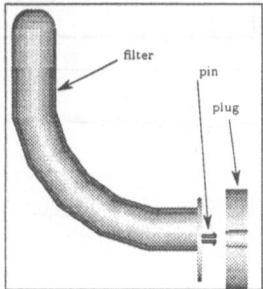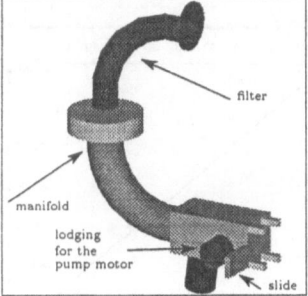

Figure 2: Phase 1. The plug, the filter and the manifold.

2.3.1 Assembly with a guarded move strategy

The plug can be grasped either along its diameter or along its normal which yield to either an horizontal insertion axis or a vertical one. However it may be noticed that the second way is more appropriate in the sense that the forces acting on the plug will lie mostly along its normal and therefore will be counterbalanced by the whole stiffness of the robot. In a first experiment we decide to perform this assembly by using a guarded move strategy along the vertical axis. The desired force value is greater than the force necessary to deform the pins. Under these conditions the assembly will be completed (provided that the lateral misalignments are not too large) and the robot will stop when the plug reaches the front part of the filter.

Experimental results are presented in Figure 3. If we consider the plot of the force along the z axis we notice a first increase corresponding to the deformation of the pins (the necessary force is approximatively 80 N). As soon as the insertion is performed the elastic energy stored in the wrist yields to an upward-downward motion of the plug which hits the pins (time=28) causing a change of sign of the F_z signal. But the damping is strong and at time 37 there is a decrease of the force while the robot is now moving the plug toward the front part of the filter. Then the plug hits this front part (time=80) and the robot moves downward until the desired force (100 N) is obtained.

We notice that the positionning errors along the axis perpendicular to the insertion axis are approximatively 2 mm and are corrected by the passive compliance behaviour of the parallel wrist. Here the insertion axis given by the operator is quite accurate and therefore the lateral forces remain small. In that case the guarded move strategy is quite efficient : failures have been observed only when the distance between the initial position of the robot and the insertion axis was greater than 3 mm.

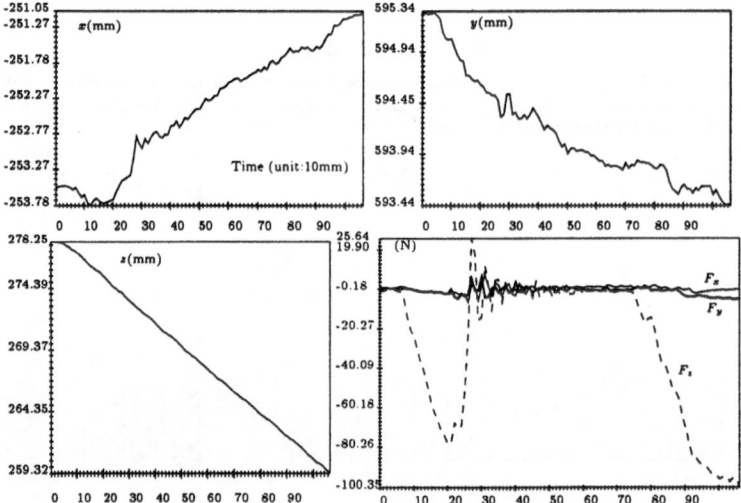

Figure 3: Phase 1, assembly of the plug and the filter. Positions of the gripper and contact forces during the assembly with passive compliance, guarded move strategy and vertical insertion axis.

2.3.2 Assembly with active compliance

For this task we may assume that the parts have the same orientation. Therefore only the x, y, z are to be controlled. The large clearance enables to use an hybrid control with a fixed compliance frame where all three axis are force-controlled with a zero desired force along x, y and a positive value along z, greater than the force necessary to deform the pins. The parameters of our controller are the desired force value and a proportional gain relating the variations of the lateral motions according to the values of the lateral forces.

Experimental results are presented in Figure 4. According to these results we notice some interesting points concerning the use of the force feedback algorithm. During the initial contact (time=0-10) small misalignments of the plug along the x, y axis are corrected. After the assembly the forces F_x, F_y are smaller than with the guarded move strategy. During the downward motion of the robot these small forces are used to slowly correct the position of the plug. At time 75 the misalignments are quite equal to zero and from this point there is no lateral motion of the plug. It must be noticed that both lateral forces are very close to zero and are better controlled around this value than with the guarded move strategy. At time 100 the plug hits the front part of the filter yielding to small variations of the position of the plug along the lateral axis. During our experiments no failure have been observed although initial errors of up to 4 mm have been introduced. As a matter of conclusion the active compliance scheme yields to a more robust controller with respect to the lateral misalignments of the plug.

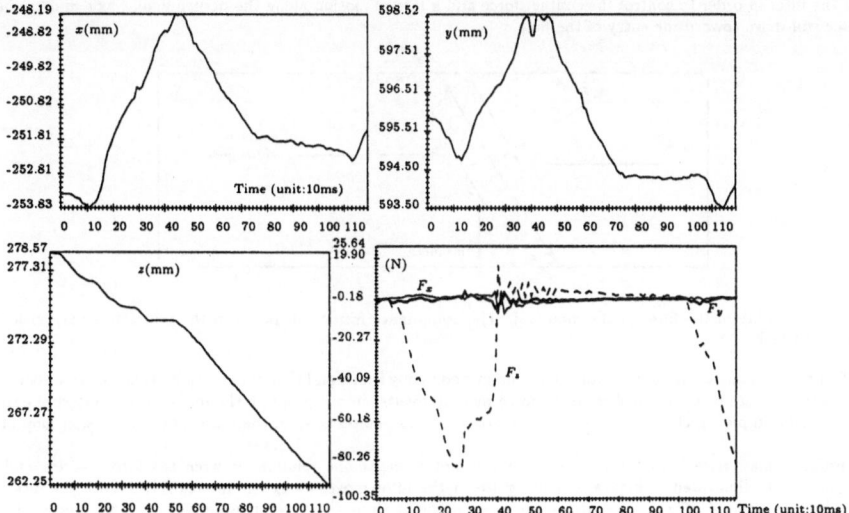

Figure 4: Phase 1, assembly of the plug and the filter. Positions of the gripper and contact forces during the assembly with active compliance, free wrist and vertical insertion axis.

2.4   Phase 2: Assembly of the filter in the manifold

The filter can be seen on Figure 2 and the manifold on Figure 1. There are large clearances between these two parts: 3.5 and 6.0 mm. If the grasping position of the filter and the position of the manifold is well known an off-line motion planning strategy may be used to determine the trajectory of the robot. We assume here that these data are not available and that the only a-priori informations are:

- an initial position of the filter such that its extremity is close from the entry of the hole in the manifold.

- a rough estimate of a direction vector enabling to put the extremity of the filter inside the hole.

Our purpose is to complete this assembly whatever is the shape of the manifold. During this task the variables to be controlled are x, y, θ where θ is the rotation angle around the z axis of the Scara robot (we assume here that the coordinates z of the robot is constant during the task). For the hybrid control scheme we may see clearly that we cannot find here a fixed compliance frame. Less obvious is that expressing the compliance frame in the tool frame may yield to a failure of the task. It would seem natural to state that in the tool frame the axis corresponding to the general axis of the filter is to be position-controlled and that the perpendicular axis is to be force-controlled. This scheme will fail when the filter hits a part of the manifold where the normal lie approximatively along the axis of the filter (Figure 5). Note that our controller will succeed in this situation: it will first cause a backward motion along the axis of the filter in order to control the contact force and a lateral motion along the perpendicular axis and gradually the filter will move toward the entry of the hole.

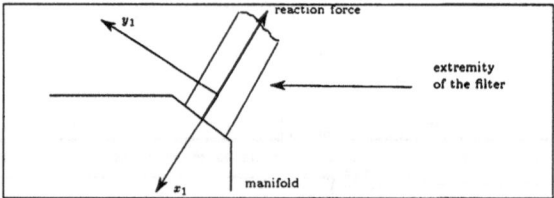

Figure 5: Assembly of the filter in the manifold. The compliance frame calculated in the tool frame may yield to a failure of the task.

In fact the compliance frame cannot be easily described in any frame and therefore we claim that the best estimation of the compliance frame is obtained from the force measurements. In our program the robot uses an internal position control loop at 10 ms which set the values of the articular coordinates used in the most internal position loop of the robot.

A problem may arise from the fact that there is not a one-to-one relation between the forces vector and the misalignments as illustrated in Figure 6. During step 1 the filter moves along the insertion axis until the filter hits the manifold at point A. The torque around the grasping point along the z axis is negative and consequently there is a negative rotation of the filter which yield to an increase of the orientation error. At this point the reaction force at A yields to a motion along the C-surface normal N. These motions yield to a second contact point B at step 2 and consequently to a change of sign of the torque. The filter rotates around the grasping point and the orientation error is partially corrected. During step 3 the torque is small but there is a reaction force yielding to the centering of the filter with a lost of contact. At step 4 there is no more contact and the grasping point moves along the insertion axis until a new contact appears.

Therefore the problem is the bad correction of the orientation during step 1 of the assembly. This is mostly due to the initial bad orientation of the filter but in any case the algorithm corrects this error in the following steps of

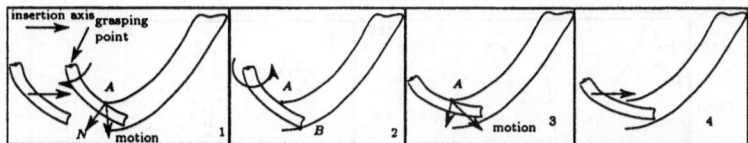

Figure 6: Assembly of the filter in the manifold. What my happen.

the assembly. For this experiment the filter is grasped on its body near its front part (our gripper does not enable to grasp the front part). The insertion is performed until the gripper hits the manifold. Figure 7 shows the record of the position of the robot, the contact forces and torque around the z axis during the assembly. We may notice on the θ record that the problem mentioned in the previous section occurs: there is a negative rotation of the wrist between time 0 and 5. After this initial error the record shows that θ is approximatively a linear function of time which means that the followed trajectory is an arc of circle as it may be expected. We may notice that a force control along the z axis may be useful because although the z force is equal to zero at the beginning of the task great variations of this force may be observed during the assembly. Our assumption is that there is a small change in the grasping position due to the great length of the filter and the torque involved at the beginning of the assembly which yields to an error on the position along the z axis.

During time 0-10 the torque is effectively positive and became then negative until time 35. At this point the torque becomes positive which means that we are in the same situation as at the beginning of the assembly. Then the torque steps back to a negative value. A similar situation occurs at time 46. At the end of the assembly (time 72) the gripper is in contact with the manifold and the torque becomes constant. Note that the filter is always in contact with the manifold (except at time 33-34 and 45). This is due to the fact that we have no a priori knowledge of the trajectory: even in the case of a lost of contact between the parts the fixed insertion axis yields to a new contact a few steps after the lost of contact.

As a matter of conclusion this experiment enables to show that the force controller is able to perform complex assembly tasks even with few informations. For the considered assembly however a reference trajectory might be useful because a reference model of the assembly enables to get a better control of the contact force.

It is possible to build a reference trajectory for this assembly from the experimental data obtained after a first experiment. To calculate this reference trajectory we consider the set of points of the data for which the torque and force are small together with the last point (at this point the assembly is performed). These sets are used as control points to calculate a spline for each of the x, y, θ variables. Figure 8 shows the reference trajectory obtained for each variable. Figures 9 shows the forces and torque measured when the gripper described the reference trajectory in open loop. Note that until the gripper hits the manifold (time 499) the forces are very small. The average values of F_x, F_y are 0.0795 N, 0.3067 N and their standard deviation is 0.293 N and 0.368 N. We see on the torque plot that at time 499 there is a one contact point between the filter and the manifold. Then the filter rotates around this point until the gripper is applied against the front of the manifold yielding to a continuous increase of the torque.

2.5 Phase 3: Assembly of the pump motor in the manifold

The lodging of the pump motor may be seen in Figure 2. The grasping is performed in such a way that the center of the gripper lie on the same vertical axis as the cylinder of the pump motor. This is a two insertions assembly task : one prismatic insertion of the motor in a slide and a cylindrical insertion. These insertions are not synchronous: the insertion in the slide occurs before the cylindrical insertion. Experimental results obtained with the force-feedback algorithm used in Phase 1 are presented in Figure 10. Note that the lateral misalignments are corrected during the insertion (between time 0-70). At time 70 the cylindrical insertion begins and causes a small disturbance on the lateral

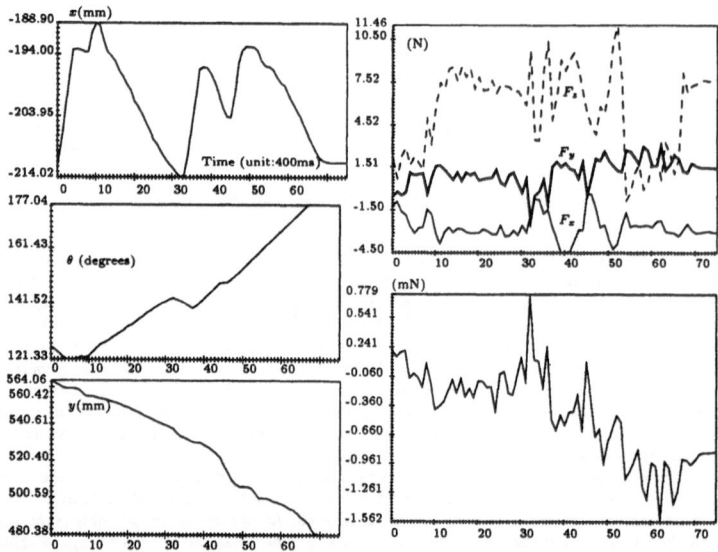

Figure 7: Phase 2: assembly of the filter in the manifold. Positions of the robot, contact forces and torque during the assembly with active compliance, free wrist and vertical insertion axis. Note that the torque is the torque exerted by the robot and therefore the opposite of the torque used by the force-feedback algorithm.

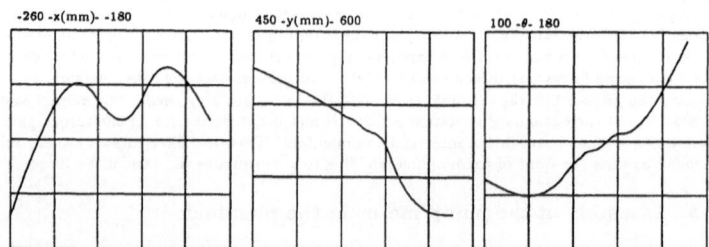

Figure 8: Phase 2: assembly of the filter in the manifold. Reference trajectory of the gripper obtained from the previous data.

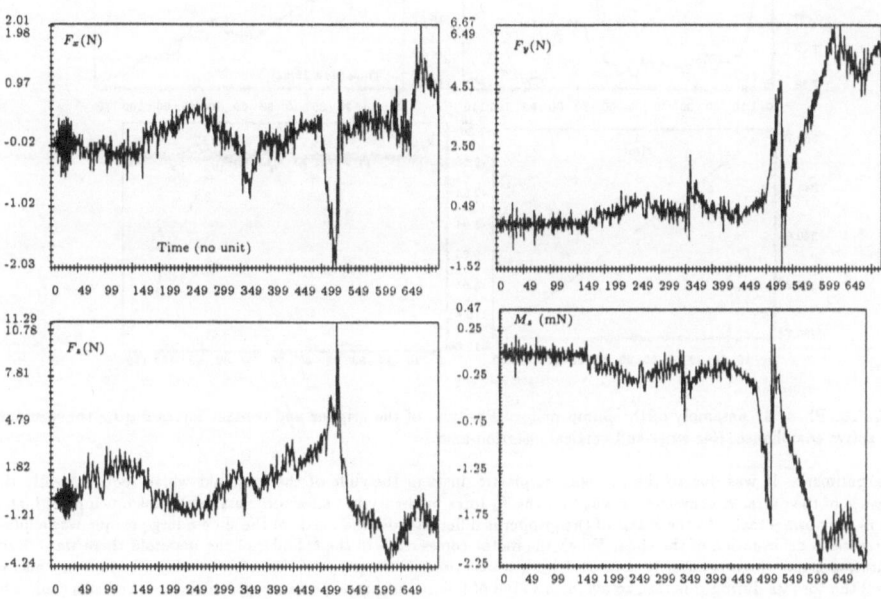

Figure 9: **Phase 2: assembly of the filter in the manifold.** Force and torque measurements obtained for the reference trajectory of the gripper followed in an open-loop mode.

position which is rapidly corrected (time 85). Note also that a large vertical axis force is necessary at the end of the task to complete the insertion (more than 10 N). As a matter of conclusion we have noticed one case of failure during

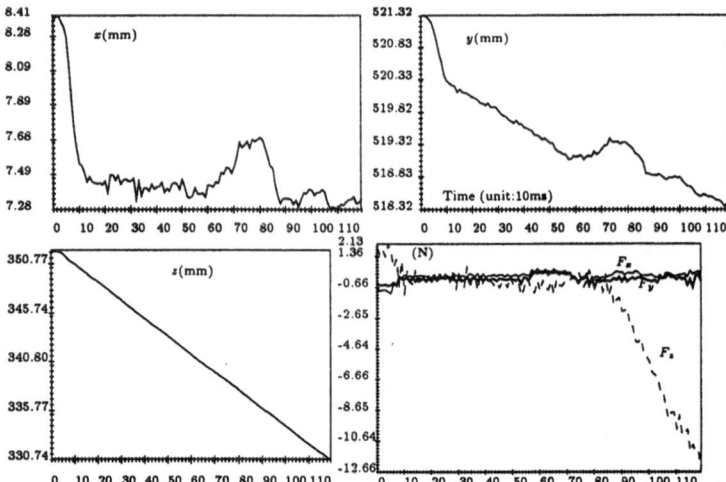

Figure 10: Phase 3: Assembly of the pump motor. Positions of the gripper and contact forces during the assembly with active compliance, free wrist and vertical insertion axis

the experiment. It was due to the presence of plastic dusts in the slide of the manifold which reduce greatly the clearance of the parts. A consequence was that the F_z force necessary to perform the assembly was much more higher than in the normal task. As the z axis of the gripper is different from the z axis of the slide a large torque was applied and creates a deformation of the slide. When the motor comes close to the cylinder of the manifold there was a large orientation error which cannot be corrected by a translation of the gripper. Therefore it may be necessary also to control the torques during this task to avoid this kind of failure but in the present state of the test bed this cannot be done. However for a more normal state of the parts the force-feedback algorithm was efficient.

2.6   Phase 4: Screwing of the plug in the manifold

In this phase the grasping position is identical to the one in Phase 1 and the insertion axis of the plug is vertical. Note that the grasping position is such that there is a small distance between the z axis of the gripper and the z axis of the plug and therefore an open loop control will be inefficient. The x, y, z axis are force controlled : the two first one are controlled through the C-surface theory and the z axis is independently controlled. The motion along the x, y axis enables to correct the misalignments of the gripper with the axis of the plug. If the forces along the x, y axis are sufficiently small a one axis proportional regulator is used to control the F_z force. If this force is close from the desired value (here 6N) a rotation of the gripper is performed. Although the range of the rotation axis of the robot is too small to fully perform the screwing it has been possible to show that the screwing was effective.

Experimental results are presented in Figure 11. We see clearly on the position records that there was first a correction of the misalignments of the parts at the beginning of the task together with a large motion along the z axis in order to obtain the desired force value (time 0-10). After that the screwing proceeds rather smoothly (the θ record shows that the screwing speed is quite constant) with a screwing rotation of 150°. Due to the distance between the z axis of the robot and the z axis of the plug we may expect the forces along the x, y axis to be theoretically a sine curve as it can be observed on the force records. Note also on this figure the record of the force F_z which value is always close from the desired value in spite of the disturbances involved by the screwing motion.

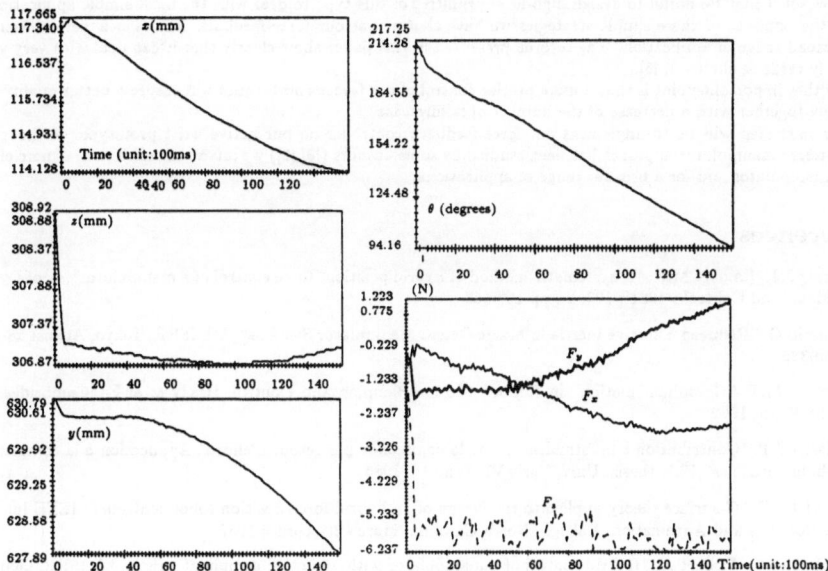

Figure 11: Phase 4: Screwing of the plug. Positions of the gripper and contact forces during the assembly with active compliance, free wrist and vertical insertion axis.

3   Conclusion

It may seem obvious that some of the assembly tasks presented in this paper can be performed in open loop (for example the assembly involved in Phase 2) but this assumption will be true only if the robotics cell is highly structured. The main problem is the grasping of the different parts involved in the assembly. We may fear that the variation in the shape of the parts and their flexibility can induce large grasping errors yielding to failure of the assemblies. Thus the need of a force sensor to detect the failure is clearly established.

In our case the test bed is not really structured: the initial position of the parts at the beginning of the assembly are determined "by hand" and this yields to rather large errors. We have also chosen to use only high-level primitives which can deal with a broad range of application and are not specific to the considered tasks. In order to use these primitives for this specific application one has to determine at most a few set of parameters : the insertion axis vector (which can be only a rough estimate) and one or two gains which are mainly related to the material of the parts and not to the assembly. We may thus expect that after a few trials a set of parameters usable for a broad range of assembly tasks will be determined.

As for the force controller we have shown that the most classical approach (i.e. guarded move strategy and hybrid control with a fixed compliance frame) may work for some of the assemblies but will fail for others. This is clearly due to the fact that the basic assumptions on which are based these approaches are not verified (especially for the Phase 2). However it may be useful to design high-level primitive of this type to deal with the most simple applications.

At the opposite of these simple strategies we have shown that our force-feedback controller is very efficient and has a broad range of application. The records presented in this paper show clearly that it can deal with very various assembly tasks as shown in [5].

Another important point is that a more precise control of the forces and torques will ensure a better quality of the assembly together with a decrease of the number of failure case.

Our next step will be to implement our force-feedback controller on our active wrist prototype. Although this macro-micro manipulator approach has been studied by some authors ([6],[2]) we intend to test it with a more efficient micro-manipulator and for a broader range of applications.

References

[1] Craig J.J., Raibert M.H. "A systematic method of hybrid position- force control of a manipulator", J. of Dyn.Sys. , Meas. and Control , 1981 , 103, 2 , pp.126-133

[2] Khatib O. "Reduced Effective Inertia in Macro/Micro Manipulator Systems", 5th ISRR, Tokyo, August 28-31,pp. 329-335.

[3] Mason M.T. "Compliant motion" in Robot , Motion-Planning and Control, Brady & al Ed.,Cambridge , The MIT Press, 1982

[4] Merlet J-P. "Contribution à la formalisation de la commande par retour d'efforts. Application à la commande de robots parallèles" PhD thesis, Univ. Paris VI, June 18, 1986

[5] Merlet J-P. "C-surface theory applied to the design of an hybrid force-position robot controller" IEEE Int. Conf. on Robotics and Automation, Raleigh, North Carolina, March 30-April 3,1987

[6] Reboulet C., Robert A. "Hybrid control of a manipulator with an active compliant wrist", 3th ISRR, Gouvieux, France, Oct. 7-11, 1985, pp.76-80

Section 9: Legged Locomotion

The first paper in this section by T. Mc Geer reports on original research on passive loco-motion, i.e., without motor control: the motion is due only to the interaction of the legs with the ground. Several models of walking machines are discussed and experimented.

The paper by A. Sano and J. Furusho presents a control method for quadrupeds at a pace gaits. The locomotion is hence dynamic since only two feet, on the same side, touch the ground together. The approach is based on a discrete-time model and experimental results show its effrectiveness.

Passive Dynamic Biped Catalogue, 1991

Tad McGeer
Vancouver, B.C., Canada
E-mail to USERTMCG@cc.sfu.ca

Symbols
(relevant formulae are referenced in parentheses)

Roman

c	axial position of mass centre (figure 12)
C	matrix of control gains (22)
I	identity matrix
g	gravitational acceleration
$\vec{H}$	angular momentum
l	leg length
m	mass
R	foot radius
r_{gyr}, r_g	radius of gyration
$\vec{S}$	stride function (1)
$\nabla \vec{S}$	jacobian of $\vec{S}$ (6)
$\nabla_1 \vec{S}$	first row of $\nabla \vec{S}$ (23)
$\nabla_c \vec{S}$	gradient of $\vec{S}$ w.r.t. $\vec{u}$ (25)
$\nabla_T \vec{S}$	gradient of $\vec{S}$ w.r.t. T_H (20)
$\nabla_{T,1}\vec{S}$	first element of $\nabla_T \vec{S}$ (23)
T	torque
$\vec{u}$	vector of control variables (25)
V	translational velocity
w	forward offset from link axis to mass centre (Δy in figure 12)
w_{hip}	hip width (figure 12)
x	coordinate normal to the ground (figure 12)
y	coordinate along the slope (figure 12)
z	eigenvalue (7); lateral coordinate (figure 12)

Greek

α_0	leg pitch angle at support transfer
β	torso inclination
γ	slope, positive downhill
γ_g	slope for gravity-powered walking
Δy	forward offset from link axis to mass centre (figure 12)
Δz	lateral offset from link axis to mass centre (figure 12)
θ	pitch angle (figure 12)
$\vec{\nu}$	start-of-step state vector (1)
σ	dimensionless pendulum frequency (9)
τ	dimensionless time $t\sqrt{g/l}$
τ_0	step period
ϕ	roll angle (figure 12)
ψ	yaw angle (figure 12)
Ω	rotation rate in pitch
ω	rotation rate

Subscripts

0	steady cycle conditions
C	stance leg
F	swing leg
H	at the hip
k	step index
T	torso or hip

466

Figure 1: A bipedal toy which walks passively down shallow slopes. Energy gained by descending the slope is balanced by energy dissipated each time the swing foot hits the ground. The illustration is adapted from McMahon [8].

Abstract

Passive dynamic bipeds walk and run by virtue of physics inherent in the interaction of their legs and the ground; they need no motor control. A diverse spectrum of passive models are now known; together they offer a lively repertoire, including locomotion at various speeds, up and down hills, in two and three dimensions, and over unevenly-spaced footholds. We review the principal passive walkers investigated to date, including experimental results and references to detailed analyses. These suggest enticing opportunities for design of efficient and dextrous walking machines.

Perhaps legged locomotion has been imposed upon the roboticist's agenda by Hollywood. After all, as any moviegoer knows, *real* robots don't have wheels. Alternatively, perhaps the practical argument is genuinely compelling: legs do promise superb all-terrain transport, although one may be sceptical about applications in the near term. Or perhaps, as for me, their attraction is simply a conundrum. Wheels, wings, skates or skis obviously constitute a sound basis for travelling hither and yon: smooth, efficient, and devoid of any hint of complexity or needless appendage. But legs! All of those bits flailing around – hardly an obvious solution, and yet in their element so much more effective than any device on file in the patent office. To anyone guided by a faith in simplicity, the challenge is disconcerting. Can so successful a scheme be as contrived as it seems? If not, where is the simplicity?

The answer, I believe, lies in *passive dynamic walking.* Through this effect legged locomotion can proceed as a passive interaction of gravity and inertia, without any motor control. The effect is demonstrated most conclusively by the walking toy in figure 1, but our own steps are guided just as surely by the same simple physics. In fact dynamics

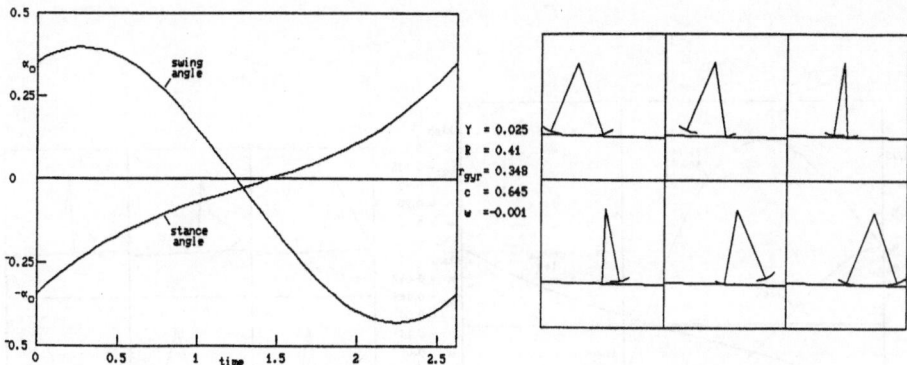

Figure 2: Passive dynamic walking cycle calculated for our first experimental biped descending a 2.5% slope. The plot shows stance and swing angles relative to the surface normal, as functions of time (made dimensionless by gravitational acceleration and leg length). The machine is modelled as two straight legs with semicircular feet and a pin joint at the hip. Parameters of the machine are tabulated in the margin.

of this sort can be seen in quite a variety of legged models, including devices having straight or knee-jointed legs (figures 2 and 3), with or without a torso (figure 4), moving in two dimensions or three (figure 5), and capable of walking or running (figure 6) or hopping (figure 7). Here we will concentrate on walking, tracing our own development from previously-reported work on simpler models through to newer results with knees and three-dimensional motion, and finally shining some light down paths for further experimental and analytical progress.

1   Straight-legged walking in 2D

Our experimental work on passive walking began with the machine shown in figure 8, which was essentially a two-dimensional version of the walking toy. We will not discuss it in detail, since a full report can be found elsewhere [3]. However a review is in order to indicate the central ideas in passive walking theory. The review begins at time zero in figure 2, when the step is just beginning and both legs are momentarily on the ground. The legs must at this instant be at equal and opposite angles ($\pm\alpha_0$), but their rotational speeds remain independently variable and so, at this instant, the model has three degrees of freedom: one angle plus two speeds. When this set is specified the subsequent motion is determined, and will proceed generally as shown in the figure. A small problem arises at midstride, when the swing foot attempts to drop below ground level (cf. figure 5) but contact can be prevented by briefly shortening the swing leg. Once the danger is past the leg can be reextended, and the motion then will continue until the legs again reach equal and opposite angles. At that point the forward foot hits the ground inelastically, dissipating some energy and changing the speeds of each leg The model is then poised to begin another step. If the speeds and angles at this instant are the same as they were at the beginning of the last step, then the model has hit upon a passively re-entrant cycle

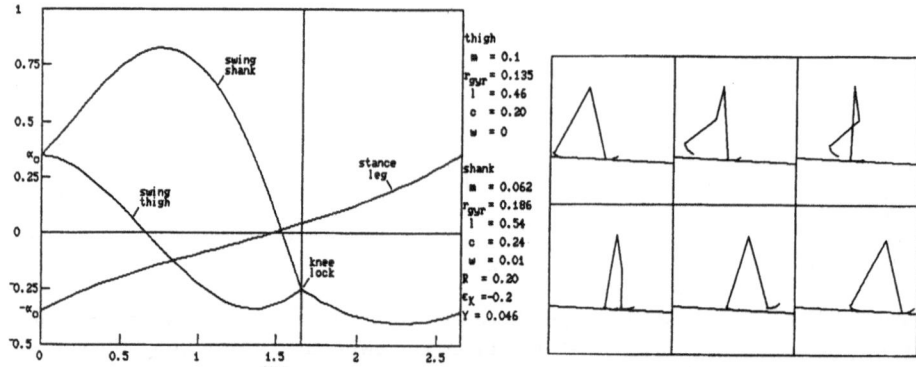

Figure 3: Passive cycle calculated for a pair of anthropomorphic legs having knees free in flexure but stopped against hyperextention. The slope is 4.6%. Model parameters are explained by McGeer [5].

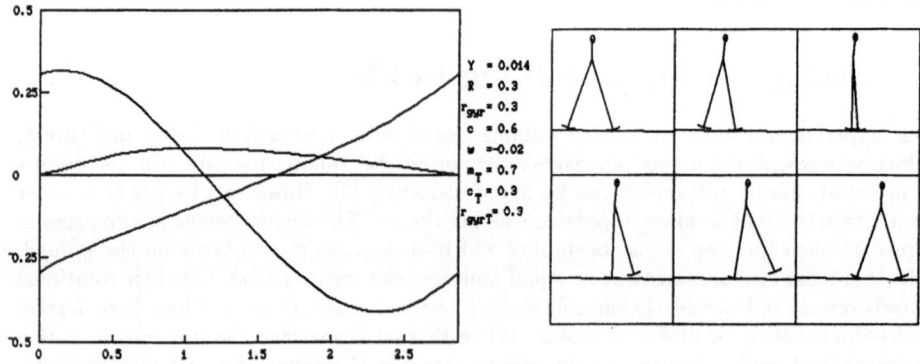

Figure 4: Walking cycle predicted for a straight-legged biped with a torso. This model leaves its legs free to swing passively, but holds its torso upright by actively controlling torque between the torso and stance leg [7].

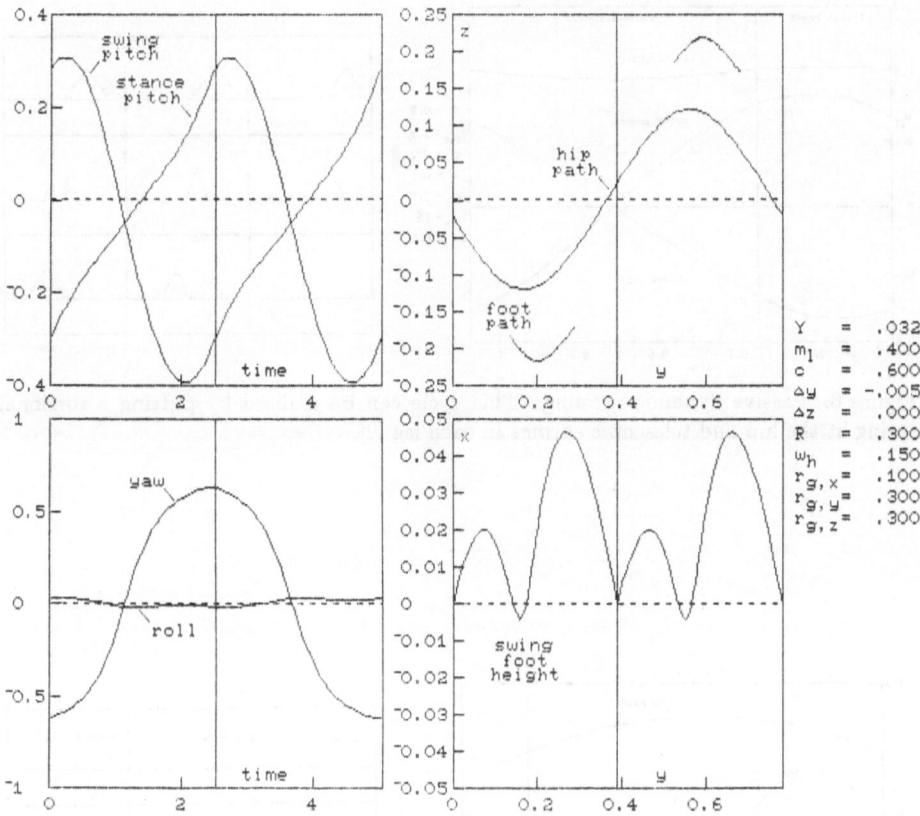

Figure 5: Three-dimensional passive cycle, as calculated for a straight-legged biped having legs separated by 15% of leg length. The slope is 3.2%. The legs' pitching motion is essentially the same as in two-dimensional passive walking, but to maintain lateral balance the model also develops a synchronised roll/yaw oscillation. In this example the yaw amplitude is substantially greater than the roll, and the cycle involves a pronounced lateral swaying of the hips (plotted in the upper right panel in units of leg length). As in the two-dimensional cycle the swing foot is calculated to pass slightly below ground level at midstride, but this problem could be remedied by knees.

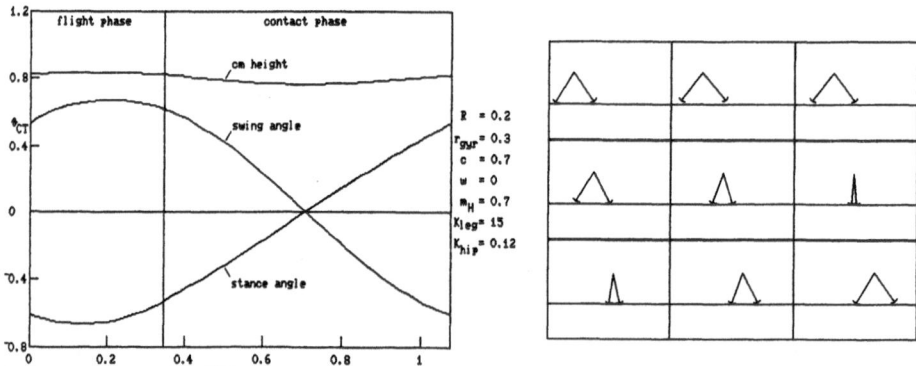

Figure 6: Passive dynamic running. This cycle can be realised by putting a torsional spring at the hip and telescopic springs in each leg [2].

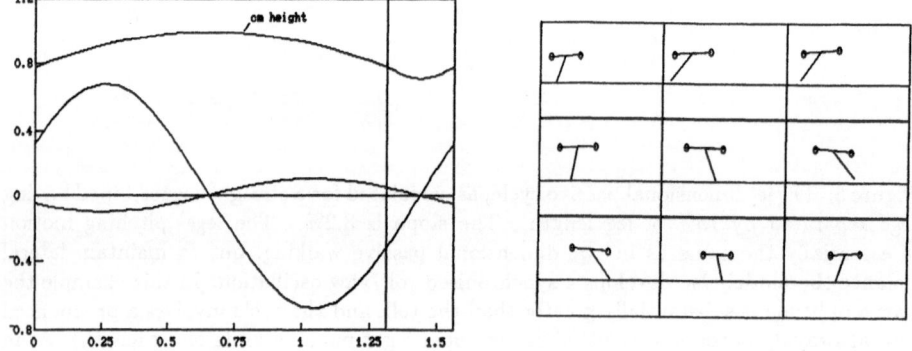

Figure 7: Passive dynamic hopping. Like passive running this cycle involves scissoring back and forth on a hip spring while bouncing on a leg spring [11].

Figure 8: Our original straight-legged biped walking passively down a 2.5% slope. Weight is 3.5 kg and leg length 50 cm. Notice that the outer legs are paired like crutches to prevent lateral tipping. A leadscrew mechanism on each leg folds the swing feet sideways for clearance at midstride.

and can keep walking indefinitely.

To explore the feasibility of such cycles we formulate a *stride function* $\vec{S}$, which will map initial conditions $\vec{\nu}$ from one stride to the next:

$$\vec{\nu}_{k+1} = \vec{S}(\vec{\nu}_k) \tag{1}$$

$\vec{S}$ can be formulated analytically using linearised equations of motion, and in past work we have taken this approach [3]. However linearisation causes some noticeable loss in accuracy for gaits of normal speed, so alternatively $\vec{S}$ can be cast as a subroutine incorporating integration of fully nonlinear dynamical equations, recognition of end-of-step contact, and calculation of speed changes due to the impulse at heel strike. Whatever the formulation, however, a cyclic gait is indicated by an argument $\vec{\nu}_0$ which maps onto itself:

$$\vec{\nu}_0 = \vec{S}(\vec{\nu}_0) \tag{2}$$

This is a set of three equations which in general cannot be solved analytically, so we resort to Newton's method. Thus we differentiate the stride function (numerically) and linearise according to

$$\vec{S}(\vec{\nu} + \Delta\vec{\nu}) \approx \vec{S}(\vec{\nu}) + \nabla\vec{S}\,\Delta\vec{\nu} \tag{3}$$

Following Newton's method an initial estimate $\vec{\nu}$ for $\vec{\nu}_0$ can be improved by iteration of

$$\Delta\vec{\nu} = \left[\mathbf{I} - \nabla\vec{S}(\vec{\nu})\right]\left(\vec{S}(\vec{\nu}) - \vec{\nu}\right) \tag{4}$$

where $\mathbf{I}$ is the identity matrix. The result of this procedure depends upon the model parameters and the initial estimate. If the parameters are such that no cyclic gait exists

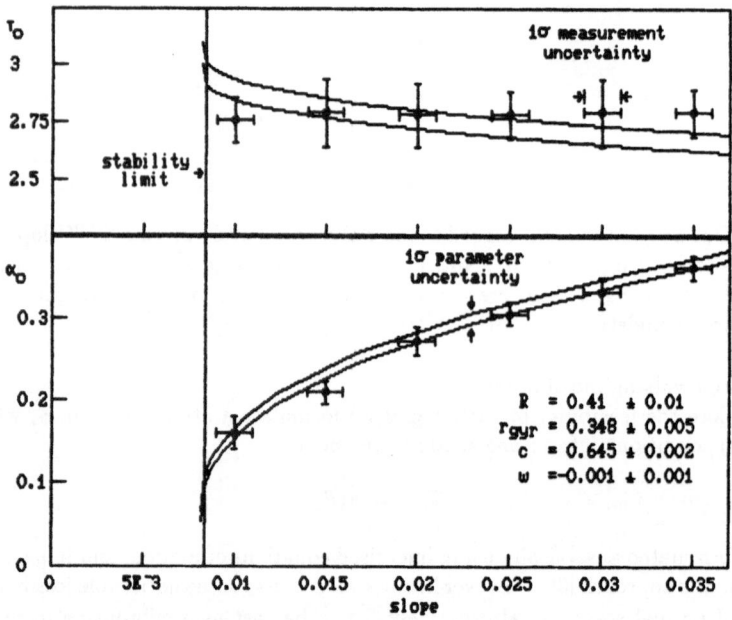

Figure 9: Analytical and experimental results for our first straight-legged biped. One time unit is 0.226 sec. Two sets of cyclic gaits can be calculated for this machine, but only the longer-period gaits could be observed quantitatively. The others were unstable. Each experimental point represents a composite of five trials on a 4 m ramp. Disagreements between the calculations and measurements are due in part to basing calculations on linearised equations of motion, and possibly in part to slightly elastic support transfer.

(*e.g.* on an uphill slope) or if the initial estimate is poor, then the iteration will quickly blow up. Otherwise, however, it will converge to one of *two* distinct solutions: that is, a model such as our experimental machine is in fact capable of *two* distinct passive gaits. Figure 9 shows the features of these gaits calculated over a range of slopes. It also shows the experimentally-measured features, which correspond to only one of the gaits. We have failed to observe the other, except over occasional fleeting steps, because it is unstable. The instability can be demonstrated using the linearised stride function. Suppose that at the start of step k the state vector is perturbed from $\vec{\nu}_0$ by $\Delta\vec{\nu}_k$. Then at the start of the next step $\Delta\vec{\nu}_{k+1}$ satisfies, according to (1) and (3),

$$\vec{\nu}_0 + \Delta\vec{\nu}_{k+1} \approx \vec{S}(\vec{\nu}_0) + \nabla\vec{S}\,\Delta\vec{\nu}_k \tag{5}$$

Cancelling the cyclic terms (2) leaves

$$\Delta\vec{\nu}_{k+1} \approx \nabla\vec{S}\,\Delta\vec{\nu}_k \tag{6}$$

This is just a set of three linear difference equations. Therefore transients following small perturbations can be analysed into three modes of the form

$$\Delta\vec{\nu}_k \sim \vec{\nu}_z\,z^k \tag{7}$$

where z is an eigenvalue of $\nabla\vec{S}$, and $\vec{\nu}_z$ is the associated eigenvector. For stability all three eigenvalues must have magnitude less than unity, with smaller magnitudes indicating that fewer strides are required to recover from a disturbance.

Table 1 lists the eigenvalues and eigenvectors calculated for the experimental machine walking on a 2.5% slope. We have given the modes names which roughly indicate their physical role: the "speed" mode is the principal means for eliminating perturbations in forward speed (reflected by its large amplitudes in Ω_C and α); the "swing" mode is the principal means for eliminating perturbations in Ω_F; and the "totter" mode is the principal means for responding to a mismatch between forward speed and step length (which normally satisfy $\alpha_0 \approx \Omega_{C_0}\tau_0/2$). More detailed discussion of them modes is offered in [3]. In the longer-period gait all three modes of the machine are stable, but in the shorter-period gait its totter mode is strongly unstable and quickly leads to toppling. With different parameter sets this problem can be tamed somewhat, but we have not found any cases in which the short-period gait is positively stable. In the long-period gait stability is much more reliable, being the rule over wide parameter variations, and holding for perturbations of useful (not merely infinitesimal) magnitude. Thus for example in experimental work we can start the machine manually, and rely on its own dynamics to handle imperfections in our technique.

2 Passive walking with knees

Our first machine left us with a lingering sense that the straight leg was not sufficiently realistic, and so in time we launched into a study of knees. We imagined a roughly anthropormorphic model, with the knee hinged freely in flexure but blocked mechanically against hyperextention. We formulated its stride function for motions of the general form shown in figure 3, with four initial conditions (one angle plus three rotational speeds), a

Table 1: Modes of the straight-legged machine on a 2.5% slope

mode	long-period gait			short-period gait		
	speed	swing	totter	speed	swing	totter
eigenvalue, z	0.692	-0.050	-0.825	0.591	0.076	9.606
α	0.653	0.136	0.993	0.391	0.191	0.643
Ω_C	0.736	0.149	0.113	0.638	0.385	0.592
Ω_F	0.179	0.979	-0.046	0.663	0.903	0.485

knee-free swing phase, an impulsive knee lock, a knee-locked swing phase, and an impulsive heel strike. We then set about searching for cycles following Newton's method. Our exploration was undertaken with fair optimism, since an earlier study of *ballistic walking* by Mochon & McMahon [9] had shown that passive dynamics could get a knee-jointed model through at least part of a walking cycle. Fortunately the promise of this work was soon borne out by our new stride function. Paired cyclic solutions emerged over a wide range of parameter variations, as in the straight-legged model. Anthropomorphic parameters proved to be particularly good, with a nicely stable long-period gait, comfortable ground clearance for the swing foot, and a positive locking torque on the stance knee maintained naturally throughout the stride. Thus the active mechanism used to clear the swing feet of our first biped now could be eliminated.

A selection of analytical results for these cycles is given in [5]. Here we add experimental data obtained with a machine called *Dynamite*. Figure 10 shows the machine, and figure 11 the calculated and measured walking behaviour. The agreement is better than in our straight-legged analysis because here the stride function was formulated using fully nonlinear dynamics. However while the solution for steady walking apparently is correct, experience suggests that the stability calculation is conservative. *Dynamite* regularly was able to walk the full length of our test ramp ($\approx$ 5 m) despite a calculated totter-mode eigenvalue of −1.5. Since both the steady-gait and stability analyses rely upon the same stride function, it is odd that there should be agreement on one and not the other; this discrepancy remains unresolved.

In the experimental work we also tested *Dynamite* with knees locked, thus reverting to the straight-legged model, and revisiting the problem of swing foot clearance. The solution for these trials was to put the machine on a chequerboard pattern of tiles. As the figure indicates gait timing and step length then proved to be quite close to those with knees-free, so the straight-legged dynamics continued to operate beneath all the new features.

Of course several important lessons had to be learned to obtain these results, and the main points to be made are as follows.

1. *Debouncing.* As the knee reextends during the latter part of the step, it makes hard contact with a mechanical stop. Once brought to rest against the stop it will stay there passively, but if it bounces it may still be flexed when the leg hits the ground. The result is usually catastrophic. To prevent bouncing and its unpleasant consequences we first tried hydraulic dampers under the kneecaps, but these proved

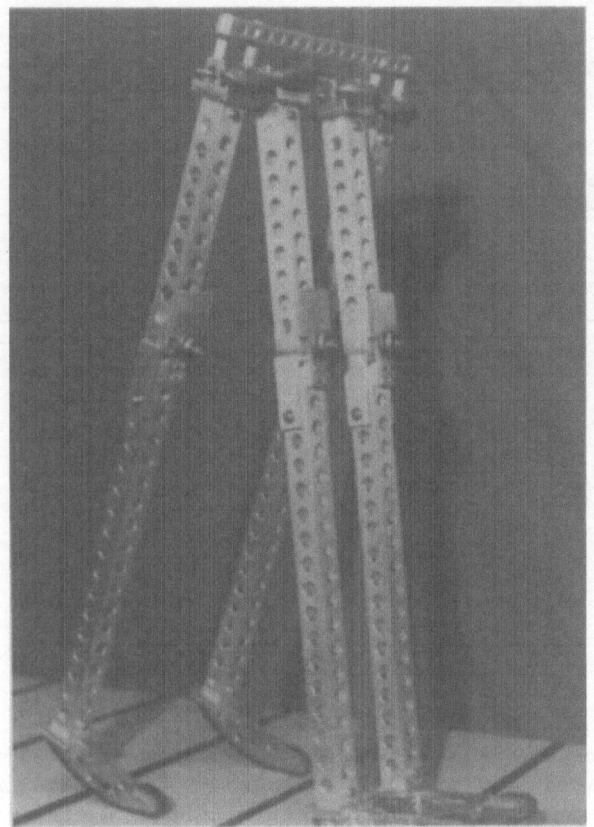

Figure 10: *Dynamite*, a two-dimensional biped with knees. Leg length is 80 cm, and mass 6.2 kg. Note the fishing weights on the hips, and on the shanks above the knees. The shank weights are attached to paddles which in this view hide the suction-cup knee debouncers.

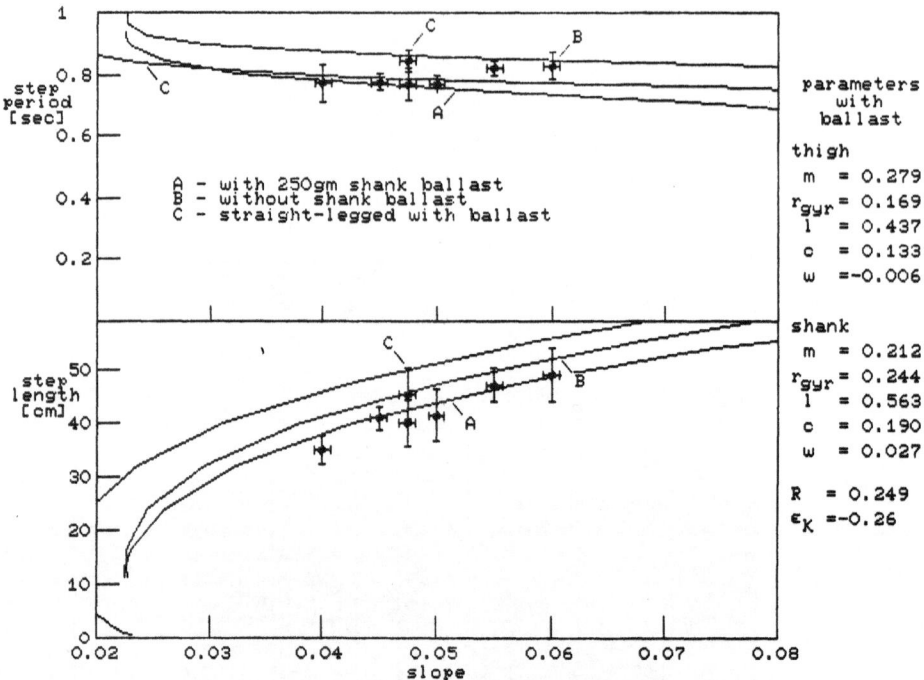

Figure 11: Walking behaviour of *Dynamite*. For the trials on 5% and shallower slopes the shanks were ballasted for improved foot clearance, as described in the text. On steeper slopes the ballast was eliminated. Each knees-free data point represents a composite of five trials, each of which included about a dozen steps. Fewer data were obtained in the straight-legged experiments. Difficulties arose because the machine had to be kept on a chequerboard pattern of elevated tiles for foot clearance, and moreover the straight-legged totter mode was calculated to be unstable. Nevertheless over many attempts we did manage a couple of straight-legged walks from start to end of the test ramp.

ineffective. Next we installed momentary mechanical catches, but these were tempermental and so not satisfactory. Finally we adopted a leaky suction cup, which proved very effective once we hit upon the right size for the leak. Sizing, however, remains tricky: the leak must be sufficiently slow to prevent ventilation of the cup during the few milliseconds of rebound, but at the same time must be sufficiently fast to prevent the cup from "grabbing" the shank when it is jerked into flexure by the impact of heel strike.

2. *Ballasting.* Impulsive energy losses, and hence the slope required for steady walking, are reduced by raising the overall mass centre of the model. *Dynamite* therefore had about 1 kg of lead across the hips. Furthermore for experiments on shallower slopes we also added 125 gm to the top of each half shank; the added inertia carried the swing leg into deeper flexure following support transfer, and so increased foot clearance. In general ballasting scales with the skeletal weight, and often carries an appreciable multiplier, so one must be careful to keep the basic structure light. *Dynamite* grew a bit out-of-hand in this respect, and so when it fell it was something of a hazard. A later machine, *Dynamite Jr.*, was smaller and caused its investigators much less apprehension on this account.

3. *Foot placement.* Whereas on the original straight-legged biped the feet were symmetric about the leg axis, on *Dynamite* they were displaced forward according to the anthropomorphic model. The reason was not just aesthetic. Forward displacement is required to put the contact force vector forward of the stance knee, and so keep it locked against its stop. Anthropomorphic sizing of the feet is also preferable, since much larger feet would reduce swing foot clearance, while much smaller feet would both increase dissipation at heel strike [3] and fail to provide locking torque unless the stance knee were allowed to hyperextend. Foot curvature is less important, but our semicircular shape is convenenient mathematically.

4. *Shank and thigh proportions.* A more subtle point of anthropomorphic design arose after our first iteration on *Dynamite* design. After a variety of computations had demonstrated that the knee-jointed model was reasonably robust we designed the experimetal machine. When assembly was complete we measured its mass distribution, and then set about calculating adjustments to be made by ballasting. The calculations revealed that no reasonable amount of ballasting would produce a satisfactory gait, that is, with stable eigenvalues, adequate foot clearance, and positive locking torque on the knee throughout the stance phase. We then systematically compared *Dynamite* with our successful anthropomorphic model, and discovered that the key difference was in the relative lengths of the shank and thigh. *Dynamite's* proportions were 50/50, but people have closer to 54/46; the shorter thigh turns out to generate substantially deeper knee flexion during the swing phase. We quickly modified *Dynamite* to 56/44, and immediately discovered a much broader range of ballasting options.

5. *Confidence in the analysis.* In responding to inquiries while preparing for experiments with *Dynamite*, we would reply quite honestly that if it didn't work we would be puzzled, but if it did work we would be amazed! When faced at the outset with

a machine whose only demonstrated abilities are very suddenly to topple face first or to buckle at the knee, analytical results often seem rather remote from the reality at hand. Under these circumstances persistence and flexibility are helpful, but one must above all maintain confidence that the analytical model is sound and that the mechanical implementation is faithful. We in particular had the problem of how to start the machine, which we tried to address through more than a month of unsuccessful tinkering with an automatic launcher. Ultimately we hit upon a manual starting technique that worked with passable reliability, and so obtained the results discussed above.

Of course once viewed in action the knee-jointed cycle, however surprising on paper, soon seems perfectly normal: seeing is believing. (Interested readers can obtain video for this purpose either from the IEEE [4] or directly from me.) Unfortunately, however, we still cannot claim good intuition about the physics of the cycle. Ballasting, for example, is guided by calculations and previous results rather than by a more holistic feeling for the dynamics. This remains a handicap, as better intuition would doubtless accelerate the evolution of improved walking models.

3   Three-dimensional walking

Let us now step out of the two-dimensional bounds of the preceding pages and consider how to operate in a three-dimensional world. As a first intermediate imagine a biped which is two-dimensional in design (that is, which has no lateral thickness and so can swing its legs past each other in the same plane) but nevertheless has freedom to move in three dimensions. Then its "three-dimensional" steady gait is the same as in 2D, and the only new complication is associated with stability – or more to the point, *instability*, since in the lateral plane the model behaves more or less as inverted pendulum. (One might hope for a "rolling coin" dynamic effect to stabilise the lateral motion. However the rolling coin owes all of its stability to spin, and a biped doesn't have the necessary store of angular momentum since its legs counter-oscillate.) Thus a small perturbation in yaw will quickly cause a steepening spiral turn followed by a sideways crash.

At this point we might observe that the 2D model also has inverted-pendulum features in the longitudinal plane, but manages to stabilise itself by periodically exchanging support between legs set some distance apart. Following this observation we might expect out-of-plane stability to benefit from lateral leg separation. Lateral separation, however, raises the question of whether and how a passive model might counter the unsupported weight of the swing leg. One can certainly attempt to reason out a scheme, but for those whose 3D intuition is unreliable there is also the option of blind recourse to numerical methods. We conceded membership in the latter category and so developed a new stride function for the 3D model sketched in figure 12. This model has *seven* start-of-step variables: yaw, roll, and stance pitch angles (swing pitch is determined by the contact condition) plus rates about the three stance leg axes, plus interleg pitch rate about the hip. The motion is calculated using fully nonlinear dynamics (which for security we derived once by hand and again using the *Autolev* program [10]). End-of-step conditions are conservation of the angular momentum vector $\vec{H}$ of the whole machine about the point of heel strike, and of $\vec{H} \cdot \hat{z}_c$ about the hip. These determine the change in rotation rates at

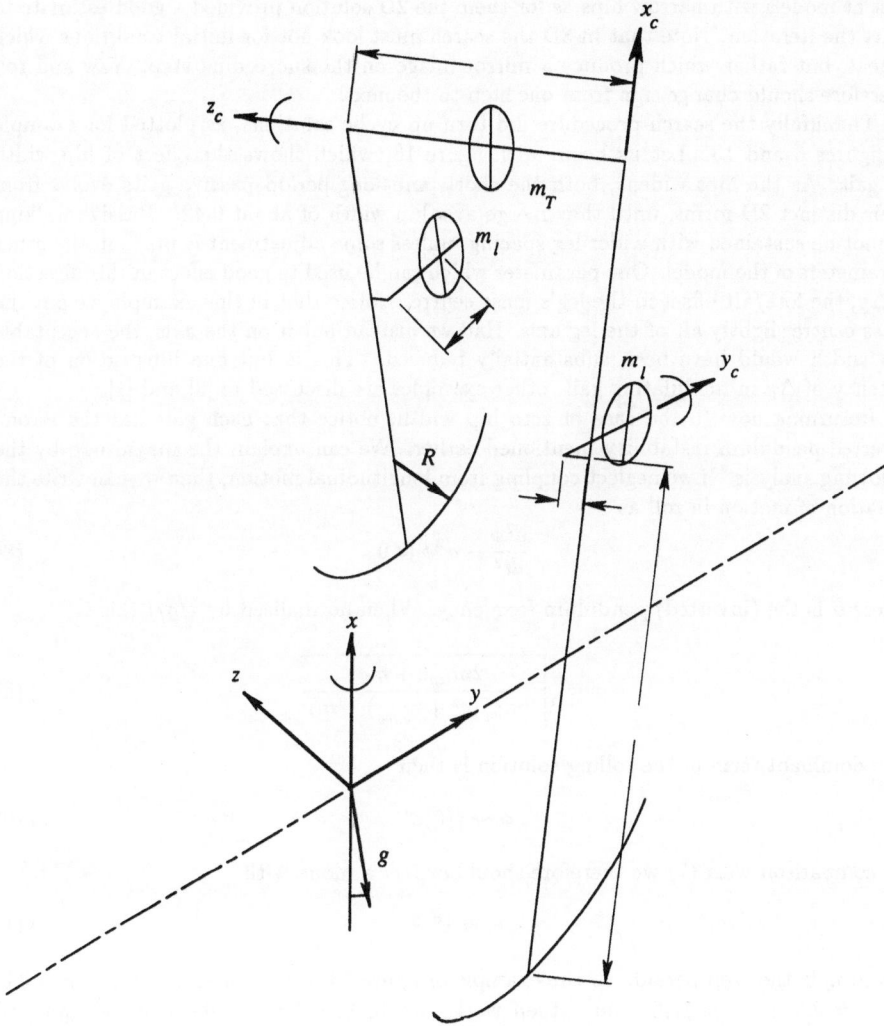

Figure 12: A three-dimensional passive biped. Like the original 2D model it has straight legs and semicircular feet. Orientation of the stance leg relative to the ground coordinates is determined by the yaw (ψ), roll (ϕ), pitch (θ) sequence of rotations about the axes indicated. The swing and stance legs have the same yaw and roll angles, but can pitch independently. Mass centres are shown for the legs and hip; those on the legs have 3D rotational inertia, while the hip axle carries only a point mass.

impact, and so together with the equations of motion complete the stride function. Once this was encoded in FORTRAN we applied the cycle-searching procedure (4), looking first at models with narrow hips as for them the 2D solution provided a good estimate to start the iteration. Note that in 3D the search must look not for initial conditions which repeat, but rather which produce a mirror image on the succeeding step. Yaw and roll therefore should change sign from one step to the next.

Thankfully the search procedure did turn up cyclic solutions, as plotted for example in figures 5 and 13. Let us begin with figure 13, which shows the effect of hip width on gait. As the hips widens, both the short- and long-period passive gaits evolve from their distinct 2D forms, until they merge at a hip width of about $0.42l$. Passive walking cannot be sustained with wider leg spacing, unless some adjustment is made in the other parameters of the model. One parameter which can be used to good effect in this direction is Δy, the fore/aft offset in the leg's mass centre. Notice that in this example we put the mass centre slightly aft of the leg axis. Had we instead put it on the axis, the acceptable hip width would have been substantially reduced. This is but one illustration of the potency of Δy in modulating gait; other examples are discussed in [3] and [1].

Returning now to the case of zero hip width, notice that each gait has the strong inverted-pendulum instability mentioned earlier. We can explain the magnitude by the following analysis. If we neglect coupling from longitudinal motion, then we can write the equation of motion in roll as

$$\frac{d^2\phi}{dt^2} - \sigma^2\phi \approx 0 \tag{8}$$

where σ is the (inverted) pendulum frequency. When normalised by $\sqrt{g/l}$ this is

$$\sigma = \sqrt{\frac{2m_{leg}c + m_T}{2m_{leg}(c^2 + r_{gyr_y}^2) + m_T}} \tag{9}$$

The dominant term in the rolling solution is then

$$\phi \sim \phi(0)e^{\sigma\tau} \tag{10}$$

By comparison with (7) we therefore should expect a mode with

$$z \approx e^{\sigma\tau_0} \tag{11}$$

where τ_0 is the step period. In the example of figure 13, $m_{leg} = 0.4$, $m_T = 0.2$, $c = 0.6$, $r_{gyr_y} = 0.3$, so σ is 1.10, and z then works out to 9.96 for the short-period gait and 20.1 for the long-period. By comparison the values calculated using (7) are about 16 and 24, respectively; the aggravated instability in the full model apparently arises from an "anti-rolling-coin" effect due to the angular momentum of the swing leg.

As anticipated this spiral instability moderates if some lateral separation is put between the legs. Unfortunately that is not to say that the instability disappears. Instead as w_{hip} increases not only does the spiral mode ultimately "bottom out" with $|z| \approx 5$, but several other modes go unstable as well. Table 2 lists these bad actors for the long-period gait with $w_{hip} = 0.15$. Furthermore explorations through parameter space have failed to turn up any set of parameters with which the situation is much better. Therefore some

Table 2: Unstable modes of example long-period gait with $w_{hip} = 0.15$

mode	spiral			directional
eigenvalue, z	-10.340	4.341	2.224	-1.066
θ_C	-0.319	0.547	0.471	-0.009
ϕ	0.592	0.166	-0.056	0.032
ψ	0.074	-0.688	-0.364	-0.999
ω_x	0.380	0.145	-0.176	-0.010
ω_y	0.608	0.223	-0.179	0.002
Ω_C	0.061	0.136	0.675	0.008
Ω_F	-0.157	0.333	0.351	0.010

variation on the model appears to be necessary for stabilisation of 3D passive walking, and we will suggest measures below.

However let us not be too hasty to dismiss the present model, for the character of its steady cycles is well worth examining. Notice first that the longitudinal motion is hardly different from that in the 2D cyclic gait (*cf.* figures 5, 2) save for the fact that in 3D the motion is somewhat accelerated. Hence the lessons of the straight-legged 2D model can be applied in the 3D case. Presumably the same would prove true for a 3D model with knees, which in particular we would expect to eliminate the problem of midstride swing clearance.

Moving on to the uniquely 3D behaviour, notice that the amplitude in roll is nearly nil while that in yaw is quite large. This combination is surprising in view of the exactly opposite strategy of our introductory toy. Its cycle develops pronounced roll but little yaw; in fact a quadrupedal cousin of the toy cannot yaw at all. The differences between the models can be ascribed to foot design. In the sagittal plane the toy's feet form a circular section centred *above* the overall mass centre. Lateral rocking then reduces to a lightly-damped pendulum oscillation, and to make the toy work well the body-rock and leg-pitch pendulums should be tuned with frequencies roughly in the ratio 3:4 [3]. The model of figure 12, on the other hand, has blade feet and so cannot rock without dissipating a lot of energy. It therefore eliminates rocking, and instead arranges for lateral balance by exploiting centrifugal effect in a turn. The balance can be demonstrated as follows. The centrifugal acceleration of each mass centre is $V\dot\psi$, V being the local speed tangent to the turn and $\dot\psi$ the yaw rate. The net centrifugal torque, when normalised by mgl, is therefore roughly

$$V\dot\psi \left(2m_{leg}c + m_T\right) \approx \frac{2\alpha_0}{\tau_0} \frac{2\Delta\psi}{\tau_0} \left(2m_{leg}c + m_T\right) \tag{12}$$

where α_0 is the excursion of the stance leg in pitch, and $\Delta\psi$ the excursion in yaw. (Note that the local speed is less than $2\alpha_0/\tau_0$ at the stance mass centre, but greater at the swing mass centre; the differences cancel.) Meanwhile the centripetal torque due to the unsupported weight of the hip and stance leg is approximately

$$m_{leg}w_{hip} + m_T\frac{w_{hip}}{2} = \left(m_{leg} + \frac{m_T}{2}\right)w_{hip}$$

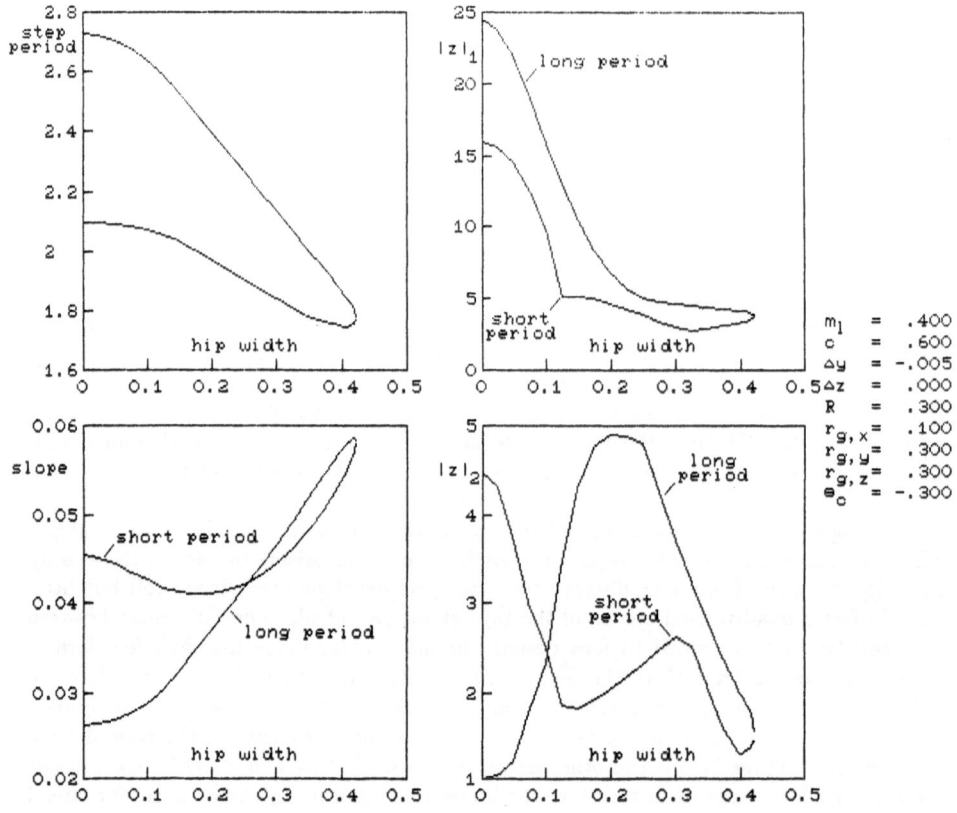

Figure 13: Step periods, equilibrium slopes, and dominant eigenvalues calculated for 3D passive bipeds walking with a stride of about 0.6 leg length (initial $\theta_C = -0.3$). Results for both short- and long-period cycles are plotted against hip width. Evidently passive walking can be maintained even with quite wide hips, although at some cost in dissipation at heel strike (hence the steepening slope). As the width increases the spiral instability is substantially moderated, but other modes go unstable, and the cyclic gaits "pinch off" before absolute stability is achieved. Parameters of the model are listed in the margin.

$$= \frac{w_{hip}}{2} \tag{13}$$

Balancing torques implies that

$$\Delta\psi \approx \frac{w_{hip}}{2} \frac{\tau_0^2}{4\alpha_0} \frac{1}{2m_{leg}c + m_T} \tag{14}$$

In the example of figure 5 this formula suggests $\Delta\psi = 0.58$, which agrees fairly well with the amplitude of 0.64 actually calculated for the cyclic gait.

It is interesting to ask how the model manages to sustain such a large yawing motion. To see the mechanism one should imagine the geometry just at the start-of-step, with the swing leg well behind the stance leg. Its unsupported weight then applies a modest rolling moment to the stance leg, but also a quite substantial yawing moment. The latter moment starts the model turning towards its centreline. The yawing torque drops to zero at midstride, and then builds in the opposite direction so that the turn straightens up by the end-of-step.

Relative to the toy our model has three apparent advantages: first, it is less tippy in roll, *i.e.* given a modest jolt with both feet on the ground, it will damp rocking cycles more quickly. Second, its feet have simpler geometry. Third, it is more robust with respect to parameter variations, because there is no requirement for tuning pitch and roll frequencies. However the model still has unsatisfactory characteristics. The yawing motion is much too exaggerated by anthropomorphic standards, and leads to excessive lateral motion of the hip. Furthermore the instabilities in the gait must somehow be controlled.

Inspection of the human model suggests how the situation can be improved. The most significant point is that we have three-degree-of-freedom hips. The rolling degree of freedom seems particularly useful, as one can quickly surmise by attempting to walk with the same constraints as our model. That is, try walking while keeping the legs as far apart at the feet as at the hips – it will feel awkward! Compare this with walking on a balance beam, with footfalls almost collinear – much easier! Evidently we use the roll freedom to reduce lateral separation between the legs, and so eliminate most of the rolling torque. Moreover the ability to modulate lateral leg separation provides a powerful control, not only for stabilisation but also for steering [12].

Hence the next recourse for the 3D model becomes evident: install roll joints at the top of each leg. These cannot be left freely hinged like the pitch joints, since if they were then the hip axle would collapse as soon as the swing leg were left unsupported. Instead stiff torsional springs must be included. We are reasonably confident that a model with joints thus stiffened will have passive gaits, but we are less sanguine about stability: it will have four modes in addition to the present model's seven, and only a committed optimist would imagine all eleven eigenvalues clustered within the unit circle! However one could stabilise the cycle by varying the equilibrium positions of the roll springs; in the next section we will outline how to develop a control law.

4 Pumped passive walking

All of the models considered thus far have been analysed under gravity power. This formulation is attractive in that it lets the models do entirely as they like, so to speak, unencumbered by any muddling of control with dynamics. However if we look carefully at how the models do as they like, we can also discover how they could do more of what we want – in particular, how they could walk uphill.

As an aid to looking carefully, imagine a reversible projector running film of a straight-legged biped (in either 2D or 3D). Run the projector in one direction, and you see the model going downhill. Run in the other, and you see it going uphill. But which direction is forward in time? You cannot say! On the one hand, you might be looking at a biped strolling downhill, dissipating energy in an impulse at each heel strike. But on the other hand, you might be looking at a biped pumping uphill, supplying energy by an impulse at each toe-off. Both interpretations are fully consistent with the equations of motion.

Of course one might protest that this sounds like a lot of smoke-and-mirrors: the biped with supposed upward mobility couldn't generate those impulses in practice, and anyway this trick wouldn't work on a knee-jointed model, which after all is the most realistic of the lot. These objections would be entirely correct. However, they also would be rather beside the point. With respect to the knee-jointed model, figure 11 indicated that its dynamics are not so unique as they first might appear. And with respect to impulses, indeed they are artificial; however, they also offer a good first approximation to what could be implemented by a machine, and to what actually is implemented (via *dorsiplantarflexion*) by people.

Thus we immediately have a solution for walking uphill: all we need do is calculate the heel-strike impulse in downhill walking, and apply the mirror image. But we are still left with the most basic problem of all: how to walk on the level? The "projector argument" suggests that the solution should be symmetric, with energy input by an impulse at toe-off and dissipated by a mirror-image impulse at heel strike. But we still don't know what impulses will work. Thus we are left again appealling to a cycle hunt via the stride function. In this case the formulation must include impulse-coupling terms as well as the heel-strike conditions and swing dynamics. With this in hand (as derived in [1]) the protocol is simply to specify an impulse (by magnitude and direction) and then apply (4) to find a cyclic gait. The results prove to be quite consient with those for passive models, but still perhaps a bit surprising at first glance.

First, for any specified impulse, cyclic gaits come in pairs, as in the gravity-powered model, but only one of the two gaits is symmetric. Second, such pairs can be generated by a wide range of impulse choices. Figure 14 illustrates these points in the form of a contour plot (for a 2D model) on axes of impulse energy and direction. For any point in this space a pair of gaits can be calculated, although in some areas (show as broken contours) one or the other is formally inadmissable because the impulse would throw the model off the ground. (Mathematically the heel strike impulse then becomes a non-physical downward pull.) The asymmetric gait in the pair turns out to have a cadence exactly twice the frequency of the leg swinging as a free pendulum. Meanwhile the symmetric gait may have a higher or lower cadence depending upon how the impulse is applied.

Figure 14 is good news for engineers aspiring to biped de ign, since its essential message is that anything goes: *any* reasonable choice of impulse will generate at least one gait and quite possibly two. Which (if either) of the two gaits develops depends upon how the biped starts, and upon stability, although with the possibility of varying impulses from one step to the next we could actively stabilise one gait and prevent development of another. We will discuss algorithms after making some further observations on the dynamics of steady

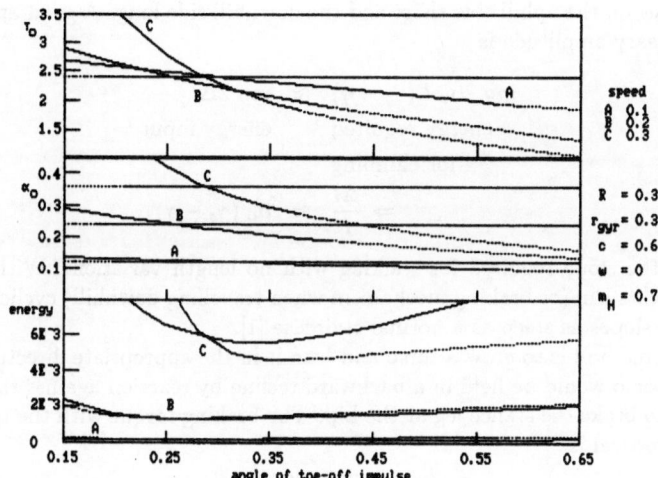

Figure 14: The energy necessary to sustain a walking cycle on level ground can be supplied by a toe-off impulse on the trailing leg. Here the bottom plot shows contours of constant walking speed as a function of impulse energy (normalised by mgl) and angle (forward from the vertical). The top plot shows the corresponding step period, and the middle plot the start-of-step leg angle. It turns out that there are two sets of solutions for each walking speed, and hence two contours.

walking.

Our earlier "projector argument" revealed one impulse for climbing uphill, but none for walking on level ground. Now having seen that there is in fact a multitude of impulses that will work for level walking, we should investigate whether the projector also failed to tell the full story for climbing. Indeed stride-function calculations reveal much more; impulses over a range of angles and energies will generate a cyclic climb. (Details are given in [6] and [1].) However it turns out that the admissable region of impulse space shrinks considerably as the slope increases, so that impulses alone ultimately are restricted only to rather shallow slopes.

For steeper slopes two options are available. The first is to vary leg length cyclically, making the leg on the uphill side short and the downhill side long. A good approximation for the necessary amplitude is

$$\underset{\substack{\text{energy required} \\ \text{for climbing}}}{mg\, 2\alpha_0\, l\, (\gamma_g - \gamma)} \approx \underset{\text{energy input}}{mg\, 2\Delta l}$$

$$\Rightarrow \frac{\Delta l}{l} \approx \alpha_0\, (\gamma_g - \gamma) \tag{15}$$

where γ_g is the slope required for walking with no length variation. With this as the pumping mechanism (or braking mechanism when travelling downhill) cyclic gaits can be sustained on slopes as steep as a normal staircase [1].

A second method is to grow a torso and lean it in the appropriate direction. Thus for descent the torso would be held in a backward recline by reaction against the stance leg; the effect is to brake the stance leg at the hip. The braking torque with the torso at angle β from the vertical is

$$T_H = -m_T g c_T \sin\beta \tag{16}$$

where the torso has mass m_T centred a distance c_T from the hip. The energy dissipated through one step is then

$$W = -\int_{-\alpha_0}^{\alpha_0} T_H d\theta_C = 2 m_T g c_T \sin\beta\, \alpha_0 \tag{17}$$

Meanwhile the excess energy gained in descending slope γ is

$$E = 2 m g l \alpha_0 (\gamma - \gamma_g) \tag{18}$$

so that the required torso recline is

$$\beta = \sin^{-1}\left(\frac{-m}{m_T}\frac{l}{c_T}(\gamma - \gamma_g)\right) \tag{19}$$

Cycle analysis for the 2D biped with torso shows that such a model can maintain stable descents on moderately steep slopes [1]. Climbing then follows by the projector method: a film of a descending biped with a backward inclination, dissipating energy via impulses and braking, could just as easily show a climbing biped with a forward bent, thrusting itself upward via impulses and hip exercise. Again both interpretations are dynamically admissable.

However a problem with relying on the torso to facilitate climbs and descents is that it isn't very powerful; with anthropomorphic values of $m_T = 0.7$ and $c_T = 0.3$, (19) dictates that the torso lean seven degrees for each degree of slope. Hence on even moderately steep slopes it runs out of leverage, and the model must revert to leg-cycling. A second problem is that the the torso is essentially an inverted pendulum, so that the torso/stance leg torque has to be adjusted actively to keep it near the desired angle. Thus for example in figure 4 the stance/torso torque is specified by a proportional-derivative feedback law on torso angle. It is interesting that human evolution has accepted this control task rather than arrange for the torso to be statically stable, as might perhaps be accomplished by moving the hips up to the shoulders. One reason might be that when standing still there would remain a control problem due to the static instability of the legs; solving that takes care of the torso as well.

5 Stability augmentation and gait modulation

Let us now take up the question of how to actively stabilise a passive gait, and how to modulate it from step to step. For a first example we shall suppose that the control is torque T_H between the legs (applied in practice by adding T_H to the stabilising torque between the stance leg and torso, and offsetting it by $-T_H$ between the swing leg and torso). We will allow T_H to change between steps, but will hold it constant through any one step. To estimate its effect we use the stride function. Thus from (5)

$$\vec{S}(\vec{\nu}_0 + \Delta\vec{\nu}_{k+1}, T_H) \approx \vec{S}(\vec{\nu}_0, T_H = 0) + \nabla\vec{S}\,\Delta\vec{\nu}_k + \nabla_T\vec{S}\,T_{Hk} \tag{20}$$

Cancelling cyclic terms as in (2) leaves

$$\Delta\vec{\nu}_{k+1} \approx \nabla\vec{S}\,\Delta\vec{\nu}_k + \nabla_T\vec{S}\,T_{Hk} \tag{21}$$

Here we have a set of linear difference equations in standard form, which invites a standard linear control law:

$$T_{Hk} = \mathbf{C}\,(\vec{\nu}_0 - \vec{\nu}_k) \tag{22}$$

Familiar methods of controller design (e.g. LQR) can now be brought to bear. (Formal controllability is unlikely ever to be a problem.) [2] gives an example of this procedure applied to active stabilisation of running. Notice that to implement the controller one need sample the state of the model only once per stride.

While mathematical methods of controller design thus come readily to hand, one can also imagine a physical approach motivated by what might be called the "stepping stone" problem. Here the objective is to precisely modulate footfalls from one step to the next, which implies that α_{k+1} is specified for all k. According to (21), the hip torque required to achieve the desired step length is

$$T_{Hk} = \frac{1}{\nabla_{T,1}\vec{S}}\left(\Delta\alpha_{k+1} - \nabla_1\vec{S}\,\Delta\vec{\nu}_k\right) \tag{23}$$

where $\nabla_{T,1}\vec{S}$ is the first element of $\nabla_T\vec{S}$, and $\nabla_1\vec{S}$ the first row of $\nabla\vec{S}$. Thus in this formulation

$$\mathbf{C} = \frac{1}{\nabla_{T,1}\vec{S}}\,\nabla_1\vec{S} \tag{24}$$

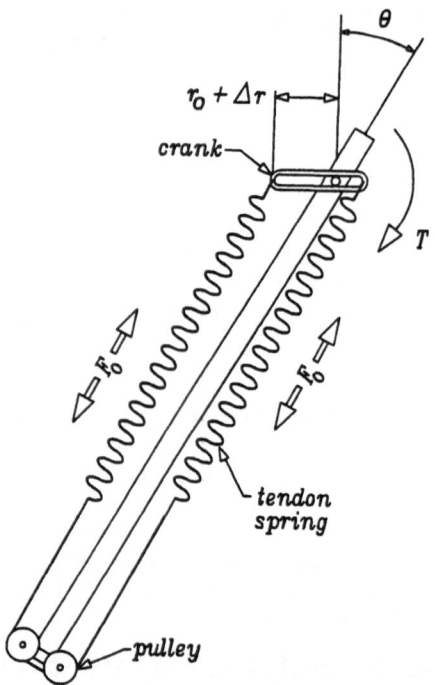

Figure 15: A hip actuator using Levered Isotonic Tendons for Human Emulation: LITHE. The pulleys maintain equal tensions in the two antagonistic tendons, while differential adjustment of their mechanical advantage determines the net torque about the crank. Elasticity in the tendons keeps the tension, and therefore the torque, constant over large swings in leg angle.

In many cases this choice is quite satisfactory. However in others it can lead to unstable modes in the remaining elements of $\vec{\nu}$. If one wants to do "stepping stone" control in these cases, one must invoke other controls – toe-off impulses, length cycling, and (in our proposed 3D development) roll-spring-equilibria – to take care of the stabilisation problem. Thus in general one could write (20) as

$$\Delta\vec{\nu}_{k+1} \approx \nabla\vec{S}\,\Delta\vec{\nu}_k + \nabla_c\vec{S}\,\vec{u} \tag{25}$$

where $\vec{u}$ includes all of the control variables. Following (23) some of these could be used to set elements of $\vec{\nu}_{k+1}$ directly (e.g. α_{k+1} for step length, ψ_{k+1} for steering). One would then be left with a system of reduced order, which could be stabilised using those variables not already "used up" for step-to-step modulation.

6 Powered walking experiments

Work on powered walking has been taken up by the Carnegie-Mellon Robotics Institute, and an experimental machine is nearing completion. It is a straight-legged 2D biped with a torso, having 1 m leg length and an anticipated mass of 20-30 kg. It will be impulse-powered by spring-loaded pistons acting between each leg and foot. These will be cocked by small motors during the stance phase, and then fired just before heel strike.

Hip actuator design has required particular care. At support transfer the stance/torso torque must be switched quickly across the hips; otherwise the suddenly unrestrained swing leg will get a big kick! The hip actuator therefore must be fast. Moreover, it must be supple: when acting on a dangling leg, it should set the equilibrium position, but not affect the pendulum frequency or damping. The geared motors or fluidic actuators used on most mechanical bipeds do not satisfy this requirement; lift one of their legs, and it will hang catatonically or, at best, grind slowly to a halt at the bottom of its swing. With joints like these the physics of passive walking are reduced to so much background noise. To preserve dynamics we have developed the so-called LITHE actuator shown in figure 15. The mechanism has proved to be quite satisfactory in balancing a leg inverted, and we look forward to equally good performance in walking experiments.

7 The collection of results

The models that we have outlined are together capable of a strong repertoire of locomotion, including walking at a range of speeds, in two or three dimensions, up and down slopes steep and shallow, and over unevenly-spaced footholds. All of this repertoire is built upon the robust foundation of the elementary passive cycle, which leads to efficient performance and simple control methods for the design engineer (and, one might imagine, a forgiving feel for a child finding his feet). In this light legs seem not so complicated after all.

At present, however, the elements that we have mentioned remain in pieces. Obviously the long-term objective is to bring them together, and the immediate opportunities for progress seem to be the following.

1. Completion of the 2D, powered, straight legged biped, and demonstration of climbs, descents, and stepping-stone control using impulse energy, torso inclination, and hip torque for pumping and modulation.

2. Design of a 2D powered biped with knees. If left without a torso this will be simpler than the current powered biped project, but more aesthetically pleasing because of the anthropomorphism. Initial analytical work is required to test impulsive pumping in the knee-jointed model (by plantarflexion), but this is likely to be successful in view of the robustness of the passive gait with respect to impulse angle and energy.

3. Analytical development of the proposed 3D model with roll springs at the hips. If this model proves to have cyclic gaits, then an experimental machine could be built (which, as mentioned earlier, would probably need active stabilisation according to the formulation of (25)). Alternatively the analytical work might be extended further by addition of knees.

Should the work on this agenda prove successful, the next step would be to make a steep-slope machine (probably with knees) pumped by leg-length cycling. That in turn would open a number of paths leading to a "full repertoire" biped, but further mapping is best left until the landscape is better known. Of course, there is also an agenda to be developed for passive *running*; figures 6 and 7 hardly begin to scratch the surface. That, however, is another story entirely.

References

[1] Tad McGeer. Dynamics and control of bipedal locomotion. 1991. Submitted to the Journal of Theoretical Biology.

[2] Tad McGeer. Passive bipedal running. *Proceedings of the Royal Society, Series B*, 1240(1297):107–134, 1990.

[3] Tad McGeer. Passive dynamic walking. *International Journal of Robotics Research*, 9(2):62–82, April 1990.

[4] Tad McGeer. Passive dynamic walking with knees. In *Video Proc. 1991 IEEE Robotics & Automation Conference*, Sacramento, California, April 1991.

[5] Tad McGeer. Passive walking with knees. In *Proc. 1990 IEEE Robotics & Automation Conference*, pages 1640–1645, Cincinnati, Ohio, May 1990.

[6] Tad McGeer. Principles of walking and running. In R. McNeill Alexander, editor, *Mechanics of Animal Locomotion*, Springer-Verlag, Berlin, 1991. (In press).

[7] Tad McGeer. *Stability and Control of Two-dimensional Biped Walking*. Technical Report CSS-IS TR 88-01, Simon Fraser University Centre for Systems Science, September 1988.

[8] Thomas A. McMahon. Mechanics of locomotion. *International Journal of Robotics Research*, 3(2):4–28, 1984.

[9] Simon Mochon and Thomas A. McMahon. Ballistic walking: an improved model. *Mathematical Biosciences*, 52:241–260, 1980.

[10] David B. Schaechter, David A. Levinson, and Thomas R. Kane. *Autolev 1.2*. OnLine Dynamics, Inc., 1605 Honfleur Drive, Sunnyvale, California, 1988.

[11] Clay Thompson and Marc Raibert. Passive dynamic running. In V. Hayward and O. Khatib, editors, *Int. Symp. of Experimental Robotics*, pages 74–83, Springer-Verlag, New York, 1989.

[12] Miles A. Townsend. Biped gait stabilisation via foot placement. *Journal of Biomechanics*, 18(1):21–38, 1985.

Realization of Dynamic Quadruped Locomotion in Pace Gait by Controlling Walking Cycle

Akihito Sano

Department of Mechanical Engineering
Gifu University
1-1 Yanagido, Gifu 501-11, JAPAN
Telephone 0582-30-1111
Fax 0582-30-1892

Junji Furusho

Department of Mechanical and Control Engineering
University of Electro-Communications
Chofu, Tokyo 182, JAPAN
Telephone 0424-83-2161
Fax 0424-84-3327

Abstract

In this study, a control method of pace gait which is often observed in middle or high-speed walking of animals is discussed. The motion in pace gait can be divided into the motion in the sagittal plane (plane vertical to the floor including the walking direction) and the motion in the lateral plane (plane vertical to the walking direction). As for the lateral motion control, a dead beat control is adopted for the purpose of adjusting the walking cycle. As for the sagittal motion control, a trajectory of the leg is designed by considering the control algorithm of the lateral plane. The effectiveness of the proposed control method was confirmed by means of computer simulation and also in walking experiments using the quadruped walking robot. Our robot COLT-3 achieved the pace walking at speed of 0.25 m/s.

1. Introduction

Recently, as one of the locomotive means, legged locomotion has been much attracted. The legged locomotion is expected to make robots maintain the facilities of a nuclear power plant or explore a planetary surface, and so forth. Many studies have been published about legged locomotion systems [1]-[6]. A pace gait considered in this study is the walking in which lateral legs form pairs and the members of a pair strike the floor in unison and they leave the floor in unison [11]. The pace gait can be observed when a camel walks in middle or high-speed [12]. On developing a good locomotion robot, a walking of animals is one of ideal models.

For the pace gait, the motion in the sagittal plane and that in the lateral plane can be approximately divided as shown in Fig.1. The control of motion in the sagittal plane is relatively easy since the fore and hind legs on the same side support the body. More particularly, the forward speed and the posture of the body in the sagittal plane can be controlled easily by articular actuators about the pitch axis. On the other hand, the motion in the lateral plane can be approximately represented as an inverted pendulum. The motion of the inverted pendulum can not directly be controlled by means of the joint torque. Therefore, a walking cycle depends on the motion in the lateral plane.

In order to realize a stable pace gait and a smooth walking, it is very important to control actively the walking cycle. Namely, the proper control of the walking cycle ensure the cooperation of the sagittal motion control and the lateral motion control. In this study, a control method of the walking cycle based on the motion control in the lateral plane is adopted. A discrete-time model is suitable for the analysis of the walking system in case of adjusting the walking cycle. By using the proposed discrete-time model, a dead beat control of the walking cycle are discussed in this paper. Moreover, a discrete-time optimal servo controller is also analyzed. The effectiveness of the proposed control method is examined by experiments with our COLT-3 walking robot.

2. Discrete-time Model

2.1 Compensation of Angular Momentum

In the stepping motion, the exchange of the supporting point causes a loss in an angular momentum about the roll axis. Therefore, unless the lost angular momentum is compensated in any way, the angular momentum of the system is attenuated so that the continuation of the stepping motion will be impossible. In this section, an indirect and efficient compensation method which utilizes an gravity effect is discussed by using Fig.2.

Figure 2(a) illustrates the motion in the lateral plane by using the inverted pendulum model. This figure shows the transition from k-1 th step to k th step. The leg-length r, r_k and r_{k+1} represent the length of the leg which was projected on the lateral plane. The leg-length can be adjusted by the bending angle of the knee joint. At first, the system is being supported by the left leg (actually by both of the fore and hind legs) in k-1 th step. The landing occurs after shortening the right leg by u_{k-1} (the leg-length changes from r to r_k). Secondly, the shortened right leg is extended up to the initial leg-length r just after the leg-support-exchange ($r_k \to r$). In k th step, the left leg is shortened by u_k similarly. Then, the supporting leg is exchanged again.

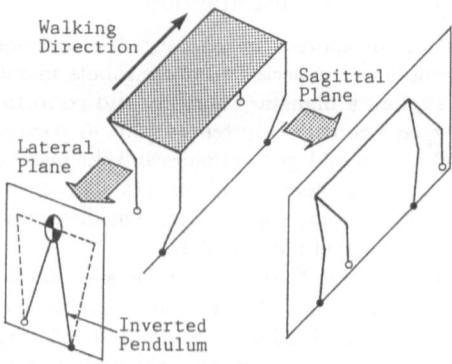

Fig.1 Pace gait

Figure 2(b) shows the trajectory of the center of gravity in k th step. The center of gravity transfers in the order of 1, 2, 3 and 4 according to the arrows in the figure. The number 1 indicates the position of the center of gravity just after the leg-support-exchange. As seen from the figure, since the period $(3 \to 4)$ when the inverted pendulum undergoes the positive moment caused by the gravity effect (the positive moment is defined as the moment which affects in the direction of rotation) is longer than the period $(2 \to 3)$ when it undergoes the negative moment, the lost angular momentum is compensated.

2.2 State Equation Using Inverted Pendulum Model

In this section, the discrete-time model for the stepping motion in the lateral plane is obtained. The relational equations are derived for every phase. The leg-length of the supporting leg is extended from r_k to r during the orbit $1 \to 2$. In this study, we assumed that this motion completes instantaneously because the behavior of the inverted pendulum is not greatly affected by the speed of the motion. In this case, according to the law of the conservation of angular momentum, the following equation holds

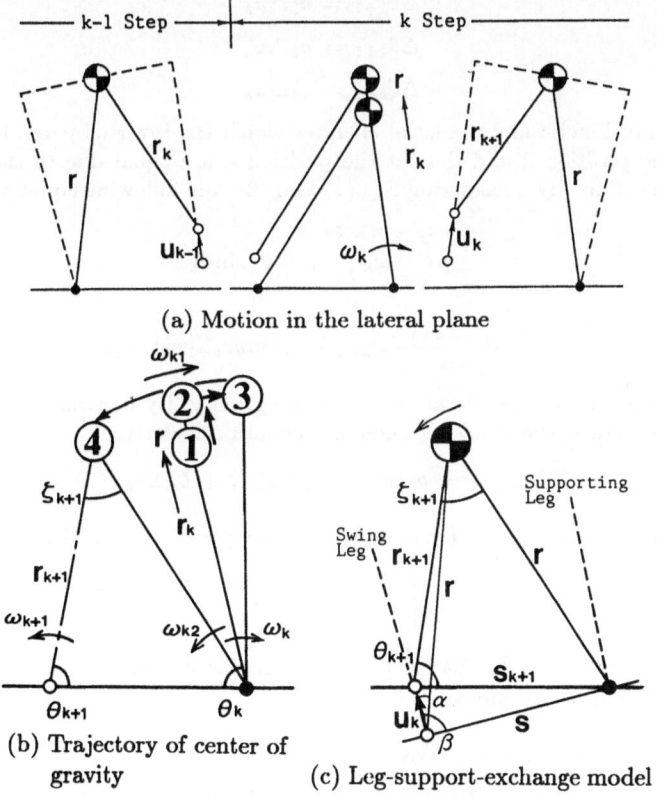

(a) Motion in the lateral plane

(b) Trajectory of center of gravity

(c) Leg-support-exchange model

Fig.2 Inverted pendulum model

$$\omega_{k1} = \frac{r_k^2}{r^2}\omega_k \tag{1}$$

where ω_k is the angular velocity before extending the supporting leg (namely just after the leg-support-exchange), and ω_{k1} is the angular velocity after extending.

Upon the leg-support-exchange which happens at the position 4, the loss of the angular velocity (loss of the angular momentum) takes place as follows:

$$\omega_{k+1} = \frac{r\cos\zeta_{k+1}}{r_{k+1}}\omega_{k2} \tag{2}$$

where ω_{k2} is the angular velocity just before the leg-support-exchange.

Figure 2(c) shows a model of the leg-support-exchange. In this figure, s_{k+1} is a stance, θ_{k+1} and ζ_{k+1} are the configuration angles of the model. When u_k slightly changed around equilibrium state $\bar{u}$, small variations $\Delta r_{k+1}, \Delta s_{k+1}, \Delta \theta_{k+1}$ and $\Delta \zeta_{k+1}$ in the vicinity of the equilibrium state are expressed as follows respectively:

$$\Delta r_{k+1} = a_1 \Delta u_k$$

$$\Delta s_{k+1} = a_2 \Delta u_k$$

$$\Delta \theta_{k+1} = a_3 \Delta u_k$$

$$\Delta \zeta_{k+1} = a_4 \Delta u_k \tag{3}$$

The sum of kinetic and potential energies which the inverted pendulum model is located at the position 2 and that at the position 4 are equal due to the law of the energy conservation. By considering Eqs.(1) and (2), the following equation holds

$$\frac{1}{2}m\left(\frac{r_k^2}{r}\omega_k\right)^2 + mgr\sin\theta_k$$

$$= \frac{1}{2}m\left(\frac{r_{k+1}}{\cos\zeta_{k+1}}\omega_{k+1}\right)^2 + mgr_{k+1}\sin\theta_{k+1} \tag{4}$$

where m is mass and g is acceleration due to the gravity. By linearizing Eq.(4) around the equilibrium state, the following difference equation is derived.

$$\Delta\omega_{k+1} = b_1\Delta\omega_k + b_2\Delta u_{k-1} + b_3\Delta u_k \tag{5}$$

By choosing $\Delta\omega_k$ and $\Delta v_k (= \Delta u_{k-1})$ as state quantities, we obtain the following state equation:

$$\begin{bmatrix} \Delta\omega_{k+1} \\ \Delta v_{k+1} \end{bmatrix} = \begin{bmatrix} b_1 & b_2 \\ 0 & 0 \end{bmatrix} \begin{bmatrix} \Delta\omega_k \\ \Delta v_k \end{bmatrix} + \begin{bmatrix} b_3 \\ 1 \end{bmatrix} \Delta u_k \tag{6}$$

The stepping motion in the lateral plane was replaced with a 2nd order difference equation with Δu_k as the input.

2.3 Relational Equation for Walking Period

In this section, a equation which expresses how the walking period T_k in k th step is affected by $\Delta\omega_k$, Δv_k and Δu_k is obtained. Let the walking period at a steady walking state be T. Figure 3 shows a motion of the inverted pendulum model which follows the extension of the supporting leg just after the leg-support-exchange, where the slant angle of the inverted pendulum model is defined as ϕ. In this case, the equation of motion is expressed as follows:

$$mr^2 \ddot{\phi}(t) - rmg \sin \phi(t) = 0 \tag{7}$$

where initial conditions are given as follows:

$$\phi(0) = \theta_k - \frac{\pi}{2}, \quad \dot{\phi}(0) = \omega_{k1} \tag{8}$$

By linearizing the above equation around the equilibrium trajectory $\bar{\phi}(t)$, the following linear time-varying system is obtained

$$\frac{d}{dt} \begin{bmatrix} \Delta\phi \\ \Delta\dot{\phi} \end{bmatrix} = \begin{bmatrix} 0 & 1 \\ \frac{g \cos \bar{\phi}(t)}{r} & 0 \end{bmatrix} \begin{bmatrix} \Delta\phi \\ \Delta\dot{\phi} \end{bmatrix} \tag{9}$$

Let us express the solution of Eq.(9) with the state transition matrix $H(t,\tau)$.

$$\begin{bmatrix} \Delta\phi(t) \\ \Delta\dot{\phi}(t) \end{bmatrix} = H(t,0) \begin{bmatrix} \Delta\phi(0) \\ \Delta\dot{\phi}(0) \end{bmatrix} \tag{10}$$

where

$$H(t,\tau) = \begin{bmatrix} h_{11}(t,\tau) & h_{12}(t,\tau) \\ h_{21}(t,\tau) & h_{22}(t,\tau) \end{bmatrix}$$

Variation of the slant angle at the time $\bar{T}$ which is caused by changes of the initial angular velocity in the k th step and the length $v_k(= u_{k-1})$ in the k-1 th step is given as follows:

$$\Delta\phi(\bar{T},0) = c_1\Delta\omega_k + c_2\Delta v_k \tag{11}$$

The slant angle $\phi(T_k)$ at the landing time $T_k(= \bar{T} + \Delta T_k)$ is expressed by the following equation.

$$\phi(T_k) = \bar{\phi}(\bar{T}) + \dot{\bar{\phi}}(\bar{T})\Delta T_k + \Delta\phi(\bar{T},0) \tag{12}$$

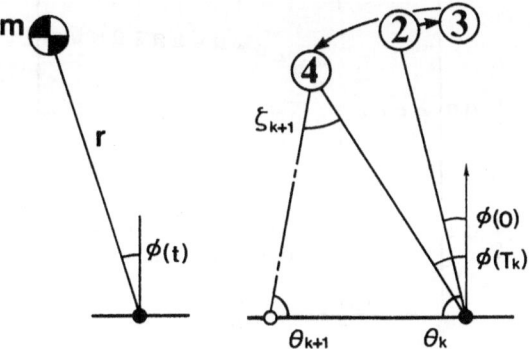

Fig.3 Inverted pendulum model after adjusting the leg-length of the supporting leg

In the above equation, higher order terms were neglected. The second term of the right member represents an effect by a small variation ΔT_k of the walking period, and the third term was given by Eq.(11). As seen from the landing configuration in the Fig.3, the $\phi(T_k)$ can be expressed as follows:

$$\phi(T_k) = \bar{\theta} + \Delta\theta_{k+1} + \bar{\zeta} + \Delta\zeta_{k+1} - \frac{\pi}{2} \tag{13}$$

By arranging with Eqs.(11), (12) and (13), we obtain the following equation.

$$\Delta T_k = d_1\Delta\omega_k + d_2\Delta v_k + d_3\Delta u_k \tag{14}$$

As seen from Eq.(14), a small variation ΔT_k of the walking period can be expressed as a linear function of $\Delta\omega_k$, Δv_k and Δu_k.

The linear discrete-time model consists of Eqs.(6) and (14) was introduced. In Fig.4, the simulation result obtained from the linear discrete-time model is shown with mark $\triangle$.This figure shows the variation of the walking period in case that the input Δu_k is 0.001m. The walking period is plotted as the ordinate, and the number of steps as the abscissa. The mark $\bigcirc$ shows the result obtained from the nonlinear continuous-time model (Eqs.(1), (2), (7) and (8)). The both are nearly in agreement for the change of the input in such a degree as shown in the figure.

3. Walking Cycle Control

Let us express Eqs.(6) and (14) as follows:

$$x_{k+1} = Ax_k + B\Delta u_k \tag{15}$$

$$\Delta T_k = Cx_k + D\Delta u_k \tag{16}$$

where

$$x_k = \begin{bmatrix} \Delta\omega_k \\ \Delta v_k \end{bmatrix}, \ A = \begin{bmatrix} b_1 & b_2 \\ 0 & 0 \end{bmatrix}$$

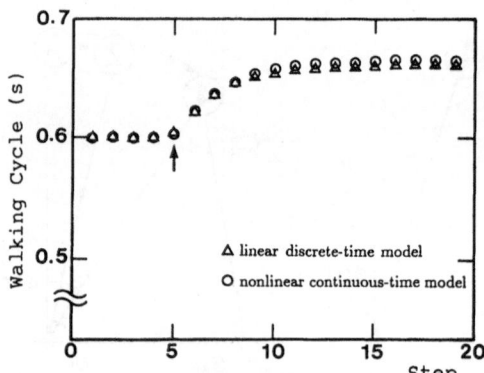

Fig.4 Variation of walking cycle in case that input Δu_k is 0.001 m

$$B = \begin{bmatrix} b_3 \\ 1 \end{bmatrix}, C = [\, d_1 \quad d_2 \,], D = d_3$$

In this section, a dead beat control and an optimal servo controller are applied in order to adjust the walking cycle.

Figure 5 shows the diagram of the controller consisting of the proportional feedback of the state and the integrated feedback of the error of walking period. At first, the parameters of the controller are designed so as to exhibit the dead beat response. Namely, it is possible for the n th system to achieve the dead beat response below n times. The obtained feedback controller is as follows:

$$\Delta u_k = \begin{bmatrix} \bar{x}_k \\ z \end{bmatrix}$$

$$K = [-0.089 - 0.568 \quad 0.046] \tag{17}$$

In Fig.6(a), the result of nonlinear simulation is shown. As seen from the figure, the walking period is nearly settling until three steps since the order of the system including an integral element is three. When the width of variation in the walking period is such a degree as this figure, the dead beat response has been realized even with the feedback controller designed by using the linear discrete-time model.

In general, it is considered that the dead beat control is sensitive to a model error. In the system of this study, the error in the leg-support-exchange model (Eq.(2)) is regarded as the most influential modeling error. Therefore, we examine in advance how the response of the system changes when the coefficient of Eq.(2) slightly varies. Let us rewrite Eq.(2) as follows:

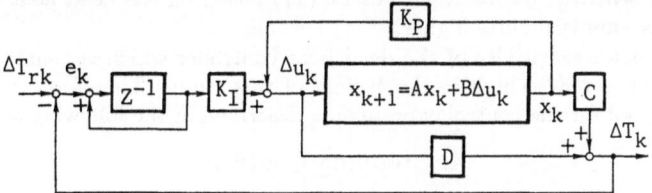

Fig.5 Structure of controller including integrator

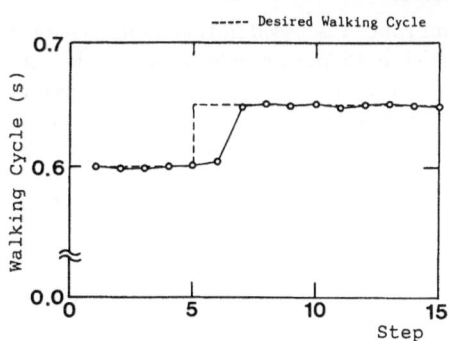

Fig.6 Variation of walking cycle at
the step response

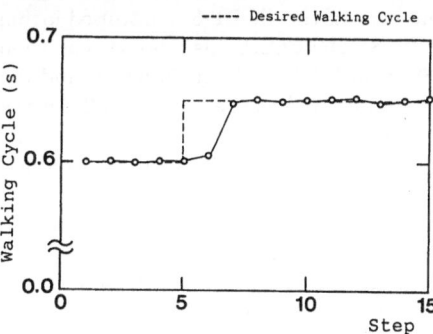

Fig.7 Sensitivity to modeling error by
changing coefficient ρ

$$\omega_{k+1} = \rho \frac{r \cos \zeta_{k+1}}{r_{k+1}} \omega_{k2} \tag{18}$$

where ρ is an variable coefficient, and above equation agrees with Eq.(2) for $\rho = 1$. Let us analyze the sensitivity to the modeling error of leg-support-exchange by changing the coefficient ρ. The dead beat controller is designed for the model with $\rho=1$. Then, the simulation is executed by using this controller and a slightly changed model ($\rho = 1 \rightarrow \rho = 0.9$). The result of the simulation is shown in Fig.7. As seen from the figure, the dead beat control has been almost achieved in spite of the veriation of the coefficient ρ.

An optimal servo controller was designed by using the following performance index.

$$J = \sum_{k=0}^{\infty} (q e_k^2 + r w_k^2) \tag{19}$$

where,

$$e_k = \Delta T_{rk} - \Delta T_k : \text{error signal}$$

$$T_{rk} : \text{reference signal of walking period}$$

$$w_k = \Delta u_{k+1} - \Delta u_k \tag{20}$$

Let us consider the selection of the weighting coefficients q and r. The weighting coefficients are chosen so as to minimize the maximum absolute value of the closed-loop eigenvalues. Then, $q=5.672$ and $r=1.0$ were obtained. The following feedback gain K was also derived.

$$K = [-0.097 \ -0.612 \ \ 0.054] \tag{21}$$

In comparison with the feedback gain of Eq.(17) based on the dead beat control, the above feedback gain is slightly higher.

The parameter sensitivity of the dead beat controller which is usually considered as a weak point is not so high in the system of this study. Therefore, the dead beat control is adopted for the control experiments described in the following section.

4. Stepping Control

4.1 COLT-3 Quadruped Locomotion Robot

One view of the COLT-3 quadruped locomotion robot discussed in this paper is shown in Fig.8. The COLT-3 is a horse type robot which is 45 kg in weight, 0.94 m in length, 0.88 m in height, 0.58 m in width and 0.52 m in leg-length. Each leg has 1 degree of freedom around the roll axis and 2 degrees of freedom around the pitch axis. In total, the robot has 12 degrees of freedom. Each articular joint is driven independently by the combination of the DC servomotor, reduction gear unit and timing belt. Tiptoes of the COLT-3 employ rubber blocks of 10 mm in diameter. Therefore, the behavior of the system during two-leg supporting phase can be regarded nearly as a kind of inverted pendulum motion.

Figure 9 is a schematic view of the actuator and sensor systems of the COLT-3 quadruped robot. An accelerometer is mounted bottom of the body in order to measure the acceleration in the walking direction, and inclinometers and angular rate sensors are mounted on the fore part of the body in order to measure angular rate and slant angle about the pitch and roll axes respectively.

In order to measure the vertical bent, strain gauges are adhered to upper and lower sides of the square pipe to which the tiptoe is fixed. In this study, the signal from these strain gauges was used to detect the landing of the swing leg. The speed sensor utilizing an ultrasonic Doppler effect was made in order to measure the body speed. The ultrasonic wave emitted from the robot is received by a microphone fixed on the floor.

4.2 Control Experiment of Stepping

The effectiveness of the proposed control method is examined by the experiments with our COLT-3. The stepping motion is achieved by using the dead beat controller. It is very difficult to measure the exact value of the $\Delta\omega_k$ because of the vibration caused by the collision phenomenon at the leg-support-exchange. Therefore, the duration $\Delta\eta_k$ after the leg-support-exchange till the angular velocity $\dot{\phi} = 0$ is adopted as a new measured value. $\Delta\eta_k$ can be expressed as linear function of $\Delta\omega_k$ and Δv_k.

$$\Delta\eta_k = f_1\Delta\omega_k + f_2\Delta v_k \tag{22}$$

Figure 10 shows the step response of the walking cycle. The walking cycle is plotted as the ordinate and the number of the steps as the abscissa. The broken line denotes the desired walking cycle, and a white circle denotes the obtained walking cycle. As seen from the figure, the walking cycle is controlled very well. The small vibration in the walking cycle is considered to be caused by mechanical factors like the backlash of the system. As a result, the walking cycle control by the dead beat controller was proved to be good effective.

5. Pace Walking Control

The trajectory of the leg in the sagittal plane is designed carefully by taking the motion control in the lateral plane into consideration. Since various external disturbances act to the robot in walking experiments, the control algorithm which is robust against the deviation of walking cycle (period of touchdown) has to be proposed. The

Fig.8 Quadruped locomotion robot
COLT-3

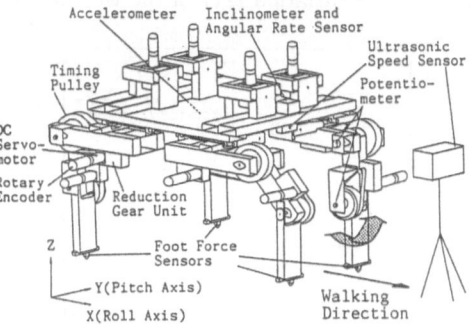

Fig.9 Actuators and sensors

action of the swing leg is divided into the three actions mentioned below. Figure 11 shows the tiptoe trajectory viewed from a coordinate system fixed on the body.

(1) Transferring action: A leg is swung forward up to a prescrived position (point C). At the beginning of this action, the leg is gradually held up while swinging back with respect to the body so that the horizontal velocity of the tiptoe becomes zero with respect to the floor $(A \rightarrow B)$. Then, the tiptoe speed with respect to the body changes from negative to positive, and the leg is swung forward and kept at the point C $(B \rightarrow C)$.

(2) Adjusting action: The position of the tiptoe is adjusted so as to realize the desired u_k. In this period, the leg is gradually swung back $(C \rightarrow D)$.

(3) Waiting action: The horizontal velocity of the tiptoe is kept zero with respect to the floor, and the robot gets ready for landing $(D \rightarrow E)$. In this period, the horizontal distance between the tiptoe of swing leg and that of supporting leg is kept constant. Further, the desired u_k has already been accomplished by the adjusting action. Accordingly, the constant stride length can be realized and the deviation in the walking cycle is permitted to some degree whenever the landing takes place in this period.

As for the trajectory of the supporting leg, the leg is rapidly extended just after the leg-support-exchange $(E \rightarrow F)$. After then, the trajectory is given so as to keep the body at a constant height $(F \rightarrow A)$. In this supporting phase (between E and A), the reference signals for the hip and knee joints of the supporting leg is generated so that the body speed is kept constant.

Figure 12 shows a photographic playback of continual dynamic walking of the COLT-3. As seen from the photographs, the walking of COLT-3 is smooth and dynamic. As a result, the lateral motion and the sagittal motion are sufficiently synchronized.

6. Conclusion

Results obtained in this study are summarized as follows:

(1) A control method of stepping in which the swing leg is shortened in order to compensate the loss of the angular momentum about roll axis was proposed. It was shown that the stepping motion can be described by the 2nd order difference equation (Eq.(6)) with Δu_k as the input.

(2) The variation ΔT_k of the walking period can be expressed by a linear function

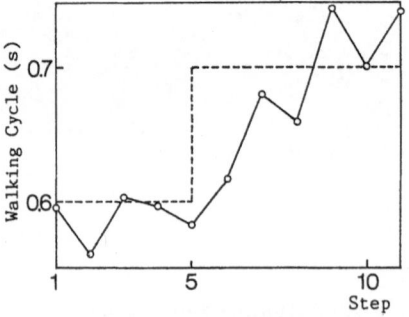

Fig.10 Experimental result of stepping motion

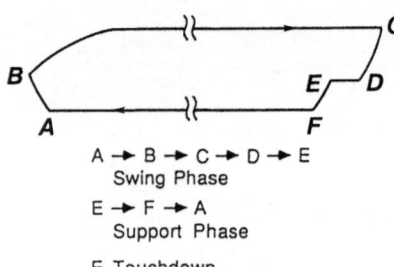

$A \rightarrow B \rightarrow C \rightarrow D \rightarrow E$
Swing Phase

$E \rightarrow F \rightarrow A$
Support Phase

E Touchdown

Fig.11 Design of leg trajectory

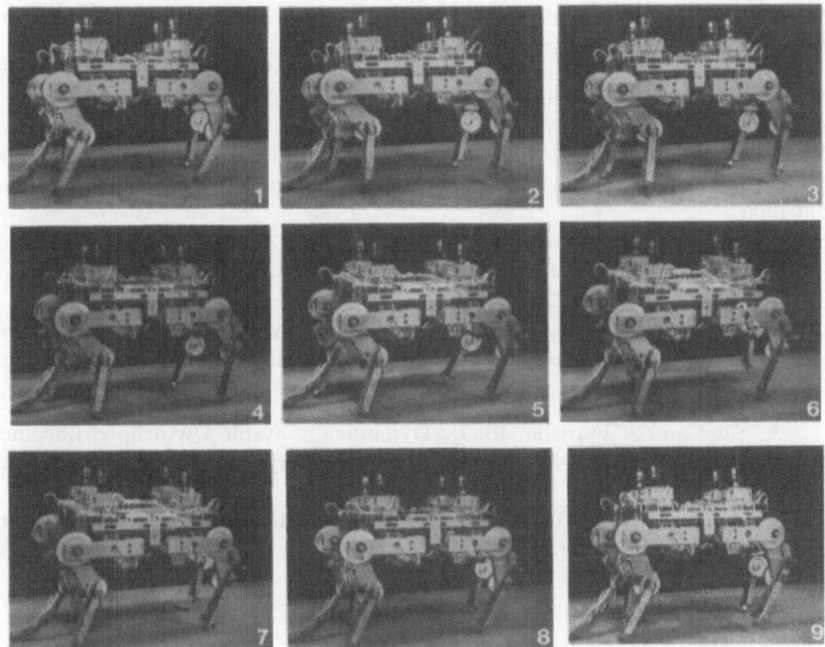

Fig.12 A photographic playback in continual dynamic pace

of the states $\Delta\omega_k$, Δv_k and the input Δu_k (Eq.(14)).

(3) The walking cycle control by the dead beat controller was discussed. It was shown that the parameter sensitivity of this method in the stepping control is not so high. Moreover, the optimal servo controller is also examined.

(4) In the stepping experiments, the proper adjustment of the walking cycle was realized by using the feedback gain of the dead beat control.

(5) The synchronization of the lateral motion and the sagittal motion was discussed, and the trajectory of the leg was designed. Finally, the smooth pace walking was realized.

REFERENCES

[1] R. B. McGhee, et al., 1968, "On the Stability Properties of Quadruped Creeping Gaits," Mathematical Biosiences, Vol.3, pp. 331-351.

[2] M. H. Raibert, 1986, Legged Robots That Balance, The MIT Press.

[3] S. M. Song and K. J. Waldron, 1989, Machines That Walk : The Adaptive Suspension Vehicle, The MIT Press.

[4] S. Hirose, 1984, "A Study of Design and Control of a Quadruped Walking Vehicle," Int. J. of Robotics Research, Vol.3, No.2, pp.113-133.

[5] H. Kimura, I. Shimoyama and H. Miura, 1988, "Dynamics in Dynamic Walk of the Quadruped Robot," (in Japanese) J. of the Robotics Society of Japan, Vol.6, No.5, pp.367-379.

[6] J. Furusho and A. Sano, 1990, "Sensor-Based Control of a Nine-Link Biped," Int. J. of Robotics Research, Vol.9, No.2, pp. 83-98.

[7] A. Sano and J. Furusho, 1990, "Realization of Natural Dynamic Walking Using The Angular Momentum Information," Proc. of 1990 IEEE Int. Conf. on Robotics and Automation, Cincinnati, pp.1476-1481.

[8] A. Sano and J. Furusho, 1988, "Analysis of Dynamic Quadruped Locomotion Based on Quasi-Angular-Momentum," (in Japanese) Trans. of SICE, Vol.24, No.12, pp.1299-1305.

[9] A. Sano, J. Furusho and A. Hashiguchi, 1991, "Basic Study of Quadruped Locomotion System with Capability of Adjusting Compliance," (in Japanese) Trans. of JSME, C, Vol.57, No.539, pp.2297-2304.

[10] J. Furusho and A. Sano, 1991, "Development of Biped Robot," Adaptability of Human Gait, ed. A. E. Patla, Elsevier Science Publishers B. V. (North-Holland), pp.277-303.

[11] A. Sano and J. Furusho, 1989, "Dynamically Stable Quadruped Locomotion (A Pace Gait in The COLT-3)," Proc. of the Int. Sympo. on Industrial Robots, pp.253-260.

[12] M. Hildebrand, 1976, "Analysis of Tetrapod Gaits: General Considerations and Symmetrical Gaits," Neural Control of Locomotion, ed. R. M. Herman et al., New York: Plenum, pp.203-236.

Section 10: Modelling and Control

Despite of all progress in computing power and software engineering involving knowledge-based techniques, robotics has its roots in mechanics and kinematics, and the actuator and joint drive problem is still one of the most central ones in the area. While hydraulic actuators -after being widely used in the early years of robotics - seemed to be obsolete for quite a while, a considerable increase in interest may be observed now due to a reduction of leakage problem, high force/torque to weight ratio and improved controllability. The paper of Boulet, Daneshmend, Hayward and Nemri drives models for fast hydraulic actuators, describes inherent characteristics e.g. concerning linearity and points out their control capabilities e.g. for achieving impedance characteristics over a wide range. Nonlinear modelling is complemented by experimental evaluations.

Many kinematic studies for robots have been done so far involving manipulability criteria and workspace considerations. In the paper of Lenarčič and Umek in contrast the human arm kinematics is investigated from the robotics viewpoint aiming mainly at corresponding workspace calculations. It outlines e.g. mutual dependence and crosstalk of joints, task specialization, joint angle limitations and aims at finding a relatively simple kinematic model based on measurements of a triple TV camera system.

Flexible links in a three-degree of freedom robot are the objects of modelling and control in a paper by Yoshikawa, Murakami and Hosoda. They derive equations of motion and state equations using Lagrange's method. Accelerometers are attached at the tips of the elastic links for observing the state variables needed. Using optimal regulator theory they show how to suppress vibrations of the arm, which has lamped masses at the tips of the light and flexible links.

The last paper by Mayeda, Maruyama, Yoshida, Ikeda and Kuwaki addresses the problem of identifying robot inertial link parameters by systematic procedures. After clarifying which parameters may be estimated from joint torque and motion data (restricted by redundancy considerations) they first present a "step by stop" method in moving single joints successively, and then compare it with a "simultaneous" technique moving all joints at the same time.

Experimental and simulation results are based on the same PUMA 560 robot.

System Identification and Modelling of a High Performance Hydraulic Actuator

Benoit Boulet, Laeeque Daneshmend, Vincent Hayward, Chafye Nemri

McGill Research Center for Intelligent Machines

McGill University

3480 University Street, Montréal, Québec, CANADA H3A 2A7

Abstract

Detailed knowledge of actuator properties is a prerequisite for advanced manipulator design and control. This paper deals with the experimental identification and modelling of the nonlinear dynamics of a high performance hydraulic actuator. Such actuators are of interest for applications which require both high power and high bandwidth. An analytical model of the system is formulated, and a software simulator implementing the force-controlled actuator model including all the nonlinear elements is shown to predict the real system's behavior quite well. The actuator properties and performance are also discussed.

1 Introduction

Hydraulic actuation used to be, and in many cases, remains the technique of choice for high performance robotic applications. However, this type of actuation is not presently receiving a great deal of attention from the robotic research community despite its often ignored advantages. This may be due, in part, to unjustified prejudice against hydraulic systems on the part of robot designers in the research community.

Hydraulic actuation is often believed to be dirty, noisy, inaccurate, inadequate for force control, complicated to use, dangerous, expensive, and hard to package. These descriptions do indeed apply to certain, general purpose, hydraulic actuators. However, hydraulic actuators specifically designed for robotics and other demanding applications, such as those discussed in this paper, overcome many of these alleged shortcomings and offer a unique set of performance characteristics.

As the objectives in advanced manipulator research become increasingly demanding, the interaction among various components of the system, and the impact of this interaction on overall manipulator performance, becomes progressively more important. This necessitates an integrated approach to manipulator design: encompassing the kinematic, structural, actuation, sensing, and control aspects of the manipulator within a unified design process. Hence, detailed knowledge of actuator properties, and the nature of the limits on actuator performance, are a prerequisite for the integrated design of advanced manipulators. Actuator characteristics are of special relevance to control law design.

This paper focuses on the modelling and system identification of one particular high performance hydraulic actuator built by ASI. A physical model is derived for this actuator, and the parameters of the various components of this model are identified experimentally. The overall force loop performance of the actuator is also investigated, and compared to the predictions of a software simulator which implements the physical model.

2 Actuator Overall Properties

With proper design, leakage has been reduced to a minimum and can be easily controlled. In addition, modern quick release flexible supply lines make connecting and disconnecting a hydraulic unit almost as easy as connecting or disconnecting an electrical component. Due to lack of space, hydraulic supplies can only be discussed briefly here. These come in many designs, some of them very compact and convenient. In our case, we used an acoustically isolated conventional supply which is not noisier than say, a ventilated back-plane chassis, and not more expensive than a bank of good quality DC motor amplifiers. The actuator itself is completely noise-free even at maximum thrust, that is 1340 N for 345 N/cm^2 (500 psi) supply. The turbulent flow is confined inside a solid metal manifold from which no audible (at least in our lab) acoustical noise can escape. This contrasts with some electro-mechanical equipment driven by switching power supplies. Also, the produced mechanical signal (force or velocity) is almost perfectly free of noise. This is typified by the sensation of smoothness when the controlled hydraulic actuator is made to interact with the experimenter's hand.

The device discussed here is a linear piston type actuator driven by an integrated high-bandwidth jet pipe suspension valve, and fitted with a force sensor. It is very compact, mechanically robust, and its mass is about .5 Kg (17 ounces). A view of the actuator without the LVDT position sensor is shown at figure 1. For a 76 mm stroke, the overall dimensions are 25 X 55 X 139 mm. Since it is a force controlled device it must include some elasticity which is almost entirely lumped in the force sensor mounted directly on the cylinder. It thus may be considered as an active instrumented structural member easily integrated in a larger assembly.

The ASI servosystem also includes a controller card which can be accessed by a host computer. The card features on-board analog linear controllers whose gains can be programmed from a host computer, allowing gain scheduling. Digital control is also possible since the valve current can be specified as desired. The system state variables can be accessed either digitally via an on-board analog to digital converter or directly by measuring the analog signals.

Force control resolution is limited by the residual solid friction forces as seen at the piston rod in closed-loop operation. Thus, resolution depends on the ability of the internal driving force to overcome these forces, and by the resolution of the sensor itself. The closed-loop force feedback gain can be fairly high, hence the effects of residual friction can be made quite small. Consequently, sensor stiffness determines the basic tradeoff between force control bandwidth and resolution. These actuators must be essentially seen as force producers due to the four-way jet pipe design of its electromagnetic valve (single stage).

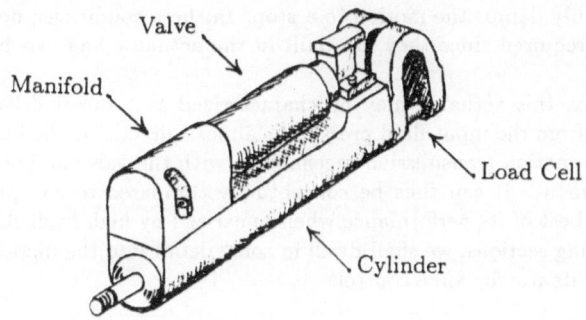

Figure 1: The ASI Hydraulic Actuator

The force output primarily results from the differential pressure across the lines leading to the chambers on each side of the piston. The pressure imbalance due to the suspension deviation is the fundamental operational mechanism. Because the valve is piggy-backed on the piston, a very direct connection between suspension deviation and force output is established.

Among the several major nonlinear characteristics of this actuator, hydraulic damping has a notable effect on performance. Hydraulic damping is a force which opposes the piston motions due to the circulation of oil through the valve orifices. For a fixed valve current that specifies a certain valve position, the effect is very small at low velocities, which make it difficult to assess, but increases faster than linearly for a certain velocity range, past which the characteristic curve tapers off. We conjecture that this effect is attributed to flow forces which become significant enough to force the opening of the valve. This phenomenon happens only when an external force applied on the piston adds up to the fluid pressure to produce higher velocities (and thus higher flow rates) than usually obtained. Thus, the force response bandwidth, kept at a maximum for small amplitude motions such as constrained or contact motions, is drastically decreased for fast motions of an inertial load in free space because the resultant velocities are in the range where the damping is exponential, enhancing stability. Hence the actuator has the intrinsic property to adapt its natural impedance characteristic to the type of tasks required in robotics. At the limit, when the fully opened valve forces maximum flow in and out the chambers, velocity saturates and is maintained constant for large variations of the disturbing load forces, as the thrust force would augment rapidly should the velocity drop. At the other end of the spectrum, when the velocity is small, the suspension deviation has a direct impact on the force output, resulting in high bandwidth force control.

High reliability is facilitated by a very small number of parts of which only two are moving parts: the bending jet pipe and the piston, not counting the LVDT position sensor. Solid friction only occurs between the piston, the rod and the cylinder in the entire assembly. The force sensor has inherent mechanical overload protection which enhances further reliability. Furthermore, elastic displacements are sensed by a noncontact Hall-effect transducer. Finally the actuator can reach its mechanical travel limit at full valve opening without incurring any damage as the oil, forced out of the vanishing chamber volume, smoothly damps the motion to a stop. In these conditions, no external mechanical stops are required since they are built-in the actuator and can be adjusted to any requirement.

In summary, this actuator may be characterized as a direct drive device since the power derived from the input fluid pressure is almost directly applied to the load without any need for a motion transmission mechanism, with the valve acting as a variable gain amplifying element. It can thus be conceptually compared to an operational amplifier producing the best of its performance when linearized by high feedback gains.

In the coming sections, we shall dwell in some detail into the modelling of this device with a view to its use for force control.

3 Physical Modelling

A "gray-box" model approach was adopted since a number of the system parameters were not known and in most cases were unavailable information. Some "reverse engineering" was performed to develop an understanding of how the system elements were designed. Our model includes linear dynamics in conjuction with nonlinear elements. These are hysteresis, static valve force characteristic, hydraulic damping and friction. These nonlinearities play an important role in the actual system and must be included if the model's predictions are to be a good approximation of the actuator's behavior. A block diagram of the closed-loop model is presented at figure 2. The linear blocks represent the valve, fluid, and force sensor dynamics, which are respectively denoted as $G(z)$, $D(z)$ and $S(s)$. Zero-order holds are used at the outputs of the discrete-time blocks but are not shown on the figure.

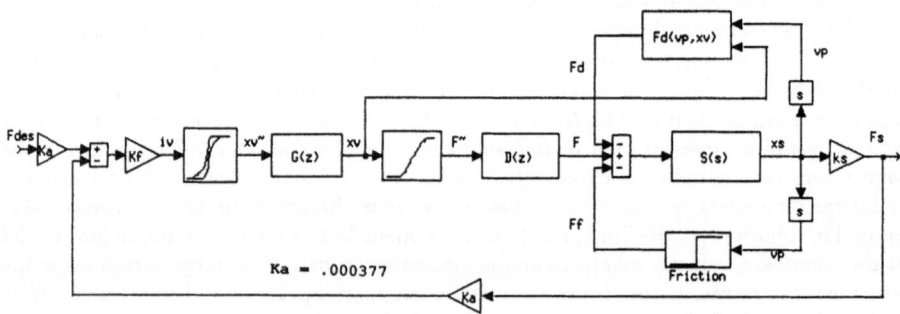

Figure 2: Block Diagram of the Closed-Loop Model

The supply pressure (345 N/cm^2) was the only available a priori information before ASI kindly agreed to provide us with proprietary information regarding the geometry of the valve. This information was needed to calculate the valve force versus the valve pipe tip position static characteristic $\tilde{F}(x_v)$ i.e. the static hydraulic force applied on the piston when it is constrained to a null velocity. All the other system parameters were unknown and had to be measured or identified.

The unknown, but measurable, model parameters were the sensor calibration, the force sensor dynamics and stiffness, the valve hysteresis characteristic, the friction characteristic and the hydraulic damping effect. The unknown, but identifiable, model parameters were the valve and fluid dynamics.

3.1 Valve Static Force Characteristic

A mathematical model of the valve static force characteristic $\tilde{F}(x_v)$ was worked out. Two assumptions were made. Firstly, the flow through the valve orifices was assumed to vary with the square root of the pressure difference across the orifices. Orifice discharge coefficients could not be measured and were estimated, based on values given in [2] (pp. 181–183). The second assumption was that direct leakage from the valve pipe tip to the return chamber is negligible. This is justified considering that even if there is some leakage, its effect should be mostly independent of x_v and should roughly be equivalent to a drop in pressure at the end of the supply line, thereby affecting only the saturation force values but not the general shape of the function. An expression was derived to calculate the steady-state force with respect to the valve pipe tip position and its characteristic is shown at figure 3. Each saturation force corresponds to the area on each side of the piston multiplied by the supply pressure.

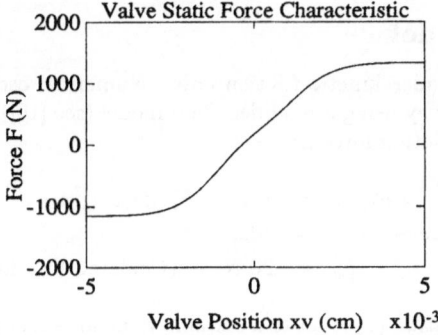

Figure 3: Valve Static Force Characteristic

It is seen that the output force is not null but has a positive bias when the valve position is zero. This is the case if the valve pipe tip position is measured from a geometrically centered origin, but an offset in valve current can be set to approximately compensate any mechanical bias such that the output force is about zero. This has the effect of shifting the valve static force characteristic and it is taken care of in the model by adding a negative offset to x_v before the function simulating the force characteristic is called.

3.2 Sensor Calibration and Dynamics

Calibration of the position sensor is performed by adjusting an offset and a gain and by measuring the piston stroke. The force sensor is calibrated similarly but its stiffness has to be measured. The force sensor essentially consists of a U-shaped piece of steel with no solid contact, thus a mass-spring-damper second-order dynamic model was chosen. We assumed that the fixture used for experimentation was perfectly rigid, although actual results showed significant bending.

3.3 Valve Hysteresis

Valve hysteresis is significant and a model was built to account for it. The model is based on a technique described in [4]. It is capable of generating minor loops from the knowledge of the major hysteresis loop. In the model, the input to the hysteresis block is the valve current i_v and the output is the DC valve pipe tip position $\tilde{x}_v$. Hysteresis output is usually chosen to be the valve's motor torque but we couldn't measure it. Hence, although the relationship between i_v and x_v would normally include the valve dynamics, we had to separate the DC hysteresis characteristic from the dynamics which relate the static and actual valve positions, $X_v(z)/\tilde{X}_v(z) = G(z)$.

The valve hysteresis is included in the system static force characteristic which can easily be measured. Friction is also included in the static force characteristic but we neglected it because of uncertainty in our friction model at very low velocities. We used the inverse of the calculated nonlinear valve static force function to obtain the lower and higher parts of the $\tilde{x}_v(i_v)$ hysteresis major loop from the DC characteristic data.

3.4 Friction Model

The friction model includes kinetic friction only. Numerical oscillation problems were avoided in the simulator by using a modified Dahl model (see [10]). The expression of the time-derivative of the friction force is:

$$\partial F_f / \partial t = \gamma (F_f - F_c \, \text{sgn}(v_p))^2 v_p, \tag{1}$$

$$\text{where} \quad \gamma = 100,$$

$$F_c = 26 \text{ N} \quad \text{(Coulomb Friction)}.$$

The parameter γ in equation (1) is set to a suitable value for a fast transient in F_f towards the Coulomb friction F_c or $-F_c$ when v_p changes sign. It should be noted that the use of this friction model which, in steady-state, is equivalent to a simple Coulomb friction model, was only intended for improving the numerical integration and not for modelling the actual Dahl effect.

3.5 Hydraulic Damping Effect

The hydraulic damping force depends on the valve pipe tip position and on the piston velocity. The family of curves used to model this effect is based on experimental data and thus it includes the flow forces acting on the valve pipe tip. Although the valve position can't be measured, we used the knowledge of the desired input currents and found the corresponding valve positions by applying these current values to the hysteresis model. The flow forces on the valve pipe and the uncertainty in the hysteresis model limit our ability to accurately predict the valve position.

3.6 Identification of Valve and Fluid Dynamics

The valve and fluid dynamics had to be identified for parametrization of the linear blocks in the model. All the linear dynamics were identified as a whole and several assumptions were made in order to be able to select the right poles and zeros for each transfer function.

It was assumed that the valve was the most restrictive limit to the open-loop bandwidth and this was based on the figures used for a similar valve in [7]. A second-order model with two distinct real poles was expected to give good results because of severe damping applied on the valve pipe tip by the fluid in the return chamber.

For the fluid dynamics, the supply and return lines were assumed to be lumped-parameter linear second-order systems. The parameters are the fluid inertia, the fluid and line compliance and the orifice resistance. The chambers on each side of the piston were assumed to be lumped-parameter first-order linear systems, the parameters being the fluid compliance and the orifice resistance. The overall fluid dynamic model order is six.

Two poles should be related to the force sensor dynamics in the identified linear transfer function which should be of the tenth order. These poles were expected to be complex and located below the force sensor's natural frequency because of the hydraulic damping effect, which is assumed to be small since the PRBS input used for identification had a low amplitude.

3.7 Actuator Model

A diagram of the physical actuator model is presented at figure 4. The dynamic and output equations relating the hydraulic force F to the sensed force F_s are:

$$F(x_v) - F_d(x_v, v_p) - F_f(v_p) = m\ddot{x}_s + b_s\dot{x}_s + k_sx_s \tag{2}$$
$$F_s = k_sx_s. \tag{3}$$

where:

$F_d(x_v, v_p)$	$\equiv$ hydraulic damping force	x_s	$\equiv$ force sensor deflection
$F_f(v_p)$	$\equiv$ friction force	m	$\equiv$ actuator mass minus piston mass
v_p	$\equiv$ piston velocity	b_s, k_s	$\equiv$ force sensor parameters

4 Experimentation

4.1 Measurement of Force Sensor Characteristics

As a first experiment, we had to measure the force sensor characteristics. We directly measured the force sensor stiffness by locking the piston to the mount and by measuring the total sensor deflection as full output force was applied in both directions. Then, by using the known saturation force values, we were able to calculate the sensor stiffness. One disadvantage of this method is that full sensor deflection probably covers a nonlinear domain of the sensed force F_s vs sensor position x_s relationship.

The force sensor impulse response was also measured by gently knocking the actuator with a piece of metal while it was held vertical. The damping factor ζ, natural frequency

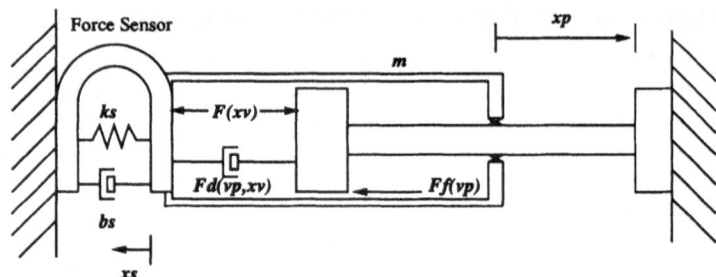

Figure 4: Actuator Model

ω_n, damping coefficient b_s and sensor stiffness k_s were then calculated and are presented at table 1. It was noted that the impulse response gave much better results than the step response because in the latter case, lateral modes were excited and masked the effect of the desired axial mode. The sensor stiffness value used in the model for simulation is the one derived from the impulse response experiment. Equation (4) is the force sensor transfer function $S(s)$ used in the model.

$$S(s) = \frac{178.6}{s^2 + 758.4s + 6712857} \tag{4}$$

actuator mass m_a		0.612 kg
actuator mass minus piston mass m		≈ 0.560 kg
sensor stiffness k_s	(direct)	43659 N/cm
	(impulse)	37592 N/cm
natural frequency $\omega_n = \sqrt{k_s/m_a}$		2478 rad/s (394 Hz)
damping factor ζ		0.14
viscous damping coefficient b_s		4.25 N/cm/s

Table 1: Measured Force Sensor Parameters

4.2 Measurement of Open-Loop Static Force Characteristic

With the piston locked to the mount, the open-loop static force characteristic was recorded while the valve current was slowly varied step by step following a triangular input. The current driver sensitivity allowed .488 mA increments in valve current. The static force characteristic and the calculated hysteresis major loop are shown at figure 5.

4.3 Measurement of Friction

Kinetic friction and stiction were measured after the oil had been taken out of the actuator (some oil was left, providing lubrication). The main disadvantage of this method is that

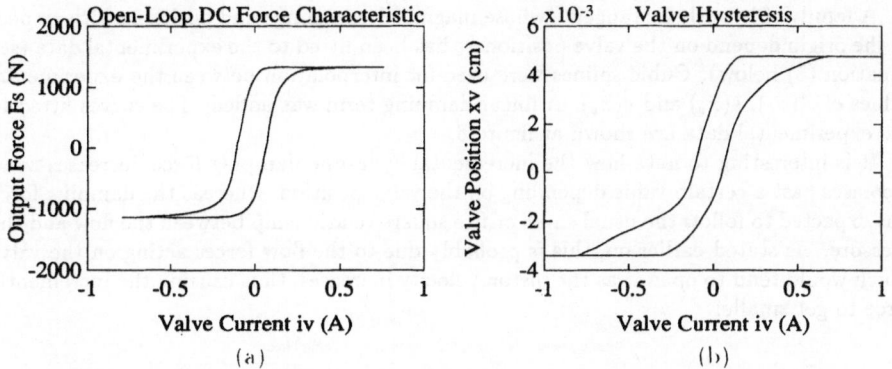

Figure 5: (a) Open-Loop Static Force Characteristic, (b) Valve Hysteresis

friction is likely to change when the pressure across the piston varies as the seal gets squeezed. *In situ* differential pressure measurements would give more accurate assessment of the phenomenon. Stiction was measured as the force at the breaking point where the piston starts moving. Coulomb friction was measured by pulling on the piston by hand and recording the force and the piston velocity. Results are shown at table 2.

stiction	(pushing on piston)	99 N
	(pulling on piston)	-54 N
Coulomb friction		$\approx$ 26 N

Table 2: Friction Measurements

4.4 Measurement of Damping Effect

For the hydraulic damping experiment, the actuator was mounted vertically such that weights could be hung from the piston (which was free to move). The procedure was as follows: we used a certain valve current as input to the open-loop system and measured the piston steady-state velocity without any load hanging to it. The corresponding steady-state force applied by the fluid pressure on the piston couldn't be measured directly but was found later by locking the piston and measuring the output force for the same input current. It is important to note that the sequence of applied currents must be the same for the hysteresis to operate in the same region. Then, for the same input current, different masses were hung to the piston and the corresponding steady-state velocities were measured. For each of these masses, the total force applied on the piston could be calculated as the sum of the measured hydraulic force and the gravitational force acting on the mass. It was assumed that the hydraulic reaction force was equal to that sum, i.e. we neglected the friction force. This procedure, which provided experimental data for one value of valve current, was repeated for different valve currents in order to be able to fit a family of curves to the data.

A family of hyperbolic tangents whose magnitudes, scalings and positions with respect to the origin depend on the valve position x_v has been fitted to the experimental data (see equation (5) below). Cubic splines were used for interpolation between the experimental values of $A(x_v)$, $s(x_v)$ and $d(x_v)$. A linear damping term was added. The curves fitted to the experimental data are shown at figure 6.

It is interesting to note how the incremental hydraulic damping force decreases as v_p increases past a certain value depending on the valve position, whereas the damping force was expected to follow the usual small orifice square relationship between the flow and the pressure. As stated earlier on, this is probably due to the flow forces acting on the valve which would tend to open it as the piston velocity increases, thus causing the incremental force to get smaller.

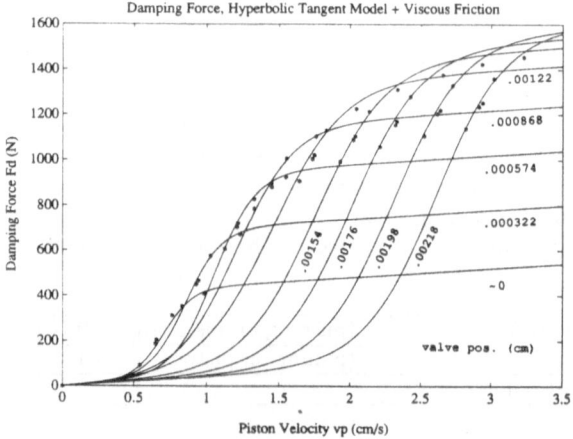

Figure 6: Experimental Hydraulic Damping Effect

$$F_d = A(\tanh(s\,(|v_p| + d)) - \tanh(sd))\,\mathrm{sgn}(v_p) + 40v_p, \qquad (5)$$
$$\text{where} \quad A = A(x_v), \quad s = s(x_v), \quad d = d(x_v).$$

The first simulations showed that the model had too much damping and a gain of .4 was added at the output of the hydraulic damping block. The model also uses bandlimited differentiation (a pole added at 40 Hz) to reduce numerical noise problems arising in the nonlinear damping loop and to improve the closed-loop response.

4.5   Identification of the Linear Part

Separate identification of the valve dynamics and the fluid dynamics was not possible; we had to identify the linear part as a whole. With the piston fixed at midstroke position, a low-amplitude PRBS input was applied to the open-loop system so that we could assume that the system was operating in a linear region. It should be noted that the fluid compliance in the cylinder depends on the piston position and reaches a maximum when the piston is at a point where both chamber volumes are equal. Therefore, the case studied here was nearly the most adverse condition to stable control when considering only the fluid dynamics (see [11], pp. 50–51). The sampling frequency was 5000 Hz.

An ARX model was estimated using a least-squares method on MATLAB^{TM} (ARX command) and the best fit was given by a tenth-order model with two delays as predicted:

$$(1 + a_1z^{-1} + a_2z^{-2} + \cdots + a_{10}z^{-10})Y(z) = (b_3z^{-3} + b_4z^{-4} + \cdots + b_{10}z^{-10})U(z), \qquad (6)$$

where $a_1 = -0.4223, a_2 = -0.3765, a_3 = -0.2802, a_4 = -0.1959, a_5 = 0.1930,$
$a_6 = 0.2234, a_7 = -0.0532, a_8 = -0.0051, a_9 = -0.0903, a_{10} = 0.0394,$
$b_3 = 0.0997, b_4 = -0.1360, b_5 = -0.0258, b_6 = 0.1094, b_7 = -0.4047,$
$b_8 = -0.1323, b_9 = -0.4253, b_{10} = -0.7851.$

The PRBS input and the system and ARX model outputs are shown respectively at figures 7 (a) and (b).The pole-zero plot of the identified model is shown in Figure 8: as can be seen, the zeros of the identified model lie outside the unit circle. This indicates that the system identification technique has yielded a non-minimum phase model. The physical system has several components which are actually distributed parameter systems, e.g. hydraulic fluid and lines, valve stem flexure, etc., and there also exist possibilities of multiple transmission paths due to the mechanics of the test set-up. Hence the non-minimum phase nature of the model appears justified. Fortunately, these non-minimum phase zeros are clustered at high frequencies. Hence controller design can be based upon frequency separation, by using an additional compensator block which filters out the high frequency behavior.

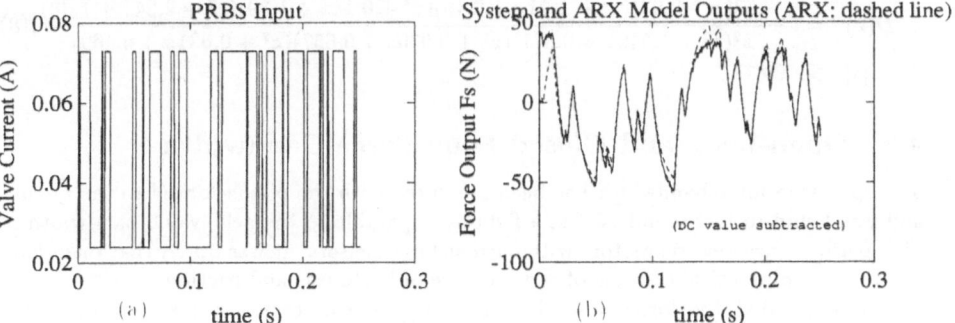

Figure 7: (a) PRBS Input, (b) System and ARX Model Outputs

4.6 Dynamics

As it was pointed out earlier on, the valve dynamics should have the lowest bandwidth and we therefore picked the only two identified real poles plus a zero at $z = 0$ for $G(z)$:

$$G(z) = \frac{0.01203z}{(z - 0.9762)(z - 0.4947)}, \qquad |z| > 0.9762 \qquad (7)$$

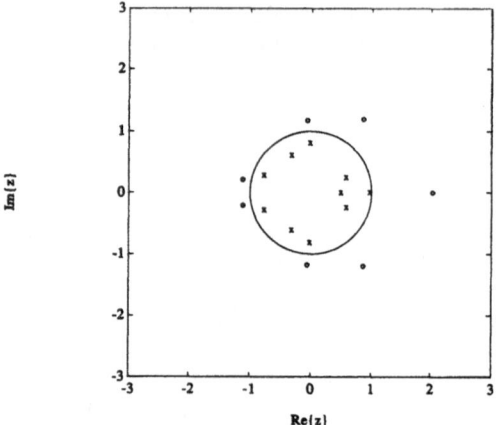

Figure 8: Poles and Zeros of Identified Transfer Function

For the fluid dynamics, we picked the three pairs of complex poles at high frequencies and the three pairs of complex non-minimum phase zeros. We also chose the only real zero at $z = 2.0282$ and placed two poles at $z = 0$ to make $D(z)$ causal (see equation (8)). These two poles get cancelled with the zero of $G(z)$ and a zero at $z = 0$ attributed to the force sensor dynamics in the identified transfer function. For the sensor dynamics, a pair of complex poles [1] around 300 Hz and one zero at $z = 0$ were disregarded.

$$D(z) = \frac{-.67(z - 2.03)(z^2 - 1.71z + 2.16)(z^2 + 0.14z + 1.39)(z^2 + 2.24z + 1.30)}{z^2(z^2 + 1.546z + 0.671)(z^2 + 0.036z + 0.657)(z^2 + 0.631z + 0.469)}, (8)$$

$$|z| > 0.8191$$

4.7 Open-Loop and Closed-Loop Force Bandwidth

The open-loop force bandwidth has been measured with the piston locked to the mount, and was found to be around 20 Hz, a figure comparable to the achievable bandwidth of high-performance electric motors with current force sensors. It was noted that open-loop control was impractical because of the presence of hysteresis and friction.

Assuming that the force closed-loop system is linear for a given amplitude of the sinusoidal input, frequency responses were experimentally obtained and are shown at figures 9 (a) and (b) for different amplitudes of the input and for a force feedback gain of 2.44. The roll-off on the magnitude Bode plot (figure 9 (a)) indicates that the system is at least of the ninth order. The closed-loop bandwidth is around 100 Hz and decreases for higher input amplitudes. This is due in part to saturating nonlinear elements in the system but also to the nonlinear hydraulic damping. The 6 dB bandwidth goes as high as 196 Hz for low-amplitude inputs.

It should be noted that an on-board lag compensator can be added so that the proportional feedback gain can be lowered to get less overshoot without compromising the

---

[1] these poles are at $z = 0.5797 + 0.2408j$ and $z = 0.5797 - 0.2408j$

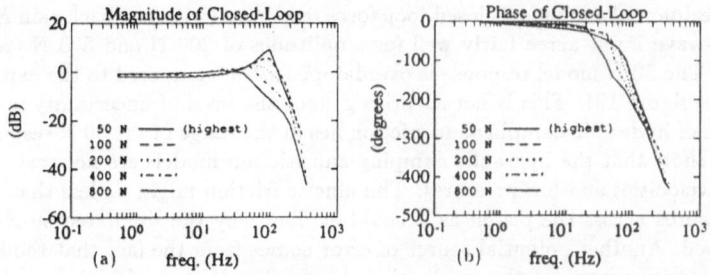

Figure 9: Closed-Loop Frequency Responses: (a) Magnitude, (b) Phase ($K_f = 2.44$)

precision at low frequencies. A limit cycle has been observed for gain values of 3.66 and higher (sustained oscillations at frequencies around 95 Hz). It was also observed that the closed-loop responses to sinusoid inputs (see figures 10 (a) and (b)) present little distortion given the degree of nonlinearity of the system. The slight distortions seen for high-amplitude, low-frequency responses (e.g. figure 10 (b)) are probably due to the piece of aluminum on which the actuator was mounted: the assembly was such that this part of the fixture bent significantly for high output forces. The fixture also had an asymmetric, nonlinear, stiffness characteristic, so that it absorbed some elastic energy from the system and then suddenly released it as it moved back and forth. This could be observed for open-loop responses as well. Another explanation would be that the flow forces acting on the valve pipe tip would slightly disturb its position, thus causing a distortion in the output force. A better experimental rig is being constructed for future experimentation.

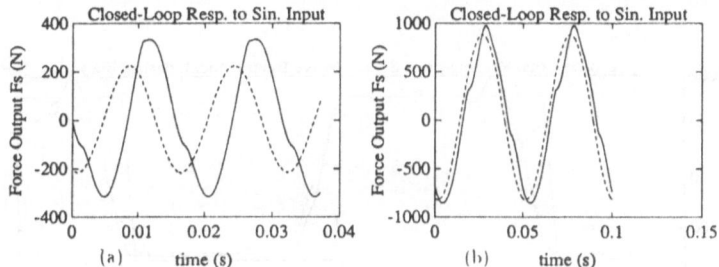

Figure 10: Closed-Loop Force Responses: (a) $f = 63$ Hz, (b) $f = 20$ Hz

5   Simulation Results

A software simulator has been built using SIMULAB^{TM}. The simulator includes discrete-time and continuous-time linear transfer functions. Zero-order holds are used at the output of the discrete-time blocks. The fifth-order Runge-Kutta integration algorithm was chosen for the simulations.

518

The experimental and model closed-loop force responses (with feedback gain $K_f = 2.44$) to a square-wave input agree fairly well for amplitudes of 200 N and 800 N (see figures 11 and 12). The 50 N model response is overdamped when compared to the experimental response (see figure 13). This is not surprising since the level of uncertainty in the combined effects of hydraulic damping and friction lies in the range of the 50 N response. The simulations show that the hydraulic damping and friction models are not really satisfactory at low velocities and low pressures. The kinetic friction might be less than expected for low pressures across the piston as it could explain why the simulated 50 N response is overdamped. Another potential source of error comes from the fact that the hysteresis model can't easily reproduce the small minor loops. The lower and higher parts used to construct the major loop were experimentally obtained and although some filtering was done on them, they are not locally perfectly smooth. Moreover, using the inverse of the valve static force function amplified these imperfections.

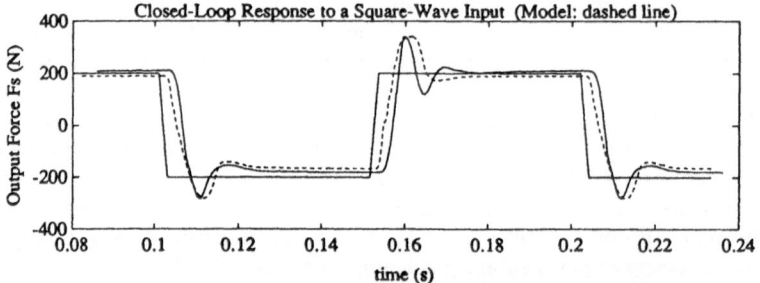

Figure 11: Closed-Loop Force Response to a 200 N Square-Wave Input ($K_f = 2.44$)

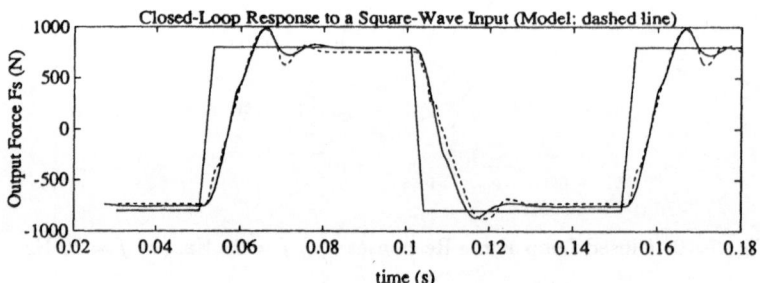

Figure 12: Closed-Loop Force Response to a 800 N Square-Wave Input ($K_f = 2.44$)

6 Conclusion

A complete nonlinear model of the high-performance ASI hydraulic servosystem has been obtained, validated and simulated. The model's ability to reproduce experimental closed-loop force responses for different amplitudes indicates that it could be a valuable tool for

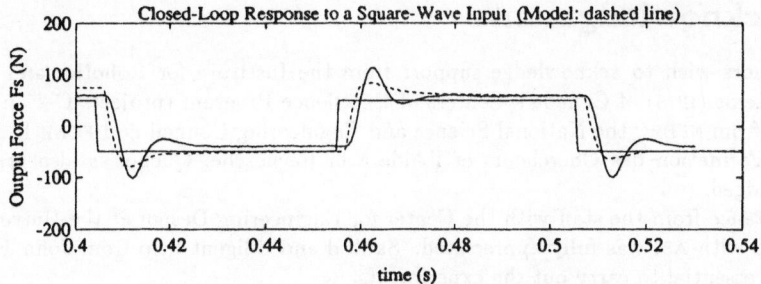

Figure 13: Closed-Loop Force Response to a 50 N Square-Wave Input ($K_f = 2.44$)

the design of better digital nonlinear force control laws. It could also be useful to explain the sytem's behavior. The model can easily be extended for simulation of position and impedance control.

The open-loop force bandwidth (20 Hz) of the hydraulic actuator is comparable to the achievable force bandwidth of high-performance electric motors while the closed-loop force bandwidth was shown to be roughly 100 Hz for $K_f = 2.44$ —much higher than any reported electric motor coupled to a force sensor of similar compliance. It must be remembered that many of the parameters reported in this paper are not intrisic to the actuator and can be modified to tradeoff various performance criteria. One of these parameters is sensor stiffness which is directly related to force control bandwidth.

The implications of the frequency response of the hydraulic actuator for control law design are encouraging: it is predominantly low-pass and the non-minimum phase zeros are clustered at high frequencies. Based on the dynamics, together with the large gains possible due to the high saturation level of the actuator, it appears ideally suited to feedback modulation of impedance over a wide range.

Further investigation into the identification of the linear dynamics is required to ascertain whether the non-minimum phase zeros in the model are artefacts of the system identification technique, or whether they can be related to specific distributed-parameter components of the overall system. Since non-minimum phase zeros place absolute limits on the achievable sensitivity minimization using feedback control, establishing the physical meaning of these zeros would be of relevance in further refining the actuator design to achieve even higher performance.

Better experiments will have to be designed for more satisfying models of the hydraulic damping and friction characteristics. The hydraulic actuator model will be used to assess the attainable range of mechanical impedance, very important for the study of antagonistic actuation. This type of actuation is required by a novel type of manipulator under construction at McGill. Some of its design features are discussed in [6]. It would be desirable to see the effect of reducing the model order by neglecting some of the fluid high frequency dynamics, and to explore digital nonlinear force and impedance control as well as dither. The system's ability to act as a force regulator while the piston is moving will also have to be assessed.

7 Acknowledgments

The authors wish to acknowledge support from the Institute for Robotics and Intelligent Systems (IRIS) of Canada's Centers of Excellence Program (projects C-2 and C-3). Funding from NSERC the National Science and Engineering Council and FCAR Les Fonds pour la Formation des Chercheurs et l'Aide à la Recherche, Québec is also gratefully acknowledged.

Assistance from the staff with the Center for Engineering Design at the University of Utah and with ASI was fully appreciated. Skillful and diligent help from John Foldvari has been essential to carry out the experiments.

References

[1] Animate Systems Incorporated (ASI) 1991. *Advanced Robotic Controller System Manual*. Salt Lake City, Utah.

[2] Blackburn, J. F., Reethof, G., Shearer, J. L. (Eds.) 1960. *Fluid Power Control*. Cambridge: The MIT Press.

[3] Bobrow, J. E., Desai, J. 1990. Modeling and Analysis of a High-Torque, Hydrostatic Actuator for Robotic Applications. *Experimental Robotics I, the First Int. Symp.* V. Hayward, O. Khatib (Eds.) Springer-Verlag, pp. 215–228.

[4] Frame, J. G., Mohan, N., Liu, T. 1982. Hysteresis Modeling in an Electromagnetic Transients Program. IEEE *Trans. on Power Apparatus and Systems* Vol. PAS-101, No. 9, September, pp. 3403–3411.

[5] Gille, J. C., Decaulne, P., Pélegrin, M. 1985. *Dynamique de la commande linéaire*. Paris: Dunod. (in french)

[6] Hayward, V. 1991. Borrowing Some Design Ideas From Biological Manipulators to Design an Artificial One. *Robots and Biological Systems,* NATO *Advanced Research Workshop*. Dario P., Sandini, G., Aebisher, P. (Eds.), Springer-Verlag, in press.

[7] McLain, T. W., Iversen, E. K., Davis, C. C., Jacobsen, S. C. 1989. Development, Simulation, and Validation of a Highly Nonlinear Hydraulic Servosystem Model. *Proc. of the 1989 American Control Conference*, AACC. Piscataway: IEEE.

[8] Shearer, J. L. 1983. Digital Simulation of a Coulomb-Damped Hydraulic Servosystem. *Trans.* ASME, *J. Dyn. Sys., Meas., Contr.* Vol. 105, December, pp. 215–221.

[9] The Math Works Inc. 1990. SIMULAB*: A Program for Simulating Dynamic Systems*. (user's guide)

[10] Threlfall, D. C. 1978. The Inclusion of Coulomb Friction in Mechanisms Programs with Particular Reference to DRAM. *Mech. and Mach. Theory* Vol. 13, pp. 475–483.

[11] Walters, R. 1967. *Hydraulic and Electro-Hydraulic Servo Systems*. Cleveland: CRC Press.

Experimental Evaluation of Human Arm Kinematics

Jadran Lenarčič and Andreja Umek
The Jožef Stefan Institute, Jamova 39, Yugoslavia

Abstract

The paper deals with the mathematical modelling of the human arm kinematics. A six-degrees-of-freedom model is obtained based on the measurement of some selective movements of the arm related to the spatial position of the wrist. Three joints are considered, the sternoclavicular joint, the glenohumeral joint and the elbow joint, while the wrist itself is not taken into account. The developed model does not correspond to the anatomical properties of the arm but approximates only its outer motion characteristics. The main advantage is its simplicity which is useful for further numerical processing, workspace calculation and evaluation.

1 Introduction

There is a variety of reasons why the investigation on the human arm kinematics from the mechanical point of view can be very useful. One of these is to apply the human arm motion properties in the design of future robot mechanisms, it can also serve in various medical applications, sports, ergonomics. A detailed kinematic study of the human arm mechanism can help, for instance, to plan and control the rehabilitation procedure of a hemiplegic arm. In robotics, the understanding of the human arm motion can give important information for the design of advanced robot manipulators and dual-arm systems, or can be used as an example for the control of redundant robot mechanisms.

The objective of the present paper is to develop a kinematic model of the human arm (without the wrist) on the basis of an experimental evaluation of some selected movements. In comparison with other investigations in this area[1, 2, 3, 4, 5], where kinematic models are obtained mainly by observation of the anatomical properties which introduces high complexity, the obtained kinematic model is basically a black-box model whose inside is a set of mathematical relations which do not neccessarily correspond to the actual motion of links and joints of the human arm. These equations approximate the output motion measured in a series of experiments, in particular the motion of the epicondylus lateralis (elbow). The main advantage of this model is its simplicity which enables further numerical processing, for example, workspace calculation and evaluation[6, 7].

In the above model, the shoulder is presented by five rotations, two in the sternoclavicular joint, three in the glenohumeral joint, and one rotation is used for the elbow flexion. The clavicular rotation, which is in reality very important degree of freedom and gives the flexibility to the shoulder, is not included in the model. Its motion is compensated by other degrees of freedom in the sternoclavicular and glenohumeral joint. Also the elbow supination is neglected, since it does not affect the position of the wrist in space. In the paper, the first sections present the measurement system and the methodology of experiments carried out on a healthy and injured human arm, then the kinematic model is developed and a discussion is given on the tasks of each particular joint.

2 Measurement Equipment and Technique

To monitor and evaluate the human arm motion characteristics, the Vicon system was used in this investigation. Three TV cameras were utilized to observe passive markers which were attached to the arm. The principle that underlies the operation of the Vicon system is embedded in the retroreflective properties of the material used for the markers. The retroreflection is property whereby a very large proportion of incident light is reflected along, or very close to, the line of incidence. The effect is that for a TV camera situated very close to the light source, a marker coated with retroreflective material will appear to be very bright as compared with other objects illuminated by the light source. Markers appear as bright spots in the TV image which is then automathically detected and digitised. Special software applied to a PDP-11 computer connected to VAX-8650 and graphic terminal Tektronix 4236 then performs all aspects of data capture and reduction to three dimensional trajectories.

For the purposes of this investigation, the Vicon system appeared sufficiently versatile as a data acquisition system, although the applied measurement technique itself presented several difficulties. Some were related to the accuracy associated with the resolution of TV cameras, their callibration and size of markers; other were related to the skin deformation (markers were attached to the skin) or to disappearance of markers from the camera's view due to the improper positioning of the cameras.

Figure 1 shows reference points on the skeleton which helped us to attach markers which served to determine the human arm motion. The marker between proc. styloideus radii and proc. styloideus ulnae represents the wrist position. Second marker is used for the elbow position at the epicondylus lateralis. In the same fashion another one is used to describe the shoulder motion by the path of the acromion with respect to the extremitas sternalis of the clavicle and two markers lie between acromion and epicondylus to signify skin deformations.

3 Analysis of Selective Movements

Previous studies related to the human arm kinematic modelling[3, 4] show that the complexity of the shoulder joints is the most critical. As long as we intend to calculate the reachable workspace of the human arm, which is related only to the space position of the reference point defined in the middle of the proc. styloideus radii and proc. styloideus ulnae, the wrist joints as well as the elbow supination/pronation can be neglected in a first approximation. Furthermore, it was shown that the effects of the scapulothoracic joint can be mathematically incorporated into the sternoclavicular and glenohumeral joint. Also the clavicular rotation in the sternoclavicular joint can be neglected.

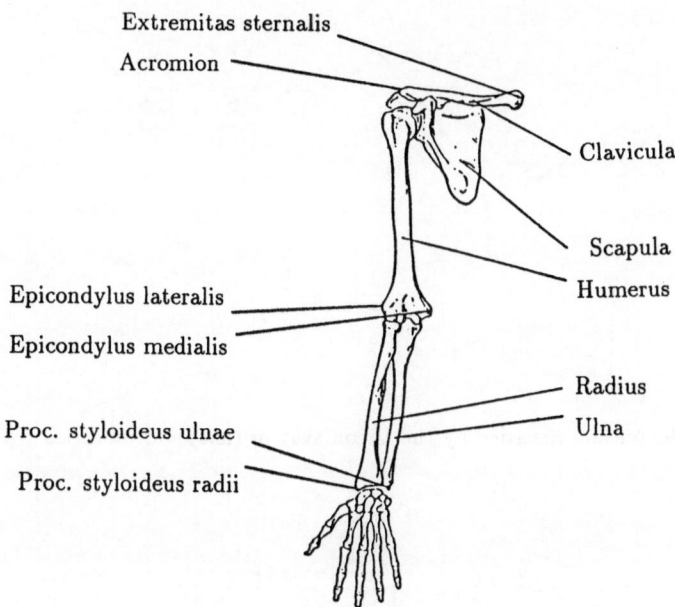

Extremitas sternalis

Acromion

Clavicula

Scapula

Humerus

Epicondylus lateralis

Epicondylus medialis

Radius

Ulna

Proc. styloideus ulnae

Proc. styloideus radii

Figure 1: Reference points for markers

Let us consider two typical movements of the shoulder, flexion/extension and abduction/adduction in the sagital and frontal plane, respectively. Figure 2 (left) shows these trajectories in the case of the humeral flexion/extension, where the first marker is attached to the acromion, the last marker to the epicondylus lateralis and two in between. The obtained trajectories do not correspond to a simple rotation about a given fixed point in the space not only because of the deformation of the skin on which markers are attached during arm movement, but also because this is a more complex movement composed of more than one simple rotation (Figure 2).

In order to calculate the instantaneous center of rotation, it is necessary to treat the recorded set of spatial points so as to obtain a smooth path transformed onto the sagital plane. The locus of the center of rotation given in Figure 2 (left), is related to the trajectory of the epicondylus. The calculated locus of the instantaneous center of rotation in case of humeral flexion movement of the arm is similar to that of humeral extension.

The same discussion can be extended to the humeral abduction/adduction movement given in Figure 3. Here the trajectories of the four markers are transformed onto the frontal plane.

Let us assume there is a point rotating along a circle. As long as the center of the circle is fixed in space, the instantaneous centre of rotation will coincide. But if the centre of the circle moves in a linear direction, the instantaneous centre of rotation will form a locus similar to those in Figure 2 and Figure 3. Based on a mathematical treatment [7], it was shown that the humeral flexion/extension movement and the abduction/adduction movement can be approximated by a combination of rotation about a fixed point and

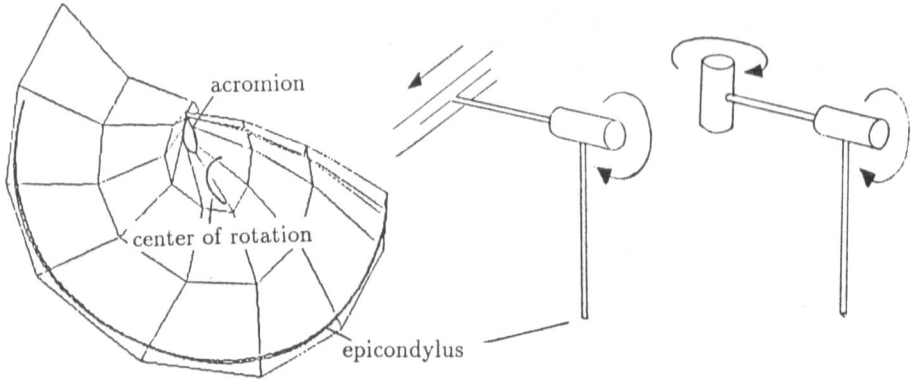

Figure 2: Flexion/extension recorded by the Vicon system (left) and modelled with two dof (right)

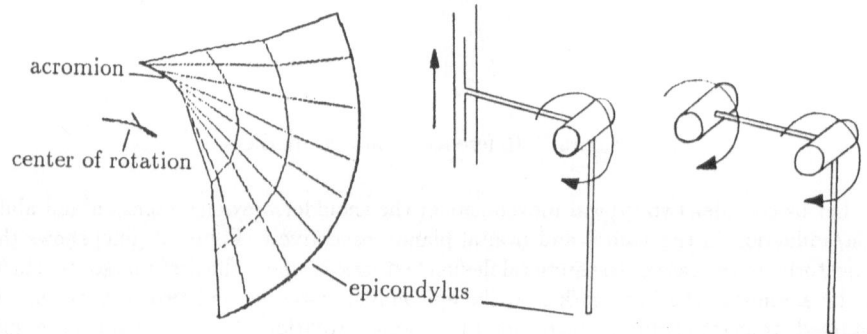

Figure 3: Abduction/adduction recorded by the Vicon system (left) and modelled with two dof (right)

a translation into a selected direction (right sides of Figure 2 and Figure 3). These translations can be substituted with adequate rotations through small angles. The horizontal translation (flexion/extension) is substituted by a rotation about a vertical axis, and the vertical translation (abduction/adduction) by a rotation about a horizontal axis. Thus, we obtain two perpendicular rotation for the flexion/extension and two parallel rotation for the abduction/adduction.

4   Complete structure of the kinematic model

The final result is that the shoulder motion is composed by five rotations, two in the sternoclavicular joint and three in the glenohumeral joint. The sternoclavicular joint contains the clavicular flexion/extension and the clavicular abduction/adduction, while the glenohumeral joint contains the humeral abduction/adduction, the humeral flex-

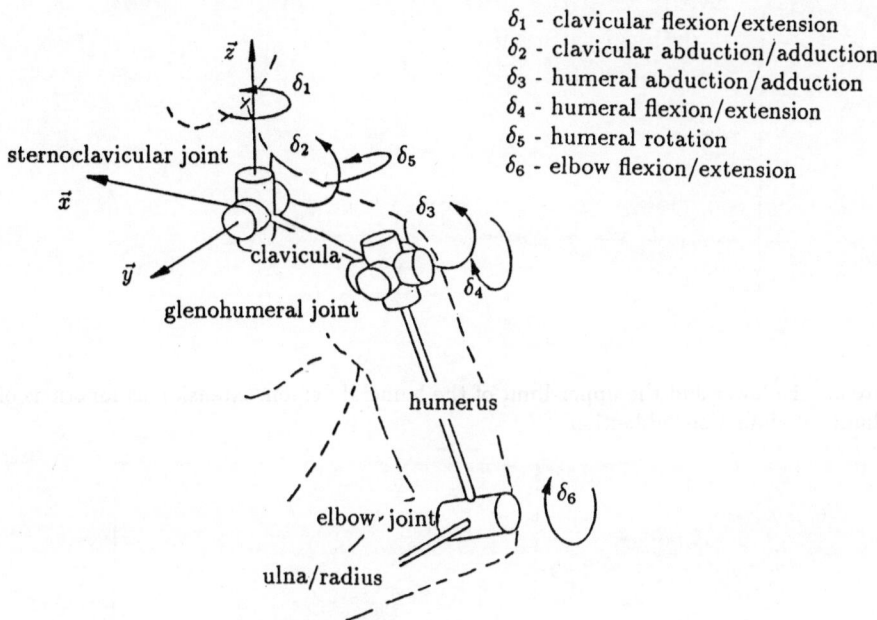

δ_1 - clavicular flexion/extension
δ_2 - clavicular abduction/adduction
δ_3 - humeral abduction/adduction
δ_4 - humeral flexion/extension
δ_5 - humeral rotation
δ_6 - elbow flexion/extension

Figure 4: A 6 dof model of the human arm

ion/extension, and the humeral rotation. Clearly, the shoulder flexion/extension is composed by the clavicular flexion/extension and humeral flexion/extension, and the shoulder abduction/adduction by the clavicular abduction/adduction and humeral abduction/adduction. The whole mechanism, which also contains the elbow flexion/extension, is presented in Figure 4.

The order of rotations influences the motion properties of the arm. In the given mechanism, the order has been selected experimentally. It is especially important in the glenohumeral joint where the ranges of motion of the included rotations are large.

5   Limitations in joint angles

In general, limitations in joint angles of the human arm depend on the values of other joint angles. In the proposed kinematic model of the arm, this is particularly evident in the humeral flexion/extension and humeral rotation. The upper and the lower limit of the humeral flexion/extension depend on the actual angle of abduction/adduction. The relationship can be specified in terms of linear functions (Figure 5) as follows

$$\delta_4 = [-43° + \delta_3/3, 153° - \delta_3/6].$$

The upper and the lower limit of the humeral rotation depend on the values of the angle of the humeral abduction/adduction and the angle of the humeral flexion/extension. The relationship is quadratic and can be specified by (Figure 6)

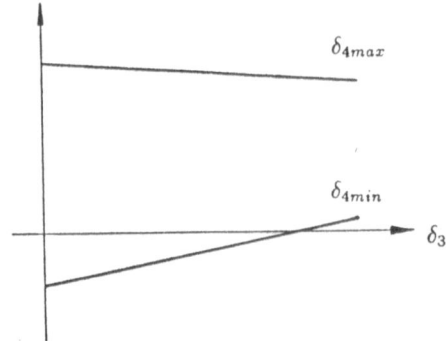

Figure 5: The lower and the upper limit of the humeral flexion/extension as functions of the humeral abduction/adduction

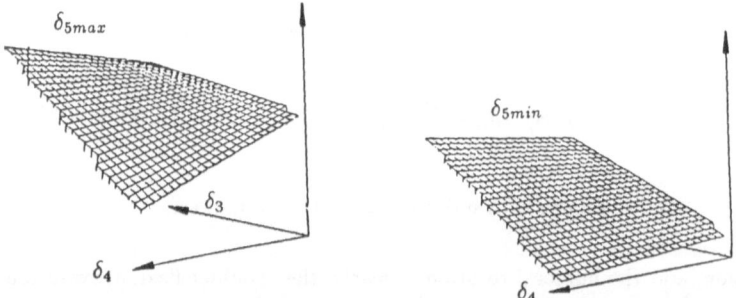

Figure 6: The lower and the upper limit of the humeral rotation as functions of the humeral abduction/adduction and flexion/extension

$$\delta_5 = [-90° + 7\delta_3/9 - \delta_4/9 + 2\delta_3\delta_4/810, 60° + 4\delta_3/9 - 5\delta_4/9 + 5\delta_3\delta_4/810].$$

Limitations on other joint angles are constant

$\delta_1 = [-17°, 17°]$
$\delta_2 = [-6°, 20°]$
$\delta_3 = [-9°, 160°]$
$\delta_6 = [-90°, 60°]$

Here all angles are given in degrees.

6   Direct and inverse kinematics

The kinematic equations established to approximate the motion properties of the human arm are relatively very simple. The mechanism contains three joints and three links whose motion can be calculated by one of the widely known procedures of vector and matrix algebra using either orthogonal rotation transformations or homogeneous transformations. The equations can be expressed in a recursive form and for a given combination of joint coordinates the position and orientation, as well as velocity and acceleration, can be obtained by very few arithmetical operations.

As usual, the inverse kinematics problem is more complex, especially because of the mechanism's redundancy. It includes six degrees of freedom which are only used for positioning the wrist. For three Cartesian space coordinates of the wrist, there is an infinity of possible solutions of joint angles $\delta_1, \delta_2, ..., \delta_6$. To solve the inverse kinematics one may take advantage of any of the known methods for redundancy resolution, however, by analysing the mechanism's structure, the procedure can be simplified.

The first two joint angles δ_1 and δ_2 move in very small ranges. We can, therefore, discretize them in few values which results in a relatively small number of the acromion positions in space. By substraction of these from the given position of the wrist (proc. styloideus radii) and considering also the change of orientation in each point, we obtain the positions which must be reached by the wrist with the remaining four-degrees-of-freedom mechanism (it includes the humerus and the forearm, the glenohumeral and the elbow joint). This mechanism is still redundant, but for a given position of the wrist only one solution for the angle of the elbow flexion δ_6 is possible. It can be calculated straightforward, while the other three angles in the glenohumeral joint δ_3, δ_4, δ_5 can still appear in an infinite number of combinations. We can solve this again by discterizing the values for example, the humeral abduction/adduction δ_3 and then for each of these, and for all positions of the wrist, calculate the remaining two angles δ_4, δ_5 straightforward. Here, we must consider the dependency of joint limits on the values of other joint angles as specified in the previous section. Together with the constraints imposed by the mechanism's structure, it turns out that the possible solutions for δ_3, δ_4, δ_5 can be obtained in unexpectedly small intervals. This fact can be demostrated by the actual inner motion (change of configuration) of the human arm which is, for a fixed position of the wrist, quite limited.

7 Redundancy and other motion properties

From the viewpoint of positioning the wrist, the mechanism that possesses 6 degrees of freedom is redundant and thus it can achieve a given position of the wrist with unlimited number of configurations. Theoretically, any combination of joint angles that correspond to a given position is acceptable. In practice, however, each joint of the human arm seems to solve a predefined task.

The main task of the sternoclavicular joint is associated with the collision avoidance between the arm and the body. The ranges of motion in this joint are small and do not contribute much to the position and orientation of the wrist. On the other hand, if this joint is fixed, the motion of the remaining part of the arm is troubled by possible collisions with the body and the resulting workspace of the arm becomes much smaller than one can expect (Figure 7). Collisions between the arm and the body evidently play an important role in the human arm motion characteristics and reachability. Therefore, the two degrees of freedom introduced in order to model the sternoclavicular joint are necessary not only

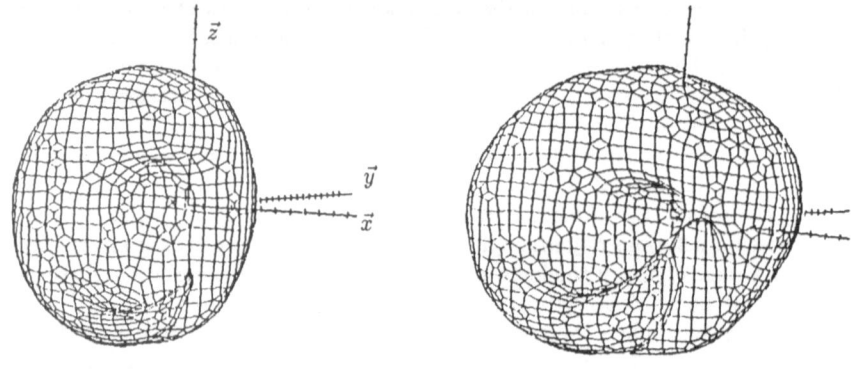

Figure 7: The human arm workspace when $\delta_A = \delta_F = 0$ (left) compared with the normal-one (right)

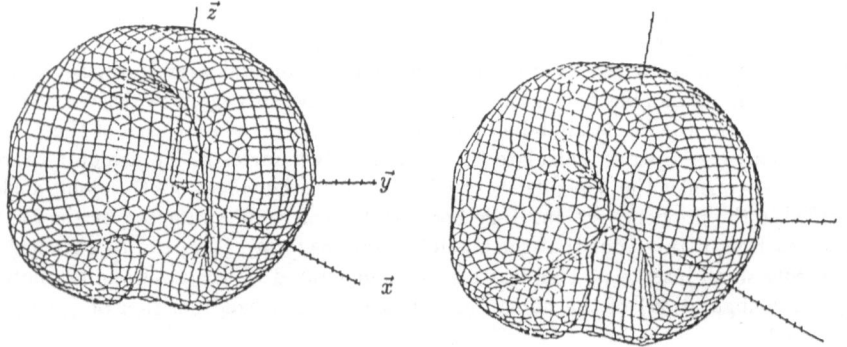

Figure 8: The obtained workspace when the elbow flexion is decreased (left) compared with the normal-one (right)

to describe the complex motion of the shoulder, but also to properly present the human arm workspace.

Basically, three degrees of freedom in the glenohumeral joint and one in the elbow joint are utilized for positioning the wrist. This mechanism is still redundant with infinite number of configurations for a given position. However, only one solution inside the limits can be found for the elbow angle. Its task is certainly associated to the reachability of the arm. In terms of the workspace properties, this rotation affects its thickness (Figure 8). If the elbow joint is fixed (as well as the sternoclavicular joint) the workspace will assume the form of a two-dimensional curve (sphere) and its volume will be zero.

The main task of the remaining three degrees of freedom in the glenohumeral joint, is the pointing operation. Once the elbow angle has been selected, the wrist can be moved on the previously mentioned sphere which radius is defined by the elbow angle. There is an infinity of possible combinations for the glenohumeral degrees of freedom corresponding to a given point on this sphere, however, the solutions can be found only inside intervals of values which differ from one point to another. It can be shown that by changing the configuration in one point the human arm can achieve different values for manipulability or kinematic index[7].

Also an interesting fact is that the limitations in joint angles do not permit collisions between various elements of the mechanism. Moreover, if we enlarge the ranges of motion, for instance, introduce a hyper-extension of the elbow, the workspace properties do not crucially change in comparison to those of the original. The "hyper" ranges of joint angles do not affect the reachability of the arm but imply collisions and other problems in a complex motion of the arm. From this point of view the kinematic structure of the mechanism and the size of the links and joint limits are optimum.

8   Conclusions

A mathematical model of the human arm kinematics is developed based on experimental evaluation of selective movements in the sagital, frontal, and horizontal plane. The developed model contains six revolute degrees of freedom for positioning the wrist and represents a redundant mechanism where each joint is dedicated to a specific task. The sternoclavicular joint facilitates the collision avoidance between the arm and the body, the glenohumeral joint effects the pointing operation of the wrist, and the elbow joint affects the reachability.

The experiments are made with three TV cameras which monitor a series of markers attached at the arm in predifined reference points. The human arm motion is thus specified in terms of trajectories effected by the markers in space. These trajectories are then numerically processed and studied in order to select the most appropriate kinematic structure of the mechanism.

The mathematical model developed in this paper does not consider the anatomical properties of the arm. It only specifies the motion of some reference points on the skeleton, especially the epicondylus sternalis and the proc.styloideus ulnae/radii. The included three joints and three links represent a ruogh approximation of the human arm. Some degrees of freedom are not taken into account, since their contribution from the positioning point of view can be added to the motion of other degrees of freedom. The main advantage of this model is its simplicity which enables complex numerical computations in the human arm motion analysis. Although the anatomical phenomena cannot be described directly by the model, it gives quite an accurate figure of the motion properties by which it may explain many kinematic particularities of the human arm.

References

[1] A.E. Engin, R.D. Peindl, "On the Biomechanics of Human Shoulder Complex - I. Kinematics for Determination of the Shoulder Complex Sinus", J. of Biomechanics, Vol.20, No.2, pp.103-117, 1987

[2] R.D. Peindl, A.E. Engin, "On the Biomechanics of Human Shoulder Complex - II. Passive Resistive Properties Beyond the Shoulder Complex Sinus", J. of Biomechanics, Vol.20, No.2, pp.119-134, 1987

[3] Z. Dvir, N. Berme, "The Shoulder Complex in Elevation on the Arm: A Mechanism Approach", J. of Biomechanics, Vol.11, pp.219-225, 1987

[4] J.E. Wood, S.G. Meek, S.C. Jacobsen, "Quantitation of Human Shoulder Anatomy for Prosthetic Arm Control - II. Anatomy Matrices, J. of Biomechanics, Vol.22, No.4, pp.309-325, 1989

[5] C. Hogfors, G. Sigholm, P. Herberts, "Biomechanical Model of Human Shoulder - I. Elements", J. of Biomechanics, Vol.20, No.2, pp.157-166, 1987

[6] J. Lenarčič, A. Umek, S. Savič, "Considerations on Human Arm Workspace and Manipulability", Proc. of the 2nd Workshop on Advances in Robot Kinematics, Linz, Austria, September 1990

[7] A. Umek, J. Lenarčič, "Recent Results in Evaluation of Human Arm Workspace", Proc. '91 ICAR, Pisa, Italy, June 1991

[8] G.L. Kinzel, L.J. Gutkowski, "Joint Models, Degrees of Freedom and Anatomical Motion Measurements", J. of Biomechanical Eng., Vol.105, pp.55-62, 1983

[9] V.T. Inman, M. Saunders, L.C. Abbott, "Observation on the Function of the Shoulder Joint", J. of Bone and Joint Surgery, Vol.42, No.1, pp.1-30, 1944

Modeling and Control of a Three Degree of Freedom Manipulator with Two Flexible Links

Tsuneo Yoshikawa*   Hiroki Murakami#   Koh Hosoda*

* Department of Mechanical Engineering
Faculty of Engineering, Kyoto University
Kyoto 606, Japan

Ishikawajima-Harima Heavy Industries Co., Ltd.
1-15, Toyosu 3-chome
Kotoku, Tokyo 135, Japan

Abstract

In this paper, we propose a 3D mass/spring model to use for real-time control of a robot with two flexible links moving in three dimensions. Based on this model, we derive the equation of motion and the state equation of a 3D flexible robot using Lagrange's method. We then show that accelerometers attached to the tips of elastic links are adequate for observing the state variables that represent flexibility in the model. Finally, we describe a controller designed by applying optimal regulator theory, and demonstrate with experimental results that the proposed controller can suppress vibrations of the arm.

1 Introduction

Recently, a need has developed for light and fast industrial robots. For use in space, long arms are needed that are light in comparison to their load. Such arms are aptly named flexible arms because their elasticity demands compensation for the arm deformations and vibrations.

In general, a mathematical model is required for analyzing or controlling a flexible manipulator. Previous studies have mostly considered a single flexible link or two flexible links moving in a plane [1]-[6] , and only a few have referred to three-dimensional (3D) motion of an arm with two flexible links [7] .

If a robot has large lumped masses at the tips of the links, and if the link masses themselves are small enough in comparison to the lumped masses that they can be neglected, then a simple lumped-mass/spring model will adequately describe the elasticity of a flexible arm manipulator. Sakawa et al. [5] have modeled the elasticity of a whole robot with one lumped mass with springs at the end tip; this way of modeling, however, does not allow description of any change in feedback gain due to change in configuration.

This paper describes a model that is useful for real-time control of a robot with two flexible links moving in three dimensions. First, we describe each flexible link using a 3D spring/mass model; then, we model the whole arm by connecting the models of the individual links, so as to treat change in feedback gain easily. Using this model, we derive the equation of motion and the state equation of a 3D flexible robot with Lagrange's method. We then show that accelerometers attached to the tips of elastic links are adequate for observing the state variables that represent flexibility in the model. There are some previous reports that use accelerometers to suppress the vibration [5][8] . Finally, we describe a controller designed by applying optimal regulator theory, and demonstrate with experimental results that the proposed controller can suppress vibrations of the arm.

2 Modeling of a Flexible Link

2.1 Assumption for Modeling

Consider the flexible arm shown in **Fig.1**. Links 2 and 3 are elastic and have lumped masses m_2, m_3 at each tip . Link 2 also has a counter-balance with mass m_c. Each joint is driven by a D.C. motor.

We make the following assumptions to build a model of the arm:

A1. Link 1 is rigid.

A2. m_2, m_3 and m_c can be considered to be point masses.

A3. The masses of links 2 and 3 are negligible in comparison to m_2 and m_3.

A4. Links 2 and 3 are slender, homogeneous and can be regarded to be simple beams.

A5. Frictions of motors and joints are assumed to be negligible.

Using these assumptions, we make the 3D spring/mass model shown in **Fig.2**, in which we model the flexibility as a spring system in three dimensions. The tip mass is considered to be connected to the 3D spring system.

2.2 Coordinate Frames

The coordinate frames are shown in **Fig.3**. Let O_0, the origin of the base frame Σ_0, be the intersection of joint axes 1 and 2, let Z_0 axis be joint axis 1, and let X_0 and Y_0 axes form a right-hand coordinate system. This system is an inertial system. The frame Σ_1 (fixed to link 1) is defined so that it coincides with Σ_0 when the first joint angle $\theta_1 = 0$. Frame Σ_2 has the same origin as Σ_0 and Σ_1. The X_2 axis is tangent to the long axis of link 2, the Z_2 axis is along the axis of joint 2, and the Y_2 axis completes a right-hand coordinate system. Frame Σ_3 is fixed to link 3, with its origin O_3 at the point where link 2 meets joint axis 3 when link 2 is not deformed.

To describe the flexibility of links 2 and 3, we define tip coordinate frames Σ_{ei}. These frames are each fixed to the tip of link i in such a way that the directions of their axes coincide with those of Σ_i when no link deformations exist.

2.3   Relation between Force and Deformation

Vectors d_i and ϕ_i are, respectively, the displacement and angular deformations of frame Σ_{ei} with respect to frame Σ_i:

$$d_i \;\; \overset{\Delta}{=} [l_i + \delta_{ix} \;\; \delta_{iy} \;\; \delta_{iz}] \quad (i = 2, 3), \tag{1}$$

$$\phi_i \;\; \overset{\Delta}{=} [\phi_{ix} \;\; \phi_{iy} \;\; \phi_{iz}] \quad (i = 2, 3), \tag{2}$$

where l_i is the length of link i, and δ_i and ϕ_i are, respectively, the deformation and angular deformation due to link elasticity.

Using the proposed model, the relation between the force and deformation is given by

$$\begin{bmatrix} f_i \\ n_i \end{bmatrix} = \begin{bmatrix} K_{i1} & K_{i3} \\ K_{i3}{}^T & K_{i2} \end{bmatrix} \begin{bmatrix} \delta_i \\ \phi_i \end{bmatrix}, \tag{3}$$

which is commonly used in structural engineering, where f_i and n_i are the force and moment to the tip of link i with respect to Σ_i, and K_{ij} is given by

$$K_{i1} = diag \left[A_i E_i / l_i \quad 12 E_i I_i / l_i^3 \quad 12 E_i I_i / l_i^3 \right]$$

$$\overset{\Delta}{=} diag \left[K_{ia} \quad K_{ib} \quad K_{ib} \right],$$

$$K_{i2} = diag \left[G_i \hat{I}_i / l_i \quad 4 E_i I_i / l_i \quad 4 E_i I_i / l_i \right]$$

$$\overset{\Delta}{=} diag \left[K_{ic} \quad K_{id} \quad K_{id} \right],$$

$$K_{i3} = \begin{bmatrix} 0 & 0 & 0 \\ 0 & 0 & -6 E_i I_i / l_i^2 \\ 0 & 6 E_i I_i / l_i^2 & 0 \end{bmatrix} \overset{\Delta}{=} \begin{bmatrix} 0 & 0 & 0 \\ 0 & 0 & K_{ie} \\ 0 & -K_{ie} & 0 \end{bmatrix},$$

where E_i is the Young's modulus of link i, G_i is the modulus of transverse of elasticity of link i, A_i is the cross-sectional area of link i, l_i is the length of link i, I_i is the area moment of inertia of link i, and $\hat{I}_i$ is the polar moment of inertia of area of link i.

3   Kinematics of the Flexible Arm

3.1   Rotation Matrix

Let the angle of joint i be θ_i and let the rotation matrix from the i th frame to the j th frame be $^j R_i$. Then,

$$^0 R_1 = \begin{bmatrix} \cos \theta_1 & -\sin \theta_1 & 0 \\ \sin \theta_1 & \cos \theta_1 & 0 \\ 0 & 0 & 1 \end{bmatrix},$$

$$^1 R_2 = \begin{bmatrix} \cos \theta_2 & -\sin \theta_2 & 0 \\ 0 & 0 & -1 \\ \sin \theta_2 & \cos \theta_2 & 0 \end{bmatrix},$$

$$^{e2} R_3 = \begin{bmatrix} \cos \theta_3 & -\sin \theta_3 & 0 \\ \sin \theta_3 & \cos \theta_3 & 0 \\ 0 & 0 & 1 \end{bmatrix}. \tag{4}$$

By neglecting the second-order terms, the rotation matrix from Σ_{e2} to Σ_2, $^2R_{e2}$, is

$$^2R_{e2} = \begin{bmatrix} 1 & -\phi_{2z} & \phi_{2y} \\ \phi_{2z} & 1 & -\phi_{2x} \\ -\phi_{2y} & \phi_{2x} & 1 \end{bmatrix}. \tag{5}$$

Further, ^0T_i is defined as the rotation matrix from Σ_0 to Σ_i:

$$^0T_1 = {}^0R_1, \tag{6}$$

$$^0T_2 = {}^0R_1{}^1R_2, \tag{7}$$

$$^0T_3 = {}^0R_1{}^1R_2{}^2R_{e2}{}^{e2}R_3. \tag{8}$$

3.2 Motions of Tip Lumped Masses

The position vector r_i $(i = 2, 3)$ of tip mass m_i with respect to Σ_0 is given by

$$r_i = \sum_{j=2}^{i} {}^0T_j d_j. \tag{9}$$

The position vector of m_c, r_c, is given by

$$r_c = {}^0T_2 d_c, \tag{10}$$

where d_c is the position vector of m_c with respect to Σ_2; $d_c = [-l_c \ 0 \ 0]$.

The velocity and acceleration of r_i are obtained by differentiating eq. (9):

$$\dot{r}_i = \sum_{j=2}^{i} \{(\sum_{k=1}^{3} U_{jk}\dot{\theta}_k + U_j^*)d_j + {}^0T_j\dot{d}_j\}, \tag{11}$$

$$\ddot{r}_i = \sum_{j=2}^{i} [\{\sum_{k=1}^{3}(\sum_{l=j}^{3} U_{jkl}\dot{\theta}_k\dot{\theta}_l + 2U_{jk}^*\dot{\theta}_k + U_{jk}\ddot{\theta}_k) + U_j^{**}\}d_j$$

$$+ \ 2(\sum_{k=j}^{3} U_{jk}\dot{\theta}_k + U_j^*)\dot{d}_j + {}^0T_j\ddot{d}_j], \tag{12}$$

where

$$U_{jk} = \frac{\partial}{\partial\theta_k}{}^0T_j,$$

$$U_{jkl} = \frac{\partial}{\partial\theta_l}U_{jk},$$

$$U_2^* = 0 \ , \ U_3^* = {}^0R_1^1R_2(\frac{d}{dt}{}^2R_{e2})^{e2}R_3, \tag{13}$$

$$U_2^{**} = 0 \ , \ U_3^{**} = {}^0R_1^1R_2(\frac{d^2}{dt^2}{}^2R_{e2})^{e2}R_3,$$

$$U_{2k}^* = 0 \ , \ U_{3k}^* = \frac{\partial}{\partial\theta_k}U_3^*,$$

The first subscript j of U shows that U_j is a matrix obtained by differentiating ^0T_j. The second and third subscripts k and l indicate partial differentiations with respect to θ_k and θ_l, respectively. The superscript $*$ represents differentiation of $^2R_{e2}$ with respect to time.

The velocity and acceleration of r_c are given by

$$\dot{r}_c = \sum_{j=1}^{2} (U_{2j}\dot{\theta}_j) d_c, \tag{14}$$

$$\ddot{r}_c = \sum_{j=1}^{2} (\sum_{k=1}^{2} U_{2jk}\dot{\theta}_j\dot{\theta}_k + U_{2j}\ddot{\theta}_j) d_c. \tag{15}$$

4 Dynamics of the Flexible Arm

4.1 Kinetic Energy

From the assumptions **A1** $\sim$ **A5**, the total kinetic energy of the robot consists of the kinetic energy of lumped masses and rotational energy of motors. Letting E_{ki} be the kinetic energy of link i,

$$E_{k1} = \frac{1}{2}I_{m1}\dot{\theta}_1^2, \tag{16}$$

$$E_{k2} = \frac{1}{2}m_2\dot{r}_2^T\dot{r}_2 + \frac{1}{2}m_c\dot{r}_c^T\dot{r}_c + \frac{1}{2}I_{m2}\dot{\theta}_2^2, \tag{17}$$

$$E_{k3} = \frac{1}{2}m_3\dot{r}_3^T\dot{r}_3 + \frac{1}{2}I_{m3}\dot{\theta}_3^2, \tag{18}$$

where I_{mi} is the rotational inertia of the i th motor (I_{m1} includes the inertia of link 1 about the Z_1 axis). The superscript T represents the matrix or vector transpose.

The total kinetic energy is

$$E_k = \sum_{i=1}^{3} E_{ki}. \tag{19}$$

4.2 Potential Energy

From eq. (3), the potential energy of elasticity of link i E_{pi} $(i = 2, 3)$ is

$$E_{pi} = \frac{1}{2} \begin{bmatrix} \delta_i^T & \phi_i^T \end{bmatrix} \begin{bmatrix} K_{i1} & K_{i3} \\ K_{i3}^T & K_{i2} \end{bmatrix} \begin{bmatrix} \delta_i \\ \phi_i \end{bmatrix}. \tag{20}$$

The potential energy from gravity, E_{pg}, is

$$E_{pg} = m_2 g^T r_2 + m_3 g^T r_3 + m_c g^T r_c, \tag{21}$$

where

$$g = \begin{bmatrix} 0 & 0 & g_0 \end{bmatrix}^T \qquad (g_0 \text{ is the acceleration due to gravity}).$$

Thus, the total potential energy E_p is

$$E_p = \sum_{i=2}^{3} E_{pi} + E_{pg}. \tag{22}$$

4.3   Equation of Motion of the Robot

Using Lagrange's method, we now derive the equation of motion of the robot. The Lagrange function L is

$$L = E_k - E_p. \tag{23}$$

Using the joint displacements θ_i, δ_i, and ϕ_i as generalized coordinates, the equation of motion for displacement θ_i is

$$\frac{d}{dt}(\frac{\partial L}{\partial \dot{\theta}_i}) - \frac{\partial L}{\partial \theta_i} = \tau_i \qquad (i=1,2,3), \tag{24}$$

where τ_i is the torque which works on joint i. From eq. (11) we get

$$\tau_i = \sum_{j=2}^{3}\{m_j(\ddot{r}_j + g)^T(\sum_{k=1}^{j} U_{ki}d_k)\} + m_c(\ddot{r}_c + g)^T(U_{2i}d_c) + I_{mi}\ddot{\theta}_i, \tag{25}$$

where we define $m_1 = 0$ for notational convenience.

The Lagrange equations of motion for deformations δ_i and angular deformations ϕ_i are

$$\frac{d}{dt}(\frac{\partial L}{\partial \dot{\delta}_i}) - \frac{\partial L}{\partial \delta_i} = 0, \tag{26}$$

$$\frac{d}{dt}(\frac{\partial L}{\partial \dot{\phi}_i}) - \frac{\partial L}{\partial \phi_i} = 0. \tag{27}$$

Therefore, we can derive equation of motion as follows:

$$m_3{}^0T_2{}^T(\ddot{r}_3 + g) + m_2{}^0T_2{}^T(\ddot{r}_2 + g) + K_{21}\delta_2 + K_{23}\phi_2 = 0, \tag{28}$$

$$m_3{}^0T_3{}^T(\ddot{r}_3 + g) + K_{31}\delta_3 + K_{33}\phi_3 = 0, \tag{29}$$

$$m_3V_3{}^T(\ddot{r}_3 + g) + K_{23}{}^T\delta_2 + K_{22}\phi_2 = 0, \tag{30}$$

$$K_{33}{}^T\delta_3 + K_{32}\phi_3 = 0, \tag{31}$$

where V_3 is the matrix defined as

$$V_3 = \frac{\partial}{\partial \dot{\phi}_2{}^T}(U_3{}^*d_3). \tag{32}$$

In assumption A2 of section 2.1, we assumed that the tip masses are point masses, so we can express ϕ as a function of θ and δ from eqs. (25),(28)– (32). From eq. (31),

$$\phi_3 = -K_{32}{}^{-1}K_{33}{}^T\delta_3. \tag{33}$$

From eq. (29), (30), and (33) ϕ_2 is given by

$$\phi_2 = K_{22}{}^{-1}V_3{}^T{}^0T_3(K_{31} - K_{33}K_{32}{}^{-1}K_{33}{}^T)\delta_3 - K_{22}{}^{-1}K_{23}{}^T\delta_2. \tag{34}$$

Differentiating eqs. (33) and (34) with respect to time, we can find $\dot{\phi}_i$ and $\ddot{\phi}_i$. Substituting them into eqs. (25),(28), and (29), we can obtain the equation of motion independent of ϕ. We do not show the final form of the equation of motion; although the solution is straightforward, it is very lengthy.

5 Control Design

5.1 Derivation of State Equation

Because the equation of motion that we have obtained is too complex to use to derive the state equation for designing a controller, we make the following assumptions to simplify it:

B1. The aim is to control the vibration around the desired fixed point, and so we assume that Coriolis and centrifugal forces are negligible.

B2. The vibration displacement is small, so the terms which include δ or ϕ are negligible except spring force terms.

B3. The beam does not deform longitudinally ($\delta_x \equiv 0$).

From B1 and B2, eqs. (12) and (15) become

$$\ddot{r}_2 = U_{21}\bar{d}_2\ddot{\theta}_1 + U_{22}\bar{d}_2\ddot{\theta}_2 + {}^0T_2\ddot{\delta}_2, \tag{35}$$

$$\ddot{r}_3 = (U_{21}\bar{d}_2 + \bar{U}_{31}\bar{d}_3)\ddot{\theta}_1 + (U_{22}\bar{d}_2 + \bar{U}_{32}\bar{d}_3)\ddot{\theta}_2$$
$$+ \bar{U}_{33}\bar{d}_3\ddot{\theta}_3 + V_3\ddot{\phi}_2 + {}^0T_3\ddot{\delta}_3, \tag{36}$$

$$\ddot{r}_c = U_{21}d_c\ddot{\theta}_1 + U_{22}d_c\ddot{\theta}_2, \tag{37}$$

where

$$\bar{d}_2 = [l_2\ 0\ 0]^T, \quad \bar{d}_3 = [l_3\ 0\ 0]^T,$$

$${}^0\bar{T}_3 = {}^0R_1{}^1R_2{}^2R_3, \quad \bar{U}_{3i} = \frac{\partial}{\partial\theta_i}{}^0\bar{T}_3.$$

Substituting eqs. (33)–(37) into eqs. (25),(28), and (29), and defining ξ and u as

$$\xi = [\theta_1\ \delta_{2z}\ \delta_{3z}\ \theta_2\ \theta_3\ \delta_{2y}\ \delta_{3y}]^T,$$

$$u = [\tau_1\ \tau_2\ \tau_3]^T,$$

we finally obtain the following simplified equation of motion:

$$\begin{bmatrix} M_h & 0 \\ 0 & M_v \end{bmatrix}\ddot{\xi} + \begin{bmatrix} F_h & 0 \\ 0 & F_v \end{bmatrix}\xi + \begin{bmatrix} 0 \\ G_v \end{bmatrix} = \begin{bmatrix} D_h & 0 \\ 0 & D_v \end{bmatrix}u, \tag{38}$$

$$M_h = \begin{bmatrix} (m_2l_2{}^2 + m_cl_c{}^2)\cos^2\theta_2 + m_3L_c{}^2 + I_{m1} & -m_2l_2\cos\theta_2 - m_3L_c(1 - K_{h1}l_3) \\ -m_2l_2\cos\theta_2 - m_3L_c & m_2 + m_3 - m_3l_3K_{h1} \\ -m_3L_c & m_3 - m_3l_3K_{h1} \end{bmatrix}$$
$$\begin{matrix} -m_3L_c(1 + K_{h2}l_3{}^2) \\ m_3(1 + K_{h2}l_3{}^2) \\ m_3(1 + K_{h2}l_3{}^2) \end{matrix} \Bigg],$$

$$M_v = \begin{bmatrix} m_2l_2{}^2 + m_cl_c{}^2 + m_3(L_s{}^2 + L_c{}^2) + I_{m2} & m_3l_3(l_3 + l_2\cos\theta_3) \\ m_3l_3(l_3 + l_2\cos\theta_3) & m_3l_3{}^2 + I_{m3} \\ m_2l_2 + m_3(l_2 + l_3\cos\theta_3) & m_3l_3\cos\theta_3 \\ m_3(l_3 + l_2\cos\theta_3) & m_3l_3 \end{bmatrix}$$

$$\begin{matrix} m_2l_2 + m_3l_2 + m_3l_3\{\cos\theta_3 - K_{v1}(l_3 + l_2\cos\theta_3)\} & m_3(l_3 + l_2\cos\theta_3)(1 + K_{v2}l_3{}^2) \\ m_3l_3(\cos\theta_3 - K_{v1}l_3) & m_3l_3(1 + K_{v2}l_3{}^2) \\ m_2 + m_3 - m_3l_3K_{v1}\cos\theta_3 & m_3\cos\theta_3(1 + K_{v2}l_3{}^2) \\ m_3(\cos\theta_3 - l_3K_{v1}) & m_3(1 + K_{v2}l_3{}^2) \end{matrix} \Bigg],$$

$$
F_h = \begin{bmatrix} 0 & 0 & 0 \\ 0 & K_{2b} - \frac{K_{2e}^2}{K_{2d}} & \frac{K_{2e}}{K_{2d}}(K_{3b} - \frac{K_{3e}^2}{K_{3d}})l_3\cos\theta_3 \\ 0 & 0 & K_{3b} - \frac{K_{3e}^2}{K_{3d}} \end{bmatrix},
$$

$$
F_v = \begin{bmatrix} 0 & 0 & 0 & 0 \\ 0 & 0 & 0 & 0 \\ 0 & 0 & K_{2b} - \frac{K_{2e}^2}{K_{2d}} & \frac{K_{2e}}{K_{2d}}(K_{3b} - \frac{K_{3e}^2}{K_{3d}})l_3 \\ 0 & 0 & 0 & K_{3b} - \frac{K_{3e}^2}{K_{3d}} \end{bmatrix},
$$

$$
D_h = \begin{bmatrix} 1 \\ 0 \\ 0 \end{bmatrix}, \qquad\qquad D_v = \begin{bmatrix} 1 & 0 \\ 0 & 1 \\ 0 & 0 \\ 0 & 0 \end{bmatrix},
$$

$$
G_v = \begin{bmatrix} (m_2l_2 + m_cl_c)\cos\theta_2 + m_3L_c \\ m_3l_3\cos(\theta_2 + \theta_3) \\ (m_2 + m_3)\cos\theta_2 \\ m_3\cos(\theta_2 + \theta_3) \end{bmatrix} g_0,
$$

where

$$
\begin{aligned}
L_c &= l_2\cos\theta_2 + l_3\cos(\theta_2 + \theta_3), \\
L_s &= l_2\sin\theta_2 + l_3\sin(\theta_2 + \theta_3), \\
K_{h1} &= \frac{K_{2e}}{K_{2d}}\cos\theta_3, \\
K_{h2} &= (K_{3b} - \frac{K_{3e}^2}{K_{3d}})(\frac{\sin^2\theta_3}{K_{2c}} + \frac{\cos^2\theta_3}{K_{2d}}), \\
K_{v1} &= \frac{K_{2e}}{K_{2d}}, \\
K_{v2} &= (K_{3b} - \frac{K_{3e}^2}{K_{3d}})/K_{2d}.
\end{aligned}
$$

Note that eq.(38) is divided into horizontal and vertical parts. Let the state values be

$$
x_h \triangleq \begin{bmatrix} \theta_1 & \delta_{2z} & \delta_{3z} & \dot\theta_1 & \dot\delta_{2z} & \dot\delta_{3z} \end{bmatrix}^T,
$$

$$
x_v \triangleq \begin{bmatrix} \theta_2 & \theta_3 & \delta_{2y} & \delta_{3y} & \dot\theta_2 & \dot\theta_3 & \dot\delta_{2y} & \dot\delta_{3y} \end{bmatrix}^T,
$$

and let the inputs be

$$
u_h = [\tau_1],
$$

$$
u_v = [\tau_2 \quad \tau_3]^T.
$$

Then, the expression of state equation of each part is

$$
\dot{x}_h = \begin{bmatrix} 0 & E_3 \\ -M_h^{-1}F_h & 0 \end{bmatrix} x_h + \begin{bmatrix} 0 \\ M_h^{-1}D_h \end{bmatrix} u_h, \tag{39}
$$

$$
\dot{x}_v = \begin{bmatrix} 0 & E_4 \\ -M_v^{-1}F_v & 0 \end{bmatrix} x_v + \begin{bmatrix} 0 \\ M_v^{-1}D_v \end{bmatrix} u_v + \begin{bmatrix} 0 \\ -M_v^{-1}G_v \end{bmatrix}, \tag{40}
$$

E_i's $i \times i$ unit matrix.

We now consider the compensation of gravity term G_v in eq. (40). The first and second terms of G_v can be compensated using a new input $u_v{}'$ defined as

$$u_v{}' = u_v + \left[\begin{array}{c} (m_2 l_2 + m_c l_c)\cos\theta_2 + m_3 L_c \\ m_3 l_3 \cos(\theta_2 + \theta_3) \end{array} \right] g_0. \tag{41}$$

For the third and fourth terms of G_v, we set

$$\delta_{2yd} = -\frac{m_3 \cos(\theta_2 + \theta_3) l_2{}^2 l_3 g_0}{2 E_2 I_2} - \frac{(m_2 + m_3)\cos\theta_2 l_2{}^3 g_0}{3 E_2 I_2}, \tag{42}$$

$$\delta_{3yd} = -\frac{m_3 \cos(\theta_2 + \theta_3) l_3{}^3 g_0}{3 E_3 I_3} \tag{43}$$

as the desired values of δ_{2y} and δ_{3y}.

5.2   Equation for Observation

The angle and angular velocity of joints are known from the outputs of optical encoders attached to each joint. To observe the elastic term δ, accelerometers are attached to the tips of the elastic beams along the XYZ axes of frame Σ_{ei}. Let a_i be the output of the accelerometer attached to the tip of link i. Then,

$$a_i = {}^0 T_i{}^T (\ddot{r}_i + g). \tag{44}$$

Using eqs. (28)-(31), neglecting the centrifugal and Coriolis forces according to assumption B1, we obtain the relation between the sensor output a_i and the deformation δ_i:

$$\left[\begin{array}{c} \delta_2 \\ \delta_3 \end{array} \right] = - \left[\begin{array}{cc} m_2 N_1^{-1} & m_3 N_1^{-1} 2 R_3 - m_3 N_1^{-1} N_2 N_3^{-1} \\ 0 & m_3 N_3^{-1} \end{array} \right] \left[\begin{array}{c} a_2 \\ a_3 \end{array} \right], \tag{45}$$

where

$$\begin{aligned}
N_1 &= K_{21} - K_{23} K_{22}^{-1} K_{23}{}^T, \\
N_2 &= K_{23} K_{22}^{-1} \bar{V}_3 {}^0 \bar{T}_3 (K_{31} - K_{33} K_{32}^{-1} K_{33}{}^T), \\
N_3 &= K_{31} - K_{33} K_{32}^{-1} K_{33}{}^T.
\end{aligned}$$

From assumption B3, δ_x is negligible. Eq. (45) reduces to

$$\left[\begin{array}{c} \delta_{2y} \\ \delta_{2z} \\ \delta_{3y} \\ \delta_{3z} \end{array} \right] = H \left[\begin{array}{c} a_{2y} \\ a_{2z} \\ a_{3x} \\ a_{3y} \\ a_{3z} \end{array} \right], \tag{46}$$

$$H = \left[\begin{array}{ccccc} -\frac{m_2 l_2^3}{3 E_2 I_2} & 0 & -\frac{m_3 l_2^3}{3 E_2 I_2}\sin\theta_3 & -\frac{m_3 l_2^2}{6 E_2 I_2}(2 l_2 \cos\theta_3 + 3 l_3) & 0 \\ 0 & -\frac{m_2 l_2^3}{3 E_2 I_2} & 0 & 0 & -\frac{m_3 l_2^2}{6 E_2 I_2}(2 l_2 + 3 l_3 \cos\theta_3) \\ 0 & 0 & 0 & -\frac{m_3 l_3^3}{3 E_3 I_3} & 0 \\ 0 & 0 & 0 & 0 & -\frac{m_3 l_3^3}{3 E_3 I_3} \end{array} \right].$$

We can find the deformation velocity $\dot{\delta}$ by differentiating eq. (46).

Eq. (46) is a redundant expression which has 5 inputs and 4 outputs, and so it can be reduced. For example, we can remove the a_{3x} sensor and still have all δ observable. In this paper, however, we use 5 sensors so as to obtain a more precise value of δ.

5.3 Optimal Regulator

We have formed the linear state equation (39) and (40) and we can assume that we get all of the state values. In this section, the theory of the optimal regulator is applied to design a controller.

We introduce performance indexes J_h and J_v defined by

$$J_h = \int \{(x_h - x_{hd})^T Q_h (x_h - x_{hd}) + u_h^T R_h u_h\} dt,$$

$$J_v = \int \{(x_v - x_{vd})^T Q_v (x_v - x_{vd}) + u_v^T R_v u_v\} dt, \tag{47}$$

where Q and R are suitable weight matrices, and x_{hd} and x_{vd} are the desired value of x_h and x_v given by

$$x_{hd} \triangleq \begin{bmatrix} \theta_{1d} & 0 & 0 & 0 & 0 & 0 \end{bmatrix}^T,$$

$$x_{vd} \triangleq \begin{bmatrix} \theta_{2d} & \theta_{3d} & \delta_{2yd} & \delta_{3yd} & 0 & 0 & 0 & 0 \end{bmatrix}^T.$$

The state feedback gain that makes these performance indices minimum can be obtained using the solution matrix P of the following Riccati equations:

$$P_h A_h + A_h^T P_h - P_h B_h R_h^{-1} B_h^T P_h + Q_h = 0,$$

$$P_v A_v + A_v^T P_v - P_v B_v R_v^{-1} B_v^T P_v + Q_v = 0. \tag{48}$$

The optimal inputs u_h^* and u_v^* are given by

$$u_h^* = -R_h^{-1} B_h P_h (x_{hd} - x_h),$$

$$u_v^* = -R_v^{-1} B_v P_v (x_{vd} - x_v). \tag{49}$$

Note that the optimal feedback gain in eq. (49) is a function of $\theta = [\theta_1 \ \theta_2 \ \theta_3]^T$. It is necessary to solve the Riccati equation to obtain the optimal gain for various desired arm configurations. A strategy such as table look-up of optimal gain for each given configuration is needed.

6 Experiments

6.1 Robot System

We evaluated the performance of the proposed control algorithm by comparing its results with those using PD control on each joint, for the case of PTP control.

The parameters of the 3D flexible robot shown in **Fig.1** are given in **Table 1**. Link 2 is elastic, with a lumped mass of 2.1kg at its tip;link 3 is also elastic, with 0.25kg at its tip. Further, link 2 has a 4.2kg counterbalance -13mm from O_2. Each joint of the robot has a DC servo motor that drives the joint through a harmonic-drive gear. The gear ratios of 1st, 2nd, and 3rd harmonic-drive are 101, 100, and 110. The outputs from the optical encoders at joints are sent through a PI/O board to the host computer. Two accelerometers are attached to the tip of link 2 to measure a_{2y}, a_{2z} and three accelerometers at the tip of link 3 to measure a_{3x}, a_{3y}, and a_{3z}. The acceleration data are sent to the computer through an A/D converter. Based on these data, the host computer determines the input command to the motors.

As the host computer, we use a 32-bit personal computer (NEC PC–9801RX) with a numerical processor 80287. Assembler and C languages are used. Sampling time is set to 4.0 ms, both for PD control of each joint and for the proposed control algorithm.

6.2 Results

For PD control, we selected the gains by trial and error. For proposed scheme, the suitable weight matrices Q and R have to be selected to calculate the optimal gain of the state feedback which coincide the joint gains of proposed scheme with those of PD control.

A experiments were done whose initial and desired points of $[\theta_1, \theta_2, \theta_3]$ were:

$$[0 \; \pi/3 \; -\pi/4] \rightarrow [\pi/9 \; \pi/4 \; -\pi/2].$$

The selected gains for PD control were:

$$\tau_1 = 300.0(\theta_{1d} - \theta_1) - 59.3\dot{\theta}_1$$
$$\tau_2 = 297.3(\theta_{2d} - \theta_2) - 34.6\dot{\theta}_2.$$
$$\tau_3 = 192.9(\theta_{3d} - \theta_3) - 14.1\dot{\theta}_3$$

For the proposed scheme, the selected weights were:

$$Q_h = diag[\, 9 \times 10^4 \; 1 \times 10^6 \; 5 \times 10^6 \; 1 \; 1 \; 1 \,]$$

$$R_h = [\, 1 \,]$$

$$Q_v = diag[\, 9.5 \times 10^4 \; 8 \times 10^4 \; 8 \times 10^5 \; 8 \times 10^5 \; 10 \; 10 \; 10 \; 10 \,]$$

$$R_v = diag[\, 1 \; 2 \,]$$

Using these weights, the control scheme were:

$$\tau_1 = -300.0(\theta_1 - \theta_{1d}) - 2941.7\delta_{2z} - 365.1\delta_{3z} - 59.3\dot{\theta}_1 - 28.2\dot{\delta}_{2z} - 73.1\dot{\delta}_{3z}$$
$$\tau_2 = -297.3(\theta_2 - \theta_{2d}) + 74.5(\theta_3 - \theta_{3d}) + 555.0(\delta_{2y} - \delta_{2yd}) - 55.7(\delta_{3y} - \delta_{3yd})$$
$$\quad -34.6\dot{\theta}_2 + 3.8\dot{\theta}_3 + 11.7\dot{\delta}_{2y} + 5.1\dot{\delta}_{3y}$$
$$\tau_3 = -57.4(\theta_2 - \theta_{2d}) - 192.9(\theta_3 - \theta_{3d}) - 76.1(\delta_{2y} - \delta_{2yd}) + 444.5(\delta_{3y} - \delta_{3yd})$$
$$\quad -7.9\dot{\theta}_2 - 14.1\dot{\theta}_3 - 21.2\dot{\delta}_{2y} + 7.8\dot{\delta}_{3y}$$

Results are shown in **Figs.4–5**. The figures show the error from the desired value of θ_1, θ_2, θ_3, and the deformations $\delta_{2y} - \delta_{2yd}$, δ_{2z}, $\delta_{3y} - \delta_{3yd}$ and δ_{3z}. We can see that the proposed scheme can suppress vibrations.

7 Conclusions

In this paper, a three-dimensional spring/mass model has been proposed to describe arm flexibility. The equation of motion and the state equation have been obtained from the model under certain assumptions. Using optimal control theory, we have designed a controller for the robot that experimental results show to be effective in suppressing vibration.

The authors would like to thank Mr. Y. Kadokawa, Toyoda Automatic Loom Works, Ltd., and Mr. A. Yoshida, Recruit Co., Ltd., for their contribution in an early stage of this research.

542

References

[1] W.J. Book, O. Maizza-Neto, and D.E. Whitney. "Feedback control of two beam, two joint systems with distributed flexibility." *Trans. of ASME, J. of DSMC*, pp.424–431, 1975.

[2] R.H. Cannon and E. Schmitsz. "Initial experiments on the end-point control of a flexible one-link robot." *Int. J. of Robotics Reserch*, 3(3):pp.62–75, 1984.

[3] F. Pfeiffer. "A feedforward decoupling concept for the control of elastic robots." *J. of Robotic Systems*, 6(4):pp.407–416, 1989.

[4] I. Simoyama and H. Miura. "A dynamic model for flexible manipulator control." *J. of the Robotics Society of Japan*, 6(5):pp.72–78, 1988. In Japanese.

[5] Y.Sakawa, F. Matsuno, et al. "Modeling and vibration control of a flexible manipulator with three axes by using accelerometers." *J. of the Robotics Society of Japan*, 6(1):pp.42–51, 1988. In Japanese.

[6] M. Uchiyama et al. "Compensating control of a flexible robot arm." *J. of the Robotics Society of Japan*, 7(4):pp.20–30, 1989. In Japanese.

[7] T. Fukuda et al. "Decoupled vibration control of 3d robotic arm with flexible links." In *Proc. of the U.S.A.-Japan Symp. on Flexible Automation*, pp.415–421, 1988.

[8] S. Futami, M. Kyura, and S. Hara. "Vibration absorption control of industrial robots by acceleration feedback." In *Proc. of IEEE Int. Conf. on Industrial Electronics*, pp.299–305, 1983.

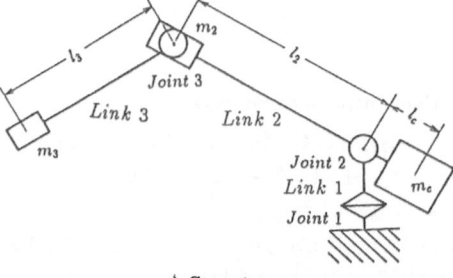

a) Structure

b) Overview

Fig.1 3-DOF flexible arm

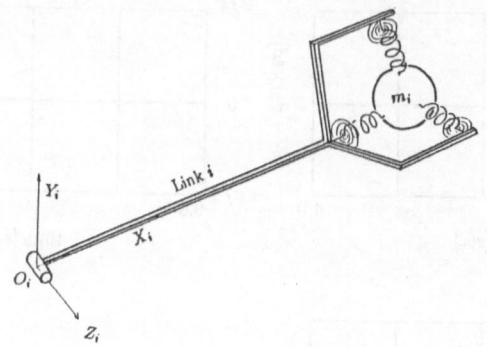

Fig.2 3D spring/mass model

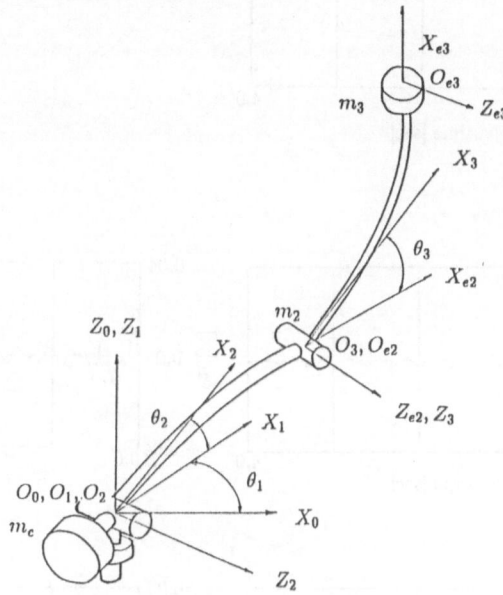

Fig.3 Coodinate frames of the flexible robot

Table 1 Parameters of links

	link 2	link 3
Length(m)	0.6	0.6
Diameter(m)	0.010	0.006
Bending rigidity($N \cdot m^2$)	101.0	13.09
Torsional rigidity($N \cdot m^2$)	77.71	10.07

544

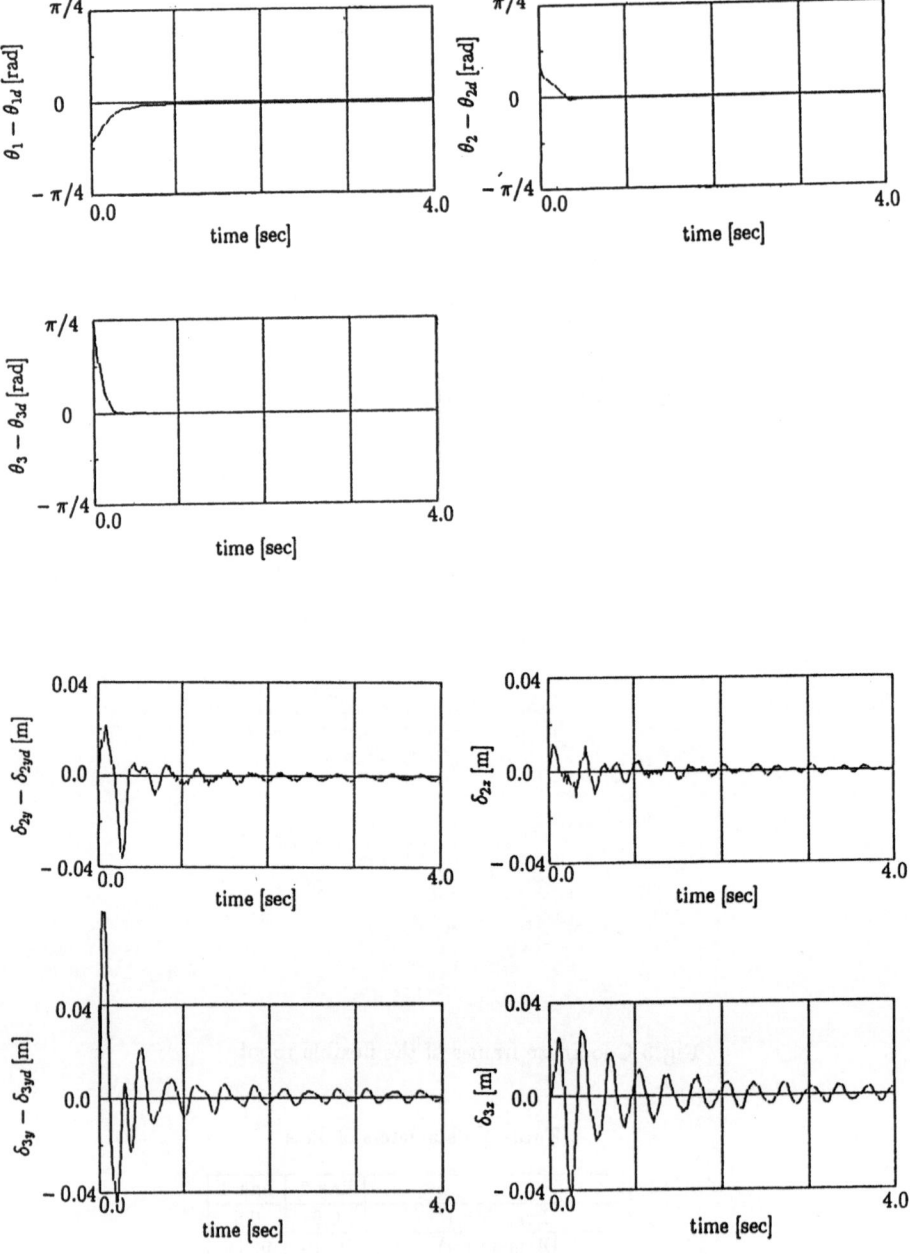

Fig.4 Experimental results (using PD control)

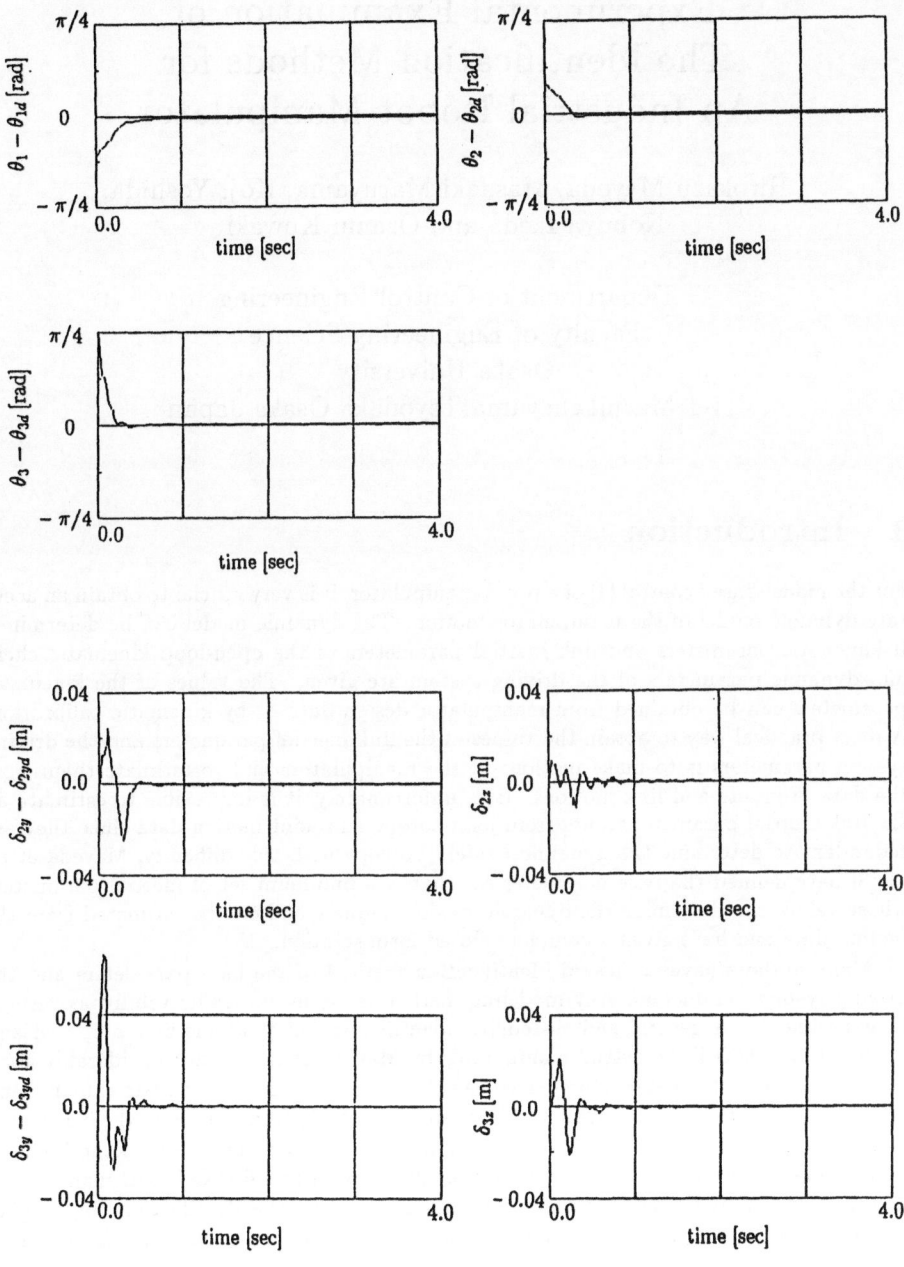

Fig.5 Experimental results (using proposed scheme)

Experimental Examination of The Identification Methods for An Industrial Robot Manipulator

Hirokazu Mayeda, Masaaki Maruyama, Koji Yoshida,
Nobuya Ikeda and Osamu Kuwaki

Department of Control Engineering
Faculty of Engineering Science
Osaka University
1-1 Machikaneyama Toyonaka Osaka Japan

1   Introduction

For the model-based control [1] of a robot manipulator, it is very crucial to obtain an accurate dynamic model of the manipulator motion. The dynamic model can be determined if kinematic parameters and link inertial parameters of the open-loop kinematic chain and dynamic parameters of the driving system are given. The values of the kinematic parameters can be obtained from manipulator design data or by kinematic calibration. A most practical way to obtain the values of the link inertial parameters and the driving system parameters is to make motions of the manipulators and to estimate them from the data of inputs and link motions. But, unfortunately, it is impossible to estimate all the link inertial parameter values from joint torque and joint motion data since they are redundant to determine the dynamic model. To cope with this difficulty, Mayeda et al. [2],[3] have defined the base parameter set that is a minimum set of inertial parameters whose values can determine the dynamic model uniquely and can be estimated from the motion data and have given a complete closed form solution.

Many authors have addressed identification method of the base parameters and the driving system parameters. Any modeling method is no use unless its validity is verified by experiments. A general and systematic identification method has first proposed and applied to a 3 D.O.F. industrial manipulator by Mayeda et al.[4], and considerably good estimation of the parameter values have been obtained. This method consists of many kind of simple test motions that move 1 or 2 joins simultaneously from the tip to the base of the manipulator and estimates the parameter values step by step. Thus this method will be called step-by-step method. The method was also applied to a 6 D.O.F. DD manipulator and the results were good [5]. DD manipulators are easier for the identification since higher accelerations are realizable and driving systems are simple and accurate. Examinations of the method on high D.O.F. industrial manipulators with gear mechanisms are left as a futher work.

There is another identification method which moves all joints simultaneously in random enough way and estimate all the parameter values using the least square method. This method may be able to be called simultaneous method. The simultaneous method has examined on a 6 D.O.F. industrial manipulators by Kawasaki et al.[6] and on a 6 D.O.F. and a 3 D.O.F. DD manipulators by Khosla[7] and Atokeson[8], respectively. All their results were reported good. Kawasaki's method taking advantage of the instrumental variable method to avoid inconsistent estimation will be more practical.

All those reported results seems good, but they applied either of the methods to their own manipulators. The two methods should be compared experimentally about same manipulators since the validities of the identification methods may seriously depend on type and D.O.F. and driving systems of manipulators. To establish practical identification methods, we need more experiments and 'know-how's about various kind of manipulators.

In this study, we first apply the step by step method to a popular 6 D.O.F. industrial manipulator PUMA 560 and show the experimental results. Next, the simultaneous method is examined on the same manipulator. Comparison of the results by the two methods will be shown and discussed.

2 The manipulator and Its Base Parameter Set

In this study, we examine the two identification methods for a popular 6 d.o.f. industrial manipulator PUMA 560. To each link i of the manipulator, a coordinate system $(o_i; \boldsymbol{x}_i, \boldsymbol{y}_i, \boldsymbol{z}_i)$ is attached in the way shown in Fig. 1. This is our convention adopted in [2]-[5]. Joint variable θ_i is the angle from $\boldsymbol{x}_{i-1}$ to $\boldsymbol{x}_i$. Let m_i and $\boldsymbol{I}_i$ be the mass and the moment of inertia tensor around o_i, and let $\boldsymbol{r}_i$ and $\boldsymbol{l}_i$ be the vectors from o_i to the center of mass of link i and o_{i+1}, respectively. We attach a load to 6-th link, which is characterized by

Approximate mass: 1.6 [kg]

Approximate location of center of mass: $[\ -7\ 7\ 11\ ]^t$ [cm]

Approximate moment of inertia matrix around o_6:

$$\begin{bmatrix} 3.67 \times 10^{-2} & 0.759 \times 10^{-2} & 1.23 \times 10^{-2} \\ 0.759 \times 10^{-2} & 3.67 \times 10^{-2} & -1.23 \times 10^{-2} \\ 1.23 \times 10^{-2} & -1.23 \times 10^{-2} & 2.87 \times 10^{-2} \end{bmatrix} [kg \cdot m^2]$$

The 6-th link and the load will be regarded as one link.

For vector $\boldsymbol{v}$ or tensor $\boldsymbol{T}$, $^i\boldsymbol{v}$, or $^i\boldsymbol{T}$ denotes the representation of $\boldsymbol{v}$ or $\boldsymbol{T}$ about $(\boldsymbol{x}_i, \boldsymbol{y}_i, \boldsymbol{z}_i)$, respectively. Then $^i\boldsymbol{l}_i$, $^i\boldsymbol{r}_i$, and $^i\boldsymbol{I}_i$ are constant vectors and a constant matrix, and they can be described as

$$^i\boldsymbol{l}_i = \begin{bmatrix} l_i^x & 0 & l_i^z \end{bmatrix}^t, \tag{1}$$

$$^i\boldsymbol{r}_i = \begin{bmatrix} r_i^x & r_i^y & r_i^z \end{bmatrix}^t, \tag{2}$$

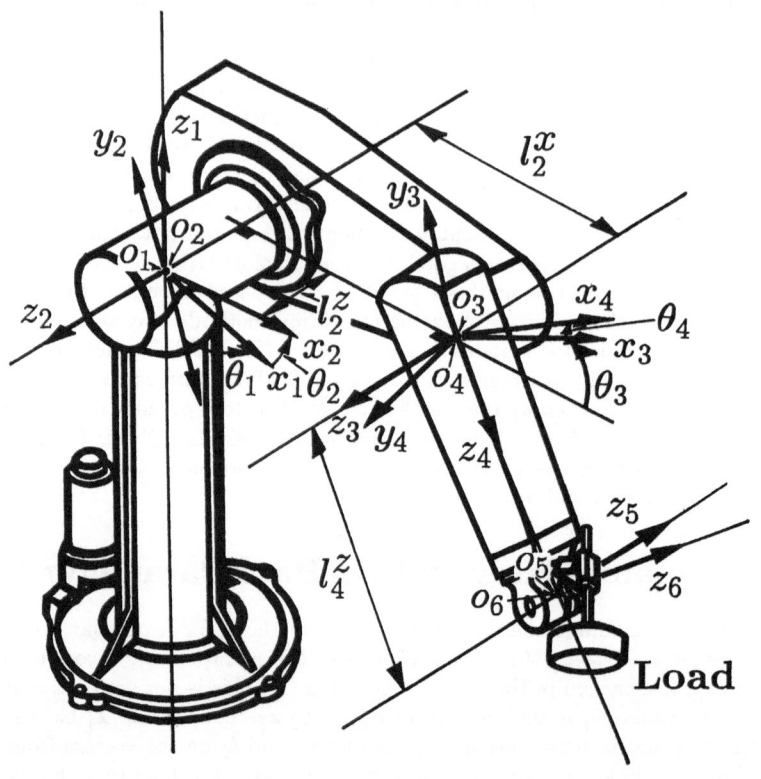

Fig. 1   PUMA 560 with Load and Its Coordinate Systems

$$
{}^i\boldsymbol{I}_i =
\begin{bmatrix}
I_i^x & I_i^{xy} & I_i^{xz} \\
I_i^{xy} & I_i^y & I_i^{yz} \\
I_i^{xz} & I_i^{yz} & I_i^z
\end{bmatrix}.
\tag{3}
$$

${}^i\boldsymbol{l}_i$ denotes the length of link i, and only $l_2^x = 43.2[cm]$, $l_2^z = -15[cm]$, $l_4^z = 43.3[cm]$ are non-zero elements. The m_i and the elements of $m_i{}^i\boldsymbol{r}_i$ and ${}^i\boldsymbol{I}_i$ for $1 > i > 6$ will be called link inertial parameters.

It is well known that the motion equation of the manipulator can be described as

$$\boldsymbol{\tau} = \boldsymbol{H}(\boldsymbol{\theta})\ddot{\boldsymbol{\theta}} + \boldsymbol{B}(\boldsymbol{\theta}, \dot{\boldsymbol{\theta}})\dot{\boldsymbol{\theta}} + \boldsymbol{G}(\boldsymbol{\theta}) \tag{4}$$

where $\boldsymbol{\tau} = [\tau_1, \cdots\cdots, \tau_6]^t$ is the joint torque vector and $\boldsymbol{\theta} = [\theta_1, \cdots\cdots, \theta_6]^t$ is the joint angle vector. If all the values of the link inertial parameters are known, we can determine the inertial term matrix $\boldsymbol{H}(\boldsymbol{\theta})$, the Coriolis and centrifugal force term matrix $\boldsymbol{B}(\boldsymbol{\theta}, \dot{\boldsymbol{\theta}})$ and the gravity term vector $\boldsymbol{G}(\boldsymbol{\theta})$ completely.

Since it is impractical to decompose the manipulator into the links, one may consider to estimate the link inertial parameter values from joint torque and joint motion data. But, unfortunately, all the link inertial parameters are redundant to determine the motion equation and it is impossible to estimate all the link inertial parameter values from the joint torque and joint motion data. Therefore we must find a minimum set of parameters, which is enough to determine the motion equation and can be estimated from the torque and motion data. Since the link inertial parameters affect the motion equation linearly, such a minimum set can consist of liner combinations of the link inertial parameters, that will be called inertial parameters. This problem has been addressed by Mayeda et al., and such a minimum set of inertial parameters is called a base parameters [2],[3].

To describe the base parameters, we introduce following notation.

$$\boldsymbol{M}_i = \sum_{j=i}^{6} m_j, \tag{5}$$

$$\boldsymbol{R}_i = \boldsymbol{M}_{i+1}\boldsymbol{l}_i + m_i\boldsymbol{r}_i, \tag{6}$$

$$\boldsymbol{J}_i = \boldsymbol{I}_i + \boldsymbol{M}_{i+1}[(\boldsymbol{l}_i \cdot \boldsymbol{l}_i)\boldsymbol{E} - \boldsymbol{l}_i \otimes \boldsymbol{l}_i], \tag{7}$$

$$^i\boldsymbol{R}_i = \left[\begin{array}{ccc} R_i^x & R_i^y & R_i^z \end{array} \right]^t, \tag{8}$$

$$^i\boldsymbol{J}_i = \left[\begin{array}{ccc} J_i^x & J_i^{xy} & J_i^{xz} \\ J_i^{xy} & J_i^y & J_i^{yz} \\ J_i^{xz} & J_i^{yz} & J_i^z \end{array} \right]. \tag{9}$$

where $\boldsymbol{E}$ is unit tensor.

Applying the results in [2] for this PUMA 560, we obtain following base parameter set consists of 36 inertial parameters.

$$J_{1z} = J_1^z + J_2^y + J_3^y + 2l_2^x R_3^z,$$

$$J_{2z} = J_2^z, \qquad\qquad J_{2(x-y)} = J_2^x - J_2^y,$$

$$J_{2xz} = J_2^{xz} - l_2^x R_3^z, \quad J_{2yz} = J_2^{yz}, \qquad\qquad J_{2xy} = J_2^{xy},$$

$$R_{2x} = R_2^y, \qquad\qquad R_{2y} = R_2^y,$$

$$J_{3z} = J_3^z + J_4^y, \qquad J_{3(x-y)} = J_3^x - J_3^y + J_4^y,$$

$$J_{3xz} = J_3^{xz}, \qquad\qquad J_{3yz} = J_3^{yz}, \qquad\qquad J_{3xy} = J_3^{xy},$$

$$R_{3x} = R_3^x, \qquad\qquad R_{3y} = R_3^y - R_4^z,$$

$$J_{4z} = J_4^z + J_5^y, \qquad J_{4(x-y)} = J_4^x - J_4^y + J_5^y,$$

$$J_{4xz} = J_4^{xz}, \qquad\qquad J_{4yz} = J_4^{yz} + l_4^z R_5^z, \qquad J_{4xy} = J_4^{xy},$$

$$R_{4x} = R_4^x, \qquad\qquad R_{4y} = R_4^y - R_5^z,$$

$$J_{5z} = J_5^z + J_6^y, \qquad J_{5(x-y)} = J_5^x - J_5^y + J_6^y,$$

$$J_{5xz} = J_5^{xz}, \qquad\qquad J_{5yz} = J_5^{yz}, \qquad\qquad J_{5xy} = J_5^{xy},$$

$$R_{5x} = R_5^x, \qquad\qquad R_{5y} = R_5^y - R_6^z,$$

$$J_{6z} = J_6^z, \qquad\qquad J_{6(x-y)} = J_6^x - J_6^y,$$

$$J_{6xz} = J_6^{xz}, \qquad\qquad J_{6yz} = J_6^{yz}, \qquad\qquad J_{6xy} = J_6^{xy},$$

$$R_{6x} = R_6^x, \qquad\qquad R_{6y} = R_6^y. \tag{10}$$

The extension of this base parameter set to general open-loop kinematic chains has been given in [3].

3   The Driving Systems

The 6 joints of the manipulator are supplied torques by 6 motors via gear mechanism. Let θ_{mi} be the rotation angle of i-th motor and $\boldsymbol{\theta}_m = [\theta_{m1}, \cdots\cdots, \theta_{m6}]^t$. Then the joint angles are related to the motor angles by the gear mechanism as

$$\boldsymbol{\theta} = \boldsymbol{K}\boldsymbol{\theta}_m \tag{11}$$

where

$$\boldsymbol{K} = \begin{bmatrix} k_1 & 0 & 0 & & & \\ 0 & k_2 & 0 & & \mathbf{0} & \\ 0 & 0 & k_3 & & & \\ & & & k_4 & 0 & 0 \\ & \mathbf{0} & & k_{5,4} & k_5 & 0 \\ & & & k_{6,4} & k_{6,5} & k_6 \end{bmatrix}, \tag{12}$$

$$k_1 = 1.60 \times 10^{-2}, \quad k_2 = 0.926 \times 10^{-2}, \quad k_3 = 1.86 \times 10^{-2},$$
$$k_4 = -1.32 \times 10^{-2}, \quad k_5 = -1.39 \times 10^{-2}, \quad k_6 = -1.30 \times 10^{-2},$$
$$k_{5,4} = 0.18 \times 10^{-3}, \quad k_{6,4} = 0.14 \times 10^{-3}, \quad k_{6,5} = 2.51 \times 10^{-3}.$$

The gear mechanism for the last 3 D.O.F. of the wrist is sophisticated, and motors 4, 5 and 6 have some interactions.

The dynamic models of the driving systems will be described as

$$\tau'_{mi} = \tau_{mi} - h_i \ddot{\theta}_{mi} - b_i \dot{\theta}_{mi} - c_i sgn \dot{\theta}_{mi} \tag{13}$$

for $1 < i < 6$ where τ_{mi} is i-th motor torque, τ'_{mi} is transmitted torque from i-th motor to gear mechanism, and h_i, b_i and c_i are the moment of inertia, the viscous friction coefficient and the Coulomb friction coefficient around i-th motor axis, respectively. The inertia, the viscous friction and the Coulomb friction of the gear mechanism are supposed to be concentrated around the motor axes. Let $\boldsymbol{\tau}'_m = [\tau'_{m1}, \cdots\cdots, \tau'_{m6}]^t$. Then

$$\boldsymbol{\tau}'_m = \boldsymbol{K}^t \boldsymbol{\tau}. \tag{14}$$

Combining (4), (13) and (14), we can obtain the dynamic model of the manipulator. h_1 and h_2 appear in the dynamic model always in the forms of

$$\boldsymbol{J}_{1z} + (k_1)^{-2} h_1, \tag{15}$$

$$\boldsymbol{J}_{2z} + (k_2)^{-2} h_2, \tag{16}$$

respectively. Therefore we can consider (15) and (16) as base parameters and abuse $\boldsymbol{J}_{1z}$ and $\boldsymbol{J}_{2z}$ to denote them, respectively. The parameters h_i, b_i and c_i for $1 < i < 6$

except for h_1 and h_2 will be called driving system parameters. To determine the dynamic model of the PUMA 560, we have to estimate all the values of the 36 base parameters and the 16 driving system parameters. Those 52 parameters will be called model parameters, and column vector of the 52 model parameters will be denoted by p.

It is easily shown [4] that all the values of the model parameters from motor torque and motor rotation data. In this manipulators, each motor current and rotation angle are measurable by an equipment and an encoder, respectively. Thus we can obtain the motor torque and motor rotation data and can estimate the value of the model parameters.

4   Identification by Step-by-step Method

In this section, we experiment the step-by-step method to estimate the values of the model parameters. This method consists of 3 types of simple test motions, as explained below, such that we only need to move one or two joints simultaneously, freezing the rest of the joint. The model parameters are divided into a certain number of subgroups, and values of model parameters in each sub-group are estimated from data of the test motions, use being made of formerly estimated model parameter values.

The extension of this method to general open-loop kinematic chains is given in [9]

4.1   Static test

If the manipulator stands still, it is easily derived that

$$\tau_i = -g \cdot (z_i \times \sum_{j=i}^{6} R_j)$$

$$= {}^i g^x R_{iy} - {}^i g^y R_{ix} + d_i$$

(17)

where g is the gravity vector, ${}^i g = [{}^i g^x \; {}^i g^y \; {}^i g^z]^t$ and d_i is a term figured out from R_{sx} and R_{sy} for $i + 1 < s < 6$. Using (17) for more than two angles of θ_i, we can estimate R_{ix} and R_{iy} provided that z_i can be not parallel to g and τ_i can be estimated from τ_{mi}. To avoid the effect of the Coulomb static friction around motor axis, we change τ_{mi} gradually and measure τ_{mi}^+ and τ_{mi}^- at the instance when θ_{mi} begins to move $+$ and $-$ directions, respectively. Then τ_{mi} is estimated as $\tau_{mi} = 1/2(\tau_{mi}^+ - \tau_{mi}^-)$.

Performing this test from $i = 6$ to 2, we can estimate R_{sx} and R_{sy} in (10). For $6 < i < 4$, certain modifications are required to estimate τ_i because of the interactions due to the gear mechanism. Using the results obtained here, we can compensate the gravity term in(4). So we omit the gravity term in later discussions.

4.2   Constant velocity motion test

Make i-th motor rotate in constant angular velocity, freezing other motors. Then, neglecting off-diagonal element of K since their effects are very small, we can derive from (13) that

$$\tau_{mi} = b_i \dot{\theta}_{mi} + c_i sgn \dot{\theta}_{mi}. \qquad (18)$$

If we realize this motion for more than two angular velocities, it is easy to estimate b_i and c_i from (18).

Performing this test for every motor, b_i and c_i for $1 < i < 6$ can be estimated. By compensation, we omit the viscous friction and the Coulomb friction in later discussion.

4.3 Accelerated motion test

Make an accelerated rotations about 5-th joint for three different θ_{6s} freezing the other joint. The motion equation about 5-th joint is easily derived as

$$(k_6^2 \boldsymbol{J6z} + h_6)\ddot{\theta}_{m6} = \tau_{m6} \qquad (19)$$

From this we can directly estimate $(k_6^2 \boldsymbol{J6z} + h_6)$.

Next, make accelerated rotations about 5-th joint for three different θ_{6s} freezing the other joints. The motion equation about 5-th joint is easily derived as

$$[k_5^2 \boldsymbol{J5z} + h_5 + k_{6,5}^2 k_6^{-2} h_6 + k_5^2(\boldsymbol{J6}(\boldsymbol{x} - \boldsymbol{y})\sin\theta_6 + 2\boldsymbol{J6xy}\sin\theta_6\cos\theta_6)]\ddot{\theta}_{m5}$$
$$\qquad\qquad (20)$$
$$= -\tau_{m5} - k_{6,5}k_6^{-1}\tau_{m6}.$$

By solving linear equations obtained from (20) for three different θ_{6s}, we can estimate $\boldsymbol{J6}(\boldsymbol{x} - \boldsymbol{y})$, $\boldsymbol{J6xy}$ and $k_5^2 \boldsymbol{J5z} + h_5 + h_6^2 k_6^{-2} h_6$.

Next, make an accelerated rotation about 6-th and 5-th joints simultaneously freezing the other joints. The motion equation about 6-th joint is easily derived as

$$k_6[k_5(\boldsymbol{J6xz}\sin\theta_6 + \boldsymbol{J6yz}\cos\theta_6) + k_{6,5}\boldsymbol{J6z}]\ddot{\theta}_{m5}$$

$$+(k_6^2\boldsymbol{J6z} + h_6)\ddot{\theta}_{m6} - k_6 k_5^2[\boldsymbol{J6}(\boldsymbol{x} - \boldsymbol{y})\sin\theta_6 + 2\boldsymbol{J6xy}(\cos^2\theta_6 - 1)]\ddot{\theta}_{m5} \qquad (21)$$

$$= \tau_{m6}.$$

Since $k_6^2 \boldsymbol{J6z} + h_6$, $\boldsymbol{J6}(\boldsymbol{x} - \boldsymbol{y})$ and $\boldsymbol{J6xy}$ have been already estimated by solving linear equations obtained from (21) for three different θ_{6s}, we can estimate $\boldsymbol{J6xz}$, $\boldsymbol{J6yz}$ and $\boldsymbol{J6z}$, and hence h_6.

Integrating both sides of (19),(20) and (21), we can avoid to use angular acceleration data.

Continuing same kind test motions for the rest of joints, we can estimate all the rested model parameters.

The estimated values of the model parameter by the step-by-step method are shown in Table. 1.

Table 1: Model Parameter Values by Step-by-step Method

Parameter	Value		Parameter	Value	
c_1	1.42×10^{-1}	[N $\cdot$ m]	J_{1z}	5.31	[kg $\cdot$m^2]
c_2	1.25×10^{-1}	[N $\cdot$ m]	J_{2z}	6.30	[kg $\cdot$m^2]
c_3	1.24×10^{-1}	[N $\cdot$ m]	J_{3z}	1.11	[kg $\cdot$m^2]
c_4	2.36×10^{-2}	[N $\cdot$ m]	J_{4z}	4.12×10^{-2}	[kg $\cdot$m^2]
c_5	1.43×10^{-2}	[N $\cdot$ m]	J_{5z}	3.81×10^{-2}	[kg $\cdot$m^2]
c_6	2.35×10^{-2}	[N $\cdot$ m]	J_{6z}	4.82×10^{-2}	[kg $\cdot$m^2]
b_1	8.63×10^{-4}	[N $\cdot$ m $\cdot$ s]	$J_2(x-y)$	-2.58	[kg $\cdot$m^2]
b_2	4.00×10^{-4}	[N $\cdot$ m $\cdot$ s]	$J_3(x-y)$	2.11×10^{-1}	[kg $\cdot$m^2]
b_3	3.67×10^{-4}	[N $\cdot$ m $\cdot$ s]	$J_4(x-y)$	2.41×10^{-2}	[kg $\cdot$m^2]
b_4	5.55×10^{-5}	[N $\cdot$ m $\cdot$ s]	$J_5(x-y)$	3.42×10^{-2}	[kg $\cdot$m^2]
b_5	4.21×10^{-5}	[N $\cdot$ m $\cdot$ s]	$J_6(x-y)$	-3.85×10^{-3}	[kg $\cdot$m^2]
b_6	5.65×10^{-5}	[N $\cdot$ m $\cdot$ s]	J_{2xy}	1.05×10^{-1}	[kg $\cdot$m^2]
h_3	3.12×10^{-4}	[kg $\cdot$m^2]	J_{3xy}	2.12×10^{-2}	[kg $\cdot$m^2]
h_4	2.30×10^{-5}	[kg $\cdot$m^2]	J_{4xy}	-7.57×10^{-3}	[kg $\cdot$m^2]
h_5	3.73×10^{-5}	[kg $\cdot$m^2]	J_{5xy}	5.63×10^{-2}	[kg $\cdot$m^2]
h_6	3.16×10^{-5}	[kg $\cdot$m^2]	J_{6xy}	-3.24×10^{-2}	[kg $\cdot$m^2]
R_{2x}	7.27	[kg $\cdot$ m]	J_{2xz}	-2.38×10^{-1}	[kg $\cdot$m^2]
R_{3x}	1.85×10^{-1}	[kg $\cdot$ m]	J_{3xz}	-4.85×10^{-1}	[kg $\cdot$m^2]
R_{4x}	1.08×10^{-2}	[kg $\cdot$ m]	J_{4xz}	-3.58×10^{-2}	[kg $\cdot$m^2]
R_{5x}	-1.16×10^{-3}	[kg $\cdot$ m]	J_{5xz}	1.41×10^{-2}	[kg $\cdot$m^2]
R_{6x}	-9.86×10^{-2}	[kg $\cdot$ m]	J_{6xz}	4.47×10^{-2}	[kg $\cdot$m^2]
R_{2y}	5.24×10^{-1}	[kg $\cdot$ m]	J_{2yz}	2.12×10^{-1}	[kg $\cdot$m^2]
R_{3y}	-2.35	[kg $\cdot$ m]	J_{3yz}	3.07×10^{-2}	[kg $\cdot$m^2]
R_{4y}	-8.60×10^{-3}	[kg $\cdot$ m]	J_{4yz}	-1.78×10^{-2}	[kg $\cdot$m^2]
R_{5y}	-2.65×10^{-1}	[kg $\cdot$ m]	J_{5yz}	2.28×10^{-2}	[kg $\cdot$m^2]
R_{6y}	1.16×10^{-1}	[kg $\cdot$ m]	J_{6yz}	-2.08×10^{-3}	[kg $\cdot$m^2]

5 Identification by Simultaneous Method

Since the model parameters affect linearly to the motor torques, (4), (13)and (14) can be modified as

$$\tau_m = \boldsymbol{\Phi}(\ddot{\boldsymbol{\theta}}, \dot{\boldsymbol{\theta}}, \boldsymbol{\theta}, sgn\dot{\boldsymbol{\theta}})\boldsymbol{p}. \tag{22}$$

Making random enough accelerated rotations for all the joints simultaneously and applying the sampled data to (22), we can estimate $\boldsymbol{p}$ by least square method.

The estimated values of the model parameters are shown in Table. 2.

Table 2: Model Parameter Values by Simultaneous Method

Parameter	Value		Parameter	Value	
c_1	1.30×10^{-1}	[N · m]	J_{1z}	3.92	[kg ·m²]
c_2	1.24×10^{-1}	[N · m]	J_{2z}	6.87	[kg ·m²]
c_3	1.15×10^{-1}	[N · m]	J_{3z}	4.47×10^{-1}	[kg ·m²]
c_4	2.30×10^{-2}	[N · m]	J_{4z}	3.38×10^{-2}	[kg ·m²]
c_5	1.51×10^{-2}	[N · m]	J_{5z}	-1.82×10^{-1}	[kg ·m²]
c_6	1.90×10^{-2}	[N · m]	J_{6z}	-3.89×10^{-2}	[kg ·m²]
b_1	9.64×10^{-4}	[N · m · s]	$J_2(x-y)$	-1.76	[kg ·m²]
b_2	3.79×10^{-4}	[N · m · s]	$J_3(x-y)$	1.99×10^{-1}	[kg ·m²]
b_3	4.38×10^{-4}	[N · m · s]	$J_4(x-y)$	4.58×10^{-2}	[kg ·m²]
b_4	6.71×10^{-5}	[N · m · s]	$J_5(x-y)$	-1.02×10^{-1}	[kg ·m²]
b_5	2.63×10^{-5}	[N · m · s]	$J_6(x-y)$	2.21×10^{-2}	[kg ·m²]
b_6	8.29×10^{-5}	[N · m · s]	J_{2xy}	2.26×10^{-1}	[kg ·m²]
h_3	4.44×10^{-4}	[kg ·m²]	J_{3xy}	1.66×10^{-1}	[kg ·m²]
h_4	3.63×10^{-5}	[kg ·m²]	J_{4xy}	-6.60×10^{-2}	[kg ·m²]
h_5	6.14×10^{-5}	[kg ·m²]	J_{5xy}	2.63×10^{-2}	[kg ·m²]
h_6	4.13×10^{-5}	[kg ·m²]	J_{6xy}	-3.03×10^{-3}	[kg ·m²]
R_{2x}	7.79	[kg · m]	J_{2xz}	3.91×10^{-1}	[kg ·m²]
R_{3x}	8.58×10^{-2}	[kg · m]	J_{3xz}	-5.58×10^{-3}	[kg ·m²]
R_{4x}	1.26×10^{-2}	[kg · m]	J_{4xz}	9.83×10^{-3}	[kg ·m²]
R_{5x}	-2.77×10^{-3}	[kg · m]	J_{5xz}	-1.64×10^{-2}	[kg ·m²]
R_{6x}	-1.21×10^{-1}	[kg · m]	J_{6xz}	1.38×10^{-2}	[kg ·m²]
R_{2y}	4.05×10^{-1}	[kg · m]	J_{2yz}	3.88×10^{-1}	[kg ·m²]
R_{3y}	-2.42	[kg · m]	J_{3yz}	-1.22×10^{-1}	[kg ·m²]
R_{4y}	-1.58×10^{-2}	[kg · m]	J_{4yz}	-7.94×10^{-2}	[kg ·m²]
R_{5y}	-2.77×10^{-1}	[kg · m]	J_{5yz}	6.33×10^{-3}	[kg ·m²]
R_{6y}	1.16×10^{-1}	[kg · m]	J_{6yz}	-4.86×10^{-3}	[kg ·m²]

6 Discussion and Comparison of The Two Methods

The test motion of the step-by-step method are very simple, and it is easy to understand how the target parameters of the test affect the motor torque. Hence programming for data processing is simple and of small size, and we can contrive good test motions for the estimations. The estimated values of subgrouped parameters in each motion test are examined by a small size simulation, and we can improve the test motion and the accuracy of the estimation. On the other hand, this method requires a number of test motion, and the estimation error of a test affect the accuracy of later estimations. The latter problem has been slightly improved by Ozaki et al.[5].

The simultaneous method is simple to understand and requires only one random enough test motion. But, the programing of the iteration algorithm of the least square estimation is very time consuming for the high D.O.F manipulator. The most serious drawback of this method is difficulty of convergence judgement and huge number of the iterations for the convergence. It took 27 hours for 90,000 iterations by SUN SPARK station IPC for almost all of the parameter values to converge. Still, a few parameters related to the wrist could not be convinced to have converged. It is difficult to find a good test motion for the fast convergence. In this method we have to use the angular acceleration data. PUMA 560 is equipped an encoder only for each joint, and the angular accelerations are obtained by differentiating the angular velocity data. These contaminated data seem to cause partially the slow convergence and the inaccuracy of the estimation.

We have assumed that b_i and c_i are constant and do not depend on direction of the rotation. But it is not true in the industrial manipulator with the gear mechanism. Especially, the PUMA 560 has very sophisticated gear mechanism without brake in the wrist. Those factors limit the accuracy of the estimations.

To evaluate the accuracy of the estimated parameter values, we have imposed a torque on each joint of the manipulator simultaneously and made an accelerated motion. The measured trajectories of the joint angles and angular velocities are compared with those obtained by simulations where the same torque data and estimated parameter values by the step-by-step method or the simultaneous method are used. The results are shown in Fig.3 - Fig.5. We can conclude from these results that the step-by-step method is more accurate way to estimate the dynamic models of industrial manipulator than the simultaneous method. For the purpose of controlling industrial manipulators, the estimated parameter values obtained by the both methods can be said considerably accurate.

7   Conclusion

The step-by-step method and the simultaneous method have been experimentally examined to estimate the dynamic model of PUMA 560. The merits and demerits of the both method are discussed and compared. To evaluate the accuracy of the estimated parameter values, the joint angle trajectories of a real motion and the simulated motions have been compared. From those results it can be said the step-by-step method is more accurate way to estimate the model parameter values than the simultaneous method. The estimated parameter values obtained by the both method will be accurate for the purpose of model based control of the industrial manipulator.

To establish the modeling method for the model based control of industrial manipulators, more experimental works on various types of manipulators are required. In further studies along this line, we will confont the necessity to take into account of compliance of the joints.

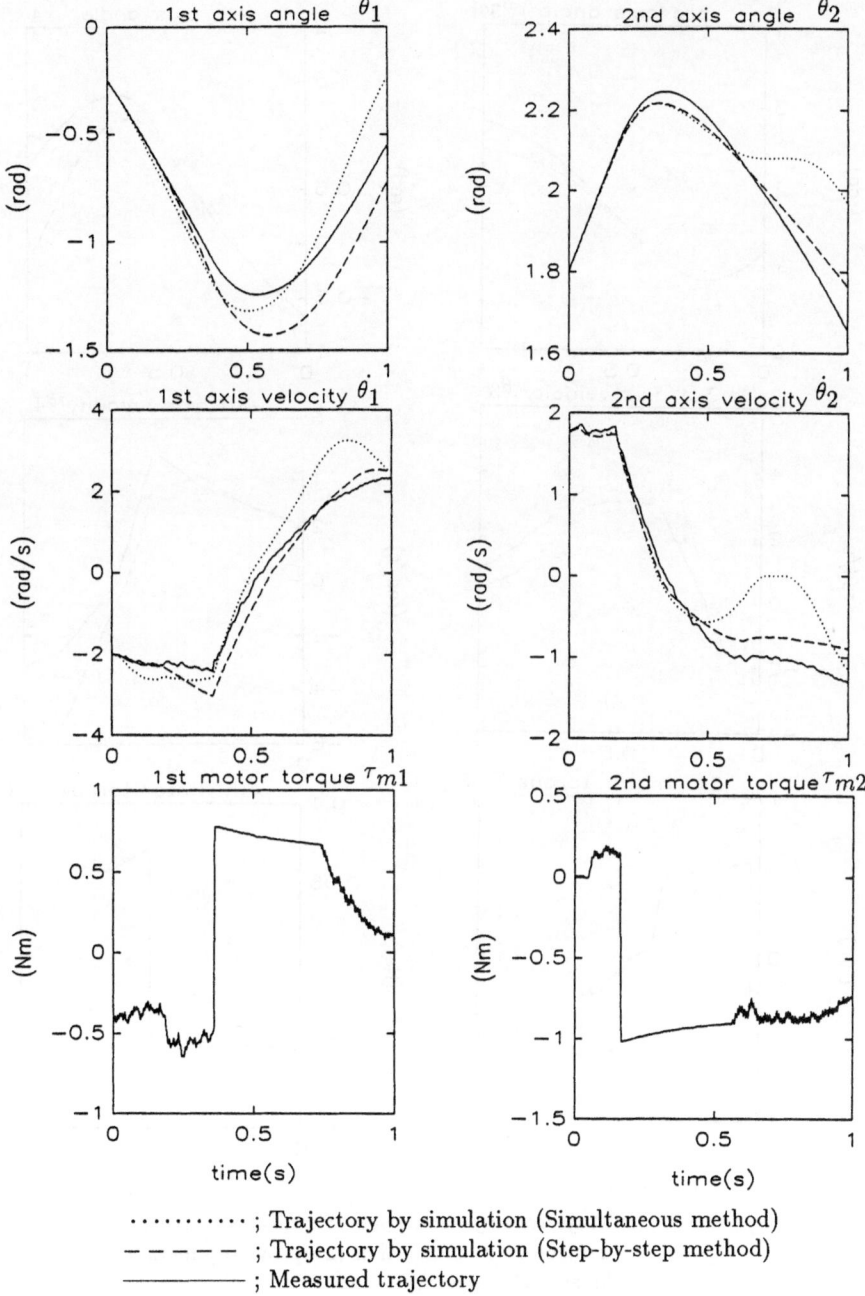

.... ; Trajectory by simulation (Simultaneous method)
– – – – – ; Trajectory by simulation (Step-by-step method)
——————— ; Measured trajectory

Fig. 2   Trajectories of Joint 1 and 2 angles

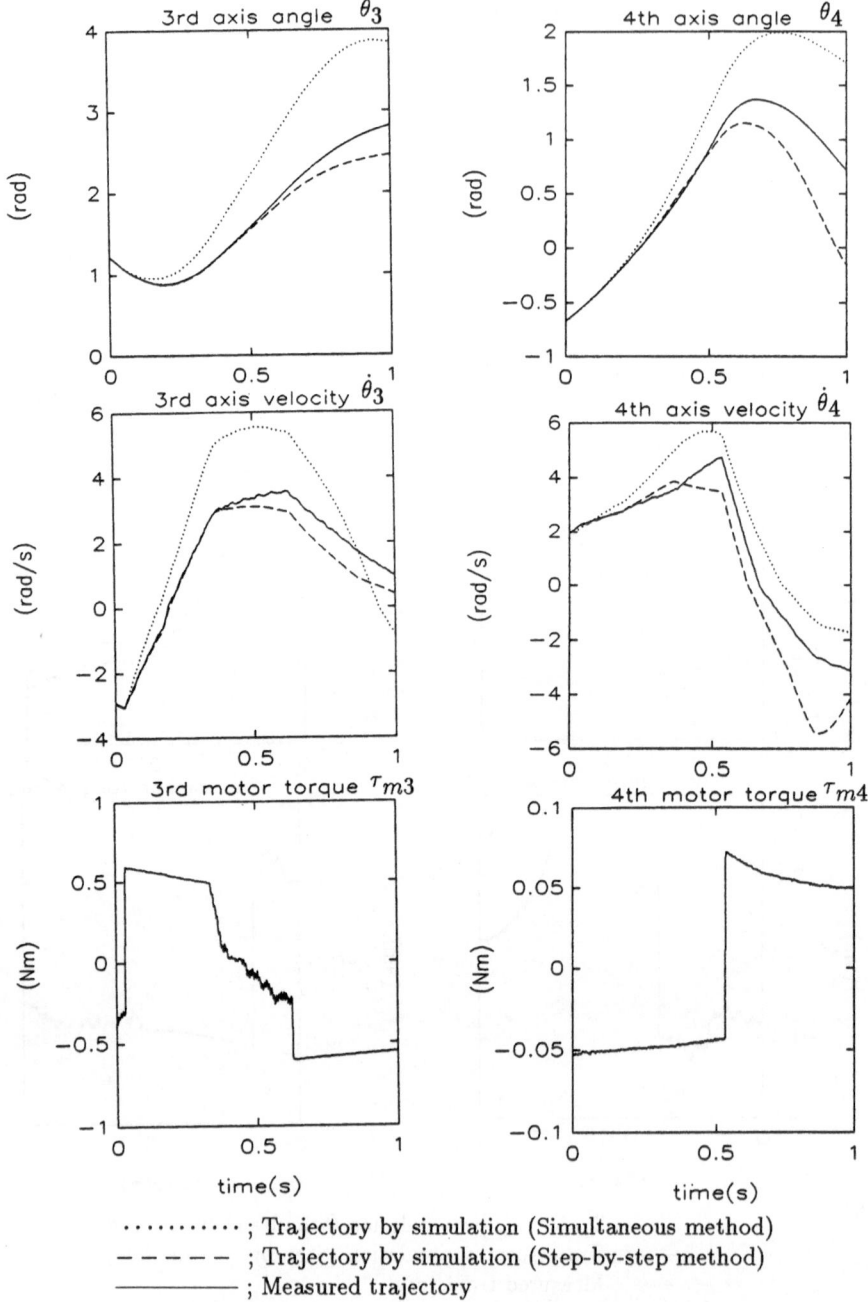

.... ; Trajectory by simulation (Simultaneous method)

− − − − − ; Trajectory by simulation (Step-by-step method)

————— ; Measured trajectory

Fig. 3   Trajectories of Joint 3 and 4 angles

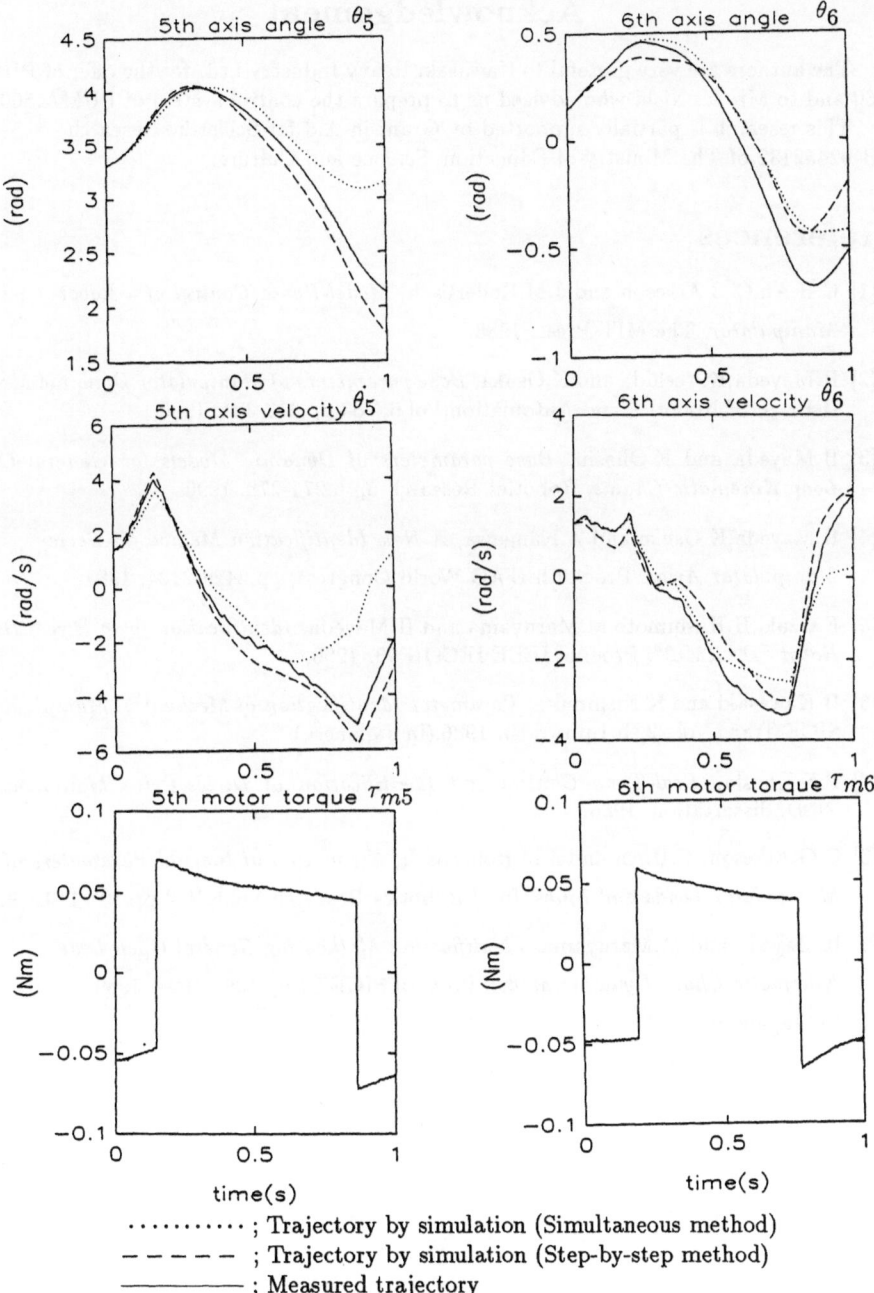

.... ; Trajectory by simulation (Simultaneous method)
− − − − − ; Trajectory by simulation (Step-by-step method)
————— ; Measured trajectory

Fig. 4   Trajectories of Joint 5 and 6 angles

Acknowledgement

The authors are very grateful to Kawasaki Heavy Industry Ltd. for the offer of PUMA 560 and to Mr. Y. Nishi who advised us to prepare the control system of PUMA 560.

This research is partially supported by Grant-in-Aid for Scientific Research (B)02452132 of The Ministry of Education, Science and Culture.

References

[1] C.H.An,C.G.Atkeson and J.M.Hollerbach, *Model-Based Control of a Robot Manipulator.* The MIT Press. 1988.

[2] H.Mayeda,K.Yoshida and K.Osuka, *Base parameters of Manipulator Dynamic Model.* IEEE J. of Robotics and Automation,Vol.6,No3,pp.312-321. 1990.

[3] H.Mayeda and K.Ohashi, *Base parameters of Dynamic Models for General Open Loop Kinematic Chains,* Robotics Research 5,pp.271-278. 1990.

[4] H.Mayeda,K.Osuka and A.Kangawa, *A New Identification Method for Serial Manipulator Arms.* Proc. 9th IFAC World Congress, pp.2429-2424. 1984.

[5] F.Ozaki,H.Hashimoto,M.Maruyama and H.Mayeda, *Identification for a Direct Drive Robot "DARM-2".* Proc. of IEEE IECON'90. 1990.

[6] H.Kawasaki and K.Nishimura, *Parameter Identification of Mechanical Manipulators.* SICE Trans.Vol.22,No1,pp.76-83. 1986.(in Japanese).

[7] P.K.Khosla, *Real-Time Control and Identification of Direct-Drive Manipulators.* Ph.D. dissertation. 1986.

[8] C.G.Atkeson, C.H.An and J.M.Hollerbach, *Estimation of Inertial Parameters of Manipulator Loads and Links.* Int.J.Robotics Research,Vol.5,No3,pp.101-119. 1986.

[9] H.Mayeda and M.Maruyama, *Identification Method for General Open-Loop Kinematic Chain Dynamic Model.* Proc. of SICE'89,pp.1091-1094. 1989. (in Japanese).

Lecture Notes in Control and Information Sciences

Edited by M. Thoma

1989–1993 Published Titles:

Vol. 135: Nijmeijer, Hendrik; Schumacher, Johannes M. (Eds.)
Three Decades of Mathematical System Theory. A Collection of Surveys at the Occasion of the 50th Birthday of Jan C. Willems.
562 pp. 1989 [3-540-51605-0]

Vol. 136: Zabczyk, Jerzy W. (Ed.)
Stochastic Systems and Optimization. Proceedings of the 6th IFIP WG 7.1 Working Conference, Warsaw, Poland, September 12-16, 1988.
374 pp. 1989 [3-540-51619-0]

Vol. 137: Shah, Sirish L.; Dumont, Guy (Eds.)
Adaptive Control Strategies for Industrial Use. Proceedings of a Workshop held in Kananaskis, Canada, 1988.
360 pp. 1989 [3-540-51869-X]

Vol. 138: McFarlane, Duncan C.; Glover, Keith
Robust Controller Design Using Normalized Coprime Factor Plant Descriptions.
206 pp. 1990 [3-540-51851-7]

Vol. 139: Hayward, Vincent; Khatib, Oussama (Eds.)
Experimental Robotics I. The First International Symposium, Montreal, June 19-21, 1989.
613 pp. 1990 [3-540-52182-8]

Vol. 140: Gajic, Zoran; Petkovski, Djordjija; Shen, Xuemin (Eds.)
Singularly Perturbed and Weakly Coupled Linear Control Systems. A Recursive Approach.
202 pp. 1990 [3-540-52333-2]

Vol. 141: Gutman, Shaul
Root Clustering in Parameter Space.
153 pp. 1990 [3-540-52361-8]

Vol. 142: Gündes, A. Nazli; Desoer, Charles A.
Algebraic Theory of Linear Feedback Systems with Full and Decentralized Compensators.
176 pp. 1990 [3-540-52476-2]

Vol. 143: Sebastian, H.-J.; Tammer, K. (Eds.)
System Modelling and Optimizaton. Proceedings of the 14th IFIP Conference, Leipzig, GDR, July 3-7, 1989.
960 pp. 1990 [3-540-52659-5]

Vol. 144: Bensoussan, A.; Lions, J.L. (Eds.)
Analysis and Optimization of Systems. Proceedings of the 9th International Conference. Antibes, June 12-15, 1990.
992 pp. 1990 [3-540-52630-7]

Vol. 145: Subrahmanyam, M. Bala
Optimal Control with a Worst-Case Performance Criterion and Applications.
133 pp. 1990 [3-540-52822-9]

Vol. 146: Mustafa, Denis; Glover, Keith
Minimum Entropy H Control.
144 pp. 1990 [3-540-52947-0]

Vol. 147: Zolesio, J.P. (Ed.)
Stabilization of Flexible Structures. Third Working Conference, Montpellier, France, January 1989.
327 pp. 1991 [3-540-53161-0]

Vol. 148: Not published

Vol. 149: Hoffmann, Karl H; Krabs, Werner (Eds.)
Optimal Control of Partial Differential Equations. Proceedings of IFIP WG 7.2 - International Conference. Irsee, April, 9-12, 1990.
245 pp. 1991 [3-540-53591-8]

Vol. 150: Habets, Luc C.
Robust Stabilization in the Gap-topology.
126 pp. 1991 [3-540-53466-0]

Vol. 151: Skowronski, J.M.; Flashner, H.; Guttalu, R.S. (Eds.)
Mechanics and Control. Proceedings of the 3rd Workshop on Control Mechanics, in Honor of the 65th Birthday of George Leitmann, January 22-24, 1990, University of Southern California.
497 pp. 1991 [3-540-53517-9]

Vol. 152: Aplevich, J. Dwight
Implicit Linear Systems.
176 pp. 1991 [3-540-53537-3]

Vol. 153: Hajek, Otomar
Control Theory in the Plane.
269 pp. 1991 [3-540-53553-5]

Vol. 154: Kurzhanski, Alexander; Laseicka, Irena (Eds.)
Modelling and Inverse Problems of Control for Distributed Parameter Systems. Proceedings of IFIP WG 7.2 - IIASA Conference, Laxenburg, Austria, July 1989.
170 pp. 1991 [3-540-53583-7]

Vol. 155: Bouvet, Michel; Bienvenu, Georges (Eds.)
High-Resolution Methods in Underwater Acoustics.
244 pp. 1991 [3-540-53716-3]

Vol. 156: Hämäläinen, Raimo P.; Ehtamo, Harri K. (Eds.)
Differential Games - Developments in Modelling and Computation. Proceedings of the Fourth International Symposium on Differential Games and Applications, August 9-10, 1990, Helsinki University of Technology, Finland.
292 pp. 1991 [3-540-53787-2]

Vol. 157: Hämäläinen, Raimo P.; Ehtamo, Harri K. (Eds.)
Dynamic Games in Economic Analysis. Proceedings of the Fourth International Symposium on Differential Games and Applications. August 9-10, 1990, Helsinki University of Technology, Finland.
311 pp. 1991 [3-540-53785-6]

Vol. 158: Warwick, Kevin; Karny, Miroslav; Halouskova, Alena (Eds.)
Advanced Methods in Adaptive Control for Industrial Applications.
331 pp. 1991 [3-540-53835-6]

Vol. 159: Li, Xunjing; Yong, Jiongmin (Eds.)
Control Theory of Distributed Parameter Systems and Applications. Proceedings of the IFIP WG 7.2 Working Conference, Shanghai, China, May 6-9, 1990.
219 pp. 1991 [3-540-53894-1]

Vol. 160: Kokotovic, Petar V. (Ed.)
Foundations of Adaptive Control.
525 pp. 1991 [3-540-54020-2]

Vol. 161: Gerencser, L.; Caines, P.E. (Eds.)
Topics in Stochastic Systems: Modelling, Estimation and Adaptive Control.
1991 [3-540-54133-0]

Vol. 162: Canudas de Wit, C. (Ed.)
Advanced Robot Control. Proceedings of the International Workshop on Nonlinear and Adaptive Control: Issues in Robotics, Grenoble, France, November 21-23, 1990.
Approx. 330 pp. 1991 [3-540-54169-1]

Vol. 163: Mehrmann, Volker L.
The Autonomous Linear Quadratic Control Problem. Theory and Numerical Solution.
177 pp. 1991 [3-540-54170-5]

Vol. 164: Lasiecka, Irena; Triggiani, Roberto
Differential and Algebraic Riccati Equations with Application to Boundary/Point Control Problems: Continuous Theory and Approximation Theory.
160 pp. 1991 [3-540-54339-2]

Vol. 165: Jacob, Gerard; Lamnabhi-Lagarrigue, F. (Eds.)
Algebraic Computing in Control. Proceedings of the First European Conference, Paris, March 13-15, 1991.
384 pp. 1991 [3-540-54408-9]

Vol. 166: Wegen, Leonardus L. van der
Local Disturbance Decoupling with Stability for Nonlinear Systems.
135 pp. 1991 [3-540-54543-3]

Vol. 167: Rao, Ming
Integrated System for Intelligent Control.
133 pp. 1992 [3-540-54913-7]

Vol. 168: Dorato, Peter; Fortuna, Luigi;
Muscato, Giovanni
Robust Control for Unstructured Perturbations:
An Introduction.
118 pp. 1992 [3-540-54920-X]

Vol. 169: Kuntzevich, Vsevolod M.; Lychak,
Michael
Guaranteed Estimates, Adaptation and
Robustness in Control Systems.
209 pp. 1992 [3-540-54925-0]

Vol. 170: Skowronski, Janislaw M.; Flashner,
Henryk; Guttalu, Ramesh S. (Eds.)
Mechanics and Control. Proceedings of the 4th
Workshop on Control Mechanics, January
21-23, 1991, University of Southern
California, USA.
302 pp. 1992 [3-540-54954-4]

Vol. 171: Stefanidis, P.; Paplinski, A.P.;
Gibbard, M.J.
Numerical Operations with Polynomial
Matrices: Application to Multi-Variable
Dynamic Compensator Design.
206 pp. 1992 [3-540-54992-7]

Vol. 172: Tolle, H.; Ersü, E.
Neurocontrol: Learning Control Systems
Inspired by Neuronal Architectures and Human
Problem Solving Strategies.
220 pp. 1992 [3-540-55057-7]

Vol. 173: Krabs, W.
On Moment Theory and Controllability of
Non-Dimensional Vibrating Systems and
Heating Processes.
174 pp. 1992 [3-540-55102-6]

Vol. 174: Beulens, A.J. (Ed.)
Optimization-Based Computer-Aided Modelling
and Design. Proceedings of the First Working
Conference of the New IFIP TC 7.6 Working
Group, The Hague, The Netherlands, 1991.
268 pp. 1992 [3-540-55135-2]

Vol. 175: Rogers, E.T.A.; Owens, D.H.
Stability Analysis for Linear Repetitive
Processes.
197 pp. 1992 [3-540-55264-2]

Vol. 176: Rozovskii, B.L.; Sowers, R.B. (Eds.)
Stochastic Partial Differential Equations and
their Applications. Proceedings of IFIP WG 7.1
International Conference, June 6-8, 1991,
University of North Carolina at Charlotte, USA.
251 pp. 1992 [3-540-55292-8]

Vol. 177: Karatzas, I.; Ocone, D. (Eds.)
Applied Stochastic Analysis. Proceedings of a
US-French Workshop, Rutgers University, New
Brunswick, N.J., April 29-May 2, 1991.
317 pp. 1992 [3-540-55296-0]

Vol. 178: Zolésio, J.P. (Ed.)
Boundary Control and Boundary Variation.
Proceedings of IFIP WG 7.2 Conference,
Sophia- Antipolis,France, October 15-17,
1990.
392 pp. 1992 [3-540-55351-7]

Vol. 179: Jiang, Z.H.; Schaufelberger, W.
Block Pulse Functions and Their Applications in
Control Systems.
237 pp. 1992 [3-540-55369-X]

Vol. 180: Kall, P. (Ed.)
System Modelling and Optimization.
Proceedings of the 15th IFIP Conference,
Zurich, Switzerland, September 2-6, 1991.
969 pp. 1992 [3-540-55577-3]

Vol. 181: Drane, C.R.
Positioning Systems - A Unified Approach.
168 pp. 1992 [3-540-55850-0]

Vol. 182: Hagenauer, J. (Ed.)
Advanced Methods for Satellite and Deep
Space Communications. Proceedings of an
International Seminar Organized by Deutsche
Forschungsanstalt für Luft-und Raumfahrt
(DLR), Bonn, Germany, September 1992.
196 pp. 1992 [3-540-55851-9]

Vol. 183: Hosoe, S. (Ed.)
Robust Control. Proceesings of a Workshop
held in Tokyo, Japan, June 23-24, 1991.
225 pp. 1992 [3-540-55961-2]

Vol. 184: Duncan, T.E.; Pasik-Duncan, B. (Eds.)
Stochastic Theory and Adaptive Control. Proceedings of a Workshop held in Lawrence, Kansas, September 26-28, 1991.
500 pages. 1992 [3-540-55962-0]

Vol. 185: Curtain, R.F. (Ed.); Bensoussan, A.; Lions, J.L.(Honorary Eds.)
Analysis and Optimization of Systems: State and Frequency Domain Approaches for Infinite-Dimensional Systems. Proceedings of the 10th International Conference, Sophia-Antipolis, France, June 9-12, 1992.
648 pp. 1993 [3-540-56155-2]

Vol. 186: Sreenath, N.
Systems Representation of Global Climate Change Models. Foundation for a Systems Science Approach.
288 pp. 1993 [3-540-19824-5]

Vol. 187: Morecki, A.; Bianchi, G.; Jaworeck, K. (Eds.)
RoManSy 9: Proceedings of the Ninth CISM-IFToMM Symposium on Theory and Practice of Robots and Manipulators.
476 pp. 1993 [3-540-19834-2]

Vol. 188: Naidu, D. Subbaram
Aeroassisted Orbital Transfer: Guidance and Control Strategies. [*In production*]
200 pp approx. 1993 [3-540-19819-9]

Vol.189: Ilchmann, A.
Non-Identifier-Based High-Gain Adaptive Control.
220 pp. 1993 [3-540-19845-8]